INORGANIC AND ORGANOMETALLIC TRANSITION METAL COMPLEXES WITH BIOLOGICAL MOLECULES AND LIVING CELLS

INORGANIC AND ORGANOMETALLIC TRANSITION METAL COMPLEXES WITH BIOLOGICAL MOLECULES AND LIVING CELLS

Edited by

KENNETH KAM-WING LO
City University of Hong Kong
Hong Kong

ELSEVIER

ACADEMIC PRESS
An imprint of Elsevier
elsevier.com

Academic Press is an imprint of Elsevier
125 London Wall, London EC2Y 5AS, United Kingdom
525 B Street, Suite 1800, San Diego, CA 92101-4495, United States
50 Hampshire Street, 5th Floor, Cambridge, MA 02139, United States
The Boulevard, Langford Lane, Kidlington, Oxford OX5 1GB, United Kingdom

Notices

Knowledge and best practice in this field are constantly changing. As new research and experience broaden our understanding, changes in research methods, professional practices, or medical treatment may become necessary.

Practitioners and researchers must always rely on their own experience and knowledge in evaluating and using any information, methods, compounds, or experiments described herein. In using such information or methods they should be mindful of their own safety and the safety of others, including parties for whom they have a professional responsibility.

To the fullest extent of the law, neither the Publisher nor the authors, contributors, or editors, assume any liability for any injury and/or damage to persons or property as a matter of products liability, negligence or otherwise, or from any use or operation of any methods, products, instructions, or ideas contained in the material herein.

Library of Congress Cataloging-in-Publication Data
A catalog record for this book is available from the Library of Congress

British Library Cataloguing-in-Publication Data
A catalogue record for this book is available from the British Library

ISBN: 978-0-12-803814-7

Publisher: John Fedor
Acquisition Editor: Katey Birtcher
Editorial Project Manager: Jill Cetel
Production Project Manager: Paul Prasad Chandramohan
Designer: Mark Rogers

Typeset by TNQ Books and Journals

CONTENTS

LIST OF CONTRIBUTORS

W.H. Ang
National University of Singapore, Singapore; NUS Graduate School for Integrative Science and Engineering, Singapore

C.S. Burke
Dublin City University, Dublin, Ireland

A. Byrne
Dublin City University, Dublin, Ireland

D.G. Calatayud
University of Bath, Bath, United Kingdom; University of Oxford, Oxford, United Kingdom

Y.-Y. Chang
The University of Hong Kong, Hong Kong, P.R. China

M.J. Chow
National University of Singapore, Singapore

F. Cortezon-Tamarit
University of Bath, Bath, United Kingdom

C. Dolan
Dublin City University, Dublin, Ireland

H. Ge
University of Bath, Bath, United Kingdom

C.G. Hartinger
University of Auckland, Auckland, New Zealand

H.U. Holtkamp
University of Auckland, Auckland, New Zealand

T.E. Keyes
Dublin City University, Dublin, Ireland

H. Li
The University of Hong Kong, Hong Kong, P.R. China

S.M. Meier
University of Vienna, Vienna, Austria

V. Mirabello
University of Bath, Bath, United Kingdom

I. Ott
Technische Universität Braunschweig, Braunschweig, Germany

S.I. Pascu
University of Bath, Bath, United Kingdom

U. Schatzschneider
Julius-Maximilians-Universität Würzburg, Würzburg, Germany

P. Scott
University of Warwick, Coventry, United Kingdom

D.H. Simpson
University of Warwick, Coventry, United Kingdom

M.P. Sullivan
University of Auckland, Auckland, New Zealand

H. Sun
The University of Hong Kong, Hong Kong, P.R. China

J. Tang
Peking University, Beijing, P.R. China

M.B.M. Theobald
University of Oxford, Oxford, United Kingdom

H.-Y. Yin
Peking University, Beijing, P.R. China

J.-L. Zhang
Peking University, Beijing, P.R. China

K.Y. Zhang
Nanjing University of Posts and Telecommunications (NUPT), Nanjing, China

PREFACE

Inorganic and organometallic transition metal complexes display diverse chemical and physical properties as a result of the variety of organic ligands combined with the rich electronic properties of d-block metal centers. Many of these properties enable complexes to form interesting noncovalent and covalent interactions with biomolecules such as nucleic acids, carbohydrates, lipids, peptides, and proteins. In fact, the long-standing interest in understanding these interactions arises in part because of their importance in the development of transition metal-based therapeutic and diagnostic reagents. There have been a growing number of studies into the cellular uptake and cytotoxic activity of transition metal complexes, with an emphasis on their use as bioimaging reagents, intracellular sensors, bioconjugation agents, enzyme inhibitors, and anticancer drugs.

This book brings together contributions from a group of chemists who provide an overview on the fundamentals and application potential of transition metal complexes in chemical biology, medicinal chemistry, and pharmaceutical chemistry. An overview of the topics covered by each contributor is highlighted below:

The first chapter by Professor Jun-Long Zhang is on the applications of zinc(II) complexes as biological imaging reagents to visualize molecular events in live cells. It highlights the use of zinc(II)–salen complexes as organelle-selective imaging probes, hydrogen peroxide sensors, and indicators for autophagy.

This is followed by Professor Tia Keyes's review of the design of luminescent transition metal complex–peptide conjugates and their use in live cell sensing and bioimaging. Of particular note is the description of cellular uptake efficiency and cytotoxicity of cell-penetrating peptides modified with ruthenium(II), osmium(II), and iridium(III) complexes and the organelle-selective delivery of signaling peptide-complex conjugates.

Professor Kenneth Yin Zhang explains the use of the photophysical properties of luminescent rhenium(I) and iridium(III) complexes as intracellular chemosensors for disease-related intracellular molecules such as reactive oxygen, nitrogen species, and biothiols, and discusses the potential applications of these complexes as photodynamic therapeutics.

Professor Wee Han Ang's contribution covers the use of the ruthenium(II)-arene scaffold as a building block for new anticancer compounds, focusing on their structure–property relationship and biomolecular interactions. In addition, their potential as anticancer drugs and their mechanisms of action are described.

Professor Ingo Ott reviews the structures and medicinal chemistry of transition metal *N*-heterocyclic carbene (NHC) complexes and also complexes that contain biologically active or naturally occurring compounds, additional metal centers, or "nonclassical" NHC ligands.

Professor Ulrich Schatzschneider introduces the endogenous production pathways and therapeutic potential of carbon monoxide, nitric oxide, and hydrogen sulfide. Probes and delivery systems of these small molecules are summarized.

Professor Peter Scott focuses on prokaryotes, describing the development of transition metal complexes as chemotherapeutic antimicrobial agents and their microbial cellular uptake, biological targets, and mechanisms of action.

Professor Sofia Pascu describes bioimaging probes constructed from a hybrid of transition metal complexes and carbon nanomaterial scaffolds and reviews approaches used to study the delivery, cytotoxicity, and uptake mechanisms of bioimaging probes assembled on single-walled carbon nanotube scaffolds.

Professor Hongzhe Sun introduces immobilized metal affinity chromatography (IMAC) and summarizes applications of IMAC in metalloproteomics and phosphoproteomics, with a focus on the modes of action of metal-based drugs and the role of phosphorylated proteins in multiple cellular processes.

Finally, Professor Christian Hartinger reviews analytic methods used to examine the protein-binding behavior, cellular accumulation, and tissue distribution of inorganic and organometallic compounds that have reached an advanced preclinical development stage, are undergoing clinical trials, or are already in clinical use.

I am most grateful to all of the authors for their extraordinary efforts and excellent contributions, and strongly believe that this collection of outstanding articles highlights the important role of transition metal complexes in different areas of biological science. The book will be useful to a broad spectrum of readers ranging from those who are new to the field to experts working at the interface of chemistry, biology, and medical science.

Kenneth Kam-Wing Lo
Hong Kong, August 2016

CHAPTER ONE

Luminescent Zinc Complexes as Bioprobes for Imaging Molecular Events in Live Cells

J. Tang, H.-Y. Yin, J.-L. Zhang
Peking University, Beijing, P.R. China

Contents

Inorganic and Organometallic Transition Metal Complexes with Biological Molecules and Living Cells
ISBN 978-0-12-803814-7
http://dx.doi.org/10.1016/B978-0-12-803814-7.00001-0

1. INTRODUCTION

Introducing coordination chemistry to biology attracts increasing attention and plays a critical role in the burgeoning field of chemical biology.[1] Because naturally occurring metal ions have been well recognized to impart utility to biology, nonnative or synthetic metal complexes, which might expand the functional roles of inorganic elements not found naturally, can also be considered to interrogate or manipulate biology. This stimulates a variety of biological applications of coordination compounds, arising from their unique electronic, magnetic, chemical, and photophysical properties.[2] In particular, along with rapid development of fluorescence microscopy techniques, luminescent transition metal complexes become an emerging class of fluorescent labels in molecular imaging; notwithstanding organic fluorophores still take a predominant role in the library of commercial cellular labels. Compared to traditional organic fluorophores, the general chemical properties of metals including tunable charge, versatile coordination sphere and thus geometry, and the potential reactivities as Lewis acids and redox centers should not be overlooked, other than the intriguing photophysical properties of the formed metal complexes such as high luminescence, large Stokes shifts and long lifetimes. This renders luminescent transition metal complexes particular utilities for biological imaging even for monitoring molecular events occurring in biological processes.[1a,3] Over the past 20 years, there has been great progress in designing luminescent metal complexes as labels or sensors, especially second and third row transition metals with d^6, d^8, and d^{10} electronic configuration such as Ir(III), Ru(II), Os(II), Re(I), Pt(II), Pd(II), Ag(I), Au(I) as well as lanthanide-based complexes, for bioimaging applications either in vitro or in vivo. Several excellent reviews[1a,3c,4] have summarized the chemical design and applications of transition metals in luminescence cellular and biomolecular imaging. Despite the tremendous progress made in this fast-growing and fascinating research field, several limitations of such metal complexes still remain including: (1) intrinsic cytotoxicity from heavy metals; (2) photocytotoxicity caused by singlet oxygen sensitization upon light irradiation; (3) excitation wavelength is far from the "biological optical window" (600–1300 nm); and (4) unclear relationship between the structure and cellular compartmentalization because of complicated variable coordination spheres.

To address these issues, applying the first row transition elements biologically available provides an opportunity to further design luminescent labels with low cytotoxicity,[5] although some features such as large Stokes shifts and long lifetimes belonging to second and third row transition metals would be sacrificed.[3c,4a–e] Moreover, for the relatively predictable electronic structures and excited states, their photophysical properties and chemical structures could probably be rationalized. Furthermore, inspired by coordination spheres around the native metal cofactors in natural proteins or enzymes, such fluorophores may exhibit metalloenyzmes or proteins-like reactivity and selectivity. On the other hand, because of continuing interest in detecting intracellular biological metal ions such as Zn(II), Cu(I), etc., there are huge amounts of fluorogenic probes with specific metal binding motifs, which also provide a chemical basis to construct luminescent transition metal labels or sensors in biological imaging.[6] Therefore these new, fascinating transition metal complexes are very appealing and can be advantageously used for molecular imaging purposes. Among the first row transition metals, luminescent zinc complexes, which emerged as promising fluorescent labels or sensors, became powerful tools to detect molecular events or bioanalytes in biological processes, although they have been much less studied compared to second and third row transition metals and lanthanide complexes. Several excellent reviews for transition metals in bioimaging have mentioned zinc complexes[1a,3c,7]; however, there is no review specially focused on the luminescent zinc complexes. In this context, we will describe the principles and trends involved in the design, functionality, and reactivity of zinc complexes and hope to inspire further research in biological imaging (Fig. 1.1).

2. GENERAL CONSIDERATIONS

2.1 Basic Properties of Zinc(II)

Zinc is a group 12 metal with an electron configuration of $[Ar]3d^{10}4s^2$. Loss of the outer shell 4s electrons yields a bare divalent zinc cation (Zn^{2+}) with the electronic configuration $[Ar]3d^{10}$. The closed-shell d^{10} electronic configuration endows Zn^{2+} with the following features: (1) no geometry-dependent ligand-field stabilization energy[8]; (2) a typical borderline Lewis acid[8a,9]; (3) redox inertness and divalent zinc is the dominant oxidation state[8c,10]; (4) diamagnetic and no d–d transition.

2.2 Coordination Chemistry

The typical coordination geometries of zinc complexes are tetra-coordinated tetrahedral, penta-coordinated trigonal bipyramidal, penta-coordinated square pyramidal, and hexa-coordinated octahedral (Fig. 1.2). These

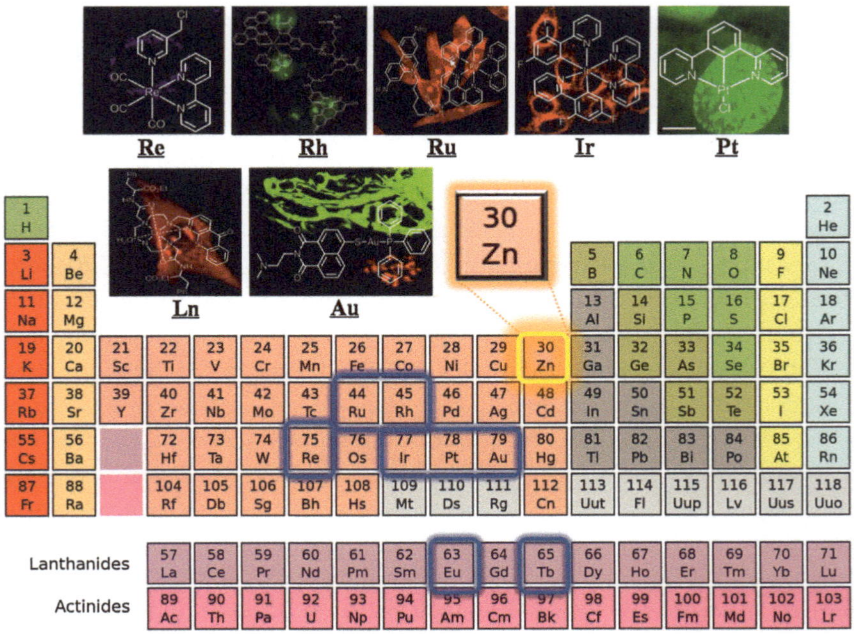

Figure 1.1 Luminescent transition metal complexes used in biological imaging.

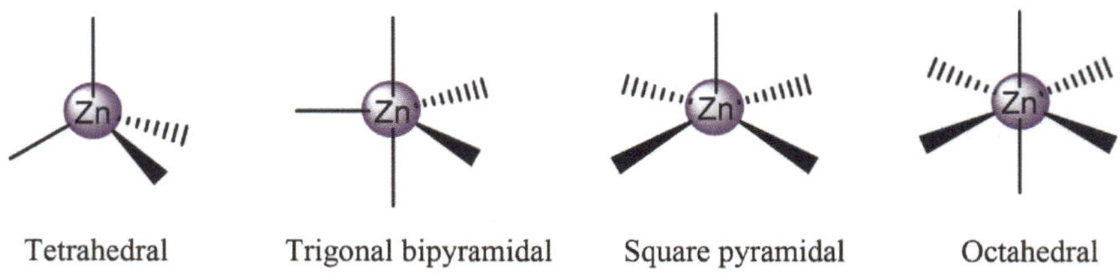

Figure 1.2 Four typical coordination geometries of zinc complexes.

configurations prevail in almost all zinc complexes, including zinc proteins existing in nature.[11] For lack of ligand-field stabilization energy, coordination geometries of zinc complexes are mainly tuned by the electrostatic and steric interactions from the ligands.

2.3 Photophysical Properties of Luminescent Zinc Complexes

To the best of our knowledge, because of the inability of the d electrons to participate in low-lying energy charge transfer or metal-centered transitions, excited states of zinc complexes are commonly ligand-centered transitions or ligand-to-ligand charge transfer (LLCT) transitions (Fig. 1.3).

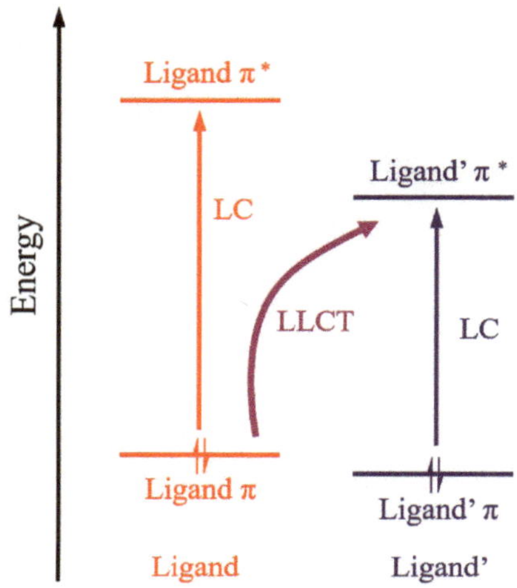

Figure 1.3 Schematic of orbitals and related electronic transitions in zinc complexes. *LC*, ligand centered; *LLCT*, ligand-to-ligand charge transfer.

Sometimes ligand-to-metal charge transfer, involving the low-lying 4s or 4p empty orbitals of the central Zn^{2+}, has been reported in the cluster-based zinc complexes.[12]

1. Ligand-centered transition
   These transitions only involve local ligand orbitals, which are almost unaffected by coordinated metal. With a closed-shell d^{10} electronic configuration, luminescence of zinc complexes mainly originates from ligand-centered transition and can be fine-tuned by elaborate modulation of the ligands' electronic states.[12,13]
2. Ligand-to-ligand charge transfer transition
   LLCT refers to the electron transfer occurring between two orbitals predominantly localized on different ligands to the same metal center. In most cases, however, the transitions are not involved in a single mechanism, but in a multiple mechanism.

Generally, compared to the free ligands, zinc complexes show stronger fluorescence. This is attributed to the enhancement of the rigidity of the fluorescent ligands by coordination to Zn^{2+}, reducing energy loss via bond vibration. For this reason, these ligands have been widely used for selectively detecting Zn^{2+} based on the emission changes. Zn^{2+} also tunes "off/on" fluorescence via photoinduced electron transfer (PET) pathways.[14] The strategy is depicted in Fig. 1.4 and a somewhat more quantitative explanation from the frontier molecular orbital (MO) energies is given in Fig. 1.5.

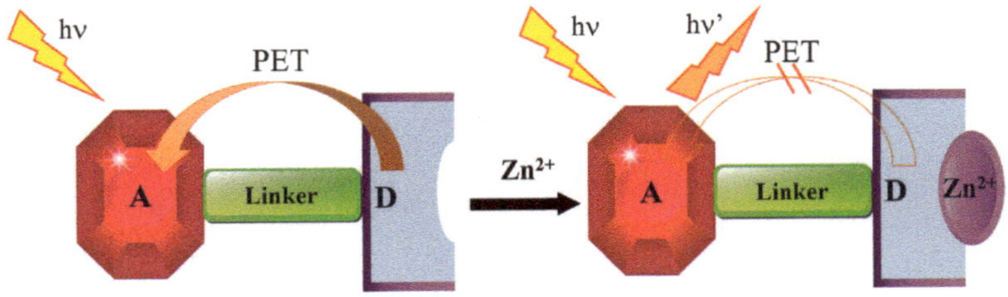

Figure 1.4 Zn^{2+} sensing by fluorescent photoinduced electron transfer (PET) indicators.

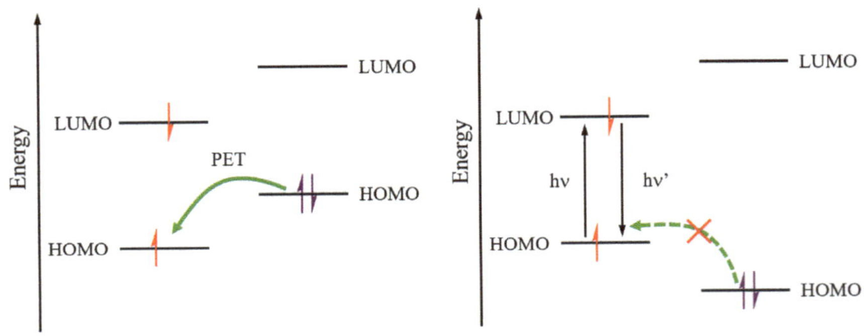

Figure 1.5 Energy diagram of photoinduced electron transfer (PET)-based sensing. *HOMO*, highest occupied molecular orbital; *hv*, electromagnetic radiation with specific wavelengths; *LUMO*, lowest unoccupied molecular orbital.

The PET process occurs because of the electron transfer from the donor, such as the lone pair of electrons on the nitrogen atom of benzylic amines, to the acceptor with the vacancy in a ground state orbital. Binding to Zn^{2+} via the lone pair electrons on the donor would obviously perturb the PET process and restore the fluorescence. Popularly, inhibition of PET by coordination to d^{10} Zn^{2+} is another predominant avenue to sense the traces of Zn^{2+} in several biological and environmental specimens.[15]

Moreover, Zn^{2+} could enhance two-photon absorption of the ligands. Tian et al.[16] reported the two-photon absorption cross-section of a β-diketone ligand (**HL**) to be 11.94 GM, and the corresponding property of the zinc complex **Zn(II)L$_2$(H$_2$O)$_2$** increases to 179.40 GM (Fig. 1.6). Combining with the results of time-dependent density functional theory calculations, it may be ascribed to complexation with Zn^{2+}, which enhances the electron-acceptor character of the central β-diketonate group and thus converts the ligand to a more strongly polarized donor–π–acceptor unit, leading to the zinc complex with dramatically enhanced two-photon absorption.

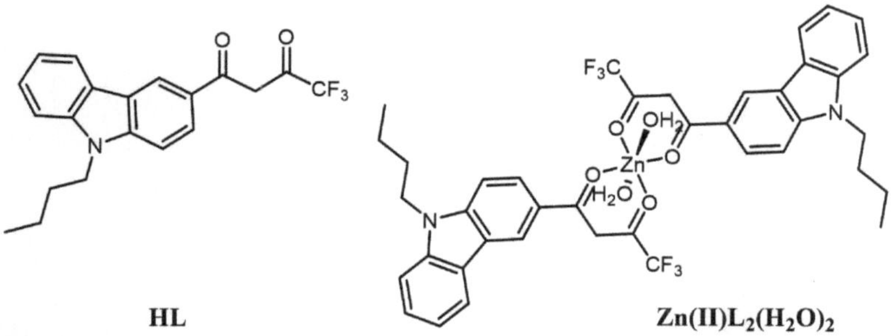

HL Zn(II)$L_2(H_2O)_2$

Figure 1.6 Structures of **HL** and **Zn(II)$L_2(H_2O)_2$** with two-proton absorption properties.

3. LUMINESCENT ZINC COMPLEXES AS BIOPROBES

Zinc is one of the most abundant transition metals in the living system and widely exists in many proteins (ca. 10% of all proteins in human body) active in metabolism, signal transduction, and gene expression. Zn^{2+} is an essential metal ion for normal biological processes, with important functions in the nervous, reproductive, and immune systems, and plays a central role in growth and development. In cell biology, Zn^{2+} is regarded as a dynamic signaling ion, forming an equilibrium within the cell or to the extracellular milieu.[9b,11a,b,17] Thus, cellular Zn^{2+} homeostasis is involved in many diseases, including degenerative diseases such as neurodegeneration (Alzheimer's disease, stroke), diabetes, cancer, and wound healing.[18] To explore the versatile functions of the micronutrient zinc, there has been a tremendous progress in new tools such as fluorescence microscopy to measure intracellular Zn^{2+} levels. Toward this goal, a lot of fluorogenic probes chelating intracellular Zn^{2+} specifically with the switched "on/off" fluorescence have been reported, which facilitates addressing the essential roles of Zn^{2+} in cellular regulation.[6i,19] On the other hand, the progress of Zn^{2+} sensors promotes the fast development of luminescent zinc coordination complexes applied in cell imaging. The earliest zinc complex as luminescent labels are Zn-DPA (DPA, dipicolylamine) complexes, which open a new access to design luminescent transition metal complexes for molecular imaging.

3.1 Zinc Complexes of DPA Ligands

3.1.1 Basic Properties of Zn-DPA Complexes

Among Zn-based biological imaging probes, Zn-DPA complexes are representative, because DPA could be regarded as a "half" part of TPEN [*N*,*N*,*N'*,*N'*-tetrakis-(2-pyridylmethyl)-ethylenediamine], which is the most widely

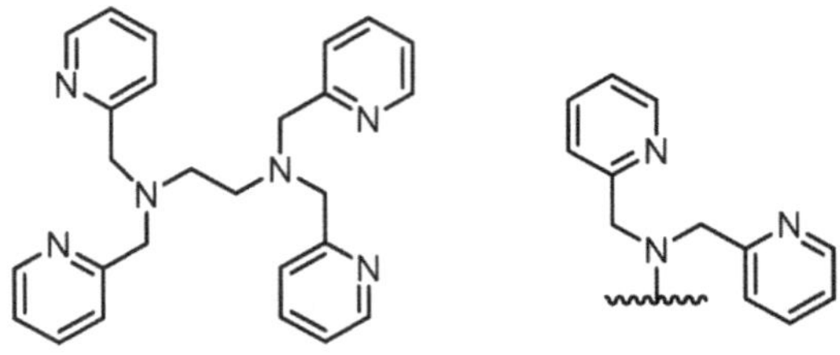

TPEN
1

DPA
2

Figure 1.7 Structures of *N,N,N′,N′*-tetrakis-(2-pyridylmethyl)-ethylenediamine (TPEN) and dipicolylamine (DPA) moiety.

used Zn^{2+} chelator with high binding affinity. Different from TPEN, forming a stable six-coordinated chelate, DPA binds Zn^{2+} of a trigonal bipyramidal geometry, and the tertiary amine could further conjugate to the fluorophore or other reporter moieties. As shown in Fig. 1.7, the electron-donating nitrogen atom on DPA often quenches the fluorescence via the PET mechanism, which could be perturbed when Zn^{2+} binds to DPA and thus switches on the fluorescence.

Zn-DPA complexes always adopt trigonal bipyramidal geometries and the rest of the coordination sites are occupied by solvent molecules generally, which could be readily replaced by anions such as phosphate. Dinuclear Zn-DPA complexes [bis(Zn-DPA)] have been reported and show improved binding ability to phosphate anions, including phosphate, pyrophosphate, ATP, and even phosphorylated peptides, proteins, and other biomolecules containing phosphate anions. Hamachi et al. first reported dinuclear Zn-DPA bioprobes (**3** and **4**) with anthracene moieties as fluorescent chemosensors to detect phosphorylated peptide in aqueous solution (Fig. 1.8).[20] This phosphate sensor also showed high selectivity toward other common anions, including carbonate, sulfate, and nitrate.[21]

3.1.2 Intracellular Phosphate Anion Probes

3.1.2.1 Polyphosphate Species

Hong et al. reported bis(Zn-DPA) bridged by a simple benzene (**5**) could selectively recognize flavin adenine dinucleotide (FAD) among flavin derivatives including flavin mononucleotide and riboflavin in water.[22] **5** has very weak fluorescence. After **5** selectively binding to phosphate anion, π–π stacking of the adenine moiety was dissociated and thus turned on the fluorescence of FAD. Notably, **5** has also been demonstrated to detect FAD in eosinophils by fluorescence microscopy and flow

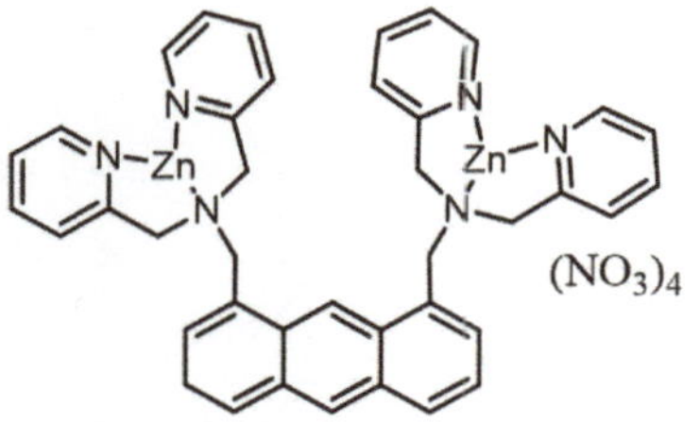

3

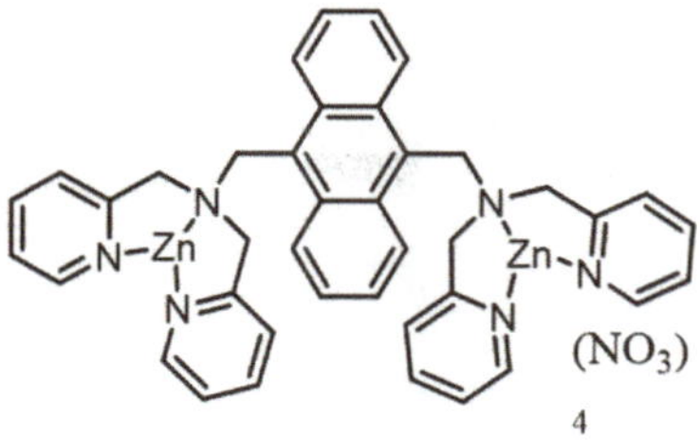

4

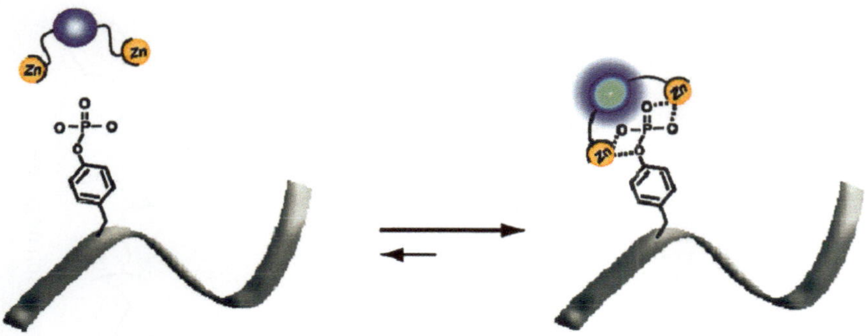

Figure 1.8 Zinc-dipicolylamine (Zn-DPA) complexes binding to phosphorylated peptide. *Reprinted with permission from Ojida, A.; Mito-Oka, Y.; Inoue, M. A.; Hamachi, I. First Artificial Receptors and Chemosensors toward Phosphorylated Peptide in Aqueous Solution.* J. Am. Chem. Soc. ***2002,*** 124 *(22), 6256–6258. Copyright 2002 American Chemistry Society.*

cytometry. Later, Hamachi et al. reported a new fluorescent chemosensor **6,** which is comprised of two Zn-DPA moieties as binding motifs and a xanthene as a fluorescent reporter.[23] They found that Zn-DPA moieties could facilitate hydroxylation of xanthene ring (Form I), forming the hydroxyl product (Form II) under neutral aqueous conditions, which is confirmed by single crystal X-ray diffraction. Form II is nonfluorescent and converts to fluorescent Form I after the addition of nucleoside polyphosphates such as ATP, which binds to the dinuclear Zn-DPA moieties and leads to dehydroxylation (Fig. 1.10). This process results in 15-fold fluorescence enhancement and a binding constant of $1 \times 10^6\,M^{-1}$ is obtained. Further imaging experiments are carried out for visualization of ATP particulate stores in Jurkat cells. It is worthy noting that this ATP sensor is pH independent and could be a general approach to detect ATP intracellularly.

For the purpose of quantitative applications, the xanthene-based bis (Zn-DPA) probe **7** for ratiometric fluorescence sensing has been reported by introduction of coumarin fluorophores to construct a fluorescence resonance energy transfer pair.[24] In this work, ratiometric fluorescence signal

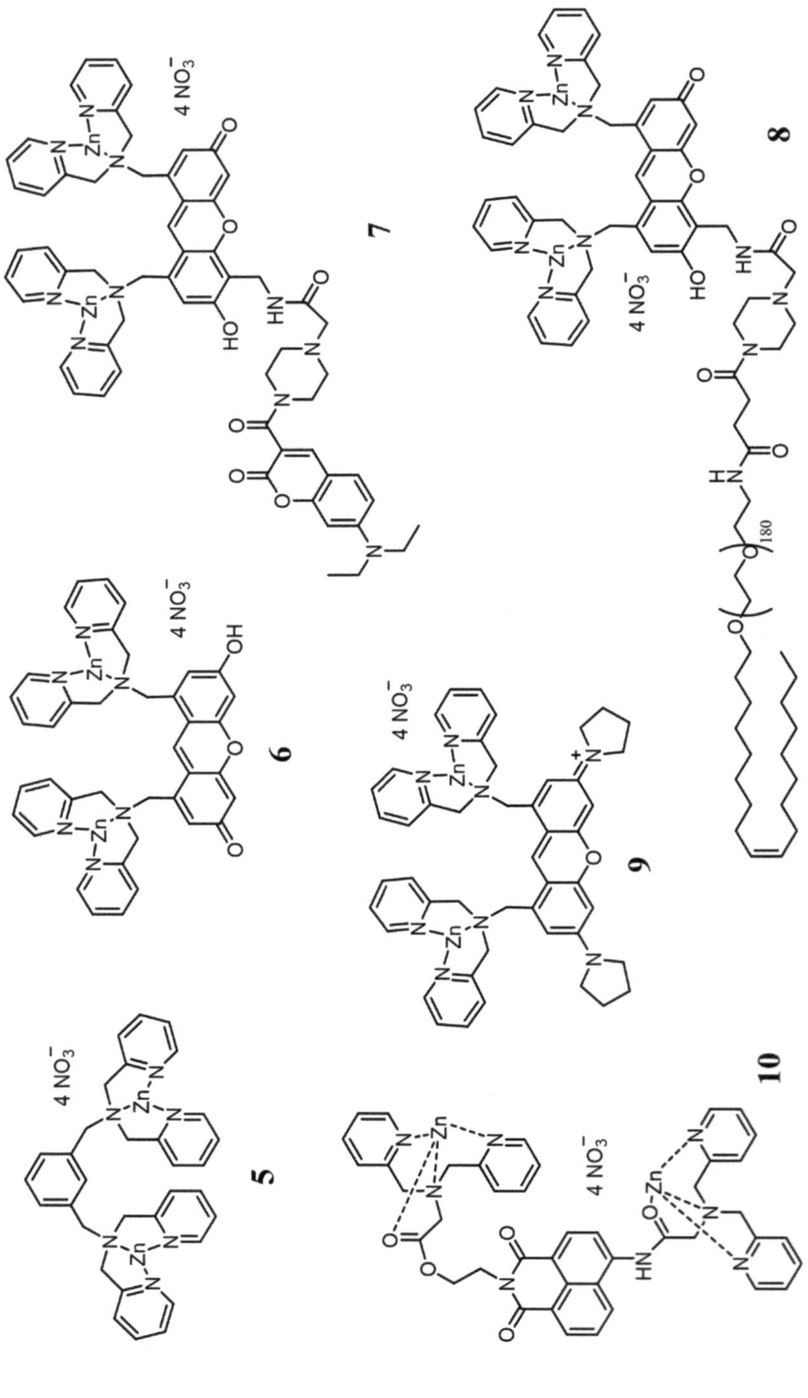

Figure 1.9 Bis(Zn-DPA)-based polyphosphate probes.

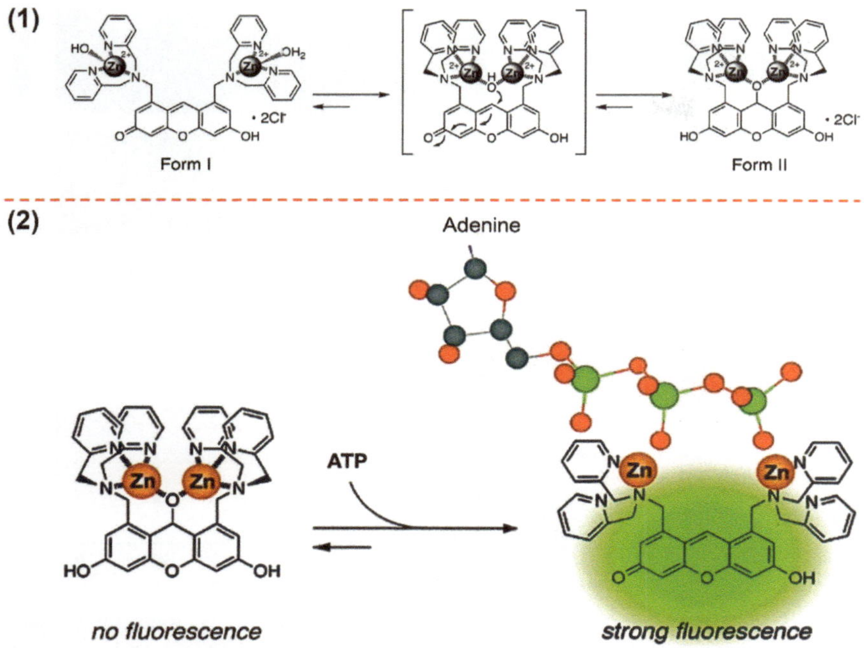

Figure 1.10 (1) Structural equilibrium of **6** under neutral aqueous conditions. (2) Mechanism of **6** probing ATP. *Reprinted with permission from Ojida, A.; Takashima, I.; Kohira, T.; Nonaka, H.; Hamachi, I. Turn-on Fluorescence Sensing of Nucleoside Polyphosphates Using a Xanthene-Based Zn(II) Complex Chemosensor.* J. Am. Chem. Soc. ***2008,*** 130 *(36), 12095–12101. Copyright 2008 American Chemical Society.*

responding to ATP concentration inside living cells could be obtained, which can be an indicator of the cellular energy level (Fig. 1.11).

The relationship between structures and organelle localizations of zinc complexes has been investigated. To image ATP on plasma membrane surfaces, a lipid chain was introduced to the xanthene ring to form probe **8**, which can detect the extracellular release of ATP during the cell necrosis process induced by streptolysin O. Meanwhile, replacing the xanthene ring with a cationic pyronin ring, **9** could stain mitochondria and probe the local change of ATP concentration during apoptosis.[25] Interestingly, these probes can be used for cell imaging and monitoring the dynamics of ATP in different subcellular organelles at the same time. In addition, Kim et al. reported a pyrophosphate-selective fluorescent chemosensor **10** with a 1,8-naphthalimide-bridged bis(Zn-DPA) structure.[26] Upon pyrophosphate binding, **10** shows fluorescence turn-off, which might be attributed to photoinduced charge transfer from negatively charged pyrophosphate to

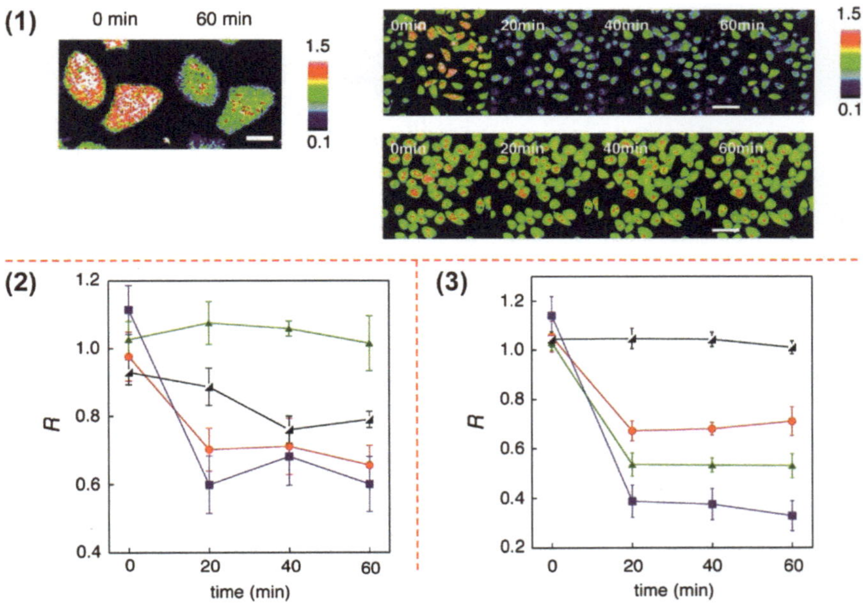

Figure 1.11 Ratiometric analysis of living cells stained with **7**. (1) Ratio images and time-lapse images of HeLa cells treated with 2-deoxyglucose (2-DG) and potassium cyanide (KCN). Average time-course plots of the emission ratios of HeLa cells (2) and HEK293 cells (3) upon treatment with KCN (green), 2-DG (red), or KCN + 2-DG (blue) or without the treatment (black). *Reprinted with permission from Kurishita, Y.; Kohira, T.; Ojida, A.; Hamachi, I. Rational Design of FRET-based Ratiometric Chemosensors for in Vitro and in Cell Fluorescence Analyses of Nucleoside Polyphosphates.* J. Am. Chem. Soc. ***2010,*** 132 *(38), 13290–13299. Copyright 2010 American Chemical Society.*

either the chelating moiety or the fluorophore. **10** has been used to detect pyrophosphate in C2C12 cells by fluorescence microscopy and displays the quenching fluorescence in the presence of pyrophosphate.

3.1.2.2 Phosphorylated Peptides and Proteins

Phosphorylation and dephosphorylation are important posttranslational modifications of native proteins, occurring site specifically on a protein surface. These biological processes play important roles in intracellular signal transduction cascades and switching the enzymatic activity. To monitor these processes in real time, the choice of fluorogenic probes with high binding affinity to phosphorylated peptides and proteins is critical. Toward this goal, Hamachi et al. developed a BODIPY-bridged bis(Zn-DPA) complex **11** for the detection of neurofibrillary tangles (NFTs) of hyperphosphorylated

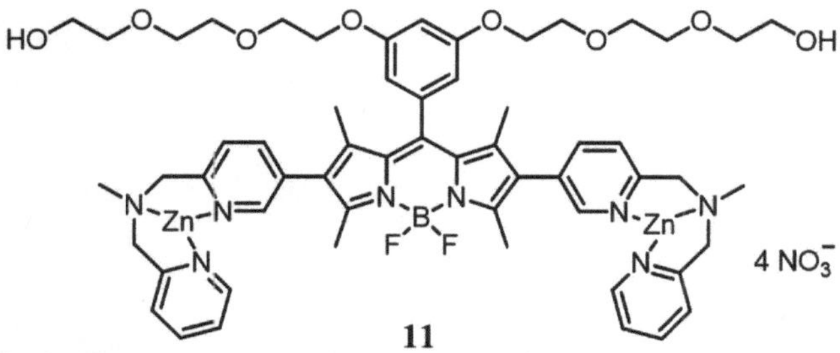

11

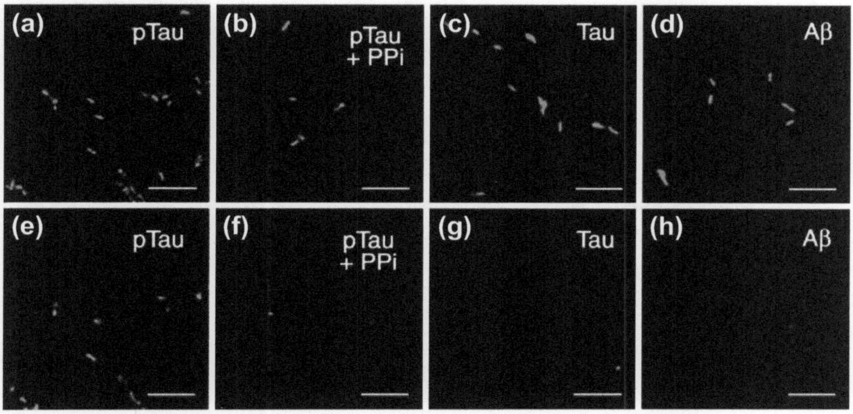

Figure 1.12 Bis(Zn-DPA)-based phosphorylated peptide probes **11** and fluorescence imaging of aggregates of tau protein (p-Tau), nonphosphorylated tau protein (n-Tau), and $A\beta_{1-42}$ stained with (a–d) thioflavin T, a nonselective fluorescent probe for β-sheet structures of protein aggregates, and (e–h) **11**. *Reprinted with permission from Ojida, A.; Sakamoto, T.; Inoue, M.-A.; Fujishima, S.-H.; Lippens, G.; Hamachi, I. Fluorescent BODIPY-Based Zn(II) Complex as a Molecular Probe for Selective Detection of Neurofibrillary Tangles in the Brains of Alzheimer's Disease Patients.* J. Am. Chem. Soc. ***2009,*** 131 *(18), 6543–6548. Copyright 2009 American Chemical Society.*

tau proteins, which is a feature of Alzheimer's disease.[27] The probe binds preferentially to phosphorylated peptides, with an association constant of about $1 \times 10^5\,M^{-1}$. A fluorescence titration experiment with the phosphorylated tau protein (p-Tau) aggregates showed the strong binding affinity of **11** to p-Tau ($EC_{50} = 9$ nM). In contrast, the weaker interaction has been found toward the nonphosphorylated tau protein (n-Tau; $EC_{50} = 80$ nM) and $A\beta_{1-42}$ fibrils ($EC_{50} = 650$ nM) (Fig. 1.12). Further histological imaging of an Alzheimer's disease patient's hippocampus tissue section demonstrated that **11** can visualize NFT deposits with switched-on fluorescence and discriminate them from the amyloid plaques.

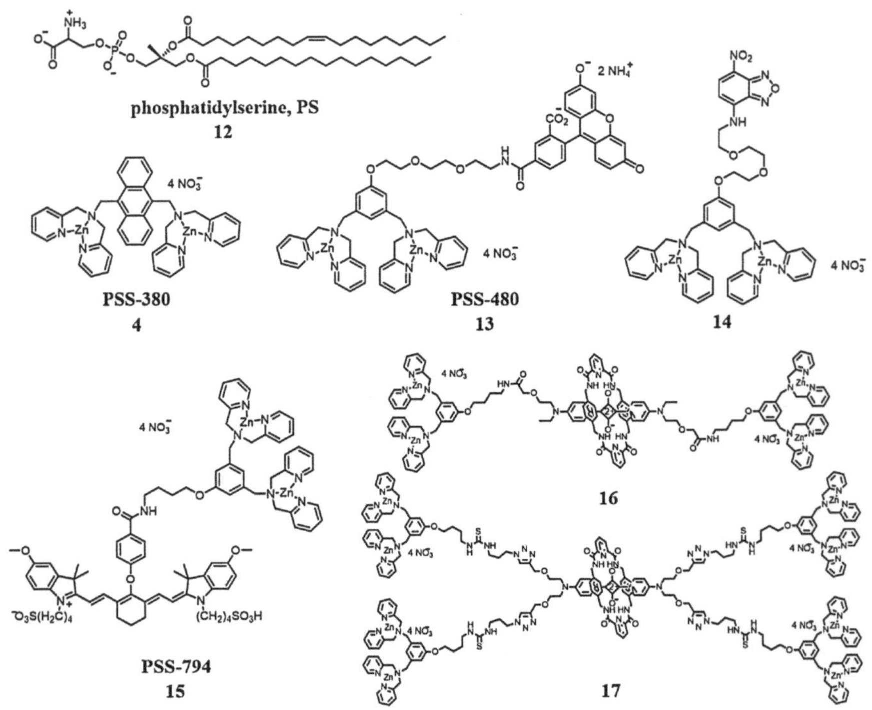

Figure 1.13 Structures of phosphatidylserine and bis(Zn-DPA)-based cell death probes.

3.1.2.3 Cell Death and Bacterial Infection

Bis(Zn-DPA) probes have been extended to detect anionic phospholipids, which are potentially useful for detecting cell apoptosis. For example, phosphatidylserine (**PS**, **12**), with an anionic phosphate as head group, is a hallmark of the early or intermediate stage of cell apoptosis and exists on the cell surface. Smith et al. first used the bis(Zn-DPA)-based **PSS-380** complex as a **PS** probe.[28] A vesicle titration experiment shows that, in the presence of anionic vesicles containing 50% **PS**, there is a 10-fold increase in fluorescence emission of **PSS-380** in contrast to no fluorescence change with zwitterionic phosphatidylcholine vesicles. Even when the concentration of **PS** is reduced to 5%, 4-fold fluorescence enhancement is still observed, indicating the high sensitivity of **PSS-380**. Living cell fluorescence imaging shows that **PSS-380** selectively stains the surface of apoptotic Jurkat cells which are pretreated with the apoptosis-inducing drug camptothecin (Fig. 1.14).[29]

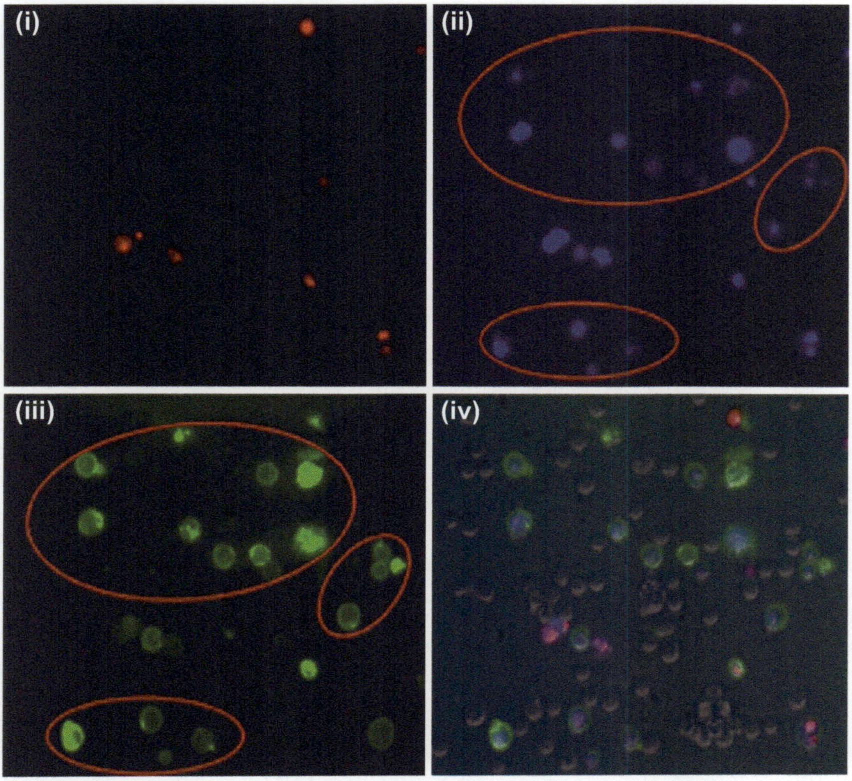

Figure 1.14 Apoptosis of Jurkat cells stained simultaneously with 7AAD (i), annexin V-FITC (ii), and **PSS-380** (iii). Image (iv) is a merge of (i–iii), and a phase-contrast image of the same cells. Apoptotic cells are highlighted by *red ovals. Reprinted with permission from Koulov, A. V.; Stucker, K. A.; Lakshmi, C.; Robinson, J. P.; Smith, B. D. Detection of Apoptotic Cells Using a Synthetic Fluorescent Sensor for Membrane Surfaces that Contain Phosphatidylserine.* Cell Death Differ. ***2003,*** 10 *(12), 1357–1359. Copyright 2003 Nature.*

As sensors with different fluorescence color can be easily designed by simply changing the reporting fluorophores, Smith et al. reported a series of different emissive bis(Zn-DPA) probes with fluorescein (**13**), 7-nitrobenz-2-oxa-1,3-diaza-4-yl (**14**), and quantum dots as the fluorescent reporters to detect **PS**.[29,30] **PSS-794**, constructed by bis (Zn-DPA) and a near infrared (NIR) carbocyanine fluorophore, was used for in vivo imaging of myopathy nude mice models (Fig. 1.15).[31] It proved to be an effective probe to detect cell death in vivo, showing major distribution at damaged muscle, while the traditional cell death probe, annexin V conjugates, showing high accumulation in the bladder in this case. **PSS-794**

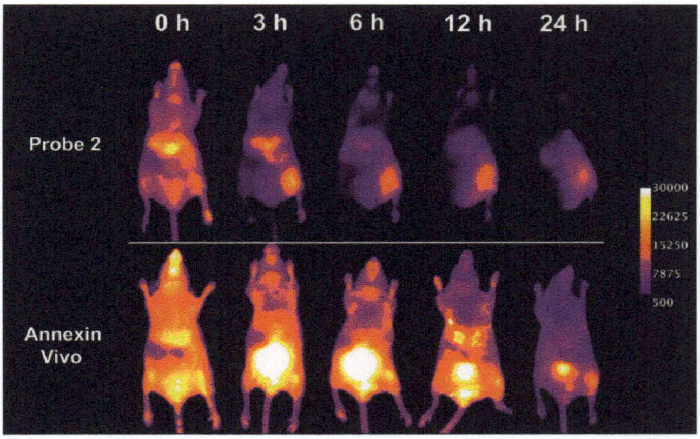

Figure 1.15 Representative near infrared (NIR) fluorescence images of a mouse treated with cell-death inducing agent in the hind leg and injected with either **PSS-794** (top row) or Annexin-Vivo 750 (bottom row) via the tail vein. *Reprinted with permission from Smith, B. A.; Gammon, S. T.; Xiao, S.; Wang, W.; Chapman, S.; McDermott, R.; Suckow, M. A.; Johnson, J. R.; Piwnica-Worms, D.; Gokel, G. W.; Smith, B. D.; Leevy, W. M. In Vivo Optical Imaging of Acute Cell Death Using a Near-infrared Fluorescent Zinc-Dipicolylamine Probe.* Mol. Pharm. ***2011***, 8 *(2), 583–590; Smith, B. A.; Xiao, S.; Wolter, W.; Wheeler, J.; Suckow, M. A.; Smith, B. D. In Vivo Targeting of Cell Death Using a Synthetic Fluorescent Molecular Probe.* Apoptosis ***2011***, 16 *(7), 722–731. Copyright 2011 American Chemical Society.*

has been also used for imaging other tissue damage, including mammary and prostate tumors, traumatic brain injury, and retinal ganglion cell degeneration.[32] Given the success of **PSS-794**, Zn-DPA complexes **16** and **17** with an increased number of DPA moieties and squaraine rotaxanes as a NIR reporter have been reported.[33] Vesicle and cell imaging studies suggested that more Zn-DPA targeting units could improve the selectivity and sensitivity as the probe accumulated more at the surface of dead and dying cells. In three separate animal models (necrotic prostate tumor rat model, thymus atrophy mouse model, and traumatic brain injury mouse model), **17** showed higher accumulation at the site of cell death than **16**, indicating that increasing the number of Zn-DPA moiety is an effective approach to enhance selectivity and sensitivity in imaging cell death processes.

Further modification of DPA structure can improve the sensitivity to **PS**. Smith et al. reported a new generation of NIR fluorescent probes such as **18** on the basis of **PSS-794** described earlier, which is optimized by using a high-throughput screening method from a library of synthetic DPA derivatives. It decreases the dissociation constant from 133 nM to 39 nM to **PS** and exhibits a 3-fold higher accumulation at diseased tissue, leading to a higher signal-to-noise ratio.[34] Bis(Zn-DPA)-based fluorescent probes have also been applied in staining bacterial membranes with anionic surfaces. Smith et al. reported that **19** shows the ability to stain the surfaces of *Escherichia coli*, *Pseudomonas aeruginosa*, and *Staphylococcus aureus* cells, which could be extended to detect bacterial infection (Fig. 1.16).[35]

3.1.3 Zn-DPA Complexes as Protein Labels

Other than phosphate sensing, Zn-DPA complexes are able to bind to histidine, aspartate, and glutamate, indicating the potential for directly labeling proteins, in a similar way to "His6 tag + Ni^{2+}". Toward this goal, Hamachi et al. reported that complex **6** and its analogs can selectively bind to the α-helix with two histidine residues at specific positions (i and $i + 4$ or $i + 7$ or $i + 11$).[36] Furthermore, another Zn-DPA complex **20** has a binding constant of $(1.1 \pm 0.1) \times 10^6\ M^{-1}$ with a tetraaspartate peptide tag, and the formed complex could be cleaved when a high concentration of pyrophosphate solution is used.[37] More importantly, **20** is orthogonal to the normally used His-6 tag and might be used to target two proteins with different fluorescence tags at the same time. Hamachi et al. demonstrated that bis(Zn-DPA) complex **22** binds to G-protein coupled receptor (GPCR) fused with a short CA6D4x2 tag (CAAAAAADDDDGDDDD) on the HEK293 cell surface (Fig. 1.18).[38] Upon binding, the reactive α-chloroacetyl moiety can react with the cysteine residue, realizing the covalent labeling (Fig. 1.18). Thus the GPCR internalization process and the following microenvironmental changes at endosome interior could be studied in living cells by fluorescence microscopy. Following this work, several zinc iminodiacetic acid (Zn-Ida) complexes (complex **23** and its analogs) with Cy5 fluorophore have been reported as binders to a distinct His-10 tag and used for selective fluorescent imaging of His-tag-fused GPCR on live cell surfaces.[39]

Aside from DPA, other Zn^{2+} chelators can also connect to a fluorescent reporter to form similar luminescent zinc complexes. For example, Zn^{2+} can selectively bind to cyclen (1,4,7,10- tetraazacyclododecane). When cyclen is connected to a fluorescent dansylamidoethyl group, the resultant zinc complex **24** is a five-coordinated, distorted square pyramidal structure and can be used as a good probe for cell apoptosis at early stages.[40] Eisenhut et al. synthesized a series of fluorescent Zn-cyclen complexes and studied their variations in the uptake level of apoptosis/necrosis cells (Fig. 1.19).[41]

In summary, the open chelation mode of Zn-DPA allows zinc to further bind to other ligands, especially anion groups such as phosphate, carboxylate, and histidine, which switches on/off the fluorescence of attached

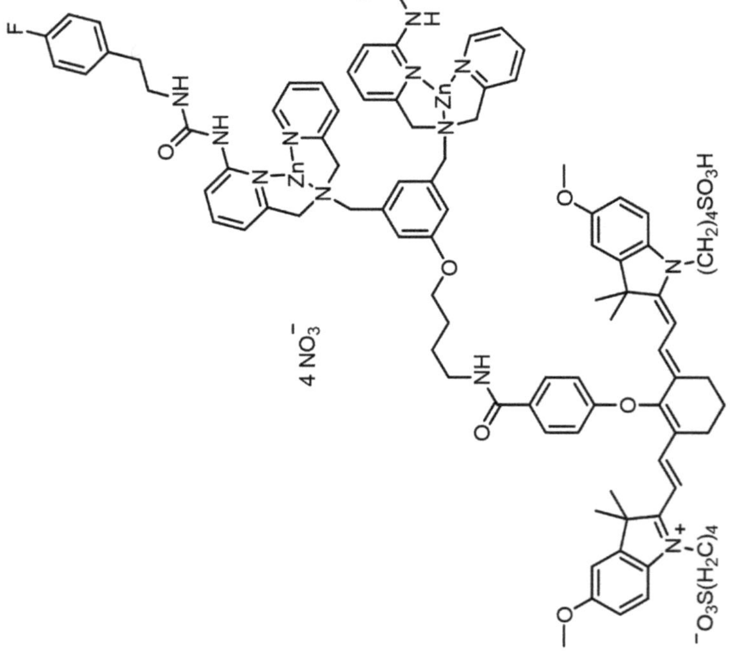

18

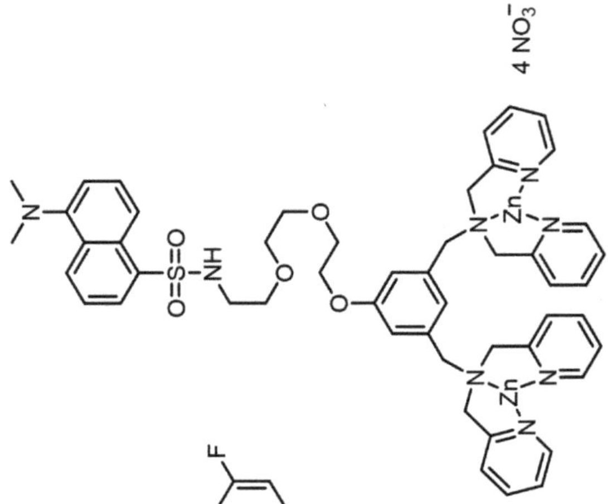

19

Figure 1.16 Structures of modified bis(Zn-DPA)-based cell-death near infrared (NIR) probe **18** and bacterial cell surface label **19**.

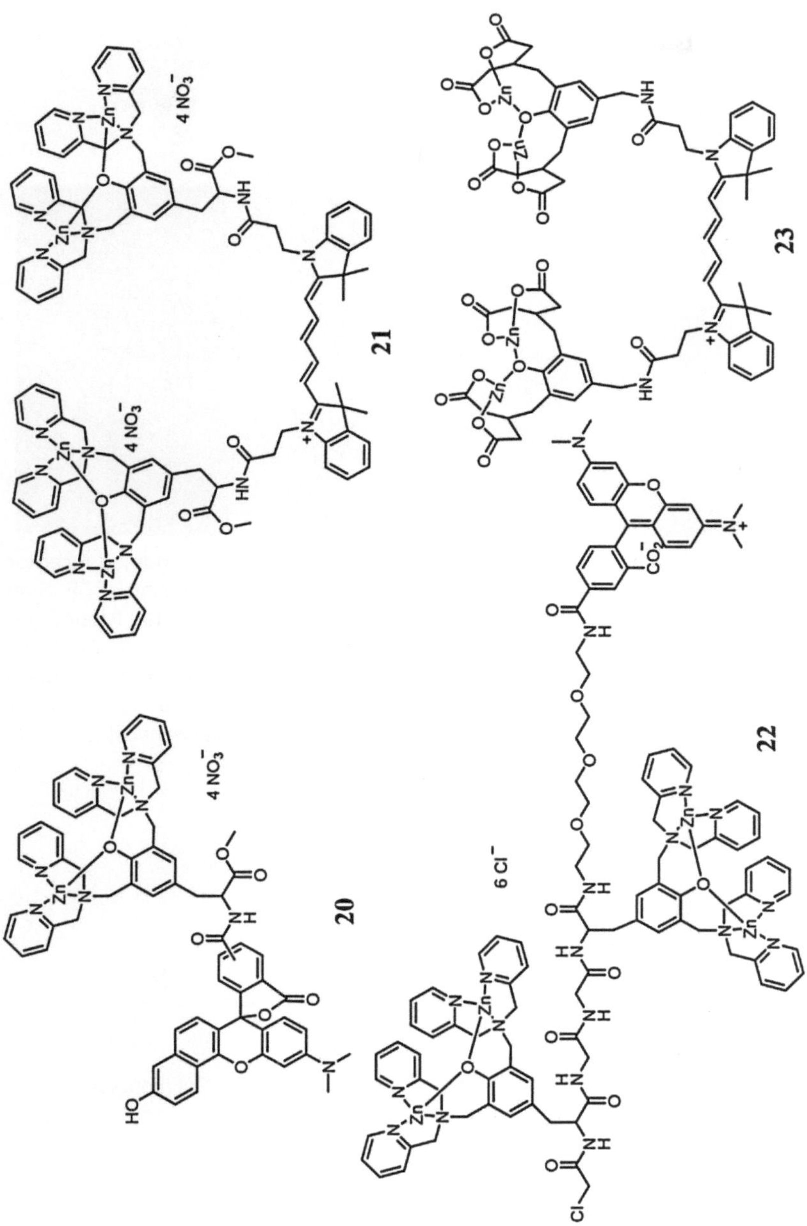

Figure 1.17 Bis(Zn-DPA)-based agents for protein labeling.

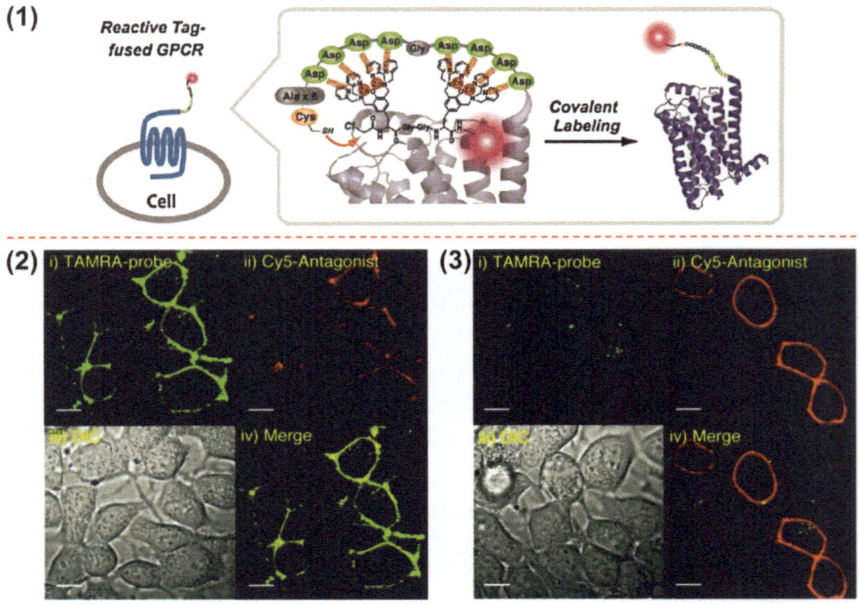

Figure 1.18 (1) Selective covalent labeling of a tag-fused G-protein coupled receptor (GPCR) on the surface of living cells using **22**. (2) and (3) Covalent labeling of GPCR on the surface of HEK293 cells with **22**. Fluorescence imaging of CA6D4x2 tag-fused GPCR (2), CA6D4x2 tag lacked GPCR (3). *Reprinted with permission from Nonaka, H.; Fujishima, S.-H.; Uchinomiya, S.-H.; Ojida, A.; Hamachi, I. Selective Covalent Labeling of Tag-Fused GPCR Proteins on Live Cell Surface With a Synthetic Probe for Their Functional Analysis.* J. Am. Chem. Soc. ***2010,*** 132 *(27), 9301–9309. Copyright 2010 American Chemical Society.*

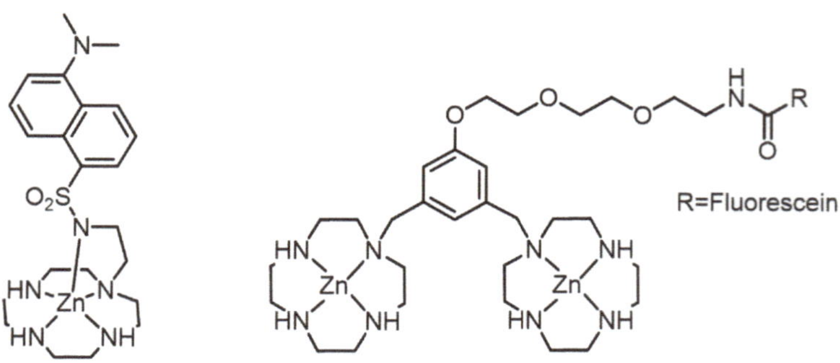

Figure 1.19 Structures of Zn-cyclen bioprobes.

fluorophores. These Zn-DPA complexes are widely studied as fluorescent bioprobes for detecting pyrophosphate species, phosphorylated peptides, and proteins in living cells, as well as cell death and bacterial infection. Importantly, Zn-DPA can be used as a binding group for tag-fused protein labeling, which would render them selective binding proteins in living cells and even whole body. However, Zn-DPA complexes do have some deficiencies. First, monomeric Zn-DPA complexes might not be stable enough and are sensitive to pH arising from the low binding affinity to Zn^{2+}. To overcome this, more than one Zn-DPA moiety was induced, but might interfere with the cellular uptake and subcellular localization. Second, these molecules were derived from the well-known Zn^{2+} sensors and constituted two parts, "fluorogenic reporter + chelator," which also increases molecular sizes. Thus direct application of luminescent zinc coordination complexes with chromophore ligands becomes an alternative approach to design zinc labels in molecular imaging.

3.2 Zinc Complexes of Thiosemicarbazone Ligands

3.2.1 Basic Properties of Zinc Thiosemicarbazone Complexes

Thiosemicarbazones are the analogs of the semicarbazides with a sulfur atom in place of the oxygen atom. They are readily prepared by the condensation reaction between a ketone (or aldehyde) and a thiosemicarbazide. These compounds are versatile chelators toward a range of metal ions, such as divalent metal ions (Cu^{2+}, Zn^{2+}) and trivalent metal ions (Co^{3+}, Fe^{3+}, Re^{3+}, Ga^{3+}, In^{3+}), with sulfur and azomethine nitrogen atoms. Biologically, these compounds have a broad spectrum of therapeutic properties with antibacterial, antimalarial, antiviral, and antitumor activities, through potentially binding to cellular copper, iron, or zinc ions.[42] However, the cellular uptake pathway and subcellular distribution are not fully investigated. Thus tracking zinc complexes of tridentate monothiosemicarbazone and tetradentate bis(thiosemicarbazone) ligands using their intrinsic fluorescence facilitates an understanding of biological activities. Importantly, zinc coordination induces the enhanced fluorescence, because the structures of zinc complexes of thiosemicarbazones become more rigid.[42a] More importantly, modification of thiosemicarbazone ligands would tune the photophysical properties of their zinc complexes (Fig. 1.20).

3.2.2 Luminescent Zinc Complexes of Tridentate Thiosemicarbazones

Triapine (3-aminopyridine-2-carboxaldehyde thiosemicarbazone) is a promising antitumor chemotherapeutic agent under phase II clinical trials.

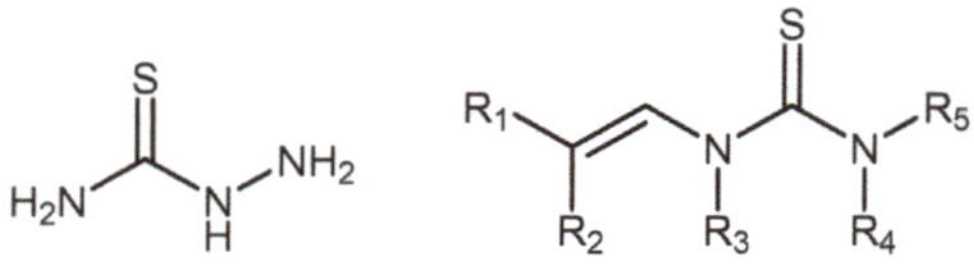

Figure 1.20 Chemical structures of thiosemicarbazide and thiosemicarbazone.

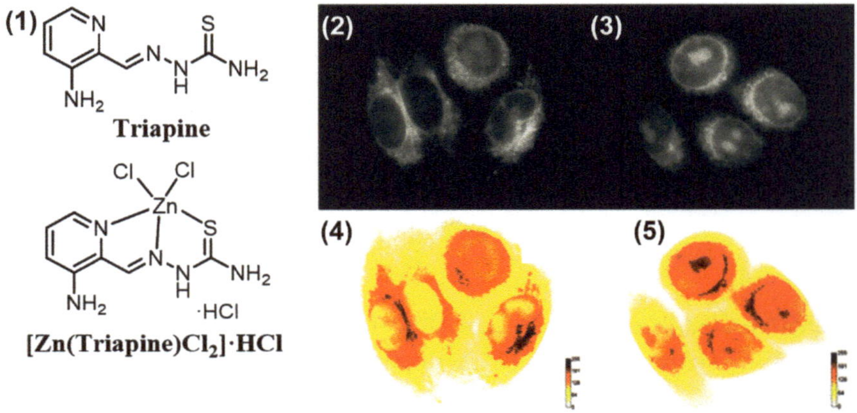

Figure 1.21 (1) Structures of **Triapine** and **Zn(Triapine)Cl$_2$**; fluorescent images of **Triapine** (2) and (4) or **Zn(Triapine)Cl$_2$** (3) and (5) in SW480 colon carcinoma cells. *Reprinted with permission from Kowol, C. R.; Trondl, R.; Arion, V. B.; Jakupec, M. A.; Lichtscheidl, I.; Keppler, B. K. Fluorescence Properties and Cellular Distribution of the Investigational Anticancer Drug Triapine (3-Aminopyridine-2-Carboxaldehyde Thiosemicarbazone) and Its Zinc(II) Complex.* Dalton Trans. ***2010***, 39 *(3), 704–706. Copyright 2010 Royal Society of Chemistry.*

To gain a deeper biological insight, Keppler et al. investigated the cellular uptake and intracellular distribution of **Triapine** and its zinc complex [**Zn(Triapine)Cl$_2$**] in living human cancer cells. Interestingly, although the cytotoxicity of **Triapine** and its zinc complex are similar in vitro, the intracellular distributions are distinctly different, as shown in Fig. 1.21. **Triapine** displays a tendency to accumulate in the nuclear membrane and the cytoplasm, whereas **Zn(Triapine)Cl$_2$** apparently stains nucleoli, which reveals the different intracellular distribution caused by the zinc coordination (Fig. 1.21).[43]

3.2.3 Luminescent Zinc Complexes of Tetradentate Thiosemicarbazones

Dilworth et al. investigated the uptake and distribution of tetradentate bis(thiosemicarbazone) complexes using fluorescence microscopy. Two fluorescent zinc bis(thiosemicarbazone) complexes, **Zn(ATSM)** and **ZnCl(ATSpyrH)**, were synthesized, which exhibited emissions in the range of 500–600 nm. Living cell imaging of their internalization in IGROV ovarian cancer cells, MCF-7 breast cancer cells, and PC-3 prostate cancer cells showed that the terminal pyrrole substituents significantly influence their intracellular distribution including nucleus and cytoplasm (Fig. 1.22).[44]

To overcome the disadvantage of weak fluorescence, Pascu et al. incorporated an aromatic ligand backbone acenaphthoquinone to replace bis-thiosemicarbazones based on **ZnCl(ATSpyrH)** systems, and prepared four zinc complexes with different exocyclic substituents (**a**, asymmetric; **b**, **c**, and **d**, symmetric). These complexes are highly cytotoxic. The cellular uptake pathways are dependent on the structural symmetry. The asymmetric complex (**a**) prefers endocytosis, whereas the symmetric derivatives (**b** and **c**) enter cells through passive diffusion. Moreover, confocal fluorescence imaging studies showed that these complexes induce rapid mitochondrial disruption.[45] To lower the cytotoxicity, Samuelson et al. reported a novel trimeric zinc complex formed by a glyoxalbis(4-methyl-4-phenyl-3-thiosemicarbazone) (**GTSCH2**) ligand, which showed high

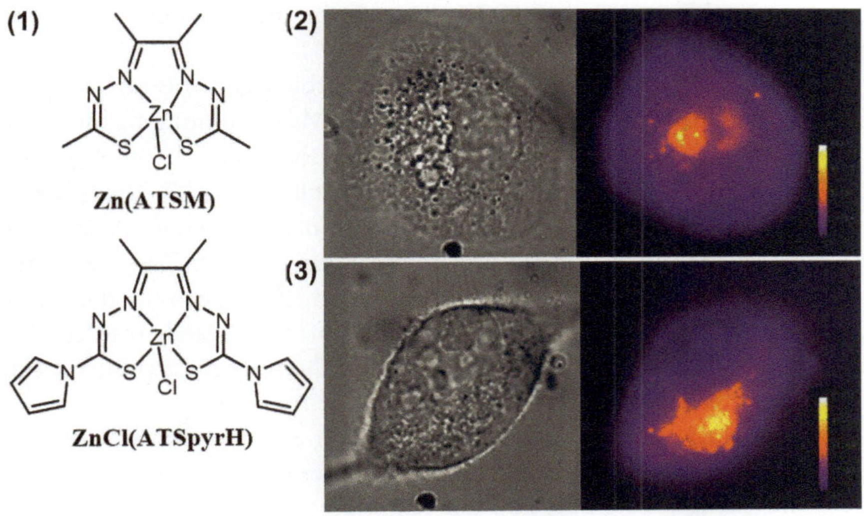

Figure 1.22 (1) Structures of **Zn(ATSM)** and **ZnCl(ATSpyrH)**; fluorescence image of **Zn(ATSM)** (2) and **ZnCl(ATSpyrH)** (3) in PC-3 prostate cancer cells. *Reprinted with permission from Cowley, A. R.; Davis, J.; Dilworth, J. R.; Donnelly, P. S.; Dobson, R.; Nightingale, A.; Peach, J. M.; Shore, B.; Kerr, D.; Seymour, L. Fluorescence Studies of the Intra-cellular Distribution of Zinc Bis(thiosemicarbazone) Complexes in Human Cancer Cells.* Chem. Commun. ***2005****, (7), 845–847. Copyright 2005 Royal Society of Chemistry.*

fluorescence and low cytotoxicity, making it a promising agent for live cell imaging (Fig. 1.23).[46]

3.3 Zinc Complexes of Salen-Type Ligands

3.3.1 Basic Properties of Znsalen Complexes

The nomenclature of salen, a typical N_2O_2 tetradentate bis-Schiff base ligand (Fig. 1.24), comes from the combination of salicylaldehyde (sal) and ethylenediamine (en). Since P. Pfeiffer first reported metal salen complexes in 1933,[47] their coordination chemistry has attracted much attention in catalytic studies, supramolecular assemblies, and materials.[48] In Znsalen complexes, Zn^{2+} locates in the N_2O_2 base plane and tends to be penta-coordinated to various Lewis bases such as neutral alcohols, carbonyls, amine, and pyridine with a square pyramid geometry.[49] In noncoordination solvents, Znsalens tend to form aggregates through intermolecular Zn···O interaction involving Lewis basic oxygen atoms (phenoxyl in another Znsalen molecule).[48c,50] The degree of aggregation is closely related to the concentration of Znsalen, steric hindrance of substituents, hydrophobic property, and polarity of the solvent (Fig. 1.25).[48c,50] Importantly, dissociating the oligomer to the monomer is accompanied by substantial changes in photophysical properties.[50,51] Thus the relationship between photophysical properties and the speciation of Znsalens offers a shortcut route for determining the aggregation process in solution.

For Znsalen, the lower-lying energy 3d orbital and the higher-lying energy 4s orbital drop Zn^{2+} out of the frontier MO, indicating that it is not involved in the related electron transition process and that the photophysical properties of Znsalen resemble those of salen ligands (**L**).[52] Thus photophysical properties of Znsalen could be finely tuned through modulation of the ligand including the conjugation degree, energy level, and the electronic effect of the substituents in the conjugated system.[52,53] Obviously, the structures of Znsalens are more rigid than those of free salen ligands; the decrease of the energy loss via bond vibration enhances fluorescence quantum yield, chemical stability, and photostability.[54]

Introducing an electron-donor (D)/electron-acceptor (A) pair to salen ligands generates intramolecular charge transfer (ICT) properties, along with the changes of absorption and emission spectra and the nonlinear optical properties. More importantly, hyperpolarization of the most important low-lying excited states reduces the loss of the nonradiative energy, enhancing the fluorescence quantum yield. Generally, when the salen ligands are anchored by a strongly electron withdrawing moiety such as

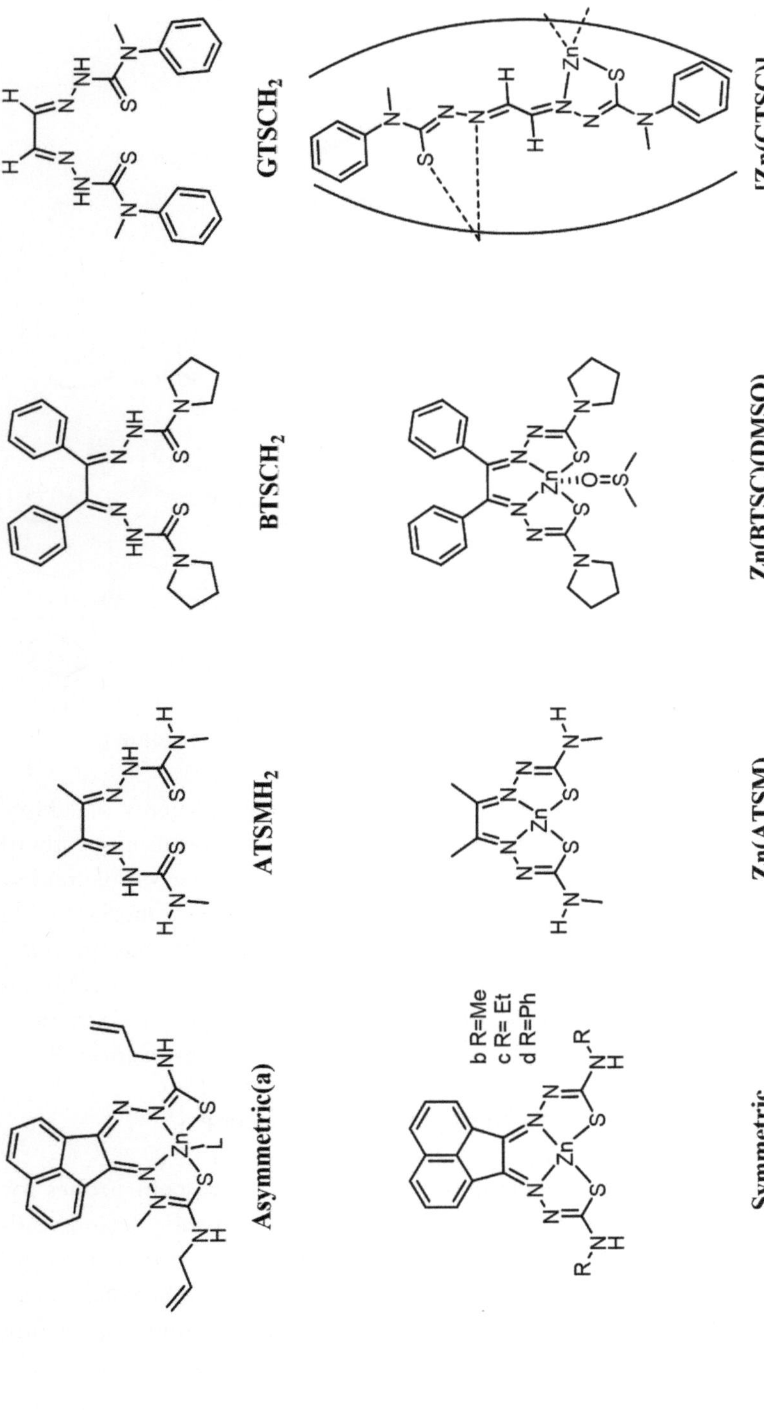

Figure 1.23 Zinc complexes of bisthiosemicarbazone ligands.

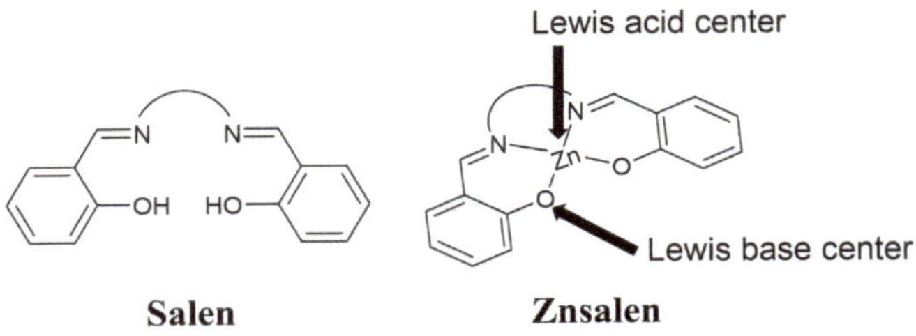

Figure 1.24 Salen and Znsalen.

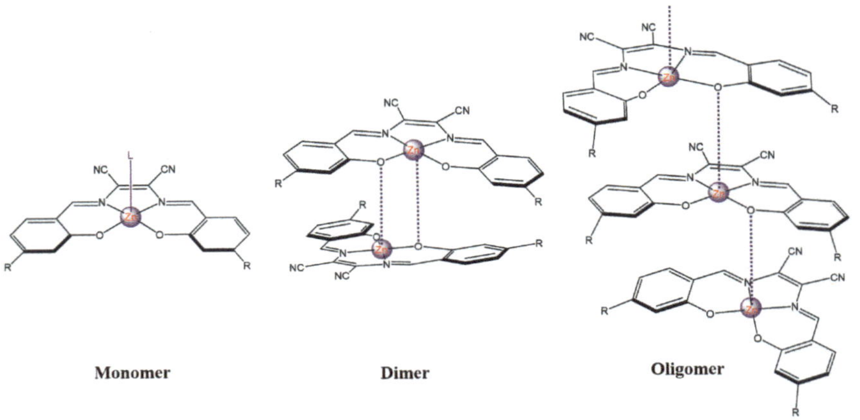

Figure 1.25 Speciation of Znsalens (monomer, dimer, or oligomer).

a cyano group (A) on diamines and dialkylamino substituents or alkoxy residues (D) on the salicylaldehyde part, the electronic absorption spectra of Znsalen would show intense ICT absorption features in the visible region (ca. 500–620 nm).[51,53b,54,55] Moreover, the electronic alteration at the 5-position of the phenoxide ring, which is involved with the large contribution to the HOMO whereas it contributes negligibly to the density of the LUMO, is also responsible for fine tuning of the emission color by introduction of electron-withdrawing or electron-donating groups at the 5-position.[56]

3.3.2 Znsalens for One- and Two-Photon Fluorescence Microscopy Imaging

Zhang et al. first reported Znsalen complexes as fluorescent probes for one- and two-photon imaging in live cells. Modulation of salen ligands with lipophilic and cationic conjugates, Znsalen complexes showed good chemo- and photo-stability, low cytotoxicity to living cells, and good subcellular selectivity, the foremost prerequisite for living cell imaging. More importantly,

Table 1.1 Znsalen Complexes and Their Subcellular Localization

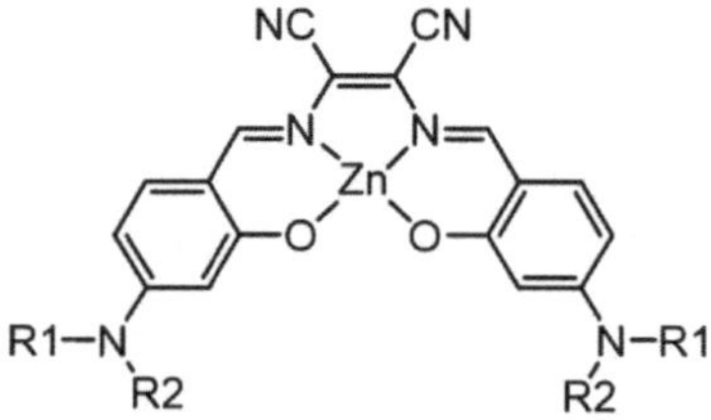

Compound	R_1	R_2	Subcellular Localization
ZnL_1	CH_2CH_3	CH_2CH_3	Cytoplasm
ZnL_2	CH_3	CH_3	Cytoplasm
ZnL_3	**R_1=R_2** =$CH_2COOC_2H_5$		**ER**
ZnL_4	**R_1**—$(CH_2)_4$—**R_2**		**ER**
ZnL_5	CH_3	$C_{14}H_{29}$	–
ZnL_6	CH_3	$CH_2CH_2CH_2Cl$	ER
ZnL_7	CH_3	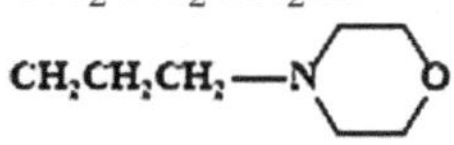	Lysosome
ZnL_8	CH_3	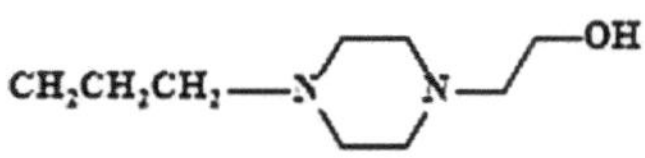	Lysosome
ZnL_9	CH_3	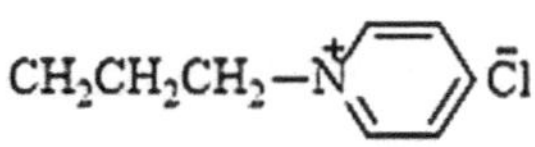	Early endosome
ZnL_{10}	CH_3	$CH_2CH_2CH_2-\dot{P}Ph_3\dot{C}l$	Early endosome

ER, endoplasmic reticulum.
Reproduced with permission from Hai, Y.; Chen, J.-J.; Zhao, P.; Lv, H.; Yu, Y.; Xu, P.; Zhang, J.-L. Luminescent Zinc Salen Complexes as Single and Two-Photon Fluorescence Subcellular Imaging Probes. *Chem. Commun.* **2011,** *47* (8), 2435–2437. Acknowledgment of copyright 2011 Royal Society of Chemistry.

anchored by cyano groups (A) on diamines and dialkylamino substituents (D) on the 4-position of the salicylaldehyde part, Znsalens show high two-photon absorption cross-sections (ca. 181–196 GM) suitable for two-photon fluorescence microscopy imaging. Except **ZnL_5**, the other nine Znsalens could be taken up efficiently by HeLa cells and localized in organelles such as lysosome, endosome, and endoplasmic reticulum (ER) (Table 1.1).[54]

Further, a Znsalen library (48 examples) (Fig. 1.26) with tunable emission was constructed as a color palette for one- and two-photon living cell imaging. Experimental and theoretical studies demonstrate that the lowest energy electronic transitions occur predominantly via salen-centered π–π* transitions, which are effective to modulate the photophysical properties. Interestingly, modulation of 4-substituents of the salicylaldehyde only changes the lipophilicity of the Znsalens and thus the subcellular localizations, and has little effect on the emission. This orthogonality is important to further design Znsalen probes in one- and two-photon imaging and multicolor living cell imaging (Fig. 1.27).[57]

3.3.3 Self-Assembly of Znsalens Influences Their Cellular Behaviors

Zhang et al. synthesized a PPh_3-derived Znsalen (**ZnL$_{11}$**) as a case study to investigate the relationship between metal induced aggregation arising from intermolecular Zn···O interaction and the biological behaviors, such as cellular uptake pathway and subcellular localization. Results from UV-vis absorption spectra, steady-state fluorescence spectra, and electronic microscopy suggest that **ZnL$_{11}$** forms aggregates through an intermolecular Zn···O bond in aqueous solution. Upon addition of pyridine, the aggregates would dissociate and disperse as monomeric species. More importantly, in the absence of pyridine, Znsalen enters the cell via a caveolae-mediated endocytosis pathway and localizes in lysosomes. Incubating with pyridine as external ligand, Znsalen could be internalized through membrane potential-dependent passive diffusion and mainly distributes in mitochondria. The distinctive results reveal that intermolecular weak interaction, like "metal induced aggregation," plays a role in tuning biological metal species and biological behavior of the complex (Fig. 1.28).[58]

Therefore modulation of the intermolecular Zn···O interaction becomes a critical issue to decipher the structural relationship to biological behaviors. Zhang et al. synthesized four cryptand-type triZnsalen complexes, which could efficiently minimize intermolecular Zn···O interaction. By way of encapsulation into poly(lactide-co-glycolide) nanoparticles, the cryptand triZnsalen displayed visible intracellular fluorescence whereas the corresponding monomeric Znsalen could not (Fig. 1.29).[59]

Very recently, Zhang et al. reported the first series of cationic sulfonium-functionalized Znsalens based on alkylation of 3-thioether or 3,5-dithioether moieties on salicylaldehydes as bioprobes for living cell imaging. The sulfonium Znsalens maintain as monomeric species in aqueous media as revealed by diffusion-ordered NMR spectroscopy and display "switched-on" fluorescence compared to the thioether analog through perturbing the PET process (Fig. 1.30). Interestingly, changing the number of sulfonium moieties could tune cell-permeability. Thus sulfonium functionalization is effective in dissociating intermolecular metal–ligand interaction and thereby modulating water solubility, photophysical properties, and even cell-permeability of fluorophores, which offers a new way to design functional metal complexes for biological studies.[60]

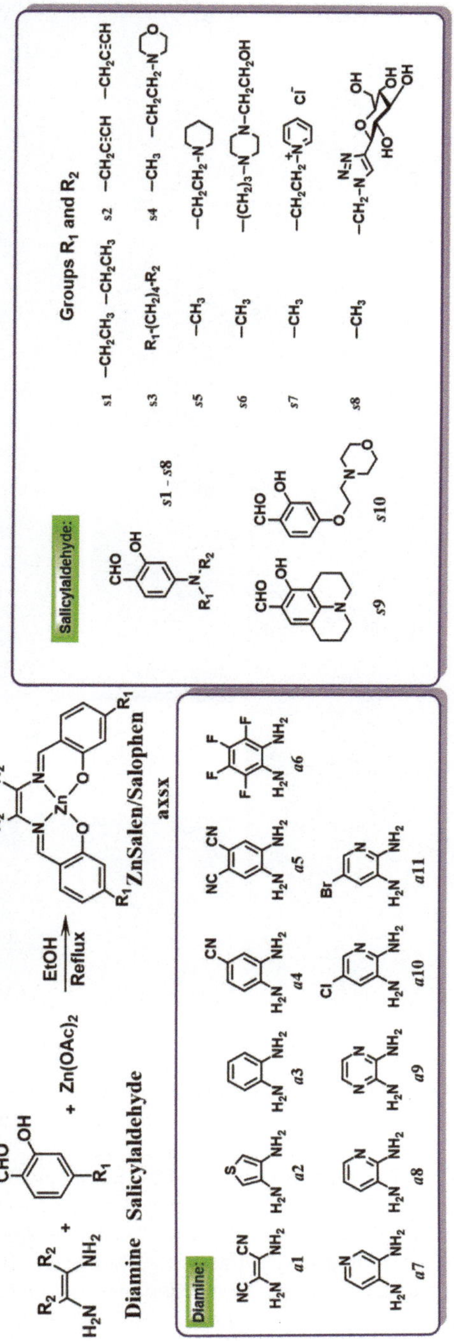

Figure 1.26 Construction of the Znsalen library.

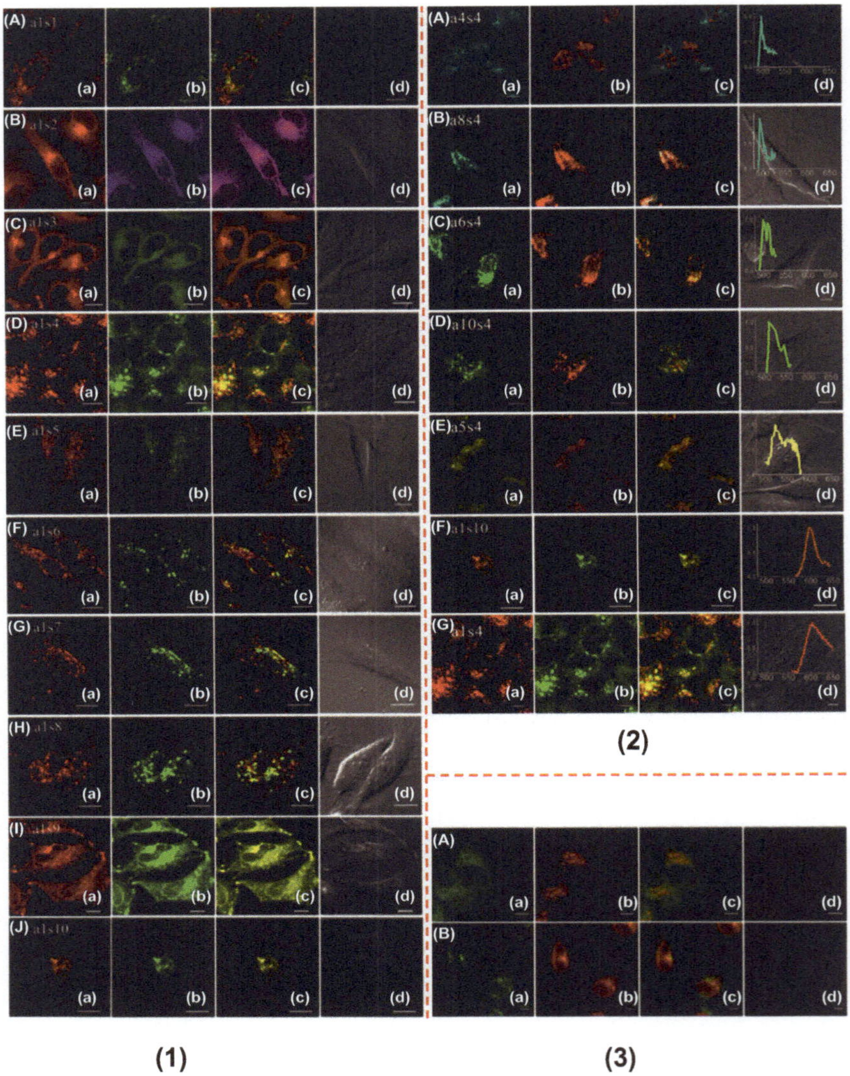

Figure 1.27 (1) and (2) fluorescence co-localization images of HeLa cells incubated with 2 μM probes in Fig. 1.26 and commercial dyes or GFP (FYVE-EGFP) (row G) and ER Tracker Blue for row B, ER Tracker Green for rows C and I, and Lysosome Tracker Green for rows A, D, E, F, H and J in (1); LysoTracker Red for Row A–Row E and LysoTracker Green for Row F and Row G in (2). (3) Two-photon multicolor images of HeLa cells. Row A: (a) a10s3 imaging of ER, green; (b) a1s4 imaging of lysosome, red; (c) merged picture of channel (a) and (b); (d) differential interference contrast (DIC); Row B: (a) a10s4 imaging of lysosome, green; (b) a1s2 imaging of ER, red; (c) merged picture of channel (a) and (b); (d) DIC. *Reprinted with permission from Xie, D.; Jing, J.; Cai, Y.-B.; Tang, J.; Chen, J.-J.; Zhang, J.-L. Construction of an Orthogonal Znsalen/Salophen Library as a Color Palette for One- and Two-Photon Live Cell Imaging.* Chem. Sci. ***2014,*** 5 *(6), 2318–2327. Acknowledgment of copyright 2014 Royal Society of Chemistry.*

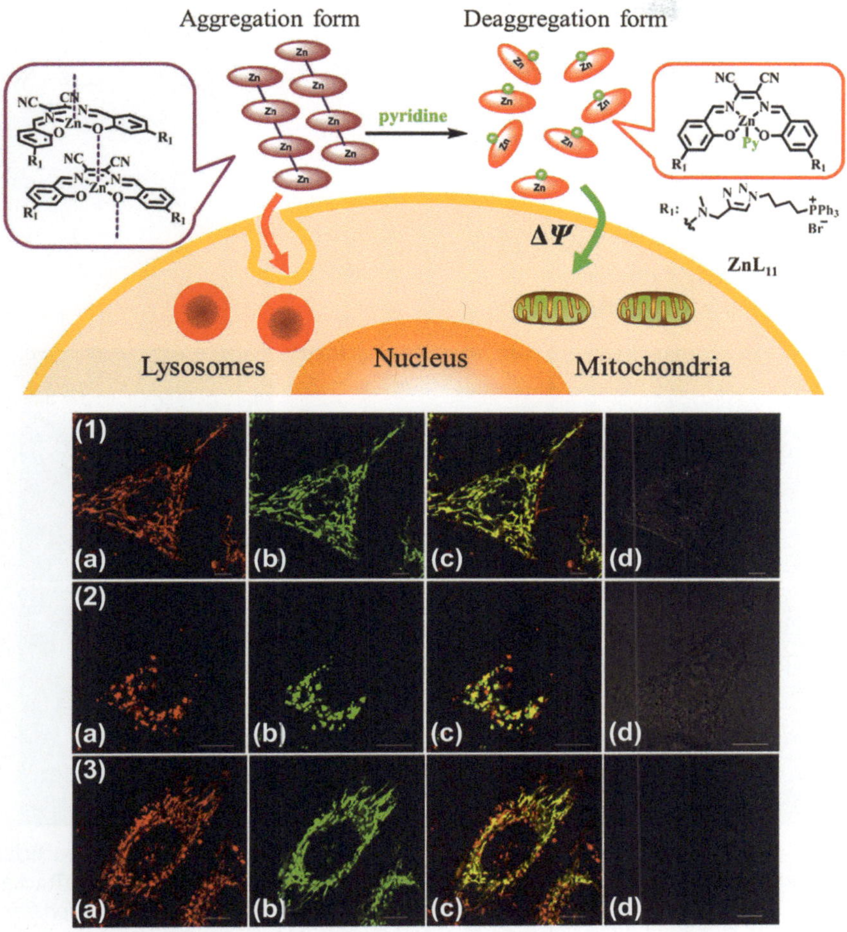

Figure 1.28 Aggregation and deaggregation forms of $\mathbf{ZnL_{11}}$ in aqueous media lead to the different cellular uptake pathways and subcellular localization. Co-localization assay in HeLa cells with (2) LysoTracker Green DND-26 and (1) and (3) MitoTracker Green: (1) $\mathbf{L_{11}}$; $\mathbf{ZnL_{11}}$ (2) in absence and (3) in presence of pyridine. *Reproduced with permission from Tang, J.; Cai, Y.-B.; Jing, J.; Zhang, J.-L. Unraveling the Correlation between Metal Induced Aggregation and Cellular Uptake/subcellular Localization of Znsalen: an Overlooked Rule for Design of Luminescent Metal Probes.* Chem. Sci. ***2015,*** 6 *(4), 2389–2397. Acknowledgment of copyright 2015 Royal Society of Chemistry.*

(1)

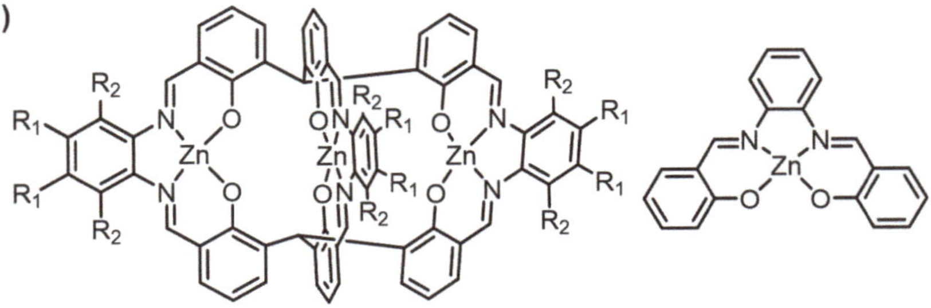

ZnL_{12} $R_1 = H, R_2 = H$;
ZnL_{13} $R_1 = OCH_2COOCH_2CH_3, R_2 = H$
ZnL_{14} $R_1 = F, R_2 = F$;
ZnL_{15} $R_1 = CN, R_2 = H$;

ZnL_{16}

(2)

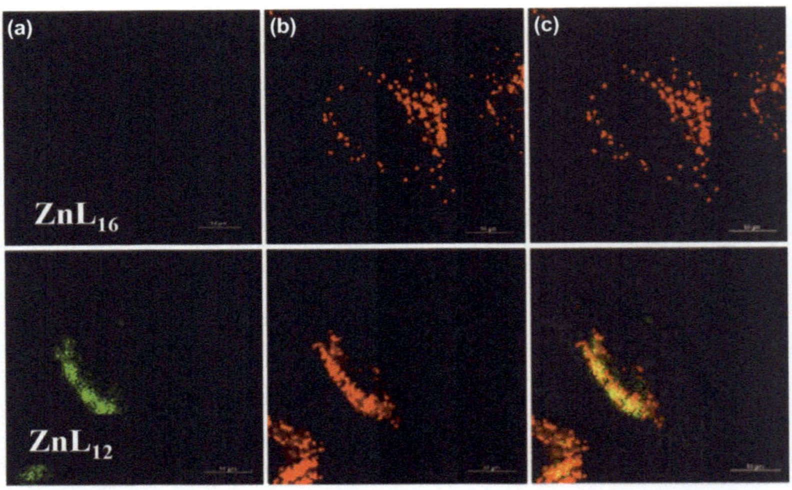

Figure 1.29 (1) Structures of **ZnL_{12-16}**. (2) Co-localization studies of the nanoparticles with LysoTracker Red: (a) images of the compounds indicated; (b) images of LysoTracker Red; (c) merged images of (a) and (b). Scale bar = 10 μm. *Reproduced with permission from Lai, J.; Ke, X.-S.; Tang, J.; Zhang, J.-L. Tris(Znsalen) Cryptand Minimizes Znsalen Aggregation Arising from Intermolecular Zn···O Interaction.* Chin. Chem. Lett. ***2015,*** 26 *(8), 937–941. Acknowledgment of copyright 2015 Chinese Chemical Society and Institute of Materia Medica, Chinese Academy of Medical Sciences.*

3.3.4 Znsalen as Photoactivatable Fluorescent Probe

Zhang et al. rationally designed effective photoactivatable fluorophore Znsalen **J-S**. Incorporating thioether moieties in the 3,3′-positions quenches the fluorescence because of a PET process. Upon one- or

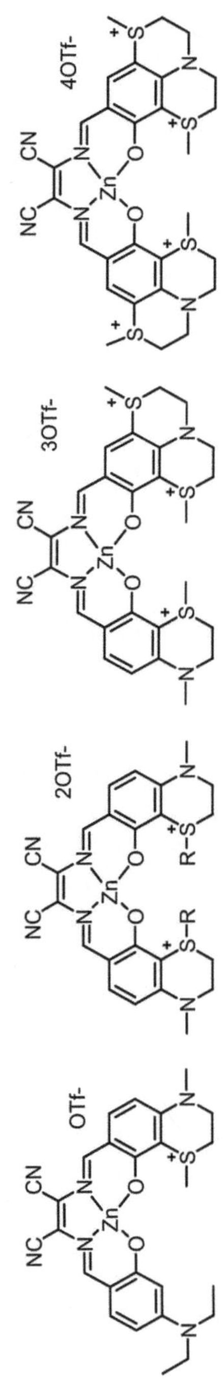

C_1Me

C_2Me R= Me
C_2Et R= Et
C_2iBu R= iBu

C_3Me

C_4Me

Figure 1.30 Structures of sulfonium Znsalens.

two-photon irradiation, the thioethers can be oxidized to sulfoxides and the fluorescence of Znsalen switches on (316- or 100-fold, respectively), because of the electron-withdrawing sulfoxides perturbing the PET process, as revealed by time-dependent density functional theory calculations. The frontier MOs involved in the vertical excitation of **J-S** and **J-SO** are presented. For **J-S**, the $S_0 \rightarrow S_1$ transition is composed of HOMO-1 → LUMO with an oscillator strength (f) of 0.0035, which is a forbidden transition ($f < 0.01$), suggesting that the S_1 state of **J-S** is a dark state. For **J-SO**, a low-lying transition-dipole-allowed S_0 to S_1 excitation (HOMO → LUMO) exhibits with an oscillator strength (f) of 0.9721, suggesting the S_1 state will be emissive. A high signal-to-noise contrast (2500-fold) was achieved by irradiation with an 840-nm laser using two-photon fluorescence microscope. **J-S** was also successfully imaged in a model living organism *Caenorhabditis elegans*, and with a ca. five times fluorescence intensity increase after one- and two-photon irradiation (Fig. 1.31).[61]

3.3.5 Znsalen as Lysosomal Hydrogen Peroxide Sensor

Hydrogen peroxide (H_2O_2) is one of the major members of reactive oxygen species (ROS) and plays essential roles as a beneficial signaling agent or a toxic hazard in physiological and pathological processes. Accumulating evidence suggests that, not only the main locales for ROS production such as mitochondria, endoplasmic reticulum, and cell membranes, but also lysosomes participate in regulating H_2O_2 flux, which determines the ultimate biological outcome of ROS and the cross-talk between subcellular regions. The main challenges for such probes include (1) the acidic conditions (pH 4.5–6.0) in lysosomes, which disfavor nucleophilic reaction-based fluorescent H_2O_2 probes and genetic encoding of fluorescent proteins, e.g., HyPer; (2) low chemoselectivity, because H_2O_2 is a mild electrophile among ROS; and (3) lack of functional groups that are capable of targeting lysosomes. Toward this goal, Zhang et al. designed the "MPO-**J-S**" probe, a novel combination of "peroxidase/fluorogenic substrate." Myeloperoxidase (MPO), which specifically converts "$H_2O_2 + Cl^-$" into hypochlorite (ClO^-), makes the original hypochlorite sensitive **J-S** "switch on" in the presence of H_2O_2 in aqueous solution under acidic conditions. It can visualize exogenous or endogenous H_2O_2 in lysosomes with high selectivity and sensitivity via one- and two-photon fluorescence microscopies (Fig. 1.32).[62]

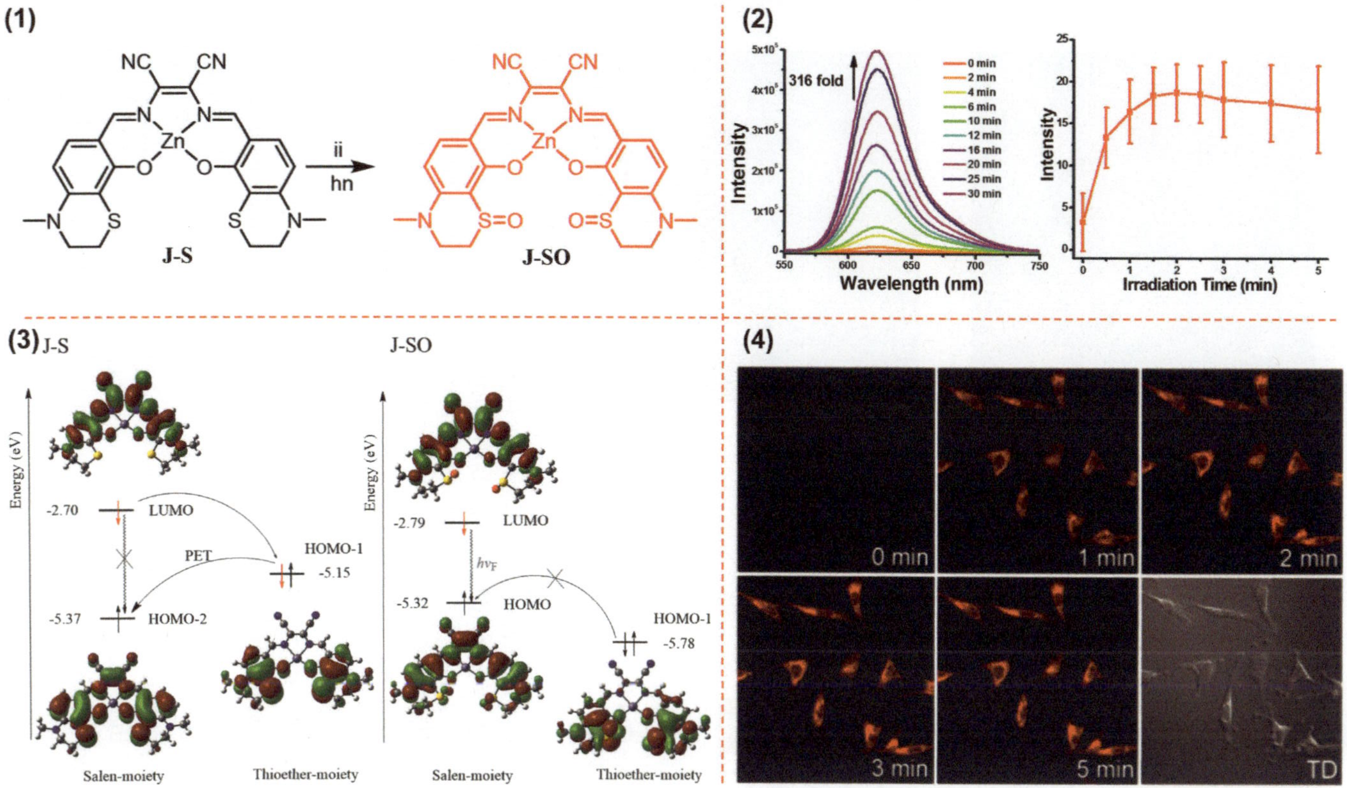

Figure 1.31 (1) Photooxidation of **J-S** to **J-SO**. (2) Spectroscopic changes of (a) UV–vis absorption and (b) emission of **J-S**, upon irradiation with a 532-nm laser (20 mW). (3) Frontier molecular orbital energy illustrations show the relative energetic dispositions of the orbitals of the two moieties in **J-S** and **J-SO**. (4) Time-course of photoactivation of **J-S** in L6 myoblasts. *PET*, photoinduced electron transfer. *Reprinted with permission from Jing, J.; Chen, J.-J.; Hai, Y.; Zhan, J.; Xu, P.; Zhang, J.-L. Rational Design of Znsalen as a Single and Two Photon Activatable Fluorophore in Living Cells.* Chem. Sci. ***2012,*** 3 *(11), 3315–3320. Acknowledgment of copyright 2012 Royal Society of Chemistry.*

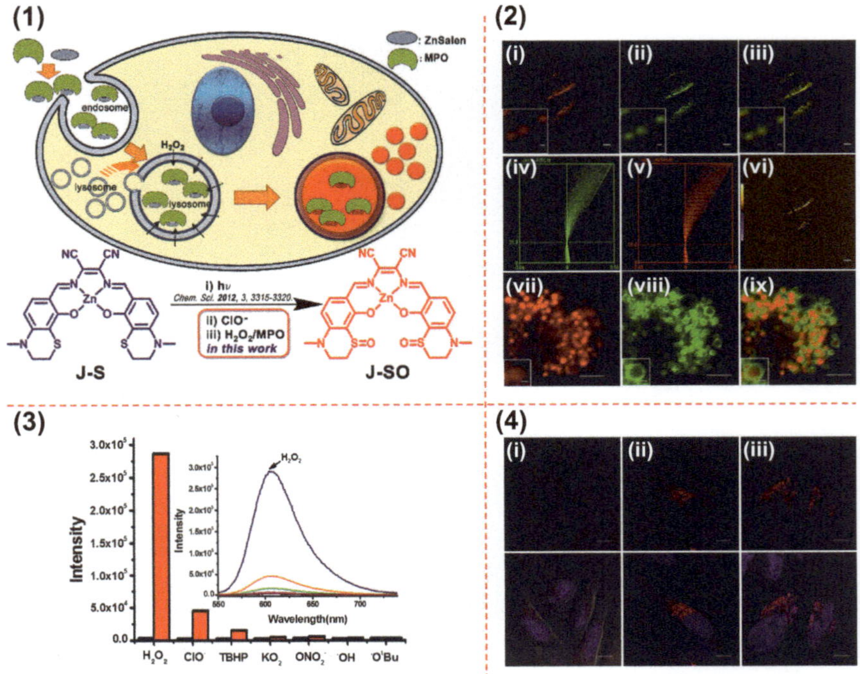

Figure 1.32 (1) Schematic of myeloperoxidase (MPO)-mediated strategy for lysosomal H_2O_2 detection; (2) **J-S** co-localization assay in HeLa cells with LysoTracker Green DND-26 and LAMP2-EGFP. (3) H_2O_2 selectivity of "MPO-**J-S**" in phosphate-buffered saline (pH 5.0) (before: black; after: red); (4) Confocal images of HeLa cells loaded with "MPO-**J-S**" under different stimulants. (i) Cells were treated with "MPO-J-S" for 12 h at 37°C. (ii) Cells loaded with "MPO-**J-S**" for 12 h were treated with H_2O_2 for 30 min. (iii) Cells loaded with "MPO-**J-S**" for 12 h were treated with phorbol myristate acetate for 1 h. *Reprinted with permission from Jing, J.; Zhang, J.-L. Combining Myeloperoxidase (MPO) with Fluorogenic Znsalen to Detect Lysosomal Hydrogen Peroxide in Live Cells.* Chem. Sci. ***2013,*** 4 *(7), 2947–2952. Acknowledgment of copyright 2013 Royal Society of Chemistry.*

3.3.6 Znsalen as Autophagy Indicator

Autophagy is the major lysosomal degradative pathway in mammalian cells and is involved in maintaining cellular homeostasis, development, and energy balance. A widely used fluorescent probe for the detection of early autophagy is GFP-LC3, a fusion of the autophagosome marker LC3 (microtubule-associated protein 1 light chain 3) and GFP (green fluorescent protein).[63] However, LC3-based probes have been reported to aggregate easily if overexpressed or coexpressed with other proteins prone to aggregation, thereby giving false-positive results.[64] Zhang et al. developed the D-glucose-attached Znsalen

(**Zn-G4**) as a sensitive and quantitative autolysosome probe for monitoring later stages of autophagy. Through introducing glucose as a hydrophilic group, the fluorescence of **Zn-G4** turned out to be closely related to the local hydrophobic environment. Upon starvation- or chemical agent-induced autophagy, the fluorescence would be "turned on" within the autolysosome or possibly the late endosome/lysosome networks. The probe can be detected by one- and two-photon imaging microscope with a high signal-to-noise ratio readout of autophagic activity. More importantly, **Zn-G4** assay proved to be convenient to perform high-throughput screening for autophagy inducers by noninvasive fluorescence detection (Fig. 1.33).[65]

3.3.7 Znsalen as Nucleic Acids Chelator

Znsalen complexes have been reported to bind to both B-DNA and G4-DNA. However, it is seldom to use the luminescence properties to design improved fluorescent probes for particular conformations of nucleic acids. Rodríguez et al. studied the intracellular behaviors of the Znsalens (**ZnL_{17} and ZnL_{18}**) that could bind to DNA in aqueous solution (Fig. 1.34).[53d] Combining with confocal co-localization imaging results, the low cytotoxicity could be ascribed to the cytosol distribution instead of the nucleus localization of the complexes, thus exerting no effect on DNA.

3.3.8 Gene Expression and Programmed cell Death Regulator

Luminescent Znsalens have also been used to regulate gene expression and programmed cell death. Zingaro et al. synthesized a set of chiral Znsalen complexes to mimic the distinctive function of zinc-finger proteins. Confocal laser scanning microscopy imaging clearly demonstrated the cellular membrane permeability of these metal complexes. The leading compound **C3** is an extremely efficient regulator of the cell cycle and anti-apoptosis genes, bcl-2, and bcl-xL. It provides a new insight into applying Znsalen derivatives as potential drugs (Fig. 1.35)[66]

3.4 Zinc Complexes of Porphyrin and Phthalocyanine Ligands

Porphyrins and phthalocyanines are highly conjugated macrocyclic ligands with interesting photochemical properties (Fig. 1.36). Their zinc complexes are widely studied and used as industrial pigments, electron transfer compounds, photosensitizers, and photodynamic therapy agents.[67] However, only a few research projects have explored their application as imaging agents.

Luedtke et al. reported a Zn-phthalocyanine complex, **Zn-DIGP**, as a high-affinity (≤2 nM) G-quadruplex probe exhibiting "turn-on"

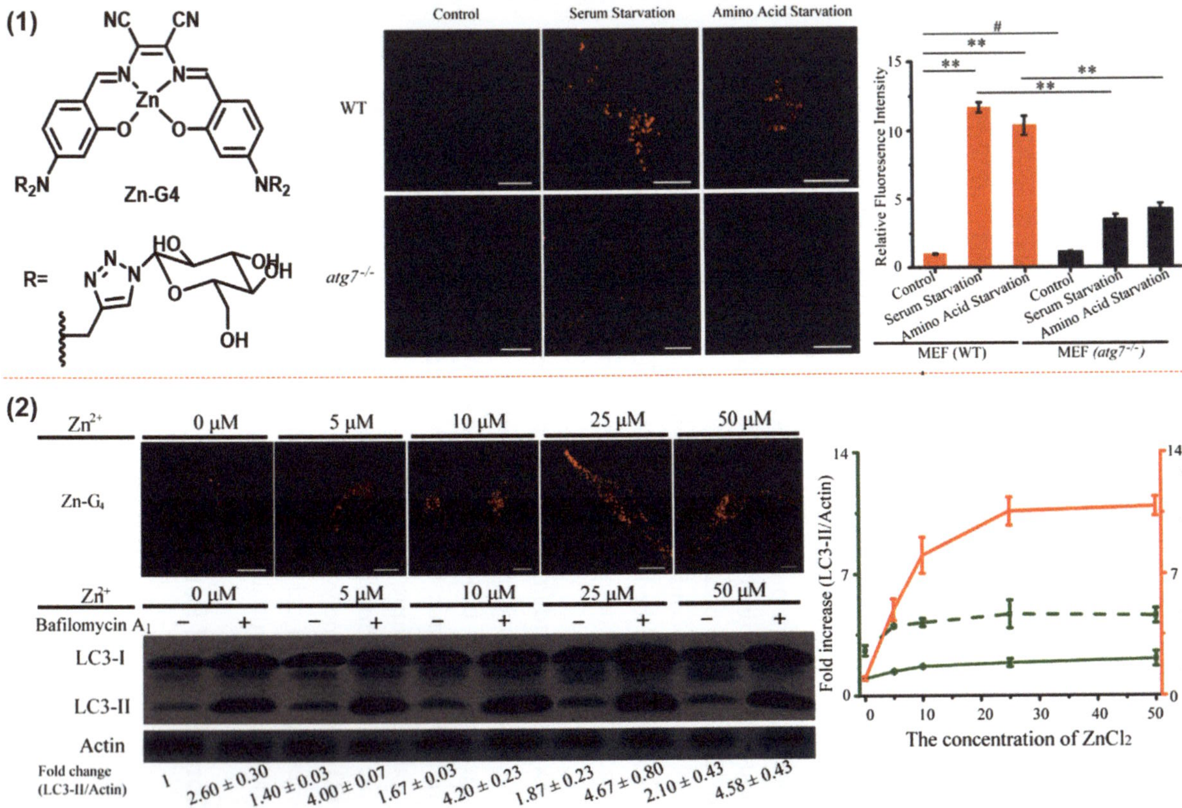

Figure 1.33 (1) Confocal images and relative fluorescence intensity of wild-type (WT) and atg7$^{-/-}$ MeF cells incubated with **Zn-G4** in complete medium (control group), in serum-free medium (serum-starvation group), or in Earle's balanced salt solution (amino acid-starvation group). Scale bar = 10 μm. (2) Left panel: confocal images of L6 myoblasts incubated with 2 μM **Zn-G4** and immunoblotting with LC3 antibodies in the presence of 0 or 100 nM bafilomycin A_1 and different concentrations of $ZnCl_2$ (0–50 μM) (scale bar = 10 μm). Right panel: comparison of the ratio of LC3-II and β-actin analyzed from immunoblotting and the relative fluorescence intensity of **Zn-G4** as analyzed from confocal images in response to various $ZnCl_2$ concentrations. (*Green solid*, $ZnCl_2$ only; *green dash*, $ZnCl_2$ plus bafilomycin A1; *red*, fluorescence intensity of **Zn-G4**). *Reprinted with permission from Chen, J.-J.; Jing, J.; Chang, H.; Rong, Y.; Hai, Y.; Tang, J.; Zhang, J.-L.; Xu, P. A Sensitive and Quantitative Autolysosome Probe for Detecting Autophagic Activity in Live and Prestained Fixed Cells. Autophagy 2013, 9 (6), 894–904. Copyright 2013 Landes Bioscience.*

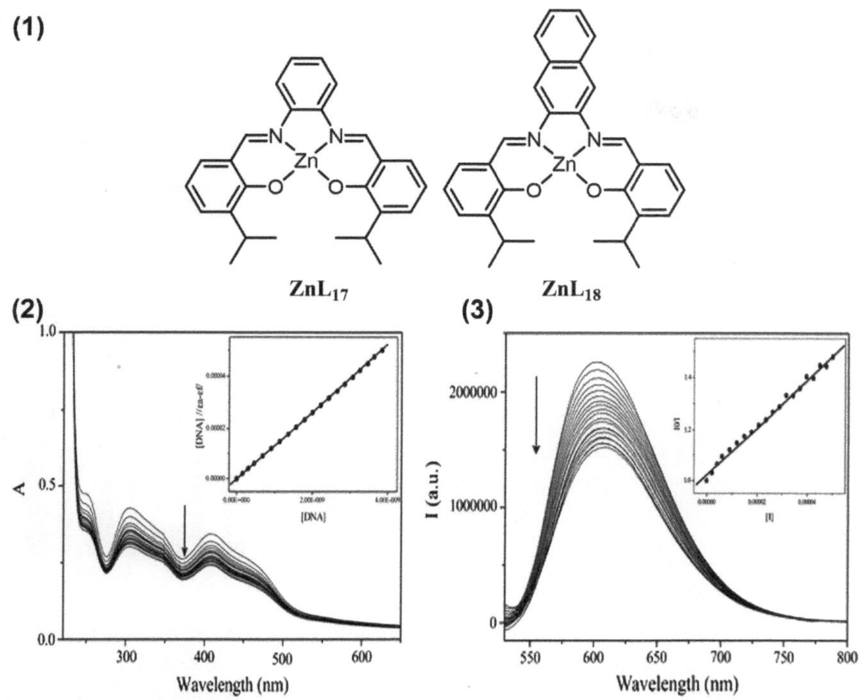

Figure 1.34 (1) Electronic absorption spectra of $\mathbf{ZnL_{17}}$ in the absence and presence of increasing amounts of DNA. Inset: plot of [DNA] versus [DNA]/($\varepsilon_a-\varepsilon_f$). (2) Emission spectra of ethidium bromide bound to DNA in the absence and presence of increasing amounts of $\mathbf{ZnL_{17}}$. Inset: Stern–Volmer plot: I_0/I versus [$\mathbf{ZnL_{17}}$]. *Reprinted with permission from Brissos, R.; Ramos, D.; Lima, J. C.; Mihan, F. Y.; Borras, M.; de Lapuente, J.; Cort, A. D.; Rodriguez, L. Luminescent Zinc Salophen Derivatives: Cytotoxicity Assessment and Action Mechanism Studies.* New J. Chem. ***2013,*** 37 *(4), 1046–1055. Copyright 2013 The Royal Society of Chemistry.*

luminescence as well as the ability to inhibit RNA expression.[68] The fluorescence of **Zn-DIGP** allows direct observation in vitro, and the cellular localization of **Zn-DIGP** in cytosol was apparently different from the distribution of common duplex DNA in nucleus, which could be easily stained by commercial probe Hoechst 33342 (Fig. 1.37). In addition, treatment of **Zn-DIGP** to SH-SY5Y neuroblastoma cells caused a rapid threefold knockdown of c-Myc RNA, which is consistent with the mechanism for G-quadruplex-mediated promoter deactivation.

Zinc porphyrins can be designed as multifunctional imaging agents. Zhang et al. reported two bimodal imaging agents consisting of a Zn(II) tetrapentafluorophenylporpholactol core and a β−substituted Gd(III)-DO3A

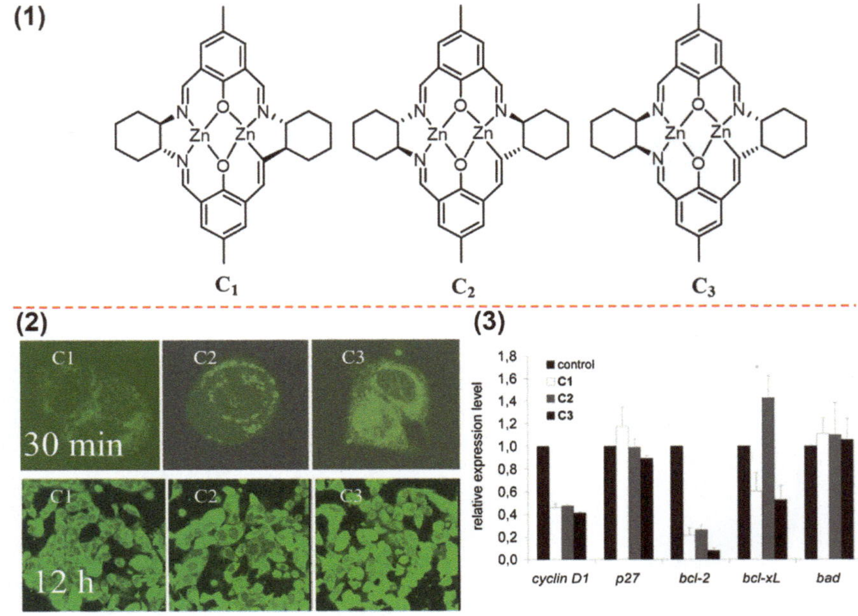

Figure 1.35 (1) Chiral zinc complexes **C1–C3**; (2) confocal fluorescence images of MCF7 cells incubated with complexes **C1–C3** for: 30 min (upper row) and 20 h (lower row); (3) the effect on relative expression of cell cycle and apoptosis-related genes. *Reprinted with permission from Gao, J.; Liu, Y. G.; Zhou, Y.; Boxer, L. M.; Woolley, F. R.; Zingaro, R. A., Artificial Zinc (II) Complexes Regulate Cell Cycle and Apoptosis-Related Genes in Tumor Cell Lines.* Chembiochem ***2007***, 8 *(3), 332–340. Copyright 2007 Wiley-VCH.*

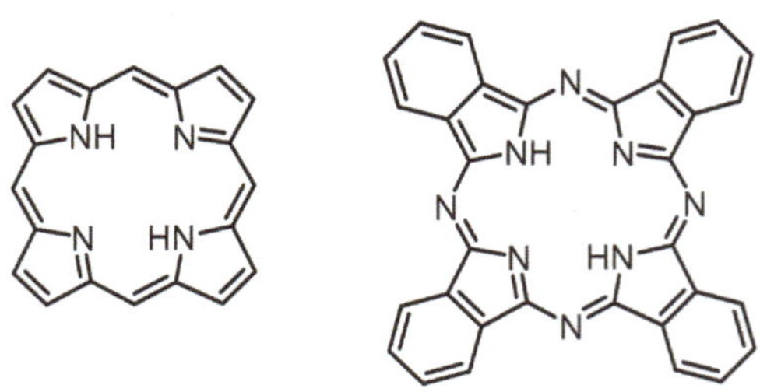

Porphyrin **Phthalocyanine**

Figure 1.36 Structures of porphyrin and phthalocyanine.

(**ZnLGd$_{539}$**) or Gd(III)-DOTA (**ZnLGd$_{595}$**) moiety (Fig. 1.38).[69] On the one hand, compared to traditional conjugation at the *meso*-position, β–modification extended the NIR absorption wavelength as well as intensity. On the other hand, these dinuclear complexes possess enhanced ionic relaxation compared to the monomeric Gd(III) MRI agents. Fluorescence imaging

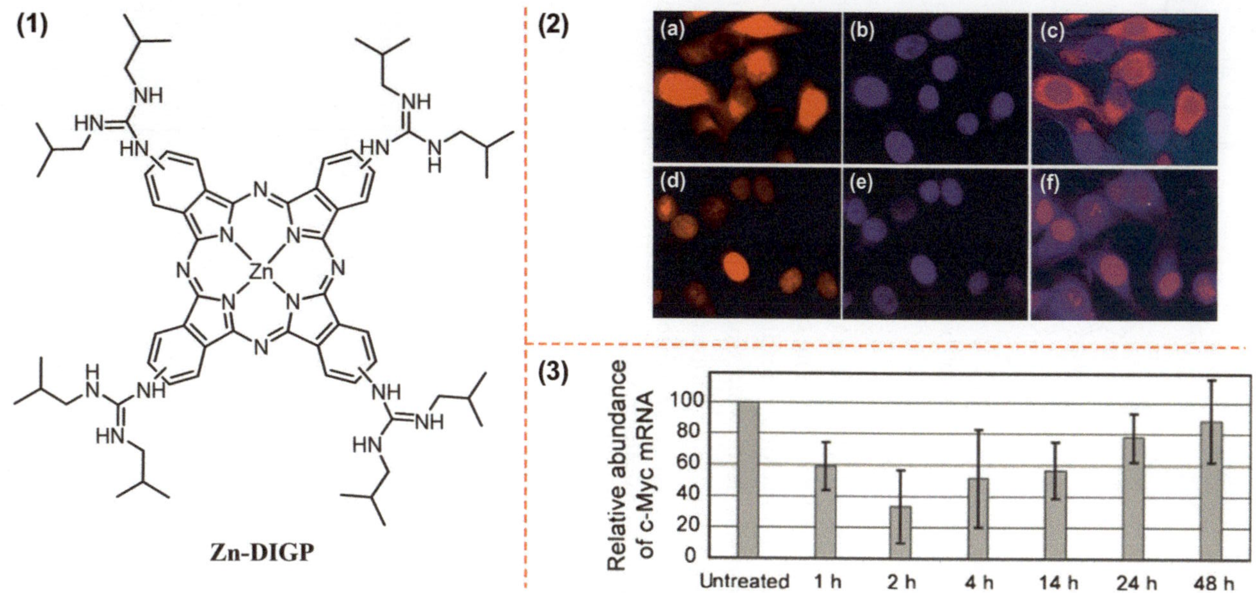

Figure 1.37 (1) Structure of **Zn-DIGP**. (2) Fixed SK-Mel-28 cells stained with **Zn-DIGP** (a–c) or **DIGP** (d–f) and Hoechst 33342. (3) Relative abundance of c-Myc mRNA extracted from SH-SY5Y neuroblastoma cells treated with 1 μM **Zn-DIGP** for 1–48 h. *Reprinted with permission from Alzeer, J.; Vummidi, B. R.; Roth, P. J.; Luedtke, N. W. Guanidinium-Modified Phthalocyanines as High-affinity G-quadruplex Fluorescent Probes and Transcriptional Regulators.* Angew. Chem. Int. Ed. Engl. ***2009,*** 48 *(49), 9362–9365. Copyright 2009 Wiley-VCH.*

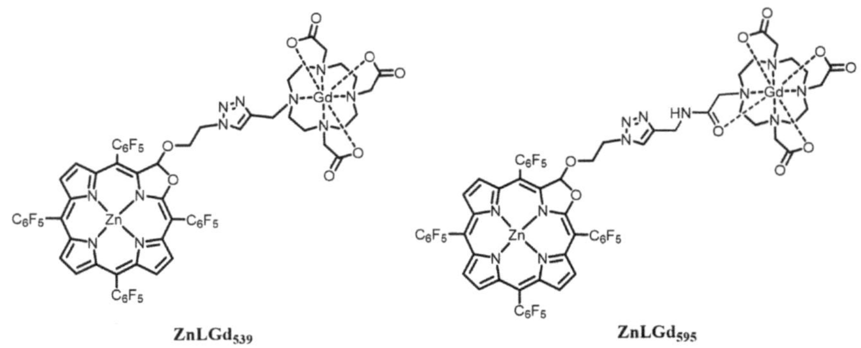

Figure 1.38 Structures of Zn-porpholactol bimodal imaging agents **ZnLGd$_{539}$** and **ZnLGd$_{595}$**.

and MRI measurements in HeLa cells demonstrate that these complexes can be used as potential bifunctional imaging contrast agents (Fig. 1.39).

A series of Zn-porpholactol β−substituted compounds were synthesized and their cellular uptake level were studied.[70] The results show that the terminal ionic groups on β−substitutions significantly influence the cellular uptake of **ZnL$_{19-24}$** and can be visualized under fluorescence microscope. Generally, **ZnL$_{19-24}$** co-localized well with the commercially available LysoTracker Green DND-26 (Fig. 1.40), and **ZnL$_{22}$** exerts photodynamic function through inducing apoptosis, including lysosome disruption, mitochondria depolarization, and nuclei condensation.

4. CONCLUSION AND OUTLOOK

In this chapter we demonstrate that luminescent zinc complexes are emerging biological probes or labels in molecular imaging. This clearly indicates that zinc coordination chemistry devotes increasing attention to the development of luminescent biological probes which are able to monitor molecular events in vitro or in vivo in real time. The advantages of the first row, biologically available metals such as low cytotoxicity and good reactivity are obvious, since they represent less concern in biological systems. We believe, however, that for many zinc complexes, high luminescence and sensitivities have still to be reached to address practical needs. A wide horizon is therefore open for further investigation and developments to meet a great demand for efficient chemosensors.

From the trend of development of luminescent Zn probes or labels, we could classify them as (1) "chelator + fluorophore", type I and (2)

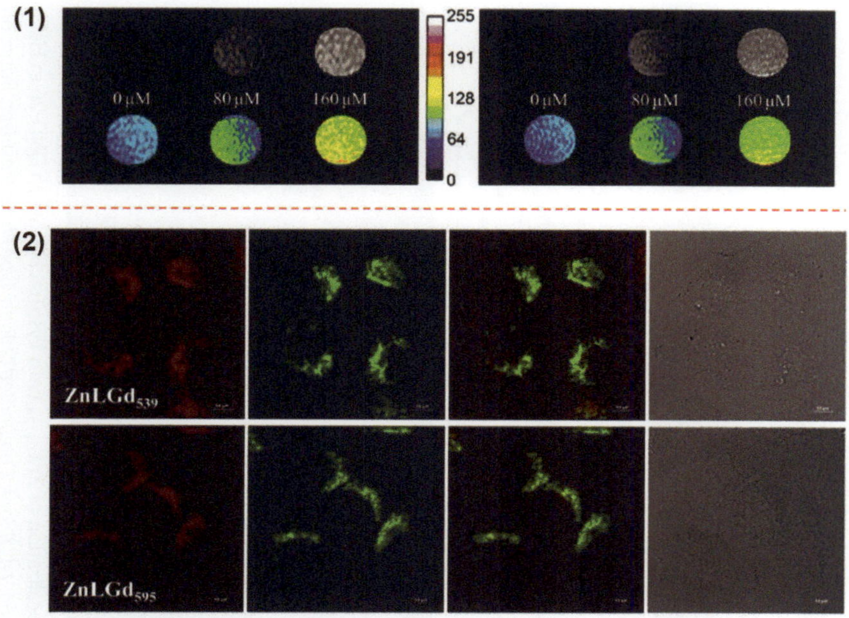

Figure 1.39 (1) T1-weighted cell phantom MR images for **$ZnLGd_{539}$** (left) and **$ZnLGd_{595}$** (right) with different concentrations. (2) Co-localization analysis of **$ZnLGd_{539}$/$ZnLGd_{595}$** and LysoTracker Green DND-26. *Reprinted with permission from Ke, X.-S.; Tang, J.; Yang, Z.-S.; Zhang, J.-L. β-Conjugation of Gadolinium(III) DOTA Complexes to Zinc(II) Porpholactol as Potential Multimodal Imaging Contrast Agents.* J. Porphyr. Phthalocyanines ***2014,*** 18 *(10 n11), 950–959. Copyright 2014 World Scientific Publishing Company.*

"Zn^{2+} + chromophore ligand", type II. We found two important issues that should be addressed in the future. The advantage of type I is reactivity. For example, the Zn-DPA complex could coordinate to anions, binding tag proteins, phosphorylated proteins, and other phosphate-containing biomolecules, which is important to monitor molecular events in biological processes. However, the molecular size of type I complexes is the disadvantage, which could interfere the biological functions or produce false signals. On the other hand, type II complexes have appropriate molecular sizes and high luminescence; however, the coordination sites are always saturated and the reactive part (or sensitive part) is dependent on ligand. This comparable sensing mechanism to organic fluorophores would diminish the advantage of coordination complexes, which is dependent on the reactivity of metal ions. For the molecular design of zinc complexes, we propose that a combination of the reactivity of the Zn^{2+} within the chromophore ligand is

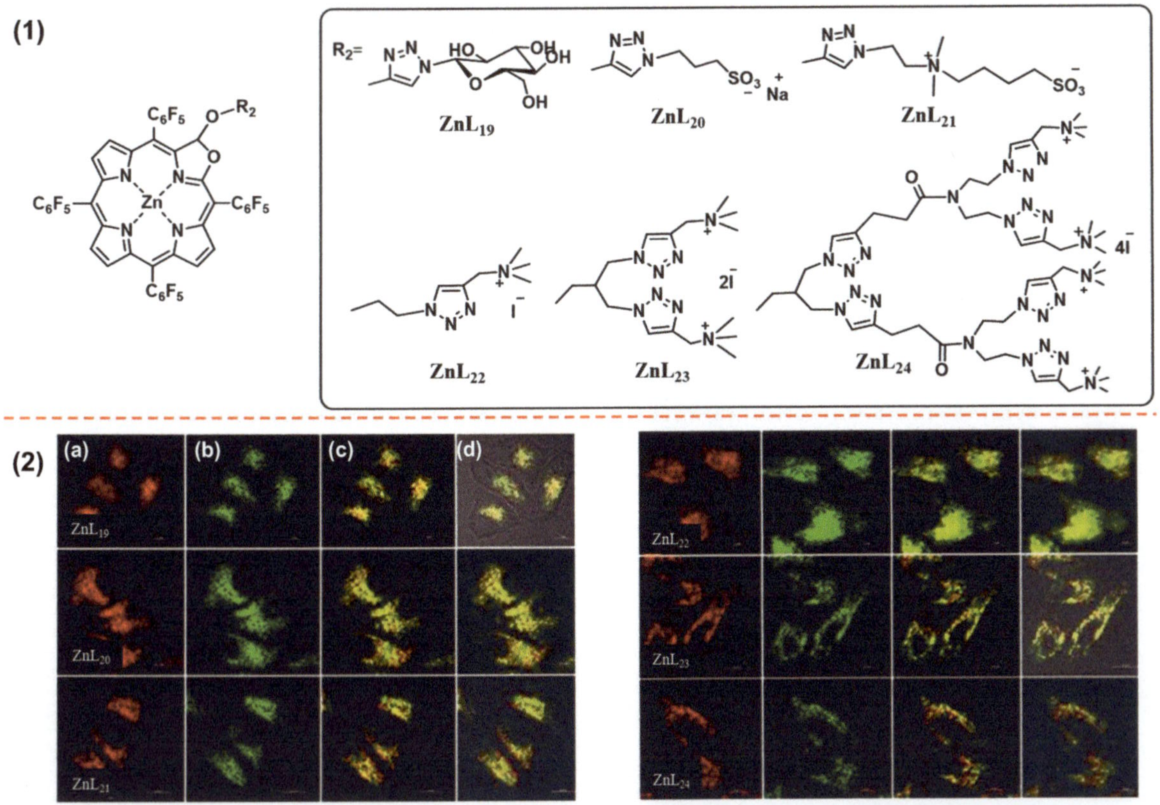

Figure 1.40 (1) Structures of Zn-porpholactol complexes **$ZnL_{19–24}$**. (2) Co-localization of **$ZnL_{19–24}$** with LysoTracker Green DND-26 in HeLa cells. *Reprinted with permission from Ke, X.-S.; Tang, J.; Chen, J.-J.; Zhou, Z.-Y.; Zhang, J.-L. β-Ionic Conjugated Chlorin-Type Photosensitizers Based on Porpholactone: Synthesis, Photophysical Properties, and Photodynamic Activity.* ChemPlusChem ***2015,*** 80 *(1), 237–252. Copyright 2015 Wiley-VCH.*

critical to impart the unique role of zinc complexes over traditional organic fluorophores.

For biological application, we are interested in developing zinc complexes for cellular metal trafficking and regulation, applying the homeostatic studies of metal ions. Intracellular metal trafficking pathways are variable and the dynamic processes are complicated. Using chromophore ligands to "light" the trafficking pathways might be an approach to understanding these biological processes. A good example is the **Zn-G4** probe, which is successfully used to monitor the later stage of autophagy and open an access to develop such probes dependent on the transformation of autolysosome to lysosome. This would be an important direction for designing metal complexes to exert the functions that traditional organic fluorophores could not or may find difficult. Finding answers to these questions will be an exciting journey into the frontier of inorganic chemistry and cell biology.

ACKNOWLEDGMENT

We thank the National Scientific Foundation of China (grants no. 20971007, 21571007, 21271013, 21321001) and National Key Basic Research Support Foundation of China (NKBRSFC) (2015CB856300) for financial support. J.-L. Zhang. gratefully thanks all of whose names appear on the reference list.

REFERENCES

1. a. Haas, K. L.; Franz, K. J. Application of Metal Coordination Chemistry to Explore and Manipulate Cell Biology. *Chem. Rev.* **2009,** *109* (10), 4921–4960.
   b. Bertini, I. *Biological Inorganic Chemistry: Structure and Reactivity;* University Science Books, 2007.
   c. Da Silva, J. F.; Williams, R. J. P. *The Biological Chemistry of the Elements: The Inorganic Chemistry of Life;* Oxford University Press, 2001.
   d. Lippard, S.; Berg, J. *Principles of Bioinorganic Chemistry;* University Science Books: Mill Valley, CA, 1994.
   e. Lippard, S. J. The Inorganic Side of Chemical Biology. *Nat. Chem. Biol.* **2006,** *2* (10), 504–507.
   f. Pyle, A. Metal Ions in the Structure and Function of RNA. *J. Biol. Inorg. Chem.* **2002,** 7 (7–8), 679–690.
2. a. Zhang, C. X.; Lippard, S. J. New Metal Complexes as Potential Therapeutics. *Curr. Opin. Chem. Biol.* **2003,** 7 (4), 481–489.
   b. Cohen, S. M. New Approaches for Medicinal Applications of Bioinorganic Chemistry. *Curr. Opin. Chem. Biol.* **2007,** *11* (2), 115–120.
   c. Fricker, S. P. Metal Based Drugs: From Serendipity to Design. *Dalton Trans.* **2007,** *43,* 4903–4917.
   d. Bruijnincx, P. C.; Sadler, P. J. New Trends for Metal Complexes With Anticancer Activity. *Curr. Opin. Chem. Biol.* **2008,** *12* (2), 197–206.
   e. Drewry, J. A.; Gunning, P. T. Recent Advances in Biosensory and Medicinal Therapeutic Applications of Zinc(II) and Copper(II) Coordination Complexes. *Coord. Chem. Rev.* **2011,** *255* (3–4), 459–472.

f. Liu, H.-K.; Sadler, P. J. Metal Complexes as DNA Intercalators. *Acc. Chem. Res.* **2011,** *44* (5), 349–359.
g. Szaciłowski, K.; Macyk, W.; Drzewiecka-Matuszek, A.; Brindell, M.; Stochel, G. Bioinorganic Photochemistry: Frontiers and Mechanisms. *Chem. Rev.* **2005,** *105* (6), 2647–2694.
h. Bünzli, J.-C. G. Lanthanide Luminescence for Biomedical Analyses and Imaging. *Chem. Rev.* **2010,** *110* (5), 2729–2755.
i. Wang, D.; Lippard, S. J. Cellular Processing of Platinum Anticancer Drugs. *Nat. Rev. Drug Discov.* **2005,** *4* (4), 307–320.

3. a. Frezza, M.; Hindo, S.; Chen, D.; Davenport, A.; Schmitt, S.; Tomco, D.; Dou, Q. P. Novel Metals and Metal Complexes as Platforms for Cancer Therapy. *Curr. Pharm. Des.* **2010,** *16* (16), 1813.
b. Meggers, E. Targeting Proteins with Metal Complexes. *Chem. Commun.* **2009,** *9,* 1001–1010.
c. Lo, K. K. -W.; Choi, A. W. -T.; Law, W. H. -T. Applications of Luminescent Inorganic and Organometallic Transition Metal Complexes as Biomolecular and Cellular Probes. *Dalton Trans.* **2012,** *41* (20), 6021–6047.
4. a. Wing-Wah Yam, V.; Kam-Wing Lo, K. Luminescent Polynuclear D10 Metal Complexes. *Chem. Soc. Rev.* **1999,** *28* (5), 323–334.
b. Thorp-Greenwood, F. L.; Balasingham, R. G.; Coogan, M. P. Organometallic Complexes ofTransition Metals in Luminescent Cell Imaging Applications. *J. Organomet. Chem.* **2012,** *714,* 12–21.
c. Ma, D.-L.; Ma, V. P. -Y.; Chan, D. S. -H.; Leung, K.-H.; He, H.-Z.; Leung, C.-H. Recent Advances in Luminescent Heavy Metal Complexes for Sensing. *Coord. Chem. Rev.* **2012,** *256* (23–24), 3087–3113.
d. Blasse, G. The Luminescence of Closed-Shell Transition-Metal Complexes. New Developments. In *Luminescence and Energy Transfer;* Springer, 1980; pp 1–41.
e. Zhao, Q.; Huang, C.; Li, F. Phosphorescent Heavy-Metal Complexes for Bioimaging. *Chem. Soc. Rev.* **2011,** *40* (5), 2508–2524.
f. Yam, V. W. -W.; Wong, K. M. -C. Luminescent Metal Complexes of D6, D8 and D10 Transition Metal Centres. *Chem. Commun.* **2011,** *47* (42), 11579–11592.
5. a. Smith, K. S.; Huyck, H. L. An Overview of the Abundance, Relative Mobility, Bioavailability, and Human Toxicity of Metals. *Environ. Geochem. Miner. Deposits* **1999,** *6,* 29–70.
b. Krishnamurti, G. S.; Naidu, R. *Chemical Speciation and Bioavailability of Trace Metals;* John Wiley and Sons, 2008.
c. House, W. A. Trace Element Bioavailability as Exemplified by Iron and Zinc. *Field Crops Res.* **1999,** *60* (1), 115–141.
6. a. Keefe, M. H.; Benkstein, K. D.; Hupp, J. T. Luminescent Sensor Molecules Based on Coordinated Metals: A Review of Recent Developments. *Coord. Chem. Rev.* **2000,** *205* (1), 201–228.
b. Prodi, L.; Bolletta, F.; Montalti, M.; Zaccheroni, N. Luminescent Chemosensors for Transition Metal Ions. *Coord. Chem. Rev.* **2000,** *205* (1), 59–83.
c. Rogers, C. W.; Wolf, M. O. Luminescent Molecular Sensors Based on Analyte Coordination to Transition-Metal Complexes. *Coord. Chem. Rev.* **2002,** *233–234,* 341–350.
d. Domaille, D. W.; Que, E. L.; Chang, C. J. Synthetic Fluorescent Sensors for Studying the Cell Biology of Metals. *Nat. Chem. Biol.* **2008,** *4* (3), 168–175.
e. Nolan, E. M.; Lippard, S. J. Tools and Tactics for the Optical Detection of Mercuric Ion. *Chem. Rev.* **2008,** *108* (9), 3443–3480.
f. Que, E. L.; Domaille, D. W.; Chang, C. J. Metals in Neurobiology: Probing Their Chemistry and Biology with Molecular Imaging. *Chem. Rev.* **2008,** *108* (5), 1517–1549.

g. Nolan, E. M.; Lippard, S. J. Small-Molecule Fluorescent Sensors for Investigating Zinc Metalloneurochemistry. *Acc. Chem. Res.* **2009,** *42* (1), 193–203.
h. Tomat, E.; Lippard, S. J. Imaging Mobile Zinc in Biology. *Curr. Opin. Chem. Biol.* **2010,** *14* (2), 225–230.
i. Zhu, L.; Yuan, Z.; Simmons, J. T.; Sreenath, K. Zn(II)-Coordination Modulated Ligand Photophysical Processes – the Development of Fluorescent Indicators for Imaging Biological Zn(II) Ions. *RSC Adv.* **2014,** *4* (39), 20398–20440.

7. a. Luo, S.; Zhang, E.; Su, Y.; Cheng, T.; Shi, C. A Review of NIR Dyes in Cancer Targeting and Imaging. *Biomaterials* **2011,** *32* (29), 7127–7138.
b. Szacilowski, K.; Macyk, W.; Drzewiecka-Matuszek, A.; Brindell, M.; Stochel, G. Bioinorganic Photochemistry: Frontiers and Mechanisms. *Chem. Rev.* **2005,** *105* (6), 2647–2694.
c. Drewry, J. A.; Gunning, P. T. Recent Advances in Biosensory and Medicinal Therapeutic Applications of Zinc(II) and Copper(II) Coordination Complexes. *Coord. Chem. Rev.* **2011,** *255* (3–4), 459–472.
d. Terenzi, A.; Lauria, A.; Almerico, A. M.; Barone, G. Zinc Complexes as Fluorescent Chemosensors for Nucleic Acids: New Perspectives for a "Boring" Element. *Dalton Trans.* **2015,** *44* (8), 3527–3535.
8. a. Müller, B.; Förster, I. Derivatives of Phosphoric and Phosphonic Acid as Corrosion Inhibitors for Zinc Pigments. *Corros. Sci.* **1996,** *38* (7), 1103–1108.
b. Koutmos, M.; Pejchal, R.; Bomer, T. M.; Matthews, R. G.; Smith, J. L.; Ludwig, M. L. Metal Active Site Elasticity Linked to Activation of Homocysteine in Methionine Synthases. *Proc. Natl. Acad. Sci. U.S.A.* **2008,** *105* (9), 3286–3291.
c. Maret, W. Molecular Aspects of Human Cellular Zinc Homeostasis: Redox Control of Zinc Potentials and Zinc Signals. *Biometals* **2009,** *22* (1), 149–157.
9. a. Walters, F. H. Design of Corrosion Inhibitors: Use of the Hard and Soft Acid-Base (HSAB) Theory. *J. Chem. Educ.* **1991,** *68* (1), 29.
b. Berg, J. M.; Shi, Y. The Galvanization of Biology: A Growing Appreciation for the Roles of Zinc. *Science* **1996,** *271* (5252), 1081.
c. Ayers, P. W. The Physical Basis of the Hard/Soft Acid/Base Principle. *Faraday Discuss.* **2007,** *135* (0), 161–190.
10. a. Maret, W. Metallothionein Redox Biology in the Cytoprotective and Cytotoxic Functions of Zinc. *Exp. Gerontol.* **2008,** *43* (5), 363–369.
b. Powell, S. R. The Antioxidant Properties of Zinc. *J. Nutr.* **2000,** *130* (5), 1447S–1454S.
11. a. Maret, W. New Perspectives of Zinc Coordination Environments in Proteins. *J. Inorg. Biochem.* **2012,** *111,* 110–116.
b. Maret, W. Zinc Coordination Environments in Proteins Determine Zinc Functions. *J. Trace Elem. Med. Biol.* **2005,** *19* (1), 7–12.
c. Vallee, B. L.; Auld, D. S. Zinc: Biological Functions and Coordination Motifs. *Acc. Chem. Res.* **1993,** *26* (10), 543–551.
12. Chen, F.; Wu, M.-F.; Liu, G.-N.; Wang, M.-S.; Zheng, F.-K.; Yang, C.; Xu, Z.-N.; Liu, Z.-F.; Guo, G.-C.; Huang, J.-S. Zinc(II) and Cadmium(II) Coordination Polymers Based on 3-(5H-tetrazolyl)benzoate Ligand with Different Coordination Modes: Hydrothermal Syntheses, Crystal Structures and Ligand-Centered Luminescence. *Eur. J. Inorg. Chem.* **2010,** *2010* (31), 4982–4991.
13. a. Gao, Y.; Wu, J.; Li, Y.; Sun, P.; Zhou, H.; Yang, J.; Zhang, S.; Jin, B.; Tian, Y. A Sulfur-Terminal Zn(II) Complex and Its Two-Photon Microscopy Biological Imaging Application. *J. Am. Chem. Soc.* **2009,** *131* (14), 5208–5213.
b. Dollberg, C. L.; Turro, C. New Quinone Diimine Complex of Zinc with pH-Dependent Emission in the Visible Region. *Inorg. Chem.* **2001,** *40* (11), 2484–2485.
c. Cheng, J.; Wei, K.; Ma, X.; Zhou, X.; Xiang, H. Synthesis and Photophysical Properties of Colorful Salen-Type Schiff Bases. *J. Phys. Chem. C* **2013,** *117* (32), 16552–16563.

14. a. Burdette, S. C.; Walkup, G. K.; Spingler, B.; Tsien, R. Y.; Lippard, S. J. Fluorescent Sensors for Zn^{2+} Based on a Fluorescein Platform: Synthesis, Properties and Intracellular Distribution. *J. Am. Chem. Soc.* **2001,** *123* (32), 7831–7841.
b. Parkesh, R.; Clive Lee, T.; Gunnlaugsson, T. Highly Selective 4-amino-1,8-naphthalimide Based Fluorescent Photoinduced Electron Transfer (PET) Chemosensors for Zn(II) under Physiological PH Conditions. *Org. Biomol. Chem.* **2007,** *5* (2), 310–317.
c. Jayabharathi, J.; Ramanathan, P.; Thanikachalam, V.; Arunpandiyan, A. Sensitive and Selective PET-Based Π-expanded Phenanthrimidazole Luminophore for Zn^{2+} Ion. *J. Fluoresc.* **2014,** *24* (3), 827–834.
15. De Silva, A. P.; Gunaratne, H. N.; Gunnlaugsson, T.; Huxley, A. J.; McCoy, C. P.; Rademacher, J. T.; Rice, T. E. Signaling Recognition Events with Fluorescent Sensors and Switches. *Chem. Rev.* **1997,** *97* (5), 1515–1566.
16. Nie, C.; Zhang, Q.; Ding, H.; Huang, B.; Wang, X.; Zhao, X.; Li, S.; Zhou, H.; Wu, J.; Tian, Y. Two Novel Six-Coordinated Cadmium(II) and Zinc(II) Complexes from Carbazate β-diketonate: Crystal Structures, Enhanced Two-Photon Absorption and Biological Imaging Application. *Dalton Trans.* **2014,** *43* (2), 599–608.
17. a. Vahrenkamp, H. Metalle in Lebensprozessen. *Chem. Unserer Zeit* **1973,** 7 (4), 97–105.
b. Tubek, S.; Grzanka, P.; Tubek, I. Role of Zinc in Hemostasis: A Review. *Biol. Trace Elem. Res.* **2008,** *121* (1), 1–8.
c. Stefanidou, M.; Maravelias, C.; Dona, A.; Spiliopoulou, C. Zinc: A Multipurpose Trace Element. *Arch. Toxicol.* **2006,** *80* (1), 1–9.
d. Salgueiro, M. J.; Zubillaga, M.; Lysionek, A.; Sarabia, M. I.; Caro, R.; De Paoli, T.; Hager, A.; Weill, R.; Boccio, J. Zinc as an Essential Micronutrient: A Review. *Nutr. Res.* **2000,** *20* (5), 737–755.
18. a. Prasad, A. S. Discovery of Human Zinc Deficiency: Its Impact on Human Health and Disease. *Adv. Nutr. Int. Rev. J.* **2013,** *4* (2), 176–190.
b. Meunier, N.; O'connor, J.; Maiani, G.; Cashman, K.; Secker, D.; Ferry, M.; Roussel, A.; Coudray, C. Importance of Zinc in the Elderly: The ZENITH Study. *Eur. J. Clin. Nutr.* **2005,** *59,* S1–S4.
c. Frederickson, C. J.; Koh, J.-Y.; Bush, A. I. The Neurobiology of Zinc in Health and Disease. *Nat. Rev. Neurosci.* **2005,** *6* (6), 449–462.
d. McCord, M. C.; Aizenman, E. The Role of Intracellular Zinc Release in Aging, Oxidative Stress, and Alzheimer's Disease. *Front. Aging Neurosci.* **2014,** *6,* 77.
e. Chasapis, C. T.; Loutsidou, A. C.; Spiliopoulou, C. A.; Stefanidou, M. E. Zinc and Human Health: An Update. *Arch. Toxicol.* **2012,** *86* (4), 521–534.
f. Loef, M.; von Stillfried, N.; Walach, H. Zinc Diet and Alzheimer's Disease: A Systematic Review. *Nutr. Neurosci.* **2012,** *15* (5), 2–12.
g. Rink, L. *Zinc in Human Health;* Ios Press, 2011.
19. a. Hirano, T.; Kikuchi, K.; Urano, Y.; Higuchi, T.; Nagano, T. Highly Zinc-Selective Fluorescent Sensor Molecules Suitable for Biological Applications. *J. Am. Chem. Soc.* **2000,** *122* (49), 12399–12400.
b. Burdette, S. C.; Walkup, G. K.; Spingler, B.; Tsien, R. Y.; Lippard, S. J. Fluorescent Sensors for Zn^{2+} Based on a Fluorescein Platform: Synthesis, Properties and Intracellular Distribution. *J. Am. Chem. Soc.* **2001,** *123* (32), 7831–7841.
c. Jiang, P.; Guo, Z. Fluorescent Detection of Zinc in Biological Systems: Recent Development on the Design of Chemosensors and Biosensors. *Coord. Chem. Rev.* **2004,** *248* (1–2), 205–229.
d. Que, E. L.; Domaille, D. W.; Chang, C. J. Metals in Neurobiology: Probing Their Chemistry and Biology with Molecular Imaging. *Chem. Rev.* **2008,** *108* (5), 1517–1549.
e. Chen, Y.; Bai, Y.; Han, Z.; He, W.; Guo, Z. Photoluminescence Imaging of Zn^{2+} in Living Systems. *Chem. Soc. Rev.* **2015,** *44* (14), 4517–4546.

20. Ojida, A.; Mito-Oka, Y.; Inoue, M. A.; Hamachi, I. First Artificial Receptors and Chemosensors toward Phosphorylated Peptide in Aqueous Solution. *J. Am. Chem. Soc.* **2002,** *124* (22), 6256–6258.
21. Han, M. S.; Kim, D. H. Naked-Eye Detection of Phosphate Ions in Water at Physiological pH: A Remarkably Selective and Easy-to-Assemble Colorimetric Phosphate-Sensing Probe. *Angew. Chem.* **2002,** *114* (20), 3963–3965.
22. Rhee, H. W.; Choi, H. Y.; Han, K.; Hong, J. I. Selective Fluorescent Detection of Flavin Adenine Dinucleotide in Human Eosinophils by Using bis(Zn^{2+}-dipicolylamine) Complex. *J. Am. Chem. Soc.* **2007,** *129* (15), 4524–4525.
23. Ojida, A.; Takashima, I.; Kohira, T.; Nonaka, H.; Hamachi, I. Turn-on Fluorescence Sensing of Nucleoside Polyphosphates Using a Xanthene-Based Zn(II) Complex Chemosensor. *J. Am. Chem. Soc.* **2008,** *130* (36), 12095–12101.
24. Kurishita, Y.; Kohira, T.; Ojida, A.; Hamachi, I. Rational Design of FRET-Based Ratiometric Chemosensors for In Vitro and in Cell Fluorescence Analyses of Nucleoside Polyphosphates. *J. Am. Chem. Soc.* **2010,** *132* (38), 13290–13299.
25. Kurishita, Y.; Kohira, T.; Ojida, A.; Hamachi, I. Organelle-Localizable Fluorescent Chemosensors for Site-Specific Multicolor Imaging of Nucleoside Polyphosphate Dynamics in Living Cells. *J. Am. Chem. Soc.* **2012,** *134* (45), 18779–18789.
26. Zhang, J. F.; Kim, S.; Han, J. H.; Lee, S.-J.; Pradhan, T.; Cao, Q. Y.; Lee, S. J.; Kang, C.; Kim, J. S. Pyrophosphate-Selective Fluorescent Chemosensor Based on 1,8-Naphthalimide–DPA–Zn(II) Complex and Its Application for Cell Imaging. *Org. Lett.* **2011,** *13* (19), 5294–5297.
27. Ojida, A.; Sakamoto, T.; Inoue, M.-A.; Fujishima, S.-H.; Lippens, G.; Hamachi, I. Fluorescent BODIPY-Based Zn(II) Complex as a Molecular Probe for Selective Detection of Neurofibrillary Tangles in the Brains of Alzheimer's Disease Patients. *J. Am. Chem. Soc.* **2009,** *131* (18), 6543–6548.
28. Koulov, A. V.; Stucker, K. A.; Lakshmi, C.; Robinson, J. P.; Smith, B. D. Detection of Apoptotic Cells Using a Synthetic Fluorescent Sensor for Membrane Surfaces that Contain Phosphatidylserine. *Cell Death Differ.* **2003,** *10* (12), 1357–1359.
29. Lakshmi, C.; Hanshaw, R. G.; Smith, B. D. Fluorophore-Linked Zinc(II)dipicolylamine Coordination Complexes as Sensors for Phosphatidylserine-Containing Membranes. *Tetrahedron* **2004,** *60* (49), 11307–11315.
30. Hanshaw, R. G.; Lakshmi, C.; Lambert, T. N.; Johnson, J. R.; Smith, B. D. Fluorescent Detection of Apoptotic Cells by Using Zinc Coordination Complexes With a Selective Affinity for Membrane Surfaces Enriched With Phosphatidylserine. *Chembiochem* **2005,** *6* (12), 2214–2220.
31. a. Smith, B. A.; Gammon, S. T.; Xiao, S.; Wang, W.; Chapman, S.; McDermott, R.; Suckow, M. A.; Johnson, J. R.; Piwnica-Worms, D.; Gokel, G. W.; Smith, B. D.; Leevy, W. M. In Vivo Optical Imaging of Acute Cell Death Using a Near-infrared Fluorescent Zinc-Dipicolylamine Probe. *Mol. Pharm.* **2011,** *8* (2), 583–590.
b. Smith, B. A.; Xiao, S.; Wolter, W.; Wheeler, J.; Suckow, M. A.; Smith, B. D. In Vivo Targeting of Cell Death Using a Synthetic Fluorescent Molecular Probe. *Apoptosis* **2011,** *16* (7), 722–731.
32. a. Smith, B. A.; Akers, W. J.; Leevy, W. M.; Lampkins, A. J.; Xiao, S.; Wolter, W.; Suckow, M. A.; Achilefu, S.; Smith, B. D. Optical Imaging of Mammary and Prostate Tumors in Living Animals Using a Synthetic Near Infrared Zinc(II)-Dipicolylamine Probe for Anionic Cell Surfaces. *J. Am. Chem. Soc.* **2010,** *132* (1), 67–69.
b. Smith, B. A.; Xie, B. W.; van Beek, E. R.; Que, I.; Blankevoort, V.; Xiao, S.; Cole, E. L.; Hoehn, M.; Kaijzel, E. L.; Lowik, C. W.; Smith, B. D. Multicolor Fluorescence Imaging of Traumatic Brain Injury in a Cryolesion Mouse Model. *ACS Chem. Neurosci.* **2012,** *3* (7), 530–537.

c. Kwong, J. M. K.; Hoang, C.; Dukes, R. T.; Yee, R. W.; Gray, B. D.; Pak, K. Y.; Caprioli, J. Bis(Zinc-Dipicolylamine), Zn-DPA, a New Marker for Apoptosis. *Invest. Ophthalmol. Vis. Sci.* **2014,** *55* (8), 4913–4921.

33. Smith, B. A.; Harmatys, K. M.; Xiao, S.; Cole, E. L.; Plaunt, A. J.; Wolter, W.; Suckow, M. A.; Smith, B. D. Enhanced Cell Death Imaging Using Multivalent Zinc(II)-bis(dipicolylamine) Fluorescent Probes. *Mol. Pharm.* **2013,** *10* (9), 3296–3303.
34. Plaunt, A. J.; Harmatys, K. M.; Wolter, W. R.; Suckow, M. A.; Smith, B. D. Library Synthesis, Screening, and Discovery of Modified Zinc(II)-bis(dipicolylamine) Probe for Enhanced Molecular Imaging of Cell Death. *Bioconjug. Chem.* **2014,** *25* (4), 724–737.
35. a. Leevy, W. M.; Johnson, J. R.; Lakshmi, C.; Morris, J.; Marquez, M.; Smith, B. D. Selective Recognition of Bacterial Membranes by Zinc(II)-Coordination Complexes. *Chem. Commun.* **2006,** *15,* 1595–1597.

    b. Leevy, W. M.; Gammon, S. T.; Jiang, H.; Johnson, J. R.; Maxwell, D. J.; Jackson, E. N.; Marquez, M.; Piwnica-Worms, D.; Smith, B. D. Optical Imaging of Bacterial Infection in Living Mice Using a Fluorescent Near-Infrared Molecular Probe. *J. Am. Chem. Soc.* **2006,** *128* (51), 16476–16477.

    c. Leevy, W. M.; Gammon, S. T.; Johnson, J. R.; Lampkins, A. J.; Jiang, H.; Marquez, M.; Piwnica-Worms, D.; Suckow, M. A.; Smith, B. D. Noninvasive Optical Imaging of *Staphylococcus aureus* Bacterial Infection in Living Mice Using a Bis-dipicolylamine-Zinc(II) Affinity Group Conjugated to a Near-Infrared Fluorophore. *Bioconjug. Chem.* **2008,** *19* (3), 686–692.

    d. White, A. G.; Fu, N.; Leevy, W. M.; Lee, J.-J.; Blasco, M. A.; Smith, B. D. Optical Imaging of Bacterial Infection in Living Mice Using Deep-Red Fluorescent Squaraine Rotaxane Probes. *Bioconjug. Chem.* **2010,** *21* (7), 1297–1304.

    e. White, A. G.; Gray, B. D.; Pak, K.Y.; Smith, B. D. Deep-Red Fluorescent Imaging Probe for Bacteria. *Bioorg. Med. Chem. Lett.* **2012,** *22* (8), 2833–2836.
36. Mito-oka, Y.; Tsukiji, S.; Hiraoka, T.; Kasagi, N.; Shinkai, S.; Hamachi, I. Zn(II) Dipicolylamine-Based Artificial Receptor as a New Entry for Surface Recognition of α-helical Peptides in Aqueous Solution. *Tetrahedron Lett.* **2001,** *42* (40), 7059–7062.
37. a. Honda, K.; Nakata, E.; Ojida, A.; Hamachi, I. Ratiometric Fluorescence Detection of a Tag Fused Protein Using the Dual-Emission Artificial Molecular Probe. *Chem. Commun.* **2006,** *38,* 4024–4026.

    b. Ojida, A.; Honda, K.; Shinmi, D.; Kiyonaka, S.; Mori, Y.; Hamachi, I. Oligo-Asp Tag/Zn(II) Complex Probe as a New Pair for Labeling and Fluorescence Imaging of Proteins. *J. Am. Chem. Soc.* **2006,** *128* (32), 10452–10459.

    c. Nonaka, H.; Tsukiji, S.; Ojida, A.; Hamachi, I. Non-enzymatic Covalent Protein Labeling Using a Reactive Tag. *J. Am. Chem. Soc.* **2007,** *129* (51), 15777–15779.
38. Nonaka, H.; Fujishima, S.-H.; Uchinomiya, S.-H.; Ojida, A.; Hamachi, I. Selective Covalent Labeling of Tag-Fused GPCR Proteins on Live Cell Surface with a Synthetic Probe for Their Functional Analysis. *J. Am. Chem. Soc.* **2010,** *132* (27), 9301–9309.
39. Fujishima, S. H.; Nonaka, H.; Uchinomiya, S. H.; Kawase, Y. A.; Ojida, A.; Hamachi, I. Design of a Multinuclear Zn(II) Complex as a New Molecular Probe for Fluorescence Imaging of His-tag Fused Proteins. *Chem. Commun.* **2012,** *48* (4), 594–596.
40. a. Koike, T.; Watanabe, T.; Aoki, S.; Kimura, E.; Shiro, M. A Novel Biomimetic Zinc(II)−Fluorophore, Dansylamidoethyl−Pendant Macrocyclic Tetraamine 1,4,7,10-Tetraazacyclododecane (Cyclen). *J. Am. Chem. Soc.* **1996,** *118* (50), 12696–12703.

    b. Kimura, E.; Aoki, S.; Kikuta, E.; Koike, T. A Macrocyclic Zinc(II) Fluorophore as a Detector of Apoptosis. *Proc. Natl. Acad. Sci. U.S.A.* **2003,** *100* (7), 3731–3736.
41. Oltmanns, D.; Zitzmann-Kolbe, S.; Mueller, A.; Bauder-Wuest, U.; Schaefer, M.; Eder, M.; Haberkorn, U.; Eisenhut, M. Zn(II)-bis(cyclen) Complexes and the Imaging of Apoptosis/Necrosis. *Bioconjug. Chem.* **2011,** *22* (12), 2611–2624.

42. a. Dilworth, J. R.; Hueting, R. Metal Complexes of Thiosemicarbazones for Imaging and Therapy. *Inorg. Chim. Acta* **2012,** *389,* 3–15.
b. Casas, J. S.; García-Tasende, M. S.; Sordo, J. Main Group Metal Complexes of Semicarbazones and Thiosemicarbazones. A Structural Review. *Coord. Chem. Rev.* **2000,** *209* (1), 197–261.
43. Kowol, C. R.; Trondl, R.; Arion, V. B.; Jakupec, M. A.; Lichtscheidl, I.; Keppler, B. K. Fluorescence Properties and Cellular Distribution of the Investigational Anticancer Drug Triapine (3-Aminopyridine-2-Carboxaldehyde Thiosemicarbazone) and Its Zinc(II) Complex. *Dalton Trans.* **2010,** *39* (3), 704–706.
44. Cowley, A. R.; Davis, J.; Dilworth, J. R.; Donnelly, P. S.; Dobson, R.; Nightingale, A.; Peach, J. M.; Shore, B.; Kerr, D.; Seymour, L. Fluorescence Studies of the Intra-cellular Distribution of Zinc Bis(thiosemicarbazone) Complexes in Human Cancer Cells. *Chem. Commun.* **2005,** *7,* 845–847.
45. a. Pascu, S. I.; Waghorn, P. A.; Conry, T. D.; Betts, H. M.; Dilworth, J. R.; Churchill, G. C.; Pokrovska, T.; Christlieb, M.; Aigbirhio, F. I.; Warren, J. E. Designing Zn(II) and Cu(II) Derivatives as Probes for In Vitro Fluorescence Imaging. *Dalton Trans.* **2007,** *43,* 4988–4997.
b. Pascu, S. I.; Waghorn, P. A.; Conry, T. D.; Lin, B.; Betts, H. M.; Dilworth, J. R.; Sim, R. B.; Churchill, G. C.; Aigbirhio, F. I.; Warren, J. E. Cellular Confocal Fluorescence Studies and Cytotoxic Activity of New Zn(II) Bis(thiosemicarbazonato) Complexes. *Dalton Trans.* **2008,** *16,* 2107–2110.
46. Dayal, D.; Palanimuthu, D.; Shinde, S. V.; Somasundaram, K.; Samuelson, A. G. A Novel Zinc Bis(thiosemicarbazone) Complex for Live Cell Imaging. *J Biol. Inorg. Chem.* **2011,** *16* (4), 621–632.
47. Pfeiffer, P.; Breith, E.; Lübbe, E.; Tsumaki, T. Tricyclische Orthokondensierte Nebenvalenzringe. *Justus Liebigs Ann. Chem.* **1933,** *503* (1), 84–130.
48. a. Wezenberg, S. J.; Kleij, A. W. Material Applications for Salen Frameworks. *Angew. Chem. Int. Ed.* **2008,** *47* (13), 2354–2364.
b. Cozzi, P. G. Metal-Salen Schiff Base Complexes in Catalysis: Practical Aspects. *Chem. Soc. Rev.* **2004,** *33* (7), 410–421.
c. Kleij, A. W. Zinc-Centred Salen Complexes: Versatile and Accessible Supramolecular Building Motifs. *Dalton Trans.* **2009,** *24,* 4635–4639.
49. Reglinski, J.; Morris, S.; Stevenson, D. E. Supporting Conformational Change at Metal Centres. Part 2: Four and Five Coordinate Geometry. *Polyhedron* **2002,** *21* (21), 2175–2182.
50. Consiglio, G.; Failla, S.; Finocchiaro, P.; Oliveri, I. P.; Bella, S. D. Aggregation Properties of Bis(salicylaldiminato)Zinc(II) Schiff-Base Complexes and Their Lewis Acidic Character. *Dalton Trans.* **2012,** *41* (2), 387–395.
51. Consiglio, G.; Failla, S.; Finocchiaro, P.; Oliveri, I. P.; Purrello, R.; Di Bella, S. Supramolecular Aggregation/Deaggregation in Amphiphilic Dipolar Schiff-Base Zinc(II) Complexes. *Inorg. Chem.* **2010,** *49* (11), 5134–5142.
52. Vladimirova, K. G.; Freidzon, A. Y.; Kotova, O. V.; Vaschenko, A. A.; Lepnev, L. S.; Bagatur'yants, A. A.; Vitukhnovskiy, A. G.; Stepanov, N. F.; Alfimov, M. V. Theoretical Study of Structure and Electronic Absorption Spectra of Some Schiff Bases and Their Zinc Complexes. *Inorg. Chem.* **2009,** *48* (23), 11123–11130.
53. a. Splan, K. E.; Massari, A. M.; Morris, G. A.; Sun, S. S.; Reina, E.; Nguyen, S. T.; Hupp, J. T. Photophysical and Energy-Transfer Properties of (Salen)zinc Complexes and Supramolecular Assemblies. *Eur. J. Inorg. Chem.* **2003,** *2003* (12), 2348–2351.
b. Lin, C.-W.; Chou, P.-T.; Liao, Y.-H.; Lin, Y.-C.; Chen, C.-T.; Chen, Y.-C.; Lai, C.-H.; Chen, B.-S.; Liu, Y.-H.; Wang, C.-C.; Ho, M.-L. Photoisomerization of a Maleonitrile-Type Salen Schiff Base and Its Application in Fine-Tuning Infinite Coordination Polymers. *Chem. Eur. J.* **2010,** *16* (12), 3770–3782.

c. Chang, K.-H.; Huang, C.-C.; Liu, Y.-H.; Hu, Y.-H.; Chou, P.-T.; Lin, Y.-C. Synthesis of Photo-luminescent Zn(II) Schiff Base Complexes and Its Derivative Containing Pd(II) Moiety. *Dalton Trans.* **2004,** *11,* 1731–1738.

d. Brissos, R.; Ramos, D.; Lima, J. C.; Mihan, F. Y.; Borras, M.; de Lapuente, J.; Cort, A. D.; Rodriguez, L. Luminescent Zinc Salophen Derivatives: Cytotoxicity Assessment and Action Mechanism Studies. *New J. Chem.* **2013,** *37* (4), 1046–1055.

54. Hai, Y.; Chen, J.-J.; Zhao, P.; Lv, H.; Yu, Y.; Xu, P.; Zhang, J.-L. Luminescent Zinc Salen Complexes as Single and Two-Photon Fluorescence Subcellular Imaging Probes. *Chem. Commun.* **2011,** *47* (8), 2435–2437.

55. a. Meng, Q.-H.; Zhou, P.; Song, F.; Wang, Y.-B.; Liu, G.-L.; Li, H. Controlled Fluorescent Properties of Zn(II) Salen-Type Complex Based on Ligand Design. *CrystEngComm* **2013,** *15* (15), 2786–2790.

b. Li, W.-H.; Zheng, G. Photoactivatable Fluorophores and Techniques for Biological Imaging Applications. *Photochem. Photobiol. Sci.* **2012,** *11* (3), 460–471.

c. Huang, S.; Zou, L.-Y.; Ren, A.-M.; Guo, J.-F.; Liu, X.-T.; Feng, J.-K.; Yang, B.-Z. Computational Design of Two-Photon Fluorescent Probes for a Zinc Ion Based on a Salen Ligand. *Inorg. Chem.* **2013,** *52* (10), 5702–5713.

56. Hwang, K. Y.; Kim, H.; Lee, Y. S.; Lee, M. H.; Do, Y. Synthesis and Properties of Salen–Aluminum Complexes as a Novel Class of Color-Tunable Luminophores. *Chem. Eur. J.* **2009,** *15* (26), 6478–6487.

57. Xie, D.; Jing, J.; Cai, Y.-B.; Tang, J.; Chen, J.-J.; Zhang, J.-L. Construction of an Orthogonal Znsalen/Salophen Library as a Colour Palette for One- and Two-Photon Live Cell Imaging. *Chem. Sci.* **2014,** *5* (6), 2318–2327.

58. Tang, J.; Cai, Y.-B.; Jing, J.; Zhang, J.-L. Unravelling the Correlation between Metal Induced Aggregation and Cellular Uptake/subcellular Localization of Znsalen: An Overlooked Rule for Design of Luminescent Metal Probes. *Chem. Sci.* **2015,** *6* (4), 2389–2397.

59. Lai, J.; Ke, X.-S.; Tang, J.; Zhang, J.-L. Tris(Znsalen) Cryptand Minimizes Znsalen Aggregation Arising from Intermolecular Zn···O Interaction. *Chin. Chem. Lett.* **2015,** *26* (8), 937–941.

60. Zhang, J.-L.; Tang, J.; Xie, D.; Yin, H.-Y.; Jing, J. Cationic Sulfonium Functionalization Renders Znsalens High Fluorescence, Good Water Solubility and Tunable Cell-Permeability. *Org. Biomol. Chem.* **2016,** *14* (13), 3360–3368.

61. Jing, J.; Chen, J.-J.; Hai, Y.; Zhan, J.; Xu, P.; Zhang, J.-L. Rational Design of ZnSalen as a Single and Two Photon Activatable Fluorophore in Living Cells. *Chem. Sci.* **2012,** *3* (11), 3315–3320.

62. Jing, J.; Zhang, J.-L. Combining Myeloperoxidase (MPO) with Fluorogenic ZnSalen to Detect Lysosomal Hydrogen Peroxide in Live Cells. *Chem. Sci.* **2013,** *4* (7), 2947–2952.

63. a. Mizushima, N.; Yoshimori, T.; Levine, B. Methods in Mammalian Autophagy Research. *Cell* **2010,** *140* (3), 313–326.

b. Kabeya, Y.; Mizushima, N.; Ueno, T.; Yamamoto, A.; Kirisako, T.; Noda, T.; Kominami, E.; Ohsumi, Y.; Yoshimori, T. LC3, a Mammalian Homologue of Yeast Apg8p, Is Localized in Autophagosome Membranes after Processing. *EMBO J.* **2000,** *19* (21), 5720–5728.

64. Kuma, A.; Matsui, M.; Mizushima, N. LC3, an Autophagosome Marker, Can Be Incorporated into Protein Aggregates Independent of Autophagy: Caution in the Interpretation of LC3 Localization. *Autophagy* **2007,** *3* (4), 323–328.

65. Chen, J.-J.; Jing, J.; Chang, H.; Rong, Y.; Hai, Y.; Tang, J.; Zhang, J.-L.; Xu, P. A Sensitive and Quantitative Autolysosome Probe for Detecting Autophagic Activity in Live and Prestained Fixed Cells. *Autophagy* **2013,** *9* (6), 894–904.

66. Gao, J.; Liu, Y. G.; Zhou, Y.; Boxer, L. M.; Woolley, F. R.; Zingaro, R. A. Artificial Zinc(II) Complexes Regulate Cell Cycle and Apoptosis-Related Genes in Tumor Cell Lines. *Chembiochem* **2007,** *8* (3), 332–340.

67. Senge, M. O.; Sergeeva, N. N. Photochemical Transformations Involving Zinc Porphyrins and Phthalocyanines. In *PATAI'S Chemistry of Functional Groups;* John Wiley & Sons, Ltd, 2009.
68. Alzeer, J.; Vummidi, B. R.; Roth, P. J.; Luedtke, N. W. Guanidinium-Modified Phthalocyanines as High-affinity G-quadruplex Fluorescent Probes and Transcriptional Regulators. *Angew. Chem. Int. Ed. Engl.* **2009,** *48* (49), 9362–9365.
69. Ke, X.-S.; Tang, J.; Yang, Z.-S.; Zhang, J.-L. β-Conjugation of Gadolinium(III) DOTA Complexes to Zinc(II) Porpholactol as Potential Multimodal Imaging Contrast Agents. *J. Porphyr. Phthalocyanines* **2014,** *18* (10n11), 950–959.
70. Ke, X.-S.; Tang, J.; Chen, J.-J.; Zhou, Z.-Y.; Zhang, J.-L. β-Ionic Conjugated Chlorin-Type Photosensitizers Based on Porpholactone: Synthesis, Photophysical Properties, and Photodynamic Activity. *ChemPlusChem* **2015,** *80* (1), 237–252.

CHAPTER TWO

Cellular Uptake and Sensing Capability of Transition Metal Peptide Conjugates

C. Dolan, C.S. Burke, A. Byrne, T.E. Keyes
Dublin City University, Dublin, Ireland

Contents

1. INTRODUCTION

The use of metal-based drugs in modern medicine is well established and accepted.[1] Indeed, photodynamic therapy employing metal complexes is applied widely in the clinic and used to treat skin-related diseases and certain cancers.[2] The application of transition metal complexes to bioimaging is much less developed but has become a focus of significant effort in recent years. Coordination compounds of ruthenium, iridium and osmium based on polypyridyl ligands have favorable photophysical and redox properties, which make them particularly amenable to imaging. These properties include synthetically tunable absorbance and emission wavelengths, increased photostability, large Stokes shifts, in addition to long-lived and often intense luminescence.

For in-cell sensing, the long-lived luminescence from ruthenium and iridium polypyridyl and related complexes is useful because it allows

Inorganic and Organometallic Transition Metal Complexes with Biological Molecules and Living Cells
ISBN 978-0-12-803814-7
http://dx.doi.org/10.1016/B978-0-12-803814-7.00002-2

for discrimination from the background autofluorescence, which is short-lived and prevalent in many biological samples. Furthermore, the long-lived excited state promotes environmental sensitivity, in particular, susceptibility to oxygen quenching, which is well known for such complexes and is facilitated by the triplet character of their excited states. Moreover, longer-lived biological processes (those on the microsecond timescale) such as membrane diffusion, protein rotation and protein folding may be studied using time-resolved luminescence anisotropy from long-lived transition metal probes.[3] The large Stokes shifts (i.e., spectral separation between their absorbance and emission maxima), frequently exhibited by transition metal luminophores, reduce the possibility of concentration quenching or self-absorption of the complex when it is present in greater than micromolar concentrations. The tunable emission of metal complexes also enables design of complexes with emission in the red to near infrared (NIR) spectral ranges to match the biological window, and away from tissue absorbance and from exciting autofluorescence of biological samples.[4,5] Finally, metal complexes tend not in general to exhibit aggregation, e.g., excimer or exciplex formation, which may occur at elevated concentrations for some organic-based molecular probes.[6] A detailed discussion on the photophysics of coordination compounds is beyond the scope of this short chapter so the reader is directed to the following literature.[7]

Confocal laser scanning microscopy is probably the most widely exploited imaging methodology applied in biology. It requires continuous laser scanning of the sample which can promote photobleaching, self-quenching and damage to the cell membrane, and so photochemical stability of the probe is particularly important.[8,9] Transition metal complexes have frequently shown improved photostability compared to conventional organic fluorophores, which is an advantage when continuous monitoring of biological samples is required, e.g., in dynamic studies. For example, Neugebauer et al. compared the photostability of a ruthenium(II) complex under imaging conditions used for human blood platelets (using 458 nm excitation wavelength).[6] They found that after 20 min of continuous imaging, the luminescent probe retained more than 50% of its initial emission intensity from within a platelet. A range of ruthenium(II),[10–15] iridium(III),[16–24] osmium(II),[25] rhenium(I),[16,26–32] and heterobimetallic[33,34] polypyridyl complexes have been explored as general probes in live cell imaging and have shown consistently good performance in this regard.

2. TRANSITION METAL COMPLEXES FOR CELL IMAGING AND SENSING

The luminescent complexes of d^6, d^8 and d^{10} transition metals have been widely studied because of their attractive and often tunable photophysical properties. One of the key drivers of this interest has been the possibility of applying luminescent transition metal complexes as sensors. The polypyridyl complexes, most notably of ruthenium(II), osmium(II), rhenium(I), rhodium(III) and iridium(III), have been widely shown to respond quantitatively to molecular oxygen and have also been studied across a range of other chemosensing domains including pH, anion and cation sensing, and in the sensing of biomolecules, such as DNA and amino acids.[35]

In spite of the tantalizing opportunities such complexes offer in terms of sensing in cells, until approximately a decade ago, reports on the use of phosphorescent transition metal complexes in bioimaging and sensing were relatively rare.[16,36] One of the earliest examples of the application of ruthenium polypyridyl complexes to cell imaging was reported by Dobrucki et al. who examined the interaction of $[Ru(bpy)_3]^{2+}$ and $[Ru(phen)_3]^{2+}$ (where bpy is 2,2′-bipyridine and phen is 1,10-phenanthroline) with macrophages.[37] It was found, consistent with our later studies, that these complexes did not penetrate the cell membrane. Although it was observed that after extensive incubation $[Ru(phen)_3]^{2+}$ was capable of binding to the membrane and was internalized after irradiation because of membrane damage.

By increasing the lipophilicity of the metal complex, Rosenzweig and coworkers observed that related ruthenium polypyridyl complexes containing pyrenyl appendages were able to permeate the membrane of macrophages and they in turn reported on intensity-based measurements of O_2 concentration.[38] As emission intensity is influenced by a range of parameters, including probe concentration, it is not a useful sensing parameter in cells, so Zhong et al. reported one of the first investigations into the sensing capability of ruthenium polypyridyl complexes using fluorescence lifetime imaging microscopy (FLIM). Here, $[Ru(bpy)_3]^{2+}$ was examined and incubated with bronchial epithelial cells. Interestingly, the complex was found to be permeable to this cell line and FLIM (under air saturated and depleted conditions) showed that the complexes retained their oxygen sensitivity within the cells.[39]

Given the sensing capabilities of ruthenium and related complexes, the relatively low level of early interest in their application to cellular imaging seems somewhat surprising, but may be attributed in part to the poor cell membrane permeability of such complexes. There has, however, been a remarkable swell of interest over the past 10 years in development of bioimaging probes based on transition metal complexes, which coincides with growing understanding of means of promoting delivery of molecular agents across cell membranes and also increasingly in targeting materials to selected organelles. These enhanced insights have emerged from the fields of pharmacy and drug delivery, and indeed in parallel with enhanced delivery of metal complexes as imaging agents, delivery of metal-based therapeutics has also been evolving.[40] A range of agents have now been identified that promote transition metal transport across the cell membrane including use of solvents or their conjugation to macromolecular structures most notably polyethylene glycol or biomolecules, such as sterols or sugars, and the focus of this chapter: peptides.

The relative delay in application of luminescent metal complexes to bioimaging may also arise from a perception that such heavy metal complexes will be cytotoxic. Certainly, this can be an issue, and indeed many of the platinum-based compounds have been deliberately designed for therapeutic applications that may lend support to this notion.[2] However, cytotoxicity is not a given. Whereas many ruthenium(II)[41] and osmium(II)[42] complexes have been engineered to exhibit potent cytotoxicity for anticancer applications, stable ruthenium(II)[43] and osmium(II)[25] polypyridyl complexes have also been shown to be noncytotoxic. Synthetic strategies that stabilize such complexes against ligand exchange and speciation in aqueous media, e.g., by modifications such as PEGylation, can be used to reduce cytotoxicity.[16,44–48] In addition, we have shown across several examples of transition metal complexes that once the lipophilicity of a complex is not high and the complex is stable, negligible cytotoxicity is observed even when such complexes are incubated with cells over windows of days.[3,6,19,25,49,50]

3. CELL-PENETRATING PEPTIDES

Peptides have emerged over the past two decades as one of the most powerful strategies for effectively targeting and delivering conjugated cargo from pharmaceuticals through imaging agents to cells and tissues.[51,52] Polypeptides have also been shown to enhance DNA binding ability,[53] and in addition have been used to deliver a wide variety of bulky exogenous cargo to living cells such as quantum dots,[54,55] nanoparticles,[56] organic fluorophores,[51,57–60] proteins,[61,62] as well as other smaller molecules.[63,64] Peptides can fulfill a range of functions in this regard, from cell-penetrating peptides (CPPs), which provide nonspecific cell permeability, to signal peptides, which provide organelle targeting or tissue-specific biomarker peptides, through peptides that combine both cell-targeting and cell-penetrating properties: cell-penetrating homing peptides.[65]

CPPs are the most widely studied class of cargo-carrying molecule to emerge in drug discovery in recent years.[66,67] They are attractive because of their low cytotoxicity, their promiscuity with regard to breadth of cell lines, their relatively short sequences (<40 amino acids), dose-dependent efficiency, and relatively little restriction as to the size or character of the cargo they carry.[68] CPPs have since moved beyond natural sequences to include libraries of CPPs that include nonprimary and unnatural amino acid sequences that have been implemented to improve cellular uptake efficiency.[68] An extensive discussion on the various CPPs available is detailed in a series of review papers,[68–70] which are recommended to the reader. For the purpose of this short chapter, however, only those most commonly conjugated to transition metal complexes to date will be considered.

CPPs vary considerably in structure although a key feature common to most examples is the presence of positively charged amino acids, particularly at the terminus of the sequence.[68,71] One of the most widely exploited CPPs is the transactivator of transcription (TAT) peptide (GRKKRRQRRRPPQ), which is derived from the transactivator protein (TAT) encoded by human immunodeficiency virus type 1.[30,58] The importance of the guanidinium groups in this peptide sequence was highlighted by Wender et al. who systematically exchanged the cationic guanidinium moieties with neutral alanine residues.[58] This led to a progressive decrease in the cellular uptake that could be directly correlated with the number of arginine residues. Therefore the large number of arginine residues in the TAT sequence led to investigations to determine if polyarginine could alone act as a CPP.

Indeed, polyarginines are now one of the most important guanidinium-rich molecular conjugates used in cell delivery. They have been shown to help overcome multidrug resistant cancer in cellular and animal models, even when the conjugated drug succumbs to resistance.[72] The implication of such a finding is significant. The conjugation of small drug molecules (Taxol, luciferin and coelenterazine H) to the octaarginine peptide, via a controlled drug release linker, improved drug administration because of substantially enhanced aqueous solubility, better localized biodistribution, and lengthened pharmacokinetics of the drugs. Importantly, the multidrug resistance described by Dubikovskaya et al. provides an alternative treatment strategy without the need to switch to another drug to circumvent the resistance.[73] Among the first peptides to be delivered with oligoarginine transporters were the RACK and cyclosporine A octapeptides, both of which progressed into clinical trials.[72]

The ability of polyarginine to diffuse across the cell membrane is critically dependent on the length of the arginine chain. A minimum chain length of Arg_5 or Arg_6 may apply depending on the cargo and Arg_8 or Arg_9 seem to show the best and most ubiquitous cargo-carrying ability.[6,74–76] As outlined later we observed that Arg_5 was incapable of transporting a ruthenium metal complex across the membrane.[6] A proviso in application of polycationic peptides as cargo carriers is that their charge can lead to cytotoxicity; indeed polyarginines alone have been reported to exhibit cytotoxicity.[77]

The need for supplementary peptide modification can arise in instances where CPPs are required to deliver very large molecular cargo, e.g., biomolecules such as nucleic acids and proteins, inside cells.[68] As tissue or in vivo delivery requires longer circulation times, the stability of the peptide, particularly toward proteolysis, needs to be considered; one strategy to achieve this is to add various hydrophilic groups to an already permeable peptide. For example, in an effort to enhance the hydrophilicity of a peptide sequence to improve its physiochemical properties, Oskolkov et al. simply replaced the isoleucine with the more hydrophilic threonine (an amino acid that is similar to isoleucine) while still preserving peptide function.[78] Further strategies for preventing proteolysis include using protease-deficient cell lines, reducing the temperature and the addition of protease inhibitors (such as phenylmethanesulfonyl fluoride).[79]

4. MECHANISM OF CPP UPTAKE

Endocytosis is broadly considered to be the main route of transmembrane transport by CPPs. This conclusion is based principally on results that have shown that CPP-led uptake is activated, i.e., temperature and energy dependent.[6,59,68,74] However, endocytosis as the sole route to CPP uptake has been called into question somewhat as some examples of temperature-independent uptake by CPPs have emerged. For example, internalization of a nickel(II) chelate of octaarginine peptide bearing a nitrilotriacetic acid ligand has been observed to occur at 4°C, a temperature at which endocytic pathways do not function.[80] In addition, direct penetration of the plasma membrane and even receptor-mediated uptake has also been reported for specific CPPs.[68] Therefore a more complex picture is emerging of the mechanism of uptake. For an in-depth discussion on the numerous potential cellular uptake mechanisms of peptides, the reader is directed to a comprehensive review by Copolovici et al.[68] However, for the purposes of this short chapter a brief synopsis of some the most commonly accepted mechanisms of CPP uptake are discussed.

Endocytosis, which encompasses phagocytosis and pinocytosis, is a highly regulated process of internalization of solutes and fluids from the extracellular matrix into the cell. It is an energy-activated process wherein the incoming molecule is engulfed by the cell plasma membrane. The membrane forms an enclosure around the material and breaks off as a vesicle that migrates into the cell where it goes to the cytoplasmic organelles, or enters the nuclear envelope.[68] Endocytosis may be further subcategorized into caveolae- and/or lipid-raft-mediated endocytosis, macropinocytosis, cholesterol-dependent, clathrin-mediated endocytosis, or caveolae- and clathrin-independent endocytosis (Fig. 2.1).[68] As pointed out by Copolovici et al., there is a practical difficulty in resolving inconsistencies in the reported mechanisms of cellular

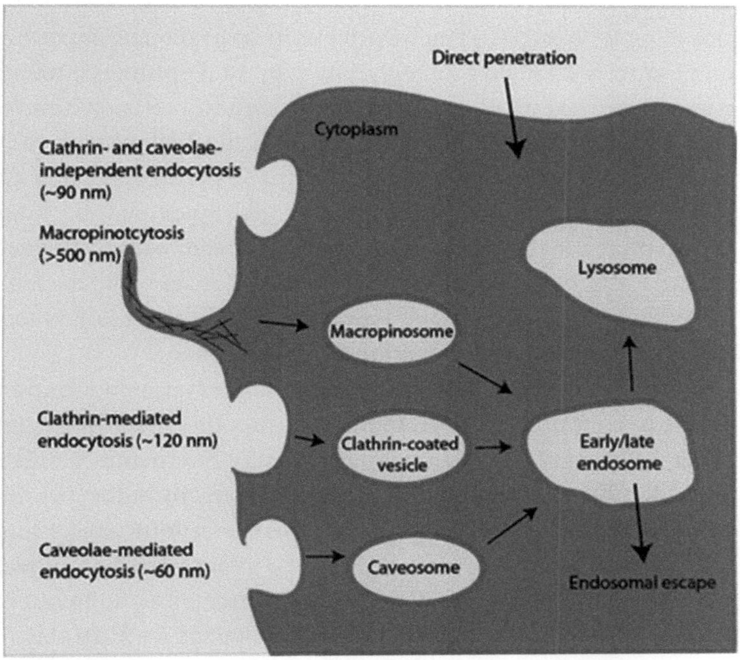

Figure 2.1 Diagram illustrating some of the uptake and release mechanisms of cell-penetrating peptides (CPPs). *Reproduced with permission from Copolovici, D.M.; Langel, K.; Eriste, E.; Langel, A.* ACS Nano ***2014***, *8, 1972–1994.*

uptake by CPPs because there are no generally accepted experimental methods to test the entry pathway by CPPs, and problems have been compounded by different experimental setups leading to contradictory results.[68]

Both lipophilicity and cationic charge of the CPP are important factors in dictating the interaction of peptides with the plasma membrane. It has been shown that as the hydrophobic interactions between CPP and membrane increase, and the number of positive charges on the peptide decreases, a reduction in transmembrane transport occurs.[68] The length of the CPP also plays a critical role; CPPs usually consist of 15–25 amino acids in length; however, as described, oligoarginine peptide sequences have also proven very efficient transporters with between seven and nine arginine residues seeming to be optimal. Moreover, the secondary structure adopted by the CPP, once in contact with the cellular plasma membrane, plays an important role in the membrane-penetrating ability of the peptide, and in turn the cytotoxicity of the CPP. The secondary structure depends mainly on the charge, hydrogen bonds, and helicity of the peptide. Several research groups have successfully determined the secondary structure of CPPs in model membranes by using circular dichroism methods with unilamellar lipid vesicles.[81]

Endocytosis is believed to begin with electrostatic interaction that occurs between negatively charged proteoglycans expressed on the cell membrane and positively charged peptides. Ligands that bind to proteoglycans may be internalized via endocytosis. For example, both the TAT protein and TAT peptide have been shown to bind strongly to heparin, a sulfated glycosaminoglycan.[68] Furthermore, negatively charged lipids may be found on outer cell membranes and therefore it may be assumed that the polypeptides recruit these lipids to create negatively charged domains on the cell's surface. Accordingly, upon membrane binding, the surface tension is reduced, which also indicates their insertion into the membrane.[68]

Many CPPs, including TAT, Antennapedia, polyarginine, transportan, MPG, and Pep-1, have been shown to migrate through the plasma membrane via at least one energy-independent pathway.[68] Numerous translocation mechanisms for different peptides have been suggested including the inverted micelle model, the carpet model (considered as the cellular uptake model at high CPP concentrations of $>10\,\mu M$), and the pore-formation model.[68] A membrane-thinning model was proposed by Herce et al. as an alternative nonendocytotic translocation mechanism for polypeptides.[82] In this model, as the peptide approaches the surface of the membrane it attracts the phosphate groups bending the dipole toward the interior of the membrane. To avert this thermodynamically unfavorable conformation the peptide inserts into the interface between the hydrophilic head groups and the carbon chains to orient away from the choline groups. As the density of peptides increases, the membrane becomes thinner and subsequently an amino acid translocates with a local bending of the membrane on both sides of the bilayer. This bending reduces the energetic barrier of the translocation and a pore forms allowing the peptide to permeate the cell membrane distal layer.

Overall, it is apparent that the mechanisms by which peptides migrate across the plasma membrane in an energy-independent mechanism are not entirely resolved. In particular, questions still remain over (1) the lack of cell specificity of CPPs because this does not correlate well with the idea that peptides could be targeting any specific cell receptor and (2) how peptides that are highly hydrophilic can translocate across the hydrophobic core of the membrane, a highly thermodynamically unfavorable prospect, through an energy independent mechanism.[82]

Finally, the uptake of a CPP and its cargo is also influenced by a number of factors including the chemical character of the cargo, differences in the physicochemical properties of the CPP, the cell line, and the concentration of CPP, all of which will have an impact on cellular uptake.[68] Nonetheless, what has been clearly demonstrated is the efficiency and wide applicability of CPPs as a transmembrane molecular cargo carrier that acts without compromising the cellular structure.

5. CPP MODIFIED TRANSITION METAL LUMINOPHORES FOR SENSING AND IMAGING

Poor aqueous solubility and low cationic charge can prohibit transition metal complex luminophores from potential driven diffusion across a cell membrane and typically complexes such as $[Ru(bpy)_3]^{2+}$ are not membrane permeable; although there have been a number of examples[11,13,83,84] of transition metal luminophores without appended moieties to promote membrane transport applied to cellular imaging. In many of these reports the complex was administered to the cells in the presence of organic solvent, in particular dimethyl sulfoxide (DMSO) or ethanol. Organic solvents can be very effective in promoting transport of the complex into the cell. Both DMSO and ethanol increase the permeability of the lipid membrane and typical concentrations used are between 1 and 5% v/v. Indeed, an important application of DMSO is in topical drug delivery where its powerful ability to increase the skin's permeability is exploited.[85] However, organic solvents can cause irreparable damage to the cell membrane and at concentrations above about 1–3% v/v can cause cell death. They can also have unpredictable effects on the cells' metabolism, e.g., DMSO is known to induce cell diffusion and cell differentiation.[85,86]

A superior approach to promoting complex permeation is to append a transporting moiety onto the complex, and ideally this is a moiety that promotes aqueous solubility as well. To this end, there have been many reports of luminescent complex probes that are conjugated to polypeptides,[19,25,87] polymers such as poly(ethylene glycol),[45,48] estradiol,[16,88] and other biomolecules,[43] in an effort to alter the membrane permeability of the dye or to specifically target them within cellular structures. These approaches have successfully rendered metal complexes water soluble, and most importantly cell permeable without the requirement for membrane-damaging permeating agents such as solvents or detergents. Such chemical modifications have been explored for many years to promote uptake and targeting in magnetic resonance contrast agents, and more recently in lanthanide probes, but they have really only become an important focus for transition metal luminophores over the past 10 years.[89,90]

Brunner and Barton reported the first application of a cargo-carrying peptide to a transition metal luminophore in 2006.[91] They described cell uptake of oligoarginine conjugated to rhodium complexes and found that the peptide was very effective in promoting complex penetration across HeLa cell membranes. Appending a fluorescein to the complex, in addition to the peptide, promoted penetration of cell membrane and nuclear envelope (Fig. 2.2).

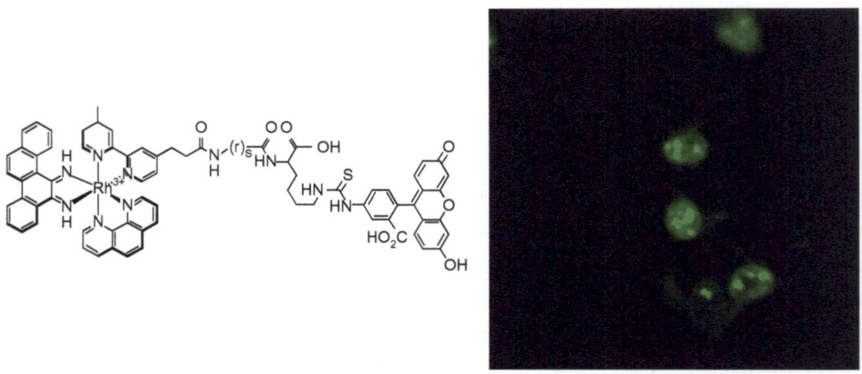

Figure 2.2 (*Left*) Schematic diagram illustrating the first application of a cargo-carrying peptide to a transition metal luminophore. (*Right*) Confocal microscopy image of HeLa cells treated with the dye peptide after incubation for 60 min with 5 μM at 37°C. The excitation wavelength was 488 nm. *Reproduced with permission from Brunner, J.; Barton, J.K.* Biochemistry (N.Y.) ***2006***, 45, *12295–12302.*

Subsequently, our group reported the first example of cell permeable ruthenium(II) polypeptide conjugates and its application to confocal fluorescent and lifetime imaging.[6] It was observed that conjugation to an octaarginine peptide promoted aqueous solubility and probe membrane permeabilization. Whereas by comparison, both the unconjugated parent complex and the analogous pentaarginine conjugated complex were membrane impermeable to both SP2 myeloma and human blood platelet membranes; emphasizing the importance of polyarginine chain length. The octaarginine modified probe transferred rapidly across the membrane of SP2 myeloma cells and permeated all but the nuclear envelope. FLIM measurements (Fig. 2.3) were used to determine the average lifetime of the dye across the various cell compartments the probe penetrated and it was found that dye residing at the plasma membrane displayed the shortest lifetimes. This was attributed to higher solubility of molecular oxygen in the cell membrane, which caused a decrease in the lifetime of the ruthenium(II) probe.

Puckett et al. also reported the application of CPPs in facilitating entry of a ruthenium(II) dye peptide into HeLa cells.[87] The parent complex used, $[Ru(phen)(bpy)(dppz)]^{2+}$ (structure shown in Fig. 2.4), contained the dppz (dipyridophenazine) ligand, which is a well-known DNA binding moiety. Ruthenium(II) complexes containing this ligand typically show light-switching behavior wherein the complex is nonemission in aqueous media, but emits on incorporation into a nonaqueous environment, such as intercalation into the DNA structure or binding to a lipid

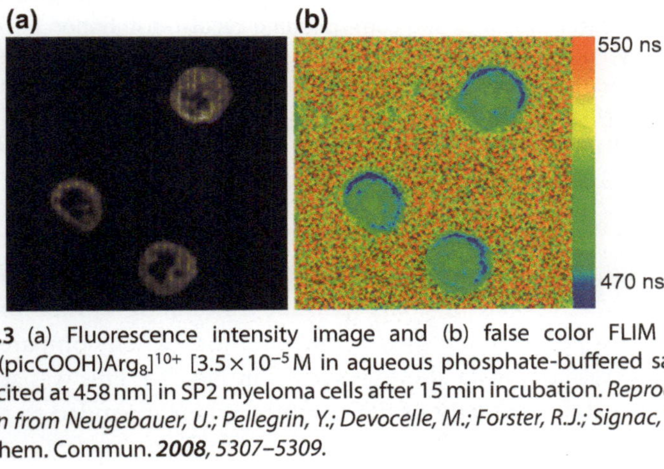

Figure 2.3 (a) Fluorescence intensity image and (b) false color FLIM image of $[Ru(bpy)_2(picCOOH)Arg_8]^{10+}$ [3.5×10^{-5} M in aqueous phosphate-buffered saline (PBS) buffer, excited at 458 nm] in SP2 myeloma cells after 15 min incubation. *Reproduced with permission from Neugebauer, U.; Pellegrin, Y.; Devocelle, M.; Forster, R.J.; Signac, W.; Moran, N.; et al.* Chem. Commun. ***2008****, 5307–5309.*

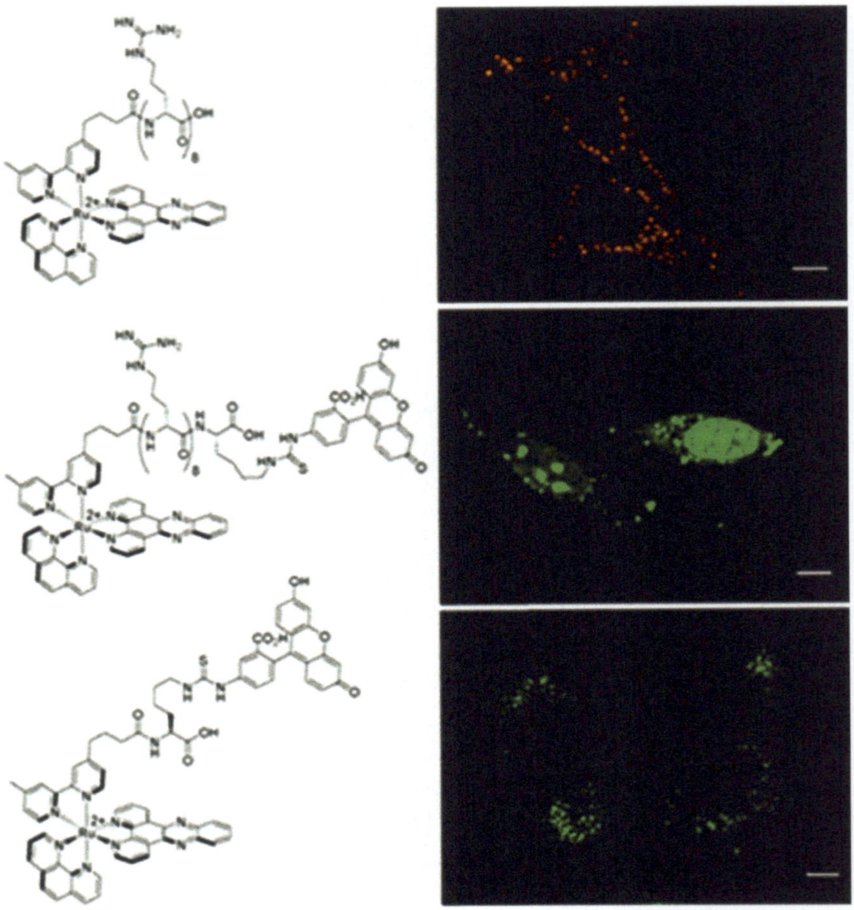

Figure 2.4 Chemical structures and corresponding cellular distribution of complexes synthesized by Puckett et al. in HeLa cells incubated at 37°C in complete medium [σ-minimum essential medium (MEM) with 10% fetal bovine serum (FBS)] then rinsed with Hanks' balanced salt solution (HBSS) solution and imaged without cell fixation. *Reproduced with permission from Puckett, C.A.; Barton, J.K.* J. Am. Chem. Soc. ***2009***, 131, *8738–8739.*

membrane.[92–94] The parent complex showed no transport across the cellular membrane prior to conjugation to the octaarginine peptide. Whereas, interestingly, the uptake of a dye peptide containing ruthenium-Arg_8-fluorescein was far greater at lower concentrations when compared to the ruthenium-Arg_8 conjugate alone, as shown in Fig. 2.4. This was attributed to the increased lipophilicity of the complex when appended to the fluorescein moiety.

Another dppz-containing metal complex conjugate [Ru(dppz)$_2$(picCOOH)Arg$_8$]$^{10+}$, shown in Fig. 2.5, was exploited for the purposes of multimodal cellular imaging.[49] As this complex exhibits the "light switch effect" (because of its dppz ancillary ligands, as described) it could be used as a membrane probe, because only the complex protected by and associated with membrane structures emits phosphorescence. Therefore confocal fluorescence imaging selectively revealed, very clearly, the membrane structures within the cell. However, as no emission was observed from the cytoplasm, luminescence could not be used to assess the true distribution of the dye within the cell. Because resonance Raman mapping does not rely on the emission of the dye, this method was employed here for the first time as a complementary technique to assess the location and distribution of the parent dye and dye conjugate. Although Raman scattering is a weak phenomenon, several orders of magnitude of signal enhancement, principally from Franck–Condon active vibrational modes associated with the absorbing chromophores within the complex, can be achieved if the exciting laser matches an absorbance of the chromophore.[95] Ruthenium polypyridyl

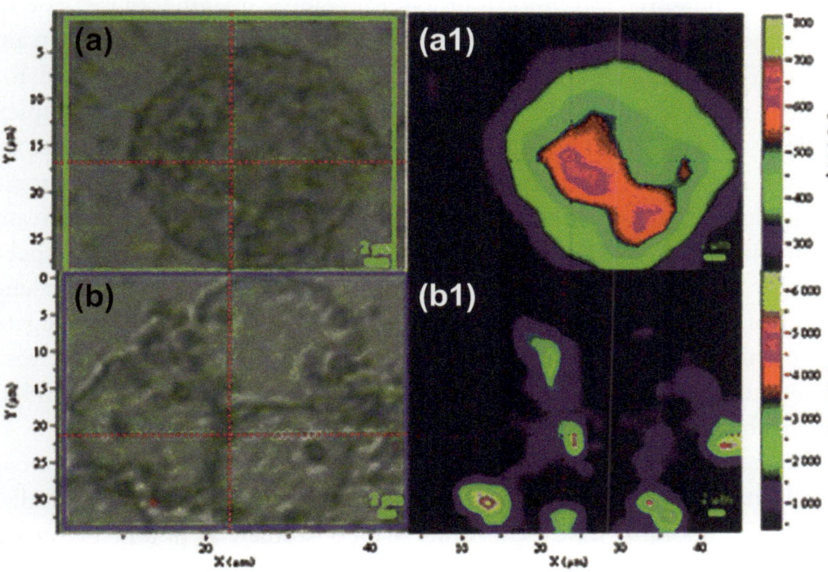

Figure 2.5 Resonance Raman intensity map of live myeloma cells after incubation with [Ru(dppz)$_2$(picCOOH)Arg$_8$]$^{10+}$ (a1) and free parent dye (b1) after excitation at 458 nm. (a) and (b) represent the bright field images of the cells in phosphate-buffered saline (PBS). *Reproduced with permission from Cosgrave, L.; Devocelle, M.; Forster, R.J.; Keyes, T.E.* Chem. Commun. ***2010***, 46, *103–105.*

complexes tend to show intense resonance Raman intensities when the exciting laser coincides with their ^{1}MLCT (metal-to-ligand charge transfer) transitions. Imaging using this signal is enabled because of the very large Stokes shifts of such complexes, which prevents interference in the resonance Raman signal from emission excited by the laser. This means that intense vibrational maps can be achieved for ruthenium polypyridyl probes at submicromolar concentrations within cells, i.e., under the same experimental conditions used for fluorescence imaging. This offers the opportunity for multimodal, multiparameter imaging where simultaneous or consecutive imaging of luminescence intensity, luminescence lifetime, and resonance Raman at a single excitation wavelength can be used to sense different elements within the cell environment. Preliminary demonstration of this approach was described for $[Ru(bpy)_2(pic\text{-}Arg_8)]^{10+}$ where confocal fluorescence imaging permitted assessment of the uptake and distribution of the dye in SP2 myeloma cells, resonance Raman mapping (which is completely independent of O_2 concentration) was used to assess pH, and the luminescence lifetime could be used to determine O_2 concentration.[96]

Related iridium(III)[19] and osmium(II)[25] peptide conjugates were also reported as potential multimodal probes for cellular imaging. In the case of the former example, the iridium(III) parent complex and conjugate were found to be cytotoxic in two different cell lines. The IC_{50} recorded for the cells in the dark was independent, in the case of the parent complex $[Ir(dfpp)_2(picCOOH)]^{+}$, of the identity of the cell, with IC_{50} values of 84.8 μM and 88 μM, respectively, for SP2 and CHO cells. The IC_{50} approximately doubled for the polyarginine conjugate and displayed a significant dependence on cell type with IC_{50} values of 35 μM and 54.1 μM, respectively, for SP2 and CHO cells. This was in spite of the fact that the concentration of the conjugate was higher in CHO. Both parent and conjugate showed high photocytotoxicity with at least an order of magnitude increase in the rate of cell death on exposure to light. Preliminary evidence pointed to a mechanism involving $^{1}O_2$ generation by the metal complex.

The least studied of the platinum group metal luminophores in the context of cell imaging are the osmium(II) polypyridyl complexes. Organometallic osmium compounds have been quite widely studied as potent cytotoxic anticancer agents.[97] Indeed, uptake of such a complex has been found to be promoted by conjugation to polyarginine by Rijt et al. who reported promising cancer cell cytotoxicity of osmium(II) arene complex [(η6-biphenyl)Os(picolinate)Cl], where polyarginine was conjugated to the picoline ligand.[98] The complex was shown to induce irreparable damage to DNA by causing a large degree of DNA unwinding. The effect of varying the polyarginine lengths (from Arg_1 to Arg_8) on the cellular uptake, toxicity, and cell distribution of the dye peptide in CHO cell lines was also explored. Their results also highlighted a strong correlation between arginine length and cellular and nuclear uptake, DNA binding, and decreased cytotoxicity. The cellular uptake and activity of the osmium functionalized polypeptides followed the order of $Arg_8 >> Arg_5 > Arg_1 \sim$ unfunctionalized.

The first report of an osmium(II) polypyridyl complex in the context of imaging also exploited a polyarginine CPP to promote uptake of the complex into CHO and SP2 myeloma cell lines.[25] The advantage of osmium(II) polypyridyl complexes over their ruthenium(II) analogs lies in their red emission maxima, which frequently extend to the NIR spectral region beyond 700 nm, which is strongly coincident with the biological optical window. Furthermore, because of their increased crystal field splitting, the osmium *d–d* states cannot be accessed by thermal excitation from the excited triplet state, which lends osmium(II) polypyridyl complexes temperature-independent photophysics and exceptional photostability even by comparison with other metal complex luminophores.[99] Such thermal stability is an important concern in cell imaging where in most instances cells are incubated at 37°C during imaging. For many ruthenium polypyridyl complexes increased temperature can lead to significant reductions in emission intensity and lifetime as well as reduced photostability of the complex. This behavior is attributed to thermal population of the e_g^* state from the triplet excited state.[100] This is particularly important in aqueous media and solutions of coordinating anions, such as Cl^- (e.g., the saline environment used for cell imaging), where the chloro or aquo products can form from ligand loss.[100] The behavior of the osmium(II) polypyridyl complex and its ruthenium analog was compared.[25] The uptake behavior was very similar, although the osmium(II) complex showed greater tendency to enter the nuclear envelope, which may indicate it is more lipophilic than the ruthenium(II) analog (Fig. 2.6). Interestingly, the osmium complex showed both superior photostability and lower cytotoxicity compared with its ruthenium equivalent. This was attributed to the remarkable stability of this complex, which was reflected by the complete absence of photobleaching even following extended periods of continuous photoirradiation. However, when compared to its ruthenium analog its luminescence was short-lived in water (20 ns), therefore rendering it insensitive to O_2. The large Stokes shift of the osmium emission here also permitted collection of Raman spectra from the probe in live cells.

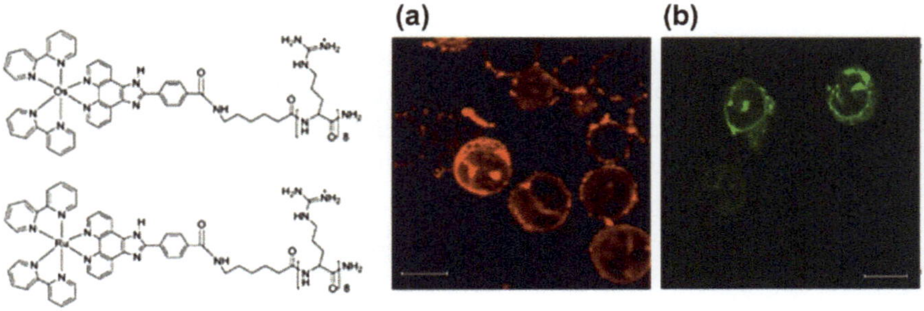

Figure 2.6 Chemical structures of osmium(II) and ruthenium(II) peptide conjugates. Cellular uptake of (a) $[Ru(bpy)_2(pic\text{-}arg_8)]^{10+}$ and (b) $[Os(bpy)_2(pic\text{-}arg_8)]^{10+}$ by live SP2 cells after 2h incubation at 37°C showing nucleoli penetration (z=2, scale bar=10 μM). The ruthenium luminescence was monitored at 630nm and the osmium at 750nm; excitation was 488nm in each case. *Reproduced with permission from Byrne, A.; Dolan, C.; Moriarty, R.D.; Martin, A.; Neugebauer, U.; Forster, R.J.; et al.* Dalton Trans. ***2015***, 44, *14323–14332.*

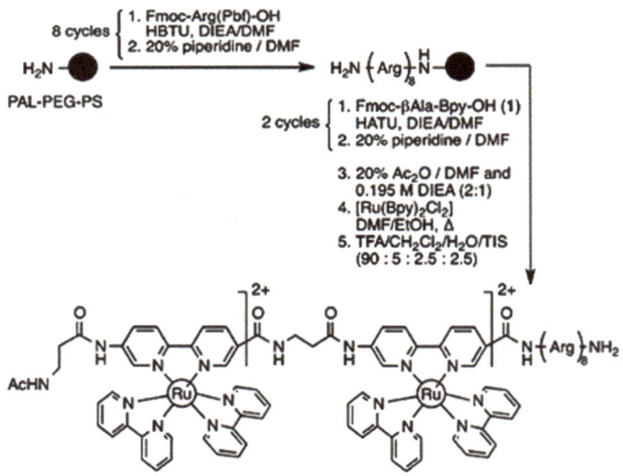

Figure 2.7 Solid-phase peptide synthesis of the octaarginine dinuclear ruthenium(II) complex. *Reproduced with permission from Gamba, I.; Salvadó, I.; Rama, G.; Bertazzon, M.; Sánchez, M.I.; Sánchez-Pedregal, V.M.; et al.* Chemistry ***2013***, 19, *13369–13375.*

Gamba et al. described a new bipyridine building block that was used for the solid-phase synthesis of a dinuclear DNA-binding ruthenium(II) metallopeptide.[101] The reaction scheme and chemical structure are shown in Fig. 2.7. A similar procedure was extended to include the synthesis of mono-, di-, and trinuclear iridium(III) metallopolymers.[76] Functionalization of the metallopolymers with an octaarginine peptide tail rendered them suitable for cellular imaging and their increased cytotoxicity was associated with strong DNA interaction and sequence selectivity. In the case of the ruthenium(II) analogs, the luminescence observed within Vero cells formed a punctuated pattern in the cytoplasm, which was attributed to the formation of endosomes arising from endocytosis as well as diffuse emission from the cell nuclei. Interestingly, the iridium(III) parent exhibited negligible cytotoxic effects, whereas the metal complexed octaarginine-metallopolymer revealed IC_{50} values comparable with cisplatin for NCIH460, A2780 cis, and MCF-7 tumor cell lines under similar experimental conditions.

6. SIGNAL PEPTIDE-MODIFIED TRANSITION METAL LUMINOPHORES

The prospect of selectively delivering an imaging or a sensing agent to individual cell organelles is particularly attractive. Small organelles can be readily distinguished by confocal microscopy once the dye is highly localized and such an approach can allow processes occurring at such sites, e.g., generation of reactive oxygen species (ROS), to be selectively monitored permitting valuable insights into cell biochemistry. One particularly useful approach in this regard is to use a signal peptide for targeting. Signal peptides (or sequences) are a broad class of peptides that operate in cells to carry out cargo delivery to organelles. Most protein synthesis occurs in the cytoplasm and the signal peptides function by temporarily [in most cases, though not for nuclear localization signal (NLS) peptides] tagging the newly formed proteins to facilitate their transport into the organelles.[102] Signal peptides are well characterized for all of the key organelles of the cell and have been widely exploited as a means of selective delivery of therapeutic agents to the organelles, e.g., drug delivery to the mitochondria[103–105] and for nonviral gene delivery to the nucleus.[106,107] The nucleus is an attractive target from the perspective of theranostics. The nuclear envelope is a double membrane that forms a continuous structure with the endoplasmic reticulum.[108] The nuclear pores permit passage of small molecules into/out of the nucleus but passage of larger molecules is mediated by nuclear localization signal sequence that recognizes importin proteins at the pore. NLS peptides are particularly synthetically accessible as they are typically less than 12 residues in length. NLS sequences that have been successfully derived from transcription factors include NF-κB, TCF1-α, TFIIC-β, and SV-40,[109,110] of which NF-κB and SV-40 have shown to enter the nucleus selectively.[50,111,112]

The interactions between metal-based luminophores, particularly the polypyridyl complexes of ruthenium, and DNA has been under intensive investigation for many years. This is driven in particular by the potential application of such materials in chemo- and photodynamic therapy.[36,113–117] Much of the insight from this work has been derived from solution-based studies of DNA-metal complex association and DNA damage. Ultimately, the extension of this work to cells and tissues requires that the complexes be directed through the highly restricted nuclear envelope. The NLS peptides hold significant promise as transporters of such complexes. Indeed, NF-κB was found to be very effective in selectively transporting ruthenium(II) complexes to the nucleus of CHO cells. As illustrated in Fig. 2.8, $[Ru(bpy)_2pic\text{-}\beta Ala\text{-}NF\kappa B]^{6+}$ selectively accumulated inside the nucleus of HeLa cells, colocalizing very accurately with a nuclear staining dye. Furthermore, in the absence of light this complex showed relatively low cytotoxicity. Interestingly, it was found that by increasing the lipophilicity of the metal complex by replacing 2,2-bipyridyl counter ligands with lipophilic 4,7-diphenyl1,10-phenanthroline (dpp) ligands, the complex could be directed to the nucleolus.[50]

Mari et al. report the conjugation of a number of ruthenium(II) polypyridyl complexes, through a photolabile linker, to an NLS peptide (NLS, L-Arg-D-Arg-L-Arg-L-Lys-$CONH_2$, Fig. 2.9).[2] In the same paper

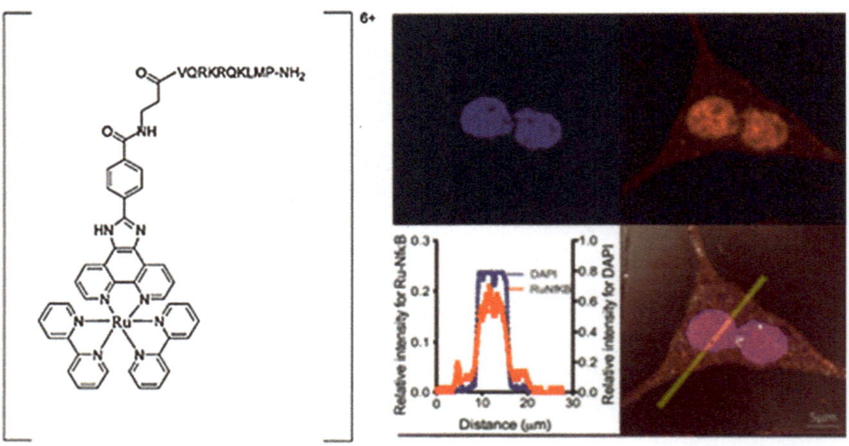

Figure 2.8 Nuclear localization of $[Ru(bpy)_2pic\text{-}\beta Ala\text{-}NF_kB]^{6+}$ and nuclear stain DAPI in CHO cells. Merged image of complex and 4′,6-diamidino-2-phenylindole (DAPI) with backscatter to demonstrate colocalization, confirmed by the corresponding cross-section and distribution plot of both complexes in the nucleus. *Reproduced with permission from Blackmore, L.; Moriarty, R.; Dolan, C.; Adamson, K.; Forster, R.J.; Devocelle, M.; et al.* Chem. Commun. ***2013***, *49, 2658–2660.*

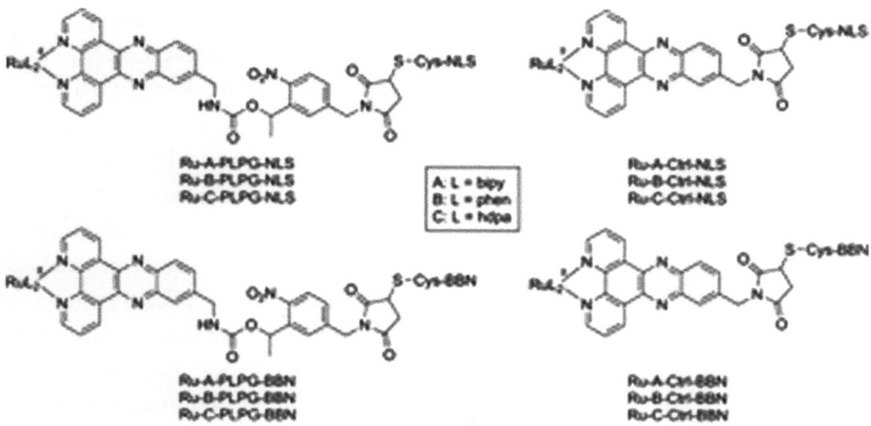

(1)

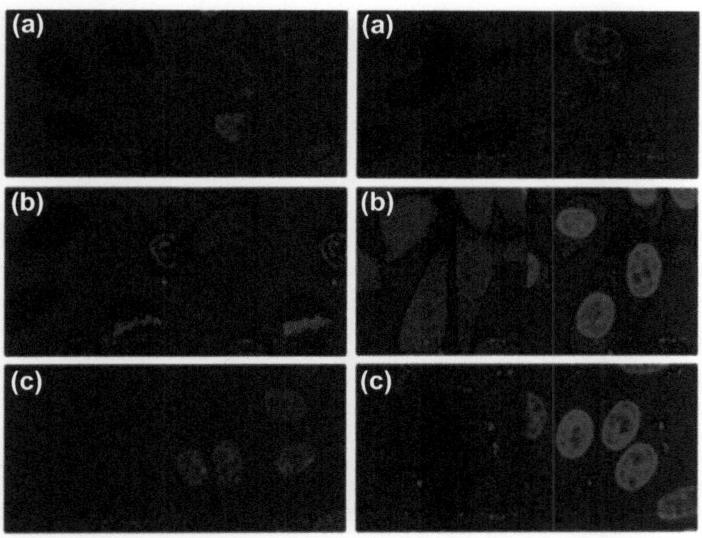

(2) (3)

Figure 2.9 (1) Structures of the light-activatable compounds conjugated to nuclear localization signal (NLS) (*top left*), Bombesin targeting peptide (BBN) (*bottom left*), the controls with NLS (*top right*), and BBN (*bottom right*). (2) Fluorescence microscopy images of HeLa cells incubated with (a) Ru-A-PLPG-NLS, (b) Ru-B-PLPG-NLS, and (c) Ru-CPLPG-NLS and (3) (a) Ru-A-PLPG-BBN, (b) Ru-B-PLPG-BBN, and (c) Ru-CPLPG-BBN. *Left*: Ru complexes (20 μm), 2 h incubation, excitation at 488 nm, emission above 600 nm; *right*: co-localization with 4′,6-diamidino-2-phenylindole (DAPI). *Reproduced with permission from Mari, C.; Pierroz, V.; Leonidova, A.; Ferrari, S.; Gasser, G.* Eur. J. Inorg. Chem. ***2015**, 3879–3891.*

they also conjugate the complexes to the Bombesin targeting peptide (BBN, L-Gln-L-Trp-L-Ala-L-Val-Gly-L-His-Leu-L-Met-$CONH_2$, Fig. 2.9), which has a very similar structure to that of the human gastrin-releasing peptide, which is overexpressed on the membrane of some cancer cells. They chose three different ancillary ligands (bipyridine, phenanthroline and dipyridylamine) to investigate the effect of lipophilicity of the complex conjugates on their uptake and cytotoxicity. Light-induced release (after 6–10 min irradiation) of the complexes was studied and toxicities were evaluated on HeLa and MRC-5 cell lines. As expected, the free complexes were characterized by poor cellular uptake and very moderate toxicity on both cell lines. However, the NLS conjugate exhibited preferential nuclear accumulation (Fig. 2.9) in the cell lines and interestingly showed no increased activity upon light irradiation. Conversely, the BBN-functionalized probes, containing phenanthroline and dipyridylamine ligands, exhibited increased cellular toxicity upon irradiation in Hela cells. This was in contrast to the MRC-5 cell line, which did not experience any toxic effect, as this particular cell line does not overexpress receptors responsible for BBN internalization. It is also noteworthy that light-stable control experiments and potential by-products displayed no toxic effect on either cell line.

Mitochondria are responsible for the cell's energy generation via ATP synthesis, supplying oxygen to the cell, and regulating apoptosis. Mitochondrial dysfunction is implicated both in disease and toxicity.[118–122] In producing ATP the mitochondria consume oxygen, therefore oxygen concentration in this organelle is a valuable measure of the cell's metabolic status reflecting both cell health or dysfunction. Mitochondria are also believed to be the primary source of ROS, in particular superoxide, within the cell[123] and ROS dysregulation is implicated in a wide range of degenerative pathologies.[119,124]

Measurement of O_2 or ROS concentrations within the mitochondria is challenging and has largely been conducted at isolated mitochondria. O_2 concentration measurement has typically relied on electrochemical measurements, using Clark cells at the isolated organelles.[125] O_2 production and ROS have been studied at isolated organelles using a range of methods including the use of probes and spin traps.[123] In situ measurement of O_2 or ROS in whole live cells is obviously preferable but use of luminescent probes in this regard is limited by the ability to direct such probes selectively across the mitochondrial double membrane. And in the case of ROS, limitations around a suitable luminescent probe remain an issue.

Consequently, mitochondria targeting peptide sequences are potentially very valuable in this regard. Signal peptides for mitochondrial targeting in nature are usually N-terminal sequences that are enriched in Arg, Ala, and Ser and poor in anionic amino acid residues. α-Helicity in the resulting structure is thought to be important in promoting transport across the mitochondrial membrane[126] and these design rules can be used as a guide in preparing synthetic mitochondrial penetrating peptide (MPP) analogs. Kelley et al. carried out a detailed systematic study of MPP sequences to evaluate the design rules around CPP sequences of value for mitochondria localization and identified a balance of charge and lipophilicity that enabled mitochondrial targeting.[127,128] They demonstrated that a sequence of eight amino acids was most successful in localizing to the mitochondria compared to chain lengths with fewer residues. Their optimal sequence (FrFKFrFK) also included D-arginine residues (r) to promote cellular stability.

Exploiting this peptide sequence we successfully demonstrated that a dimeric ruthenium(II) polypyridyl complex could be very selectively targeted to the mitochondria of mammalian cells, with even greater selectively than the commercially available MitoTracker Red, as shown in Fig. 2.10.[129] Using luminescence lifetime imaging, the dinuclear probe was shown to respond to molecular oxygen concentrations within the mitochondria. Antimycin A, a mitochondrial electron transport inhibitor, was used to switch off the mitochondrial machinery, and the ability of the probe to respond dynamically to changing O_2 concentrations within live HeLa cells was demonstrated. Using calibrated oxygen-dependent lifetime data it was possible to semiquantitatively measure O_2 concentration within the mitochondria, but it was evident that the probe also responded to elevated ROS levels within the organelle. Where ROS was produced, the probe lifetime decreased significantly below what would be expected for saturation O_2 concentrations. This example shows the potential power of combining signal peptides with transition metal molecular probes. With further development, dynamic and quantitative insights into mitochondrial behavior in live cells and tissues should be possible.

Selective mitochondrial accumulation of cytotoxic dinuclear ruthenium(II) polypyridyl complexes, without an appended peptide moiety, has also been reported by Pisani et al. (Fig. 2.11).[130] Increasing the length of the lipophilic ligand bridge resulted in significant improvements in cytotoxic properties. Furthermore, they showed for the first time that bis-ruthenium(II) complexes can selectively accumulate in the mitochondria of live cancer cells. Complexes containing the longest ligand bridge

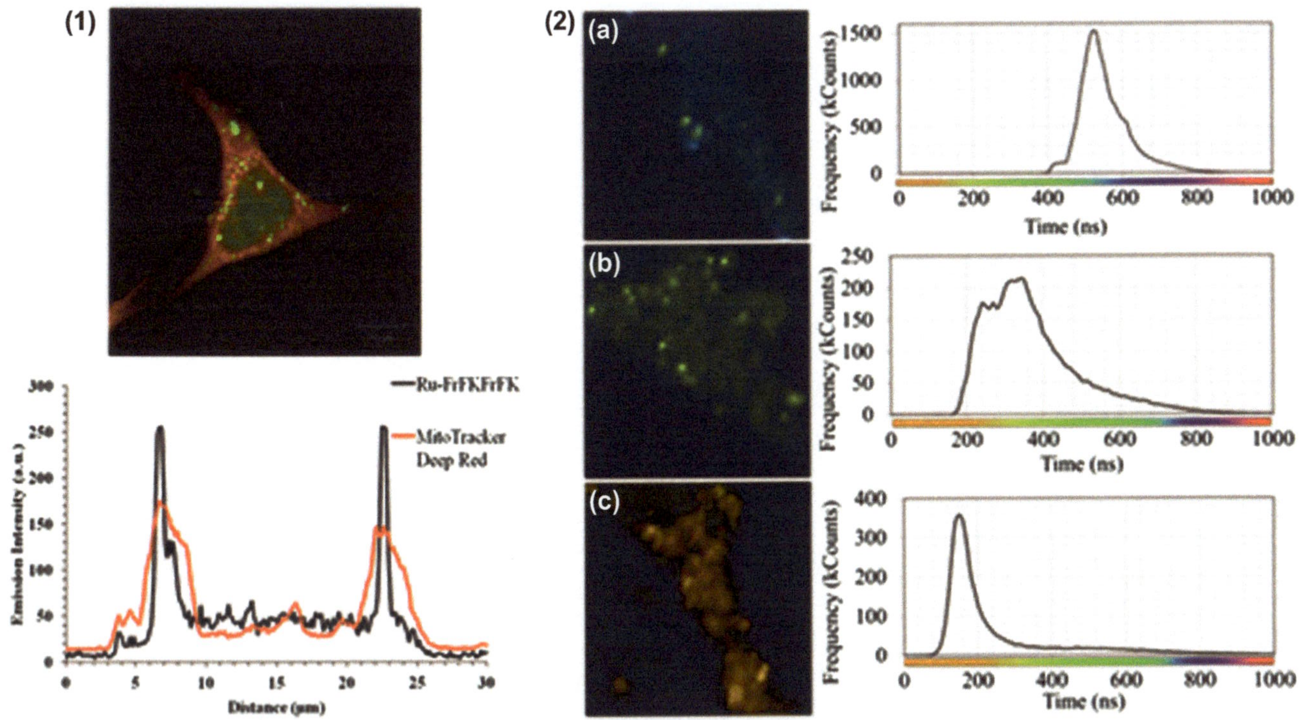

Figure 2.10 (1) Confocal image of $[Ru(bpy)_2(phen\text{-}Ar)2\text{-}FrFkFrFk]^{7+}$ in a HeLa cell, with MitoTracker Red. The cross-section corresponding plot demonstrates the collocation of the two complexes within the mitochondria. (2) Luminescence lifetime of dinuclear probe in response to the addition of Antimycin A. HeLa cells were incubated with the dye peptide for 2 h in the dark (a), and Antimycin A (200 μM/mL) was added. Cells were incubated for 10 min with Antimycin and fluorescence lifetime imaging microscopy (FLIM) recorded (b). Cells were incubated for a further 90 min, where total exposure time was 100 min, and FLIM lifetime recorded (c) ($n = 2$). FLIM was carried out by accumulating photon counts over 10 min in each case under identical, wavelength, and laser intensity conditions. *Reproduced with permission from Martin, A.; Byrne, A.; Burke, C.S.; Forster, R.J.; Keyes, T.E.* J. Am. Chem. Soc. ***2014***, 136, *15300–15309.*

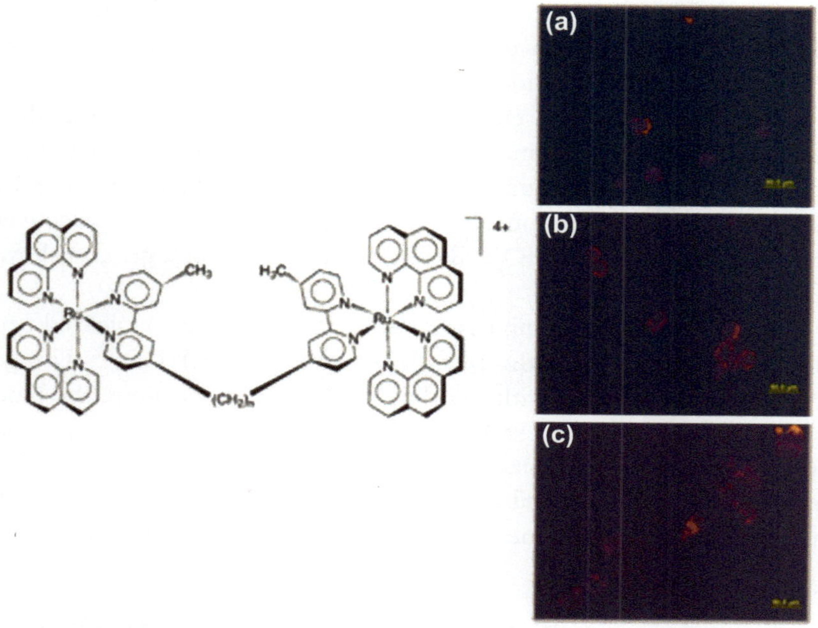

Figure 2.11 (*Left*) The chemical structure of $Rubb_n$ (where $n=2$, 5, 7, 10, 12, or 16 methylene groups in the linking chain). (*Right*) Fluorescence microscopy images of cellular uptake and organelle accumulation of $Rubb_n$. Incubation concentrations are: (a) $Rubb_5$ (50 mM), (b) $Rubb_{10}$ (25 mM), and (c) $Rubb_{16}$ (5 mM). *Reproduced with permission from Pisani, M.J.; Weber, D.K.; Heimann, K.; Collins, J.G.; Keene, F.R.* Metallomics ***2010****, 2, 393–396.*

($n=12$ and 16) exhibited increased cytotoxicity in leukemia cell lines. It was concluded that the increased lipophilicity, as a result of chain length, of the metal complex allowed for greater cellular accumulation resulting in increased cell death.

Peptide ligand sequences have also been used to target metal complexes to specific proteins within cells. Probably the most widely exploited ligand sequence in this regard is the Arg-Gly-Asp (RGD) tripeptide residue. RGD is an integrin recognition sequence. The integrins are a family of transmembrane glycoproteins that are fairly ubiquitous in mammalian cells and are involved cell adhesion, mobility, and signaling. They have been fairly widely targeted in cell imaging because of the prevalence of integrin αvβ3 at the surface of tumor cells.[131]

The RGD sequence was conjugated to two ruthenium(II) polypyridyl complexes and its association studied with platelet integrin.[3] The long-lived excited states and polarized emission of ruthenium polypyridyl complexes make them useful as sensors for study of protein binding and conformation. Here, time-resolved luminescence anisotropy was used to demonstrate conformational changes occurring at the integrin when it was activated with two chemical activators, dithiothreitol and Mn^{2+}, which are purported to induce conformational changes analogous to those occurring physiologically within the platelet during thrombosis. Furthermore, the ruthenium-RGD conjugates were found to selectively bind to the surface of CHO cells expressing platelet integrin, as shown in Fig. 2.12.

Similarly, an iridium luminophore was conjugated with a cyclized peptide sequence containing the integrin targeting motif RGD.[132] Using a series of model peptides with histidine flanking each terminus, the binding activity and reaction kinetics of various iridium-histidine cyclized peptides of different ring sizes were characterized. Ultimately, the RGD-functionalized peptide exhibited increased targeting affinity than its linear form and enhanced membrane permeability in comparison with a

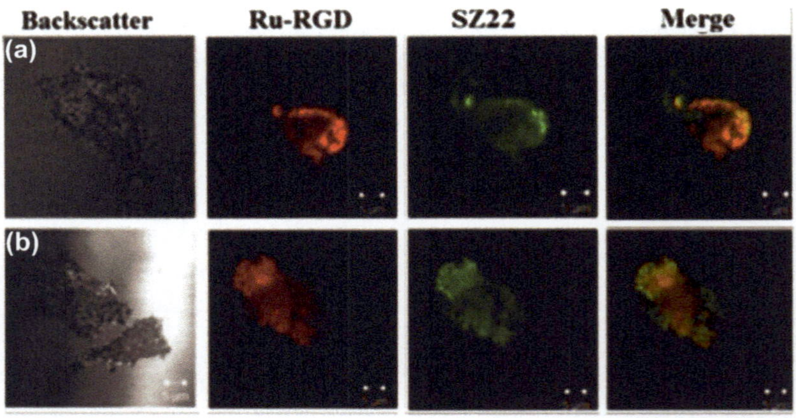

Figure 2.12 Luminescence confocal images of a transfected CHO-$\alpha_{11b}\beta_3$ cell expressing the resting form of integrin $\alpha_{11b}\beta_3$ compared to a CHO cell not expressing the integrin protein in the same image. (a) bpy-RGD and (b) dpp-RGD, Ru-RGD (red) successfully colocalized with SZ22 (green) localizing on the surface of the cell (merge, yellow). Ru-RGD was added at 70 μM concentration in phosphate-buffered saline (PBS) (pH 7.4) and incubated at 37°C for 4 h. SZ22 was added 1/250 dilution overnight at 4°C and secondary AlexaFluor488 goat antimouse was added 1/500 for 2 h at room temperature. Excited at 458 nm, ruthenium luminescence was collected with a long pass filter of 615 nm. No luminescence was observed on the cell failing to express the integrin protein. *Reproduced with permission from Adamson, K.; Dolan, C.; Moran, N.; Forster, R.J.; Keyes, T.E.* Bioconjugate Chem. ***2014**, 25, 928–944.*

fluorescein-labeled cyclic RGDyK peptide (Fig. 2.13). Furthermore, the cell death-inducing peptide, KLA-linked, iridium-HRGDH (Ir-HRGDH-KLA) showed dramatically enhanced cytotoxicity and high selectivity for cancer cells versus noncancer cells.

The peptide somatostatin is a growth hormone-inhibiting polypeptide that is interesting in imaging as receptors for this peptide have been identified in certain tumor types. Rau et al. conjugated this 14 amino acid long peptide sequence to a luminescent a ruthenium(II) polypyridyl complex, which showed receptor-mediated endocytotic uptake and also excellent cellular selectivity attributed to the targeting of the somatostatin receptors, which are upregulated at tumor cells.[113] The focus of this particular article was photodynamic

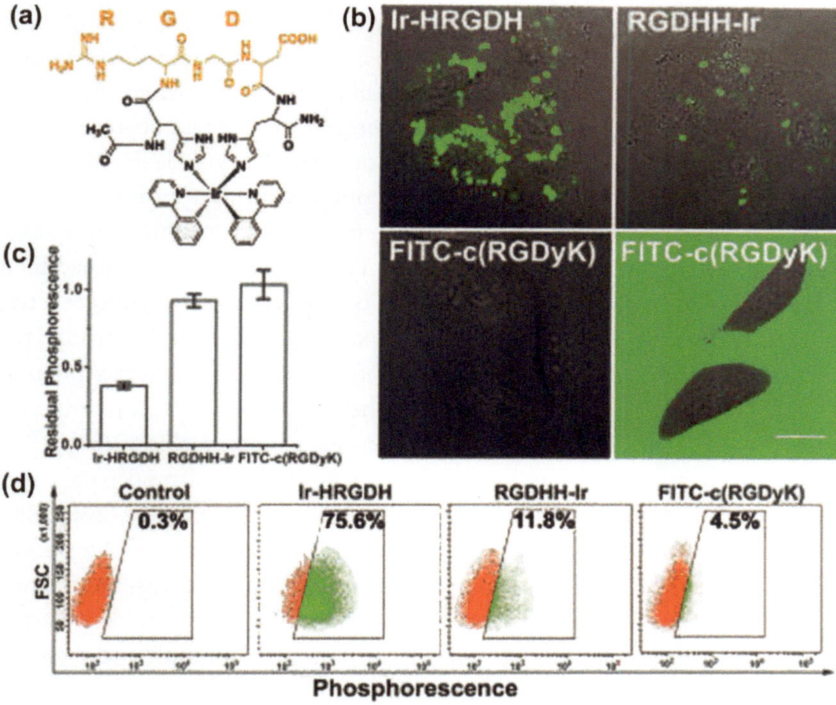

Figure 2.13 Increased cellular uptake of an iridium-peptide (Ir-HRGDH). (a) Structure illustration of Ir-HRGDH. (b–d) Cell uptake behavior study by using (b) confocal laser scanning microscopy, (c) the residual phosphorescence assay, and (d) flow cytometry in A549 cells incubated with Ir-HRGDH (10 μM, λex = 405 nm), RGDHH-Ir (10 μM, λex = 405 nm), or FITC-c(RGDyK) (10 μM, λex = 488 nm) for 24 h at 37°C. Scale bar = 20 μm. *Reproduced with permission from Ma, X.; Jia, J.; Cao, R.; Wang, X.; Fei, H.* J. Am. Chem. Soc. ***2014***, 136, *17734–17737.*

therapy rather than use as an imaging agent because this material showed strong photocytotoxicity, but low cytotoxicity in the absence of light.

Phototriggers are quite widely used in biological research and comprise a chemical entity composed of two parts: the biomolecule under interrogation and a "cage" moiety that inhibits its action.[133] Upon irradiation, the compound of interest is freed and can interact with the surrounding media. An example of such a novel phototrigger was reported by Salierno et al., who detailed the introduction of glutamate, an excitatory neurotransmitter, with a ruthenium-bipyridine core, suitable for UV, visible and IR irradiation.[133] The ability to precisely deliver glutamate is crucial to modulate the activity of neural circuits. Accordingly, an ideal way to achieve this is to use an optical photorelease of glutamate, which until now could only be done by using harmful UV radiation. Sensitization, using a ruthenium complex, led to a full family of caged compounds that could be activated with visible light (as high as 532 nm). Furthermore, the ruthenium(II) complex $[Ru(bpy)_2(PMe_3)(Glu)]$ (where PMe_3 = trimethylphosphine) showed no apparent toxicity and could be used for standard physiological experiments hours after incubation.

Tryptophan (Trp)-containing tripeptide sequences (AlaTrpAla) have also been conjugated to transition metal complex luminophores including rhodium[134] and ruthenium(II) (1,4,5,8-tetraazaphenanthrene) complexes through a variety of linker motifs.[135,136] The formation of photoadducts with Trp moieties offer many possibilities of applications for photoreactive ruthenium(II) complexes, which could be exploited for photodamaging Trp-containing peptides and proteins. A series of 1,4,5,8,9,12-hexaazatriphenylene and 1,4,5,8-tetraazaphenanthrene ruthenium(II) tripeptide complexes tethered to oligonucleotides (ODNs) was also reported (Fig. 2.14).[137,138] The ODN sequence can be synthesized as an antisense strand to a target DNA sequence. Upon binding the complementary sequence, photoirradiation can yield a permanent crosslink between the strands. Either the photocrosslinking arises from a dimerization process of two Trp units, induced by the photoreactive complex, or the photocrosslinking corresponds with the biadduct formation that links the same ruthenium(II) complex to two Trp moieties. This photoreactivity appears very promising for a wide range of applications involving either DNA or peptides and proteins.

The emergence of metal-(bio)polymer assemblies has led to the fabrication of numerous interesting and practical structures, whose size dimensions are on the micro- to nanometer scale. Metal centers may be used to facilitate structure–function relationships and tune macromolecular properties of (bio)polymer chains through a variety of binding mechanisms.[139]

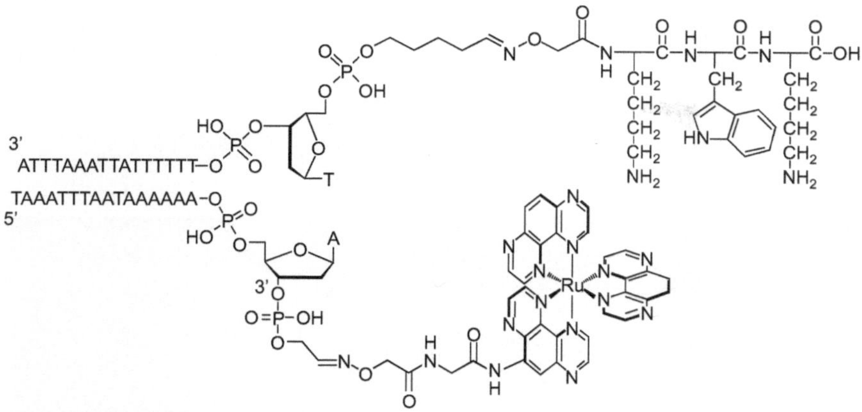

Figure 2.14 Structure of the duplex bearing the ruthenium(II) complex $[Ru(TAP)_2(phen)]^{2+}$ at the 3′ end and the tripeptide Lys-Trp-Lys at the 5′ end of the duplex without G base. *Reproduced with permission from Marcélis, L.; Ghesquière, J.; Garnir, K.; Kirsch-De Mesmaeker, A.; Moucheron, C.* Coord. Chem. Rev. ***2012***, 256, *1569–1582.*

The resulting metallopolymers may possess diverse kinetic, thermodynamic, physical, chemical, and, in particular, structural properties that derive from the presence of the metal ions.

Fatal neurodegenerative diseases such as Alzheimer's disease and prion diseases are linked to misfolding of disease-specific amyloidogenic proteins.[140] Fibrillar structure is one of the most significant assembly structures for proteins.[141] Kim et al. detail the preparation of linear and branched pyrene-labeled histidine-rich peptide conjugates that exhibited nanofibril formation in aqueous media.[141] The molecular packing differences between both peptides resulted in their specific binding affinities to metal ions. Both fibrils exhibited great affinity for Cu^{2+} metal ions as a fluorescence "turn-off" sensor but remarkably the branched nanoprobe revealed a "light-up" response, allowing for detection of Ag^{+} metal ions in HeLa cells (Fig. 2.15).

Another interesting perspective in designing metal-based therapeutic and diagnostic probes is the use of transition metals themselves in directing their cellular uptake and subcellular localization. Tang et al. demonstrated metal-induced aggregation as a result of the intermolecular interactions between the Zn-luminescent probe and water molecules.[142] Comparative studies showed that the Zn-probe undergoes cellular uptake through caveolae-mediated endocytosis and internalizes in the endosomal/lysosomal compartments, whereas the free ligands localized in the mitochondria (Fig. 2.16). Additional transmission

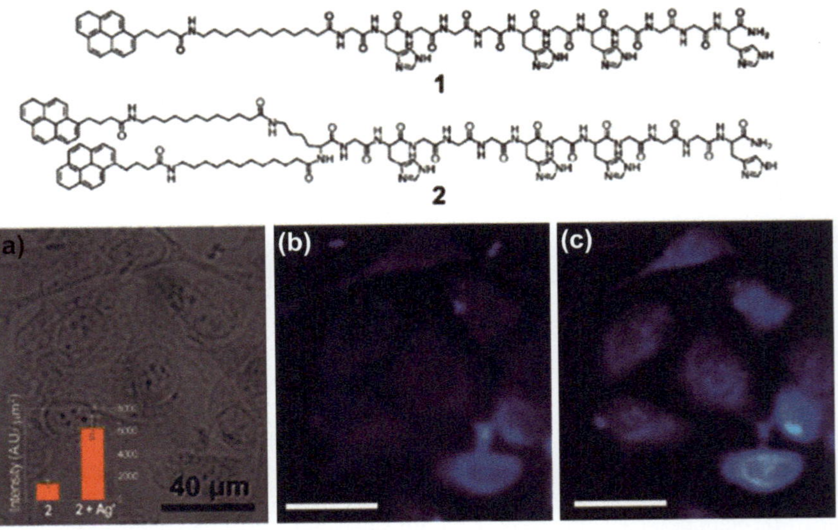

Figure 2.15 (*Top*) Molecular structures of peptide-conjugated amphiphiles. (*Bottom*) (a) Bright field (b and c) fluorescence images of (b) HeLa cells with 2 (10 mM) and (c) 2-loaded HeLa cells in the presence of $AgNO_3$ (150 mM, scale bars = 40 mm). Insert of (a): the intensity of fluorescence per area both before and after incubation with Ag^+. *Reproduced with permission from Kim, I.; Jeong, H.; Kim, Y.; Lee, N.; Huh, K.; Lee, C.; et al.*, J. Mater. Chem. B ***2014***, 2, *6478–6486.*

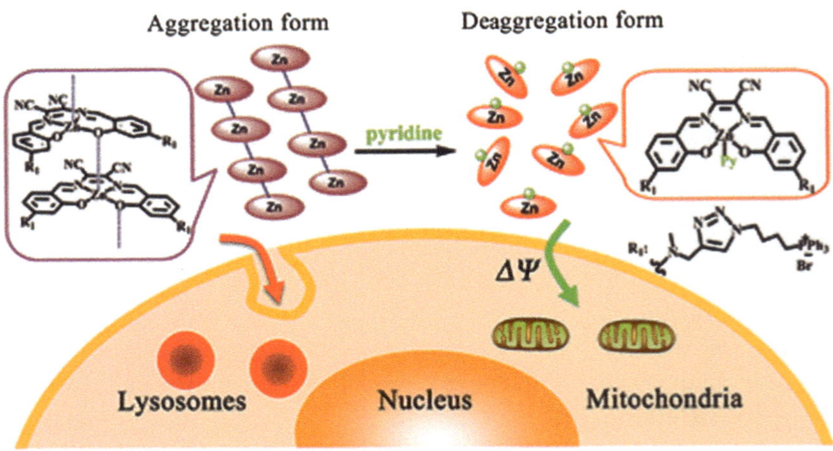

Figure 2.16 Schematic diagram illustrating how the aggregation and deaggregation forms of a Zn-based probe in aqueous media lead to different cellular uptake pathways and subcellular localization. *Reproduced with permission from Tang, J.; Cai, Y.; Jing, J.; Zhang, J.* Chem. Sci. ***2015***, 6, *2389–2397.*

electron microscopy and dynamic light-scattering studies revealed the tendency of the metal-based probe to form large-sized fibrous structures in aqueous media. Interestingly, in the presence of pyridine, the degree of aggregation could be tuned, and therefore localization of the metal probe within the cell could be controlled.

Other less frequently used transition metal signal peptide probes, not covered in extensive detail in this chapter, include those containing rhenium,[26,143,144] rhodium,[134,145,146] and nickel.[80] For a more detailed discussion on the application of the lanthanides, with particular emphasis on cell imaging, the reader is directed to the following papers.[147–151]

The incorporation of a rhenium complex into a peptide structure, where for the first time the metal complex itself was critical for receptor binding, was reported by Rosita et al..[144] Here, the authors successfully managed to create an imaging probe for targeting the growth hormone secretagog receptor. They report on the design and synthesis of ghrelin (a 28 amino acid peptide hormone) analogs through modification of the *n*-octanoyl Ser-3 side chain, with the key component being the neutral and lipophilic cyclopentadienylrhenium tricarbonyl complex which mimics the lipophilic properties of the original ghrelin side chain. They concluded that of all the conjugates analyzed, the rhenium-ghrelin surrogate was the best candidate ($IC_{50} = 35$ nM) to be developed as a cancer imaging agent.

Raszeja et al. also described the conjugation of a rhenium(I) metal complex to neurotensin, a shortened version of the native neuropeptide neurotensin.[152] Both uptake and cellular distribution of the parent and dye conjugate were compared in several cancer cell lines, namely, HeLa (cervical cancer), HT-29 (colon cancer), IMIM-PC2, and PT-45 (both pancreatic cancers). In general, both compounds were taken up well and distributed in the cytosol; however, they were impermeable to the nuclear membrane. Interestingly, a qualitative comparison between the rhenium complex and conjugate indicated that the organometallic compound was better internalized and more rapidly uptaken than the peptide conjugate, especially when considering that the rhenium complex was incubated at concentrations 10 times lower than the conjugate. It is considered that the lipophilic parent complex is most likely internalized by a mechanism of passive diffusion, whereas the more pronounced uptake differences across different cell lines can be attributed to a receptor-mediated uptake mechanism for the neurotensin conjugate.

7. CONCLUSIONS AND FUTURE OUTLOOK

Over the past decade luminescent transition metal polypyridyl complexes have begun to be very seriously considered as probes for live cell imaging and sensing. Their very attractive optical and photophysical properties, in particular their tunable optical properties and long-lived excited states, make them uniquely amenable to cell-based sensing. Advances in means of promoting cellular uptake, arising from the domains of drug delivery, have overcome a key barrier to earlier progress with metal-based luminopores; that of cell permeability. Peptides are particularly attractive for both uptake and targeting of such complexes. Initially, guanidinium-rich CPPs provided a leap forward as they both increase the aqueous solubility of the metal complex and promote efficient uptake of cargo across cellular membranes without damaging its structure. Ongoing optimization of transmembrane transporters along with growing understanding of gene sequences has enhanced molecular cargo delivery and targeting in cells across a variety of imaging, diagnostic and therapeutic applications, and these developments offer useful opportunities for directing metal luminophores. In parallel, advances in peptide chemistry, in particular solid-phase peptide synthesis, have opened up an arsenal of easily accessible peptide sequences including artificial peptides that may be tailored for cargo delivery and targeting within cells and tissues.

Sensing dynamic environmental changes inside living cells remains a significant challenge in cell biology. Whereas transition metal luminophores such as those based on the metal complexes of ruthenium, iridium, osmium, rhenium, etc. have been shown for many years to be very effective sensor probes across a range of analytes; this sensing capability is only beginning now to be extended to cells. The focus to date has principally been on O_2 sensing but the possibilities are extensive. Enabled by FLIM or for intensity-based measurements by including self-referencing via incorporation of a nonenvironmentally sensitive probe into the molecular structure, the use of such luminophores across far more diverse sensing applications is predicted.

Peptides will continue to grow as a means of delivering and targeting metal complexes in cells and tissues. Many signal peptides in their natural function are cleaved from their cargo by proteolysis once they have delivered the agent to their site of action. We predict that such biomimicry could be incorporated into metal complex delivery where on-delivery emission is switched on or off by this cleavage to enable theranostic applications of such materials. A conceptual barrier to the application of metal complexes in cell imaging is also increasingly being overcome: the expectation that heavy metals will be cytotoxic. Numerous metal complex luminophores, particularly those of Ru(II) complexes, have now been shown to exhibit low cytotoxicity and peptides can play a role in improving the toxicological profile of such metal complexes.

Finally, as some preliminary studies discussed earlier have demonstrated, the exceptionally large Stokes shifts and long lifetimes of inorganic probes offer the unique opportunity for their exploitation in multimodal or multiplexed sensing where resonance Raman and luminescence intensity and/or lifetime measurements are combined in single measurements. If this field is expanded it can lead to some very useful dynamic insights into the cell environment. Overall, the future for transition metal luminophores in imaging and sensing is very bright indeed.

ACKNOWLEDGMENTS

The authors gratefully acknowledge support of Science Foundation Ireland under Grant Nos. 15/TIDA/2895 and 14/IA/2488, and The Irish Research Council for Postgraduate Scholarships.

REFERENCES

1. Medici, S.; Peana, M.; Nurchi, V. M.; Lachowicz, J. I.; Crisponi, G.; Zoroddu, M. A. *Coord. Chem. Rev.* **2015,** *284,* 329–350.
2. Mari, C.; Pierroz, V.; Leonidova, A.; Ferrari, S.; Gasser, G. *Eur. J. Inorg. Chem.* **2015,** 3879–3891.
3. Adamson, K.; Dolan, C.; Moran, N.; Forster, R. J.; Keyes, T. E. *Bioconjugate Chem.* **2014,** *25,* 928–944.
4. Pauli, J.; Grabolle, M.; Brehm, R.; Spieles, M.; Hamann, F. M.; Wenzel, M.; et al. *Bioconjugate Chem.* **2011,** *22,* 1298–1308.
5. Welsher, K.; Sherlock, S. P.; Dai, H. *Proc. Natl. Acad. Sci. U.S.A* **2011,** *108,* 8943–8948.
6. Neugebauer, U.; Pellegrin, Y.; Devocelle, M.; Forster, R. J.; Signac, W.; Moran, N.; et al. *Chem. Commun.* **2008,** 5307–5309.
7. Balzani, V.; Bergamini, G.; Campagna, S.; Puntoriero, F. In *Photochemistry and Photophysics of Coordination Compounds: Overview and General Concepts;* Balzani, V., Campagna, S., Eds.; Springer Berlin Heidelberg, 2007; pp 1–36.
8. Nwaneshiudu, A.; Kuschal, C.; Sakamoto, F. H.; Anderson, R. R.; Schwarzenberger, K.; Young, R. C. *J. Invest. Dermatol.* **2012,** *132,* e3.
9. Lee, S.; Xie, J.; Chen, X. *Chem. Rev.* **2010,** *110,* 3087–3111.
10. Gill, M. R.; Garcia-Lara, J.; Foster, S. J.; Smythe, C.; Battaglia, G.; Thomas, J. A. *Nat. Chem.* **2009,** *1,* 662–667.
11. Ding, Y.; Wu, Q.; Zheng, K.; An, L.; Hu, X.; Mei, W. *RSC Adv.* **2015,** *5,* 63330–63337.
12. Puckett, C. A.; Barton, J. K. *J. Am. Chem. Soc.* **2007,** *129,* 46–47.
13. Xu, W.; Zuo, J.; Wang, L.; Ji, L.; Chao, H. *Chem. Commun.* **2014,** *50,* 2123–2125.
14. Chen, L.; Peng, F.; Li, G.; Jie, X.; Cai, K.; Cai, C.; et al. *J. Inorg. Biochem.* **2016,** *156,* 64–74.
15. Tang, T. S.; Yip, A. M.; Zhang, K. Y.; Liu, H.; Wu, P. L.; Li, K. F.; et al. *Chemistry* **2015,** *21,* 10729–10740.

16. Lo, K. K. *Acc. Chem. Res.* **2015,** *48,* 2985–2995.
17. Xiong, L.; Zhao, Q.; Chen, H.; Wu, Y.; Dong, Z.; Zhou, Z.; et al. *Inorg. Chem.* **2010,** *49,* 6402–6408.
18. Zhao, Q.; Yu, M.; Shi, L.; Liu, S.; Li, C.; Shi, M.; et al. *Organometallics* **2010,** *29,* 1085–1091.
19. Dolan, C.; Moriarty, R. D.; Lestini, E.; Devocelle, M.; Forster, R. J.; Keyes, T. E. *J. Inorg. Biochem.* **2013,** *119,* 65–74.
20. Mandal, S.; Poria, D. K.; Ghosh, R.; Ray, P. S.; Gupta, P. *Dalton Trans.* **2014,** *43,* 17463–17474.
21. Lee, L. C.; Lau, J. C.; Liu, H.; Lo, K. K. *Angew. Chem. Int. Ed.* **2016,** *55,* 1046–1049.
22. Zhang, K. Y.; Lo, K. K. *Inorg. Chem.* **2009,** *48,* 6011–6025.
23. Kuil, J.; Steunenberg, P.; Chin, P. T. K.; Oldenburg, J.; Jalink, K.; Velders, A. H.; et al. *ChemBioChem* **2011,** *12,* 1897–1903.
24. Wang, X.; Jia, J.; Huang, Z.; Zhou, M.; Fei, H. *Chemistry* **2011,** *17,* 8028–8032.
25. Byrne, A.; Dolan, C.; Moriarty, R. D.; Martin, A.; Neugebauer, U.; Forster, R. J.; et al. *Dalton Trans.* **2015,** *44,* 14323–14332.
26. Mari, C.; Panigati, M.; D'Alfonso, L.; Zanoni, I.; Donghi, D.; Sironi, L.; et al. *Organometallics* **2012,** *31,* 5918–5928.
27. Balasingham, R. G.; Coogan, M. P.; Thorp-Greenwood, F. *Dalton Trans.* **2011,** *40,* 11663–11674.
28. Choi, A. W.; Liu, H.; Lo, K. K. *J. Inorg. Biochem.* **2015,** *148,* 2–10.
29. Lo, K. K.; Hui, W. *Inorg. Chem.* **2005,** *44,* 1992–2002.
30. Leonidova, A.; Pierroz, V.; Adams, L. A.; Barlow, N.; Ferrari, S.; Graham, B.; et al. *ACS Med. Chem. Lett.* **2014,** *5,* 809–814.
31. Fernandez-Moreira, V.; Ortego, M. L.; Williams, C. F.; Coogan, M. P.; Villacampa, M. D.; Gimeno, M. C. *Organometallics* **2012,** *31,* 5950–5957.
32. Yin Zhang, K.; Ka-Shun Tso, K.; Louie, M.; Liu, H.; Kam-Wing Lo, K. *Organometallics* **2013,** *32,* 5098–5102.
33. Wragg, A.; Gill, M. R.; Turton, D.; Adams, H.; Roseveare, T. M.; Smythe, C.; et al. *Chemistry* **2014,** *20,* 14004–14011.
34. Wragg, A.; Gill, M. R.; McKenzie, L.; Glover, C.; Mowll, R.; Weinstein, J. A.; et al. *Chemistry* **2015,** *21,* 11865–11871.
35. Ma, D.; Ma, V. P.; Chan, D. S.; Leung, K.; He, H.; Leung, C. *Coord. Chem. Rev.* **2012,** *256,* 3087–3113.
36. Zhao, R.; Hammitt, R.; Thummel, R. P.; Liu, Y.; Turro, C.; Snapka, R. M. *Dalton Trans.* **2009,** 10926–10931.
37. Dobrucki, J. W. *J. Photochem. Photobiol. B* **2001,** *65,* 136–144.
38. Ji, J.; Rosenzweig, N.; Jones, I.; Rosenzweig, Z. *J. Biomed. Opt.* **2002,** *41,* 404–409.
39. Zhong, W.; Urayama, P.; Mycek, M. A. *J. Phys. D* **2003,** *36,* 1689.
40. Blunden, B. M.; Stenzel, M. H. *J. Chem. Technol. Biotechnol.* **2015,** *90,* 1177–1195.
41. Furrer, J.; Süss-Fink, G. *Coord. Chem. Rev.* **2016,** *309,* 36–50.
42. Peacock, A. F. A.; Habtemariam, A.; Fernandez, R.; Walland, V.; Fabbiani, F. P. A.; Parsons, S.; et al. *J. Am. Chem. Soc.* **2006,** *128,* 1739–1748.
43. Ramu, V.; Ali, F.; Taye, N.; Garai, B.; Alam, A.; Chattopadhyay, S.; et al. *J. Mater. Chem. B* **2015,** *3,* 7177–7185.
44. Samanta, D.; Kratz, K.; Zhang, X.; Emrick, T. *Macromolecules* **2008,** *41,* 530–532.
45. Li, S.; Liu, H.; Zhang, K.; Lo, K. *Chemistry* **2010,** *16,* 8329–8339.
46. Li, S. P.; Lau, C. T.; Louie, M.; Lam, Y.; Cheng, S. H.; Lo, K. K. *Biomaterials* **2013,** *34,* 7519–7532.
47. Knop, K.; Hoogenboom, R.; Fischer, D.; Schubert, U. *Angew. Chem. Int. Ed.* **2010,** *49,* 6288–6308.
48. Martin, A.; Moriarty, R. D.; Long, C.; Forster, R. J.; Keyes, T. E. *Asian J. Org. Chem.* **2013,** *2,* 763–778.

49. Cosgrave, L.; Devocelle, M.; Forster, R. J.; Keyes, T. E. *Chem. Commun.* **2010,** *46,* 103–105.
50. Blackmore, L.; Moriarty, R.; Dolan, C.; Adamson, K.; Forster, R. J.; Devocelle, M.; et al. *Chem. Commun.* **2013,** *49,* 2658–2660.
51. Wender, P. A.; Rothbard, J. B.; Jessop, T. C.; Kreider, E. L.; Wylie, B. L. *J. Am. Chem. Soc.* **2002,** *124,* 13382–13383.
52. Rothbard, J. B.; Garlington, S.; Lin, Q.; Kirschberg, T.; Kreider, E.; McGrane, P. L.; et al. *Nat. Med.* **2000,** *6,* 1253–1257.
53. Cardo, L.; Sadovnikova, V.; Phongtongpasuk, S.; Hodges, N. J.; Hannon, M. J. *Chem. Commun.* **2011,** *47,* 6575–6577.
54. Bilan, R.; Fleury, F.; Nabiev, I.; Sukhanova, A. *Bioconjugate Chem.* **2015,** *26,* 609–624.
55. Yukawa, H.; Kagami, Y.; Watanabe, M.; Oishi, K.; Miyamoto, Y.; Okamoto, Y.; et al. *Biomaterials* **2010,** *31,* 4094–4103.
56. Papst, S.; Brimble, M. A.; Evans, C. W.; Verdon, D. J.; Feisst, V.; Dunbar, P. R.; et al. *Org. Biomol. Chem.* **2015,** *13,* 6567–6572.
57. Cardo, L.; Thomas, S. G.; Mazharian, A.; Pikramenou, Z.; Rappoport, J. Z.; Hannon, M. J.; et al. *ChemBioChem* **2015,** *16,* 1680–1688.
58. Wender, P. A.; Mitchell, D. J.; Pattabiraman, K.; Pelkey, E. T.; Steinman, L.; Rothbard, J. B. *Proc. Natl. Acad. Sci. U.S.A.* **2000,** *97,* 13003–13008.
59. Fuchs, S. M.; Raines, R. T. *Biochemistry (N.Y.)* **2004,** *43,* 2438–2444.
60. Yamamoto, N.; Bryce, N. S.; Metzler-Nolte, N.; Hambley, T. W. *Bioconjugate Chem.* **2012,** *23,* 1110–1118.
61. Fawell, S.; Seery, J.; Daikh, Y.; Moore, C.; Chen, L. L.; Pepinsky, B.; et al. *Proc. Natl. Acad. Sci. U.S.A.* **1994,** *91,* 664–668.
62. Ghosh, P.; Yang, X.; Arvizo, R.; Zhu, Z.; Agasti, S. S.; Mo, Z.; et al. *J. Am. Chem. Soc.* **2010,** *132,* 2642–2645.
63. Vázquez, O.; Blanco-Canosa, J. B.; Vázquez, M. E.; Martínez-Costas, J.; Castedo, L.; Mascareñas, J. L. *ChemBioChem* **2008,** *9,* 2822–2829.
64. Cohen, J. L.; Almutairi, A.; Cohen, J. A.; Bernstein, M.; Brody, S. L.; Schuster, D. P.; et al. *Bioconjugate Chem.* **2008,** *19,* 876–881.
65. Svensen, N.; Walton, J. G. A.; Bradley, M. *Trends Pharmacol. Sci.* **2012,** *33,* 186–192.
66. Marie, E.; Sagan, S.; Cribier, S.; Tribet, C. *J. Membr. Biol.* **2014,** *247,* 861–881.
67. Liu, H.; Zeng, F.; Zhang, M.; Huang, F.; Wang, J.; Guo, J.; et al. *J. Controlled Release* **2016,** *226,* 124–137.
68. Copolovici, D. M.; Langel, K.; Eriste, E.; Langel, A. *ACS Nano* **2014,** *8,* 1972–1994.
69. Gray, B. P.; Brown, K. C. *Chem. Rev.* **2014,** *114,* 1020–1081.
70. Zorko, M.; Langel, Ü. *Adv. Drug Deliv. Rev.* **2005,** *57,* 529–545.
71. Li, Z. J.; Cho, C. H. *J. Transl. Med.* **2012,** *10,* S1.
72. Wender, P. A.; Cooley, C. B.; Geihe, E. I. *Technologies* **2012,** *9,* e49–e55.
73. Dubikovskaya, E. A.; Thorne, S. H.; Pillow, T. H.; Contag, C. H.; Wender, P. A. *Proc. Natl. Acad. Sci. U.S.A.* **2008,** *105,* 12128–12133.
74. Fuchs, S. M.; Raines, R. T. *Cell. Mol. Life Sci.* **2006,** *63,* 1819–1822.
75. Takechi, Y.; Yoshii, H.; Tanaka, M.; Kawakami, T.; Aimoto, S.; Saito, H. *Langmuir* **2011,** *27,* 7099–7107.
76. Gamba, I.; Salvado, I.; Brissos, R. F.; Gamez, P.; Brea, J.; Loza, M. I.; et al. *Chem. Commun.* **2016,** *52,* 1234–1237.
77. Jones, S. W.; Christison, R.; Bundell, K.; Voyce, C. J.; Brockbank, S. M. V.; Newham, P.; et al. *Br. J. Pharmacol* **2005,** *145,* 1093–1102.
78. Oskolkov, N.; Arukuusk, P.; Copolovici, D.; Lindberg, S.; Margus, H.; Padari, K.; et al. *Int. J. Pept. Res. Ther.* **2011,** *17,* 147–157.
79. Ryan, B. J.; Henehan, G. T. *Current Protocols in Protein Science;* John Wiley & Sons, Inc, 2001.

80. Futaki, S.; Niwa, M.; Nakase, I.; Tadokoro, A.; Zhang, Y.; Nagaoka, M.; et al. *Bioconjugate Chem.* **2004,** *15,* 475–481.
81. Dunkin, C. M.; Pokorny, A.; Almeida, P. F.; Lee, H. *J. Phys. Chem. B* **2011,** *115,* 1188–1198.
82. Herce, H. D.; Garcia, A. E. *J. Biol. Phys.* **2008,** *33,* 345–356.
83. Yoshihara, T.; Hosaka, M.; Terata, M.; Ichikawa, K.; Murayama, S.; Tanaka, A.; et al. *Anal. Chem.* **2015,** *87,* 2710–2717.
84. Zhang, K. Y.; Liu, H.; Tang, M.; Choi, A. W.; Zhu, N.; Wei, X.; et al. *Inorg. Chem.* **2015,** *54,* 6582–6593.
85. Notman, R.; Noro, M.; O'Malley, B.; Anwar, J. *J. Am. Chem. Soc.* **2006,** *128,* 13982–13983.
86. Yu, Z. W.; Quinn, P. J. *Mol. Membr. Biol.* **April–June 1998,** *15* (2), 59–68.
87. Puckett, C. A.; Barton, J. K. *J. Am. Chem. Soc.* **2009,** *131,* 8738–8739.
88. Lo, K. K.; Lee, T. K.; Lau, J. S.; Poon, W.; Cheng, S. *Inorg. Chem.* **2008,** *47,* 200–208.
89. Artemov, D.; Bhujwalla, Z. M. *Curr. Pharm. Biotechnol.* **2004,** *5,* 485–494.
90. Vithanarachchi, S. M. *Curr. Mol. Imaging* **2012,** *1,* 12–25.
91. Brunner, J.; Barton, J. K. *Biochemistry (N.Y.)* **2006,** *45,* 12295–12302.
92. Friedman, A. E.; Chambron, J. C.; Sauvage, J. P.; Turro, N. J.; Barton, J. K. *J. Am. Chem. Soc.* **1990,** *112,* 4960–4962.
93. Olson, E. J. C.; Hu, D.; Harmann, A.; Jonkman, A. M.; Arkin, M. R.; Stemp, E. D. A.; et al. *J. Am. Chem. Soc.* **1997,** *119,* 11458–11467.
94. Poynton, F. E.; Hall, J. P.; Keane, P. M.; Schwarz, C.; Sazanovich, I. V.; Towrie, M.; et al. *Chem. Sci.* **2016,** 7 *(5),* 3075–3084.
95. Chalmers, J. M.; Griffiths, P. R. *Resonance Raman Spectroscopy, Handbook of Vibrational Spectroscopy;* John-Wiley and Sons: New York, 2002; pp 534–556.
96. Neugebauer, U.; Cosgrave, L.; Pellegrin, Y.; Devocelle, M.; Forster, R. J.; Keyes, T. E. In *Proc. SPIE 8427, Biophotonics: Photonic Solutions for Better Health Care III, 84270C;* (June 1, 2012). 84270C-84270C-13.
97. van Rijt, S. H.; Mukherjee, A.; Pizarro, A. M.; Sadler, P. J. *J. Med. Chem.* **2010,** *53,* 840–849.
98. van Rijt, S. H.; Kostrhunova, H.; Brabec, V.; Sadler, P. J. *Bioconjugate Chem.* **2011,** *22,* 218–226.
99. Benniston, A. C.; Harriman, A.; Li, P.; Sams, C. A. *J. Phys. Chem. A* **2005,** *109,* 2302–2309.
100. Van Houten, J.; Watts, R. J. *J. Am. Chem. Soc.* **1976,** *98,* 4853–4858.
101. Gamba, I.; Salvadó, I.; Rama, G.; Bertazzon, M.; Sánchez, M. I.; Sánchez-Pedregal, V. M.; et al. *Chemistry* **2013,** *19,* 13369–13375.
102. Castro-Fernández, C.; Maya-Núñez, G.; Conn, P. M. *Endocr. Rev.* **2005,** *26,* 479–503.
103. Heller, A.; Brockhoff, G.; Goepferich, A. *Eur. J. Pharm. Biopharm.* **2012,** *82,* 1–18.
104. Malhi, S. S.; Murthy, R. S. R. *Expert Opin. Drug Delivery* **2012,** *9,* 909–935.
105. Murphy, M. P.; Smith, R. A. J. *Adv. Drug Delivery Rev.* **2000,** *41,* 235–250.
106. Cartier, R.; Reszka, R. *Gene Ther.* **2002,** *9,* 157–167.
107. Escriou, V.; Carrière, M.; Scherman, D.; Wils, P. *Adv. Drug Delivery Rev.* **2003,** *55,* 295–306.
108. Hetzer, M. W. The Nuclear Envelope. *Cold Spring Harb. Perspect. Biol.* **2010,** *2,* a000539.
109. Ragin, A. D.; Morgan, R. A.; Chmielewski, J. *Chem. Biol.* **2002,** *9,* 943–948.
110. Branden, L. J.; Mohamed, A. J.; Smith, C. I. E. *Nat. Biotech.* **1999,** *17,* 784–787.
111. Noor, F.; Wüstholz, A.; Kinscherf, R.; Metzler-Nolte, N. *Angew. Chem. Int. Ed.* **2005,** *44,* 2429–2432.
112. Boulikas, T. *J. Cell. Biochem.* **1994,** *55,* 32–58.
113. Wang, T.; Zabarska, N.; Wu, Y.; Lamla, M.; Fischer, S.; Monczak, K.; et al. *Chem. Commun.* **2015,** *51,* 12552–12555.
114. Elias, B.; Kirsch-De Mesmaeker, A. *Coord. Chem. Rev.* **2006,** *250,* 1627–1641.
115. Jacquet, L.; Davies, R. J.; Kirsch-De Mesmaeker, A.; Kelly, J. M. *J. Am. Chem. Soc.* **1997,** *119,* 11763–11768.
116. Puckett, C. A.; Barton, J. K. *Bioorg. Med. Chem.* **2010,** *18,* 3564–3569.

117. Tuite, E.; Lincoln, P.; Norden, B. *J. Am. Chem. Soc.* **1997,** *119,* 239–240.
118. Duchen, M. R.; Szabadkai, G. *Essays Biochem.* **2010,** *47,* 115–137.
119. Tiede, L. M.; Cook, E. A.; Morsey, B.; Fox, H. S. *Cell Death Dis.* **2011,** *2,* e246.
120. Galluzzi, L.; Larochette, N.; Zamzami, N.; Kroemer, G. *Oncogene* **2006,** *25,* 4812–4830.
121. Green, K.; Brand, M. D.; Murphy, M. P. *Diabetes* **2004,** *53,* S110–S118.
122. Frantz, M.; Wipf, P. *Environ. Mol. Mutagen* **2010,** *51,* 462–475.
123. Murphy, M. *Biochem. J.* **2008,** *417,* 1–13.
124. Valko, M.; Leibfritz, D.; Moncol, J.; Cronin, M. T. D.; Mazur, M.; Telser, J. *Int. J. Biochem. Cell Biol.* **2007,** *39,* 44–84.
125. Li, Z.; Graham, B. H. *Methods Mol. Biol.* **2012,** *837,* 63–72.
126. Emanuelsson, O.; Nielsen, H.; Brunak, S.; von Heijne, G. *J. Mol. Biol.* **2000,** *300,* 1005–1016.
127. Horton, K. L.; Stewart, K. M.; Fonseca, S. B.; Guo, Q.; Kelley, S. O. *Chem. Biol.* **2008,** *15,* 375–382.
128. Yousif, L. F.; Stewart, K. M.; Kelley, S. O. *ChemBioChem* **2009,** *10,* 1939–1950.
129. Martin, A.; Byrne, A.; Burke, C. S.; Forster, R. J.; Keyes, T. E. *J. Am. Chem. Soc.* **2014,** *136,* 15300–15309.
130. Pisani, M. J.; Weber, D. K.; Heimann, K.; Collins, J. G.; Keene, F. R. *Metallomics* **2010,** *2,* 393–396.
131. Desgrosellier, J. S.; Cheresh, D. A. *Nat. Rev. Cancer* **2010,** *10,* 9–22.
132. Ma, X.; Jia, J.; Cao, R.; Wang, X.; Fei, H. *J. Am. Chem. Soc.* **2014,** *136,* 17734–17737.
133. Salierno, M.; Marceca, E.; Peterka, D. S.; Yuste, R.; Etchenique, R. *J. Inorg. Biochem.* **2010,** *104,* 418–422.
134. Antos, J. M.; McFarland, J. M.; Iavarone, A. T.; Francis, M. B. *J. Am. Chem. Soc.* **2009,** *131,* 6301–6308.
135. Gauthier, N.; De Winter, J.; Gerbaux, P.; Moucheron, C.; Luhmer, M.; Kirsch-De Mesmaeker, A. *Inorg. Chem.* **2010,** *49,* 6796–6798.
136. Rebarz, M.; Ghesquiere, J.; Boisdenghien, A.; Defrancq, E.; Moucheron, C.; Kirsch-De Mesmaeker, A. *Inorg. Chem.* **2010,** *49,* 10867–10874.
137. Marcélis, L.; Ghesquière, J.; Garnir, K.; Kirsch-De Mesmaeker, A.; Moucheron, C. *Coord. Chem. Rev.* **2012,** *256,* 1569–1582.
138. Marcélis, L.; Surin, M.; Lartia, R.; Moucheron, C.; Defrancq, E.; Kirsch-De Mesmaeker, A. *Eur. J. Inorg. Chem.* **2014,** *2014,* 3016–3022.
139. Dmitriev, R. I.; O'Donnell, N.; Papkovsky, D. B. *Bioconjugate Chem.* **2015**.
140. Xiao, Y.; Ma, B.; McElheny, D.; Parthasarathy, S.; Long, F.; Hoshi, M.; et al. *Nat. Struct. Mol. Biol.* **2015,** *22,* 499–505.
141. Kim, I.; Jeong, H.; Kim, Y.; Lee, N.; Huh, K.; Lee, C.; et al. *J. Mater. Chem. B* **2014,** *2,* 6478–6486.
142. Tang, J.; Cai, Y.; Jing, J.; Zhang, J. *Chem. Sci.* **2015,** *6,* 2389–2397.
143. Polyakov, V.; Sharma, V.; Dahlheimer, J. L.; Pica, C. M.; Luker, G. D.; Piwnica-Worms, D. *Bioconjugate Chem.* **2000,** *11,* 762–771.
144. Rosita, D.; DeWit, M. A.; Luyt, L. G. *J. Med. Chem.* **2009,** *52,* 2196–2203.
145. Mandal, S.; Poria, D. K.; Seth, D. K.; Ray, P. S.; Gupta, P. *Polyhedron* **2014,** *73,* 12–21.
146. Sardesai, N.; Lin, S.; Zimmermann, K.; Barton, J. *Bioconjugate Chem.* **1995,** *6,* 302–312.
147. Montgomery, C. P.; Murray, B. S.; New, E. J.; Pal, R.; Parker, D. *Acc. Chem. Res.* **2009,** *42,* 925–937.
148. Parker, D. *Aust. J. Chem.* **2011,** *64,* 239–243.
149. Smith, D. G.; McMahon, B. K.; Pal, R.; Parker, D. *Chem. Commun.* **2012,** *48,* 8520–8522.
150. Rajendran, M.; Yapici, E.; Miller, L. W. *Inorg. Chem.* **2014,** *53,* 1839–1853.
151. Soulié, M.; Latzko, F.; Bourrier, E.; Placide, V.; Butler, S. J.; Pal, R.; et al. *Chemistry* **2014,** *20,* 8636–8646.
152. Raszeja, L.; Maghnouj, A.; Hahn, S.; Metzler-Nolte, N. *ChemBioChem* **2011,** *12,* 371–376.

CHAPTER THREE

Luminescent Rhenium(I) and Iridium(III) Complexes for Intracellular Labeling, Sensing, and Photodynamic Therapy Applications

K.Y. Zhang
Nanjing University of Posts and Telecommunications (NUPT), Nanjing, China

Contents

1. INTRODUCTION

Luminescent transition metal complexes have been used as an alternative to organic fluorescent dyes to label or probe biomolecules for more than 20 years because of their favorable photophysical properties, such as high luminescence quantum yields and tunable luminescence colors.[1,2] Over time, with the rapid development of luminescence imaging microscopes and techniques, the intracellular labeling and sensing performance of these complexes have attracted increasing interest.[3–7] In intracellular applications, the high photostability of transition metal complexes allows continuous exposure to irradiation and enables real-time imaging. The triplet emissive states render these complexes to show large Stokes shifts and long luminescence lifetimes. The former minimizes the aggregation-induced luminescence

Inorganic and Organometallic Transition Metal Complexes with Biological Molecules and Living Cells
ISBN 978-0-12-803814-7
http://dx.doi.org/10.1016/B978-0-12-803814-7.00003-4

quenching and the latter facilitates time-resolved imaging that analyzes the phosphorescence signal without interference from background autofluorescence. Luminescent tricarbonyl rhenium(I) polypyridine complexes and biscyclometalated iridium(III) polypyridine complexes of general formula $[Re(N^\wedge N)(CO)_3(L)]^+$ and $[Ir(N^\wedge C)_2(N^\wedge N)]^+$ exhibit high cellular uptake efficiency because of their positive formal charge and ligand-controlled high lipophilicity, which allows electrostatic attraction with negatively charged cell surface and facilitates the diffusion into cytoplasm.[3,5–7] These complexes show sensitive photophysical properties toward ligands and the surrounding microenvironment, facilitating development of sensors. Intensity-based luminescence sensing simply correlates luminescence intensity at a certain wavelength with the concentration of analytes. However, the intensity is also determined by sensor concentration, which is usually uncertain and difficult to control or measure in an intracellular or in vivo environment. The use of these complexes to develop wavelength-ratiometric sensors, which reflects the presence and the concentration of analytes by the ratio of intensities at two luminescence wavelengths, significantly improves the accuracy and precision, especially in intracellular and in vivo sensing. Upon photoexcitation, the long-lived triplet-excited states of these complexes are readily quenched by oxygen molecules, accompanied with the generation of singlet oxygen ($^1O_2$), which can cause cell death. Thus these complexes have great potential to be used as photodynamic therapy (PDT) reagents.

In this chapter, we first summarize the development of luminescent rhenium(I) and iridium(III) complexes for covalently labeling biomolecules in their native environments. The use of the complexes as intracellular chemosensors for disease-related intracellular molecules including reactive oxygen and nitrogen species are then discussed. Furthermore, the strategies to design wavelength-ratiometric intracellular sensors are introduced. Finally, since the triplet-excited states of these complexes sensitize the generation of singlet oxygen, which is one of the main inducers of cell apoptosis, the potential applications of these complexes for photodynamic therapy are discussed.

2. INTRACELLULAR LABELS

The incorporation of reactive functional groups into luminescent transition metal complexes affords phosphorescent labeling reagents that can be covalently attached to biomolecules such as proteins.[5,7] For example, aldehyde and isothiocyanate groups can react with the amino group in the side chain of lysine, and thus luminescent complexes modified with these reactive groups can be used to label lysine-containing proteins. A series of luminescent cyclometalated iridium(III) bis(quinolylbenzaldehyde) diimine complexes $[Ir(qba)_2(N^\wedge N)](PF_6)$ (**1**) have been designed and synthesized for intracellular protein labeling.[8] These complexes exhibited intense and long-lived orange–yellow emission in fluid solutions, which has been assigned to a triplet intraligand (^{3}IL)-excited state mixed with some triplet metal-to-ligand charge-transfer (^{3}MLCT) character associated with the qba ligand. After reductive amination with the lysine residues of the protein bovine serum albumin (BSA), luminescent bioconjugates iridium(III)-BSA that showed intense and long-lived triplet phosphorescence were formed, and the dye-to-protein ratios were determined to be between 0.3 and 0.9. Sodium dodecyl sulfate-polyacrylamide gel electrophoresis analysis indicated that the labeling process did not alter much the molecular weight of BSA, but a portion of protein molecules was crosslinked by the two amine-reactive aldehyde moieties of each molecule of the iridium(III) complexes. HeLa cells loaded with one of the complexes showed perinuclear staining (Fig. 3.1). A similar staining pattern was observed after the cells were fixed by methanol or paraformaldehyde. In the control experiments, the living HeLa cells showed similar intracellular staining when treated with the aldehyde-free complexes $[Ir(pq)_2(N^\wedge N)]$ (PF_6) (Hpq = 2-phenylquinoline), but the luminescence intensity was decreased dramatically upon fixation of the cells. The reason was that the aldehyde complexes were covalently attached to various biomolecules as well as intracellular structures, but the aldehyde-free complexes were washed off upon fixation of the cells with organic solvents.

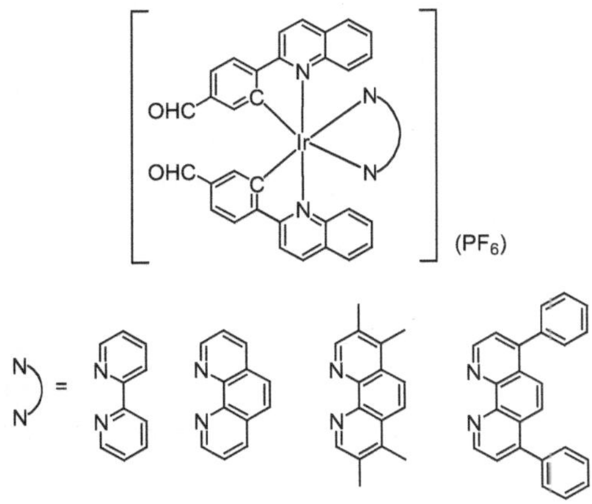

1

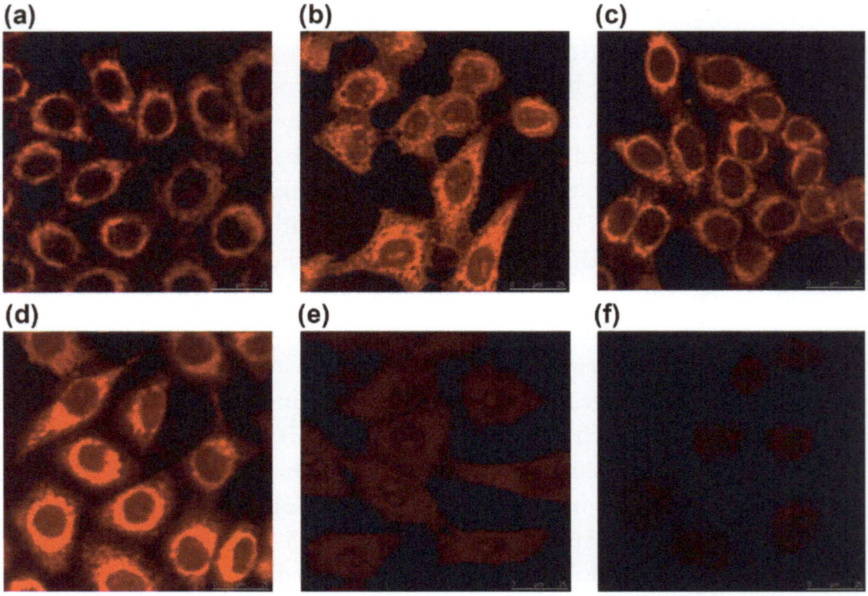

Figure 3.1 Fluorescence images of HeLa cells incubated with one of the aldehyde complexes (5 μM) at 37°C for 1 h (a) before and (b) after subsequent fixation with MeOH, or (c) fixation with paraformaldehyde, followed by extensive washing with MeOH; (d)–(f): fluorescence images of HeLa cells treated and fixed with the same procedures except that the aldehyde-free complex [Ir(pq)$_2$(Ph$_2$-phen)](PF$_6$) was used as the staining reagent. *Reproduced from Lee, P.-K.; Liu, H.-W.; Yiu, S.-M.; Louie, M.-W.; Lo, K.K.-W.* Dalton Trans. ***2011,* *40, 2180–2189 by permission of the Royal Society of Chemistry.***

Another iridium(III) complex containing an amine-targeting isothiocynate group [Ir(pq)$_2$(phen-ITC)](PF$_6$) (**2**) has been synthesized to investigate its interaction with proteins and identify its molecular targets inside living cells.[9] The luminescence of the complex in 3T3, human retinal pigmented epithelial (RPE), and HeLa cells showed nearly complete colocalization with the MitoTracker, indicating that the complex selectively labeled mitochondria in diverse living mammalian cells (Fig. 3.2). Importantly, the luminescence of the complex survived cell lysis, protein extraction, and gel electrophoresis under denaturing conditions. The conjugation of the complex with intracellular proteins was rapid at 37°C, but was completely abolished at 4°C at nonphysiological temperature or in the presence of sodium azide. Proteomic analysis indicated that many luminescently labeled proteins were mitochondrial proteins. The most highly labeled protein was VDAC1 (voltage-dependent anion channel 1), which is located on the outer mitochondrial membrane.[10] Other strongly labeled proteins included 14-3-3 protein ε (YWHAE) and peroxiredoxin 1 (Prdx1). Prolonged exposure of cells to complex **2** led to mitochondrial shortening and fragmentation because ofVDAC1 deregulation.[11] Also bioconjugation of the complex with many abundant cytoskeletal proteins, such as keratins, vimentin, and tubulin, occurred upon longer incubation.

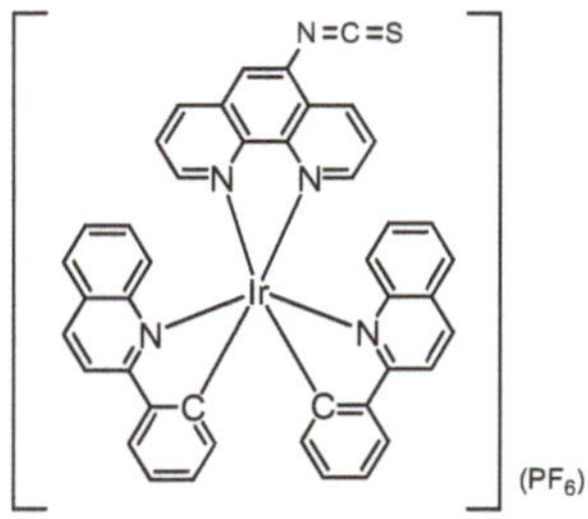

2

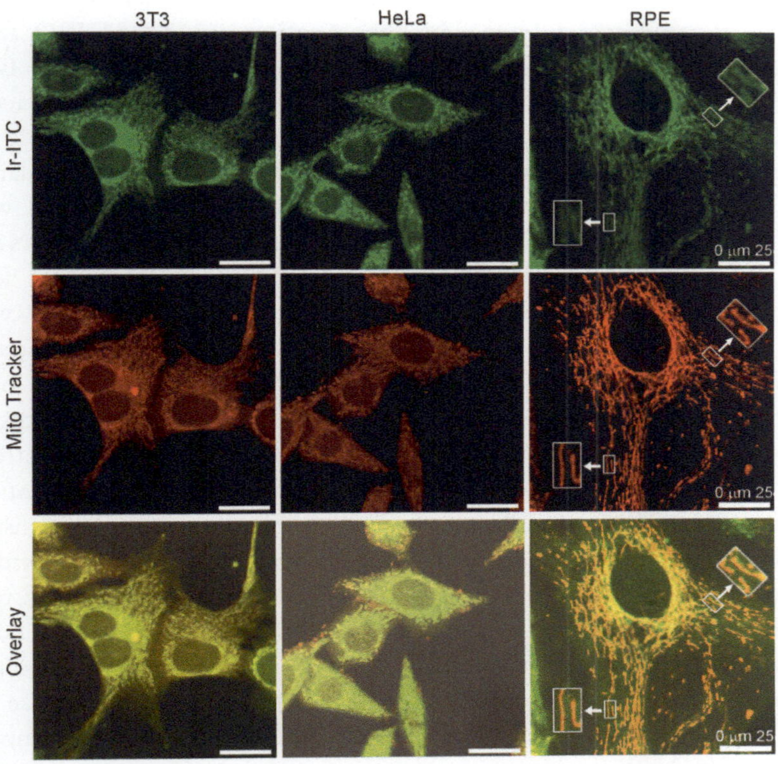

Figure 3.2 Confocal luminescence images of living 3T3, HeLa, and retinal pigmented epithelial (RPE) cells. Cells were preincubated with 50 nM MitoTracker reagent [$\lambda_{ex}=532$ nm, $\lambda_{em}=(620\pm30)$ nm, *red* (Gray in print versions)] for 15 min at 37°C, then further incubated with 5 μM complex **2** for 10 min [$\lambda_{ex}=405$ nm, $\lambda_{em}=(550\pm30)$ nm, *green* (Light gray in print versions)]. *Yellow* (White in print versions) areas indicate the colocalization of probes in mitochondria structure. Scale bar = 25 μm. *Reproduced from Wang, B.; Liang, Y.; Dong, H.; Tan, T.; Zhan, B.; Cheng, J.; Lo, K.K.-W.; Lam, Y.W.; Cheng, S.H.* ChemBioChem ***2012***, 13, *2729–2737 by permission of Wiley-VCH.*

Bioorthogonal chemistry is a versatile tool for the specific labeling and imaging of biomolecules in their native environments.[12,13] Generally, a substrate labeled with a chemical reporter, which is usually small in size and does not disturb the function and activity of the substrate, is first incorporated into living cells or organisms. Then, a bioorthogonal probe carrying a complementary functionality is used to recognize the chemical reporter and visualize the biomolecules of interest. Azide is a particularly useful chemical reporter because of its small size, nonnative nature, and inertness toward biomolecules. It readily undergoes strain-promoted alkyne–azide cycloaddition with cyclooctynes such as dibenzocyclooctyne (DIBO) without the need of a catalyst. Rhenium(I) (**3**)[14] and iridium(III) (**4**)[15] complexes containing a DIBO pendant have been developed and their labeling behavior has been examined using a model compound benzyl azide. It was found that the attachment of DIBO to the metal complexes did not reduce the reaction rate of the DIBO-azide systems. These complexes were used to label proteins including BSA, HSA, and apotransferrin (aTf) that had been modified with azidoacetic acid, affording protein conjugates that exhibited the characteristic luminescence of the corresponding complex. To perform intracellular labeling, living Chinese hamster ovary (CHO) cells were pretreated with 1,3,4,6-tetra-*O*-acetyl-*N*-azidoacetyl-D-mannosamine (Ac_4ManNAz), which was then metabolically converted to *N*-azidoacetyl sialic acids and subsequently located on the cell surface. The cells treated with the rhenium(I) DIBO complexes exhibited intense luminescence from the cell membrane and internal compartments, which has been ascribed to the bioorthogonal labeling and intracellular trafficking of the labeled glycans, because the luminescence of cells without pretreatment of Ac_4ManNAz was negligible. Unfortunately, in the case of the iridium(III) DIBO complexes, the cell staining did not show distinguishable difference when the cells were loaded with phenylpyridine and phenylquinoline complexes, whether the cells were pretreated with Ac_4ManNAz or not, because the high lipophilicity of these two complexes led to efficient nonspecific internalization into living cells. Two polar carboxyl groups were introduced into the complex to reduce the lipophilicity and inhibit the nonspecific cellular uptake of the complex. Living CHO cells treated with this complex did not show any luminescence. In contrast, incubation of Ac_4ManNAz-treated CHO cells with this complex gave intense luminescence from the cell membrane, highlighting the bioorthogonal labeling characteristics (Fig. 3.3).

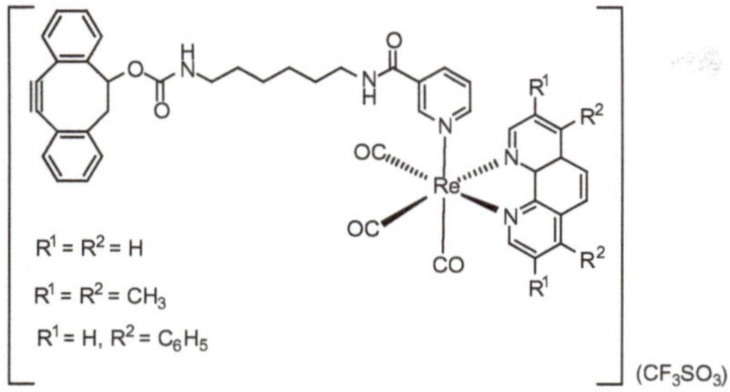

3

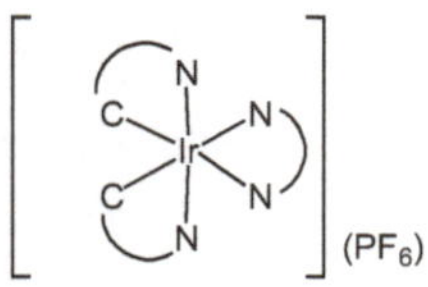

N^C = ppy, N^N = bpy-C6-DIBO

N^C = pq, N^N = bpy-C6-DIBO

N^C = ppy-COOH, N^N = bpy-TEG-DIBO

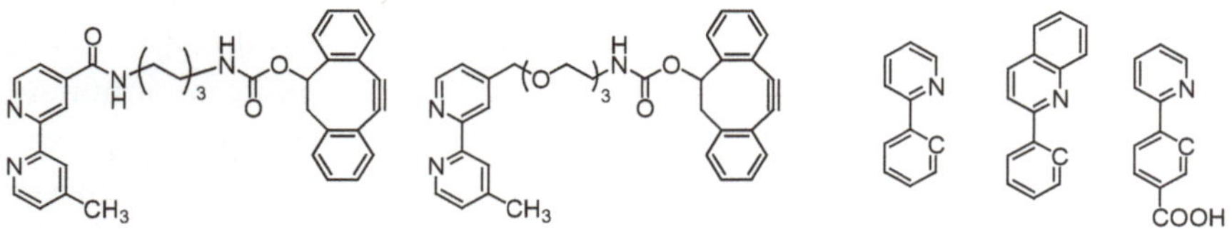

4

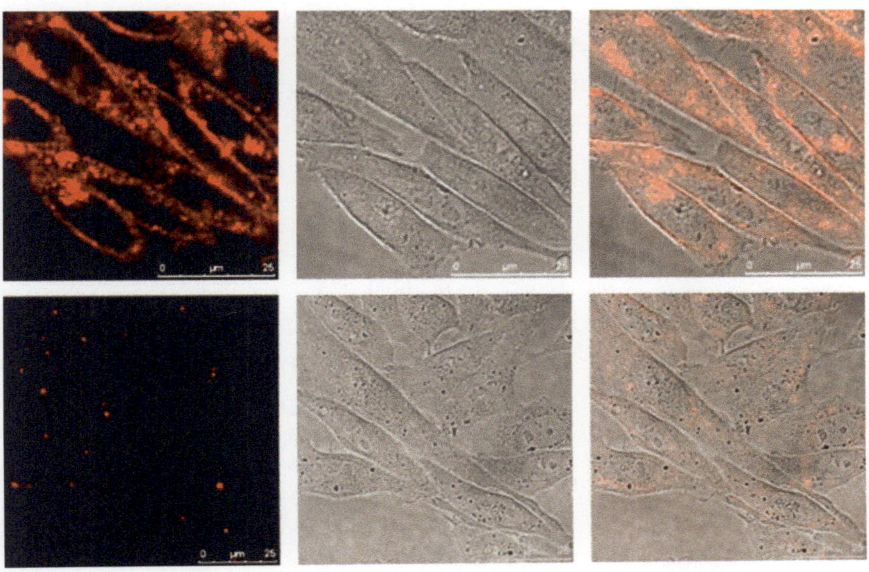

Figure 3.3 Fluorescence (*left*), brightfield (*middle*), and overlaid (*right*) confocal microscopy images of $Ac_4ManNAz$-treated (*top*) and -untreated (*bottom*) CHO cells incubated with the complex containing carboxylate groups (30 μM, 37°C, 5 h). *Reproduced from Lo, K.K.-W.; Chan, B.T.-N.; Liu, H.-W.; Zhang, K.Y.; Li, S.P.-Y.; Tang, T.S.-M.* Chem. Commun. ***2013,*** *49, 4271–4273 by permission of the Royal Society of Chemistry.*

Bioorthogonal probes that display luminescence turn-on upon labeling allow real-time monitoring of biomolecules and biological processes because the probes remain nonemissive as long as the appended bioorthogonal functional group is intact. To develop fluorogenic or phosphorogenic probes, the bioorthogonal functional group should be able to quench the emission of the luminophores, but is converted into a nonquenching molecule upon bioorthogonal labeling. Three iridium(III) complexes containing a nitrone moiety [Ir(N^C)$_2$(bpy-nitrone)](PF$_6$) (**5**) have been designed and synthesized as novel phosphorogenic bioorthogonal probes.[16] These complexes were nonemissive because of the *cis-trans* photo-isomerization of the C=N group. Upon cycloaddition reaction with a strained cyclooctyne (1*R*,8*S*,9*S*)-bicyclo[6.1.0]non-4-yn-9-ylmethanol (BCN), the photoisomerization was inhibited, leading to significant luminescence enhancement. The complexes were used to label BCN-modified proteins, and the resulting protein conjugates exhibited very significant luminescence enhancement (about 800-fold) compared to the free complexes. Although nitrones showed spin-trapping behavior and reactivity toward thiols, reactive oxygen/nitrogen species (RONS) and biological thiols did not cause luminescence response and the complexes remained intact in the presence of these molecules. Living CHO-K1 cells treated with these complexes did not display any intracellular luminescence. However, when the cells were pretreated with *N*-[(1*R*,8*S*,9*S*)-bicyclo[6.1.0]non-4-yn-9-ylmethyloxycarbonyl]-1-aminodecane (BCN-C10) followed by extensive washing, further treatment with the complexes gave rise to significant intracellular luminescence because of the specific bioorthogonal labeling inside the cells (Fig. 3.4).

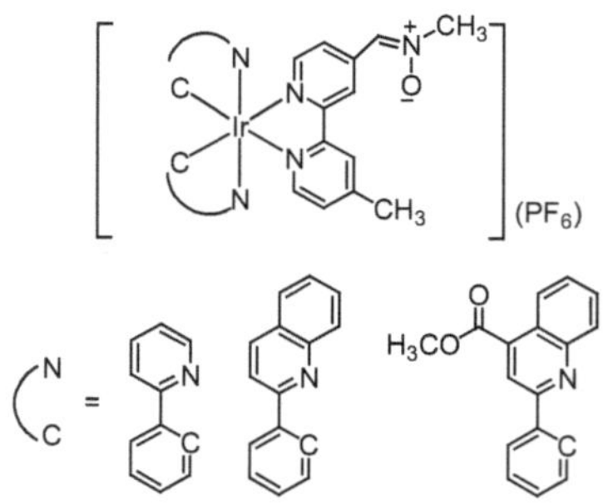

5

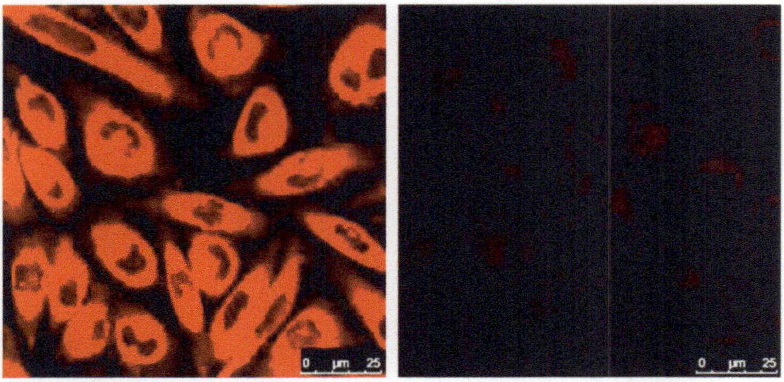

Figure 3.4 Laser-scanning confocal microscopy images of CHO-K1 cells incubated with one of the complexes **5** (5 µM) at 37°C for 1 h with (*left*) or without (*right*) pretreatment with *N*-[(1*R*,8*S*,9*S*)-bicyclo[6.1.0]non-4-yn-9-ylmethyloxycarbonyl]-1-aminodecane (BCN-C10) (125 µM) at 37°C for 30 min. Scale bar = 25 mm. *Reproduced from Lee, L.C.-C.; Lau, J.C.-W.; Liu, H.-W.; Lo, K.K.-W.* Angew. Chem. Int. Ed. ***2016,*** 55, *1046–1049 by permission of Wiley-VCH.*

1,2,4,5-Tetrazine undergoes inverse electron-demand Diels–Alder cycloaddition with dienophiles such as alkenes and alkynes with very fast reaction kinetics, high reaction selectivity, and high compatibility with water.[17,18] Importantly, it quenches many luminophores via Förster resonance-energy transfer (FRET), but can be converted into a nonquenching molecule upon reaction with a dienophile, resulting in luminescence turn-on. The mononuclear and binuclear rhenium(I) complexes $[Re(N^\wedge N)(CO)_3(py\text{-}Tz)](CF_3SO_3)$ and $[\{Re(N^\wedge N)(CO)_3\}_2(\mu\text{-}py\text{-}Tz\text{-}py)](CF_3SO_3)_2$ (**6**) exhibited weak luminescence ($\Phi_{em} < 0.05$) upon photoexcitation due to FRET-based luminescence quenching.[19] Upon reaction with BCN, the complexes showed significant luminescence enhancement by up to 180-fold. The reactivity of the binuclear complexes was much higher than that of the free ligand and the mononuclear analogs because of the strong electron-withdrawing effect of the cationic rhenium(I) polypyridine unit. One of the binuclear complexes has been used to track the intracellular distribution of the bioconjugate BSA–BCN. HeLa cells pretreated with BSA–BCN and incubated with the complex showed punctate cytoplasmic staining (Fig. 3.5). Since no intracellular luminescence was observed when unmodified BSA was used instead of BSA–BCN, the observed cytoplasmic luminescence was because of intracellular protein labeling.

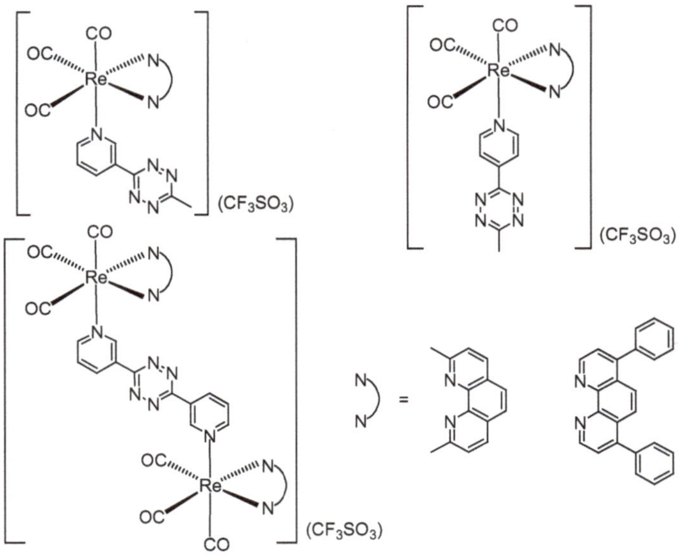

6

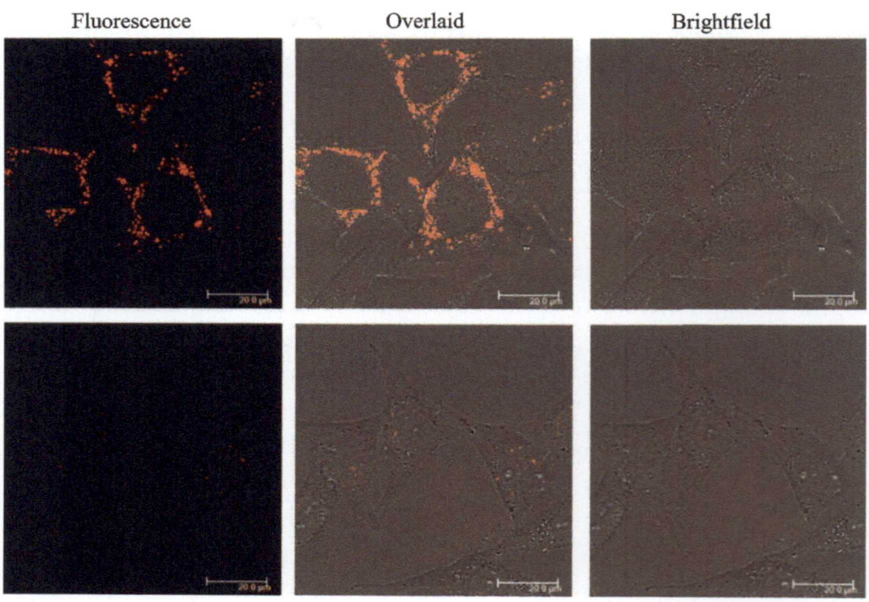

Figure 3.5 Confocal microscopy images of live HeLa cells that have been incubated with bovine serum albumin–(1*R*,8*S*,9*S*)-bicyclo[6.1.0]non-4-yn-9-ylmethanol (BSA–BCN) (*top*) and unmodified BSA (*bottom*) (7.4 μM), respectively, at 37°C for 30 min, and subsequently with one of the binuclear complexes (5 μM, 37°C, 30 min). *Reproduced from Choi, A.W.-T.; Tso, K.K.-S.; Yim, V.M.-W.; Liu, H.-W.; Lo, K.K.-W.* Chem. Commun. ***2015,*** 51, *3442–3445 by permission of the Royal Society of Chemistry.*

3. INTRACELLULAR SENSORS

Luminescent transition metal complexes have been widely used as sensors for a variety of analytes because of their sensitive phosphorescence toward ligands and surrounding microenvironment.[1–7] The detection of intracellular RONS has received particular attention as these reactive molecules are involved in many physiological and pathological processes, and implicated in carcinogenesis and neurodegenerative disorders, such as Alzheimer's disease, Parkinson's disease, multiple sclerosis, and Huntington's disease.[20,21] Nitric oxide (NO) is an important free radical endogenously produced in different mammalian cells by nitric oxide synthase, which catalyzes the conversion of L-arginine to NO and L-citrulline.[22] The rhenium(I) polypyridine complexes functionalized with a diaminoaromatic moiety $[Re(N^\wedge N)(CO)_3(py(NH_2)_2)](CF_3SO_3)$ (**7**) have been developed for intracellular NO sensing based on the reaction between electron-rich diaminoaromatic compounds and NO, yielding triazole derivatives.[23] These complexes were weakly emissive because of the quenching of the ^{3}MLCT $[d\pi(Re)\rightarrow\pi^*(N^\wedge N)]$ luminescence by the diaminoaromatic moiety via photoinduced electron transfer. Upon reaction with NO, these diamine complexes were converted to the triazole derivatives, which revealed intense luminescence upon photoexcitation. Unfortunately, when one of the complexes was used to visualize intracellular NO in HeLa cells, the complex-loaded cells remained weak emissive after addition of an exogenous NO donor, 3-(2-hydroxy-1-methyl-2-nitrosohydrazino)-*N*-methyl-1-propanamine (NOC-7). The reason was probably because of the short intracellular lifetime of NO ($t_{1/2}$ = about 5 s) and the rather low reactivity of the complex. To realize intracellular sensing, new rhenium(I) diamine complexes $[Re(N^\wedge N)(CO)_3(py\text{-}diamine)](CF_3SO_3)$ (**8**) were designed.[24] An electron-donating methoxy group was added to the diaminoaromatic ligand to increase its electron density and thus the reactivity toward NO. Also the diaminoaromatic unit was attached to the tricarbonylrhenium(I) complexes through an insulating spacer-arm to reduce the electron-withdrawing effect caused by the rhenium(I) center. HeLa cells treated with one of the complexes did not reveal significant intracellular luminescence. However, further incubation of the cells with NOC-7 led to substantial luminescence enhancement because of the conversion of the diamine complex to its triazole derivative. The detection of endogenous NO by the complex has also been demonstrated using the macrophage cell line RAW 264.7, which is known to produce NO when stimulated by endotoxin lipopolysaccharides (LPS). Upon stimulation of the complex-loaded cells by LPS, intracellular luminescence turn-on was observed. The luminescence enhancement was suppressed by addition of the NO synthase inhibitor NG-nitro-L-arginine, indicating that the increase of luminescence intensity was caused by the intracellular formation of NO.

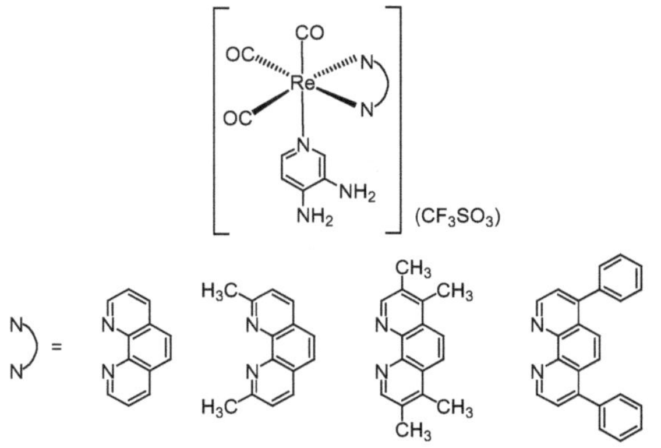

7

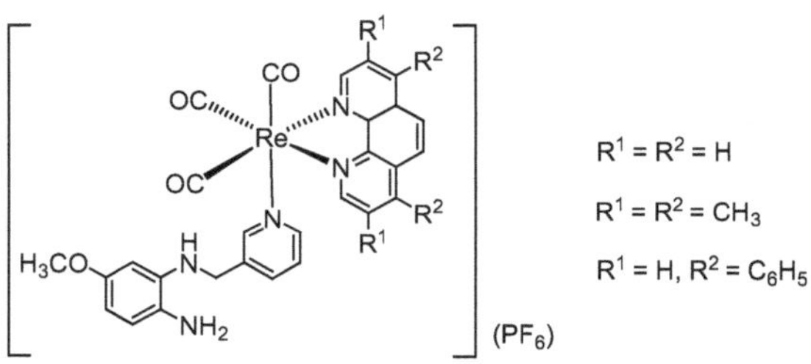

8

The methoxy-substituted diaminoaromatic moiety has been incorporated into iridium(III) complexes, affording a series of phosphorogenic intracellular NO sensors [Ir(N^C)$_2$(bpy-diamine)](PF$_6$) (**9**).[25] As expected, the complexes emitted weakly because of the intermolecular reductive quenching of the excited complexes by the diamine moiety, and exhibited luminescence enhancement in the presence of NOC-7. A similar phosphorogenic response was observed in HeLa cells when the cells were treated with NOC-7 (Fig. 3.6). The generation of the triazole products in the HeLa cells has been confirmed by electrospray ionization mass spectrometry of the cell extract. The detection of endogenous NO using the iridium(III) complexes has also been demonstrated using the macrophage cell line RAW 264.7.

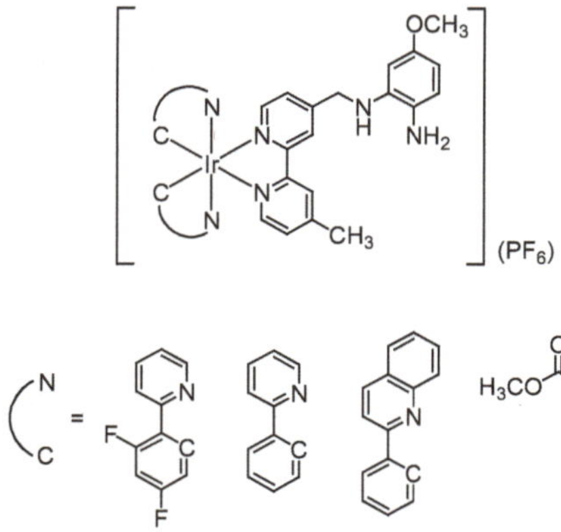

9

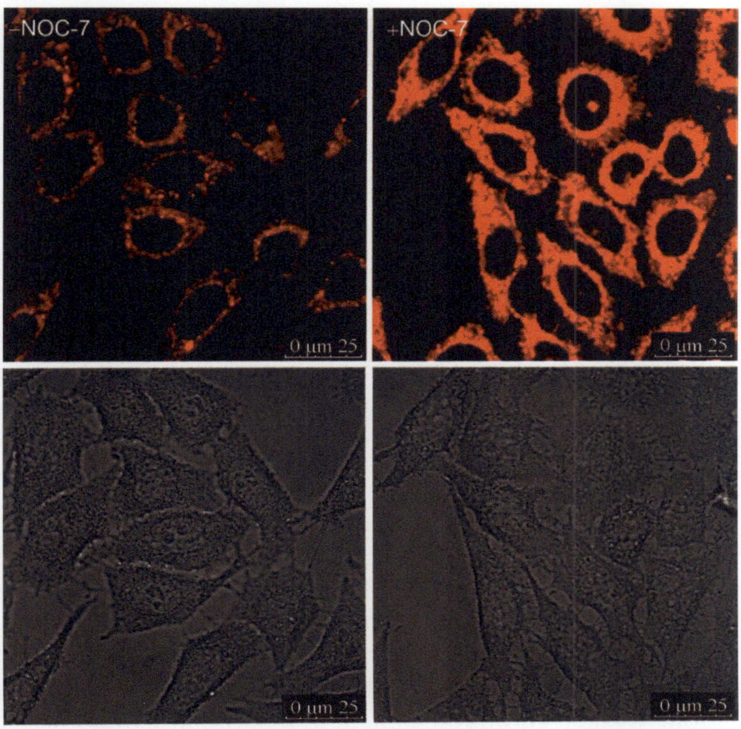

Figure 3.6 Luminescence (*top*) and brightfield (*bottom*) images of HeLa cells upon incubation with one of complexes **9** (5 μM, 1 h) at 37°C without (*left*) and with (*right*) post-treatment with 3-(2-hydroxy-1-methyl-2-nitrosohydrazino)-*N*-methyl-1-propanamine (NOC-7) (100 μM, 1 h). *Reproduced from Law, W.H.-T.; Leung, K.-K.; Lee, L.C.-C.; Poon, C.-S.; Liu, H.-W.; Lo, K.K.-W.* ChemMedChem ***2014***, 9, *1316–1329 by permission of Wiley-VCH.*

Hypochlorite (ClO^-) is another important reactive oxygen species (ROS) that plays an important role in the destruction of pathogens in the immune system.[26,27] It is endogenously produced in leukocytes by myeloperoxidase that catalyzes the oxidation of chloride to ClO^- by hydrogen peroxide.[28,29] Unregulated ClO^- production may be associated with various cardiovascular diseases, neuron degeneration, arthritis, and cancer.[30] A series of iridium(III) complexes [Ir(N^C)$_2$(bpy-dmn)] Cl (**10**) appended with a diaminomaleonitrile moiety, which acted not only as a ClO^- recognition group but also as a luminescence quencher, has been designed and synthesized.[31] The complexes were weakly emissive but displayed intense luminescence in the presence of ClO^- because of the oxidation of diaminomaleonitrile to a carboxylate group. The luminescence turn-on response is specific to ClO^- over other ROS and the luminescence colors varied from green to red depending on the cyclometalating ligands of the complexes. Upon internalization of the complexes into living HeLa and RAW 264.7 cells, the cells showed very weak luminescence. The intracellular distribution of the complexes has been analyzed by inductively coupled plasma mass spectrometry. The results indicated that most of the complexes were concentrated in the mitochondria, where most intracellular ROS are produced. Upon addition of ClO^- to the complex-loaded cells, the mitochondria were lighted up because of the oxidation of the diaminomaleonitrile to a carboxylate group. The use of the complexes to detect endogenous ClO^- in RAW 264.7 macrophage cells, which produce endogenous ClO^- when stimulated with LPS and phorbol myristate acetate, has also been demonstrated. Incubation of stimulated cells with the complexes led to intense intracellular luminescence and the staining patterns were highly colocalized with commercial MitoTrackers. In another study, the bis(phenylpyridine) iridium(III) analog [Ir(N^C)$_2$(bpy-dmn)]Cl (**11**) revealed a maximum two-photon absorption cross-section of about 78.1 GM at 750 nm.[32] This complex has been used to image ClO^- in the mitochondria of living cells (Fig. 3.7), multicellular spheroids, and zebrafish via two-photon fluorescence microscopy. This technique is very attractive because the near infrared excitation used allows deeper penetration, weaker autofluorescence, less photobleaching, and lower phototoxicity than the ultraviolet and visible light excitation.

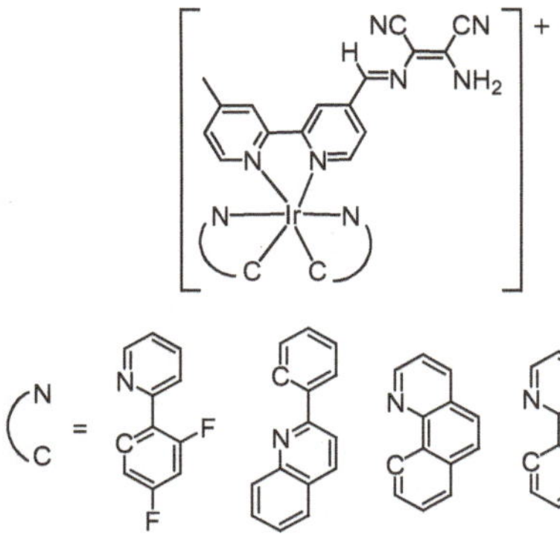

10

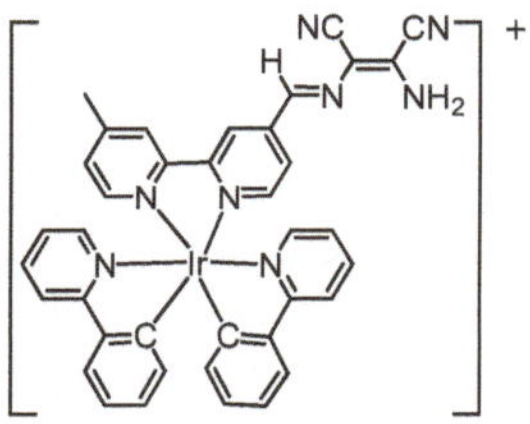

11

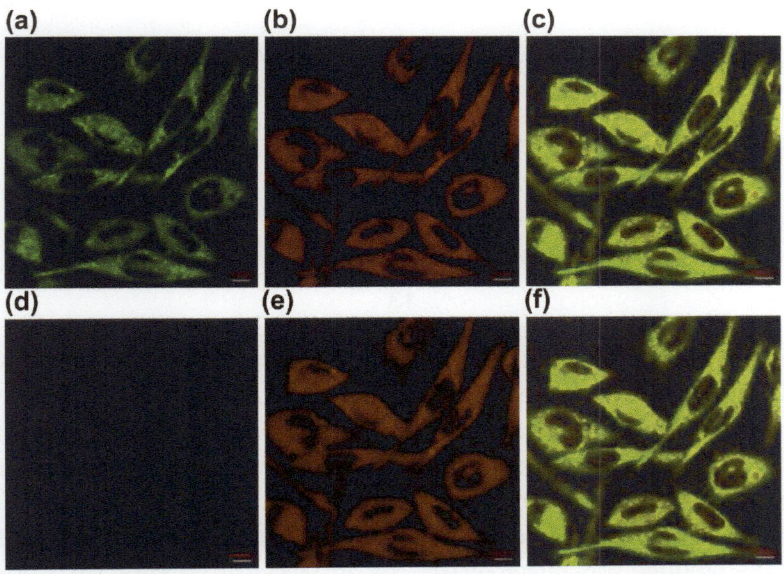

Figure 3.7 One- and two-photon microscopy imaging of exogenous ClO^- in HeLa cells. (a) One-photon microscopy image of cells incubated with MitoTracker *Green* (Light gray in print versions) for 30 min, $\lambda_{ex}=488$ nm, $\lambda_{em}=500–540$ nm; one-photon (b) and two-photon (e) microscopy images of cells pretreated with NaClO (10 μM) for 1 h and then treated with complex **11** (10 μM) for another 1 h, $\lambda_{ex}=405$ nm (one-photon)/750 nm (two-photon), $\lambda_{em}=580–620$ nm; (d) brightfield image of cells; (c) overlay image of (a), (d), and (b); (f) overlay image of (a), (d), and (e). Scale bar = 10 mm. *Reproduced from Li, G.; Lin, Q.; Sun, L.; Feng, C.; Zhang, P.; Yu, B.; Chen, Y.; Wen, Y.; Wang, H.; Ji, L.; Chao, H.* Biomaterials ***2015***, 53, *285–295 by permission of Elsevier.*

The thiol-containing amino acid cysteine (Cys) plays important roles in many physiological and pathophysiological processes in living systems.[33] It acts as a disease-associated physiological regulator, a potential neurotoxin, and a biomarker for medical conditions.[34] Different strategies have been reported to design iridium(III) complexes that display luminescence changes in response to Cys. Iridium(III) aldehyde complexes such as **12**[35] react with Cys to form cyclic thiazolidine, which usually significantly alters the luminescence properties of the complexes. Iridium(III) complexes containing a 2,4-dinitrobenzenesulfonyl (DNBS) moiety such as **13**[36] showed phosphorogenic response toward Cys because DNBS acts as a strong electron-acceptor to quench the luminescence of the complexes but its covalent linkage with the complex core is cleaved by Cys, resulting in luminescence recovery.

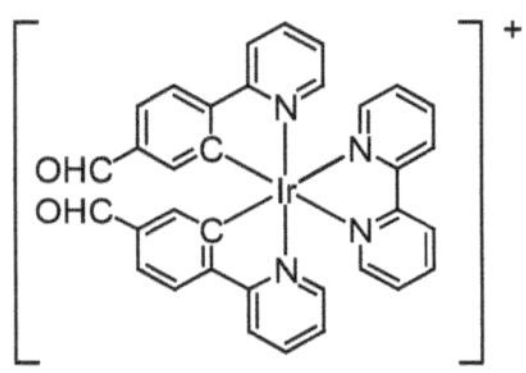

12

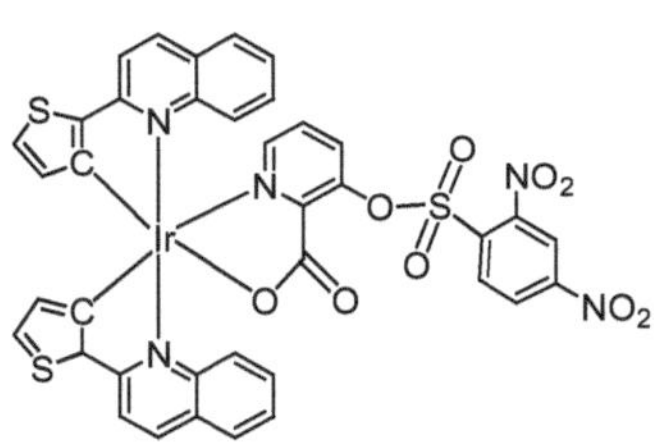

13

An iridium(III) complex (**14**) containing two α,β-unsaturated ketone groups has been reported as an intracellular sensor for Cys.[37] Complex **14** was weakly emissive, but emitted intense luminescence in the presence of Cys, because the complex and Cys underwent 1,4-addition reaction, which broke the p−π conjugation between the ketone and the cyclometalating ligands, resulting in turn-on luminescence response. The **14**-Cys_2 adduct was characterized by matrix-assisted laser desorption/ionization time-of-flight mass spectrometry. The remarkable luminescence response could be observed under either single-photon excitation at 400 nm or two-photon excitation at 800 nm, facilitating two-photon imaging of intracellular Cys. Additionally, the luminescence lifetimes of complex **14** and the adduct **14**-$(Cys)_2$ were 85 and 231 ns, respectively, which are much longer than those of fluorescent organic dyes, which are generally several nanoseconds. Thus the long-lived luminescence could be in principle separated from a strongly autofluorescent background in the time domain. The utilization of complex **14** for sensing Cys has been demonstrated in the presence highly fluorescent rhodamine B via time-gated luminescence titration. The prompt luminescence of a mixture of complex **14** and rhodamine B was dominated by the intense fluorescence of rhodamine B. Addition of Cys did not cause noticeable luminescence changes. In contrast, time-gated acquisition of the spectrum after a 100-ns delay completely removed the fluorescence of rhodamine B, and the luminescence intensity was enhanced by 3.2-fold in the presence of Cys. The intracellular Cys sensing was performed via two-photon imaging, time-gated luminescence imaging, and photoluminescence lifetime imaging microscopy (PLIM). Another iridium(III) complex containing two α,β-unsaturated ketone groups in its diimine ligand (**15**) has been applied for imaging Cys in living zebrafish.[38] Zebrafish exposed to complex **15** exhibited significant luminescence (Fig. 3.8). However, when the animal model was pretreated with *N*-methylmaleimide (NEM), which is a general trapping reagent of thiols, the emission became extremely weak. These in vivo studies showed that complex **15** could enter zebrafish and efficiently detect Cys within living organisms.

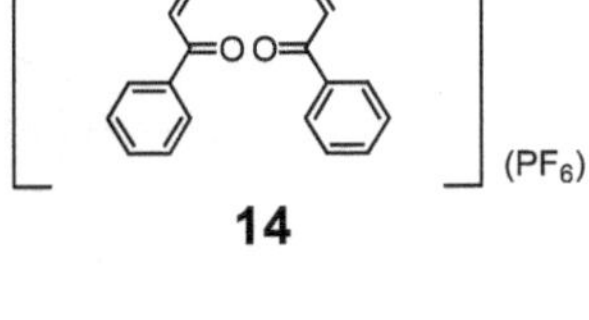

14

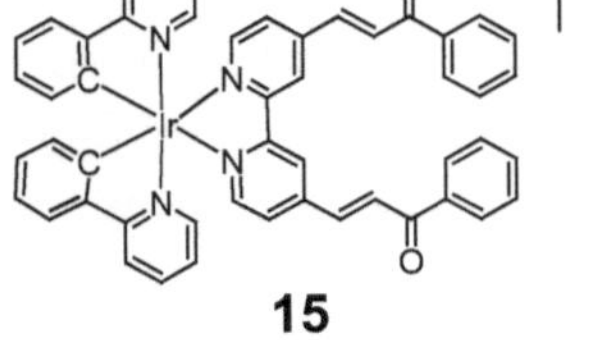

15

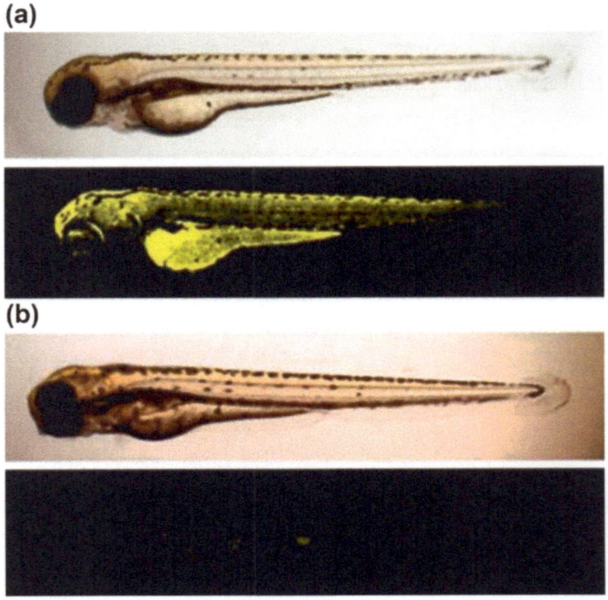

Figure 3.8 In vivo images of zebrafish treated with complex **15**. (a) Three-day-old zebrafish was incubated with 10 μM of complex **15**. (b) Three-day-old zebrafish was pre-incubated with 200 μM *N*-methylmaleimide (NEM) for 15 min and then incubated with 10 μM of complex **15** for 30 min. *Upper panels* show phase contrast images and lower panels show luminescence images. *Reproduced from Mao, Z.; Wang, M.; Liu, J.; Liu, L.-J.; Lee, S.M.-Y.; Leung, C.-H.; Ma, D.-L.* Chem. Commun. ***2016,*** *52, 4450–4453. by permission of the Royal Society of Chemistry.*

Intracellular oxygen concentration is of significant importance in the determination of physiological and pathological processes in biological systems.[39] Hypoxia, referring to inadequate supply of oxygen, is an important feature of many diseases such as solid tumors, inflammatory diseases, and cardiac ischemia.[40] Phosphorescent transition metal complexes display extremely high sensitivity toward oxygen because of the efficient quenching of their triplet emissive states by oxygen molecules. Live cells incubated with the iridium complex $[Ir(btp)_2(acac)]$ (**16**) did not show notable phosphorescence but exhibited intense cytoplasmic phosphorescence when cultured under a low oxygen condition.[41] Another iridium(III) complex (**17**) was concentrated in the nuclei upon internalization into living cells.[42] The luminescence from the nuclei was significantly enhanced when the cells were cultured under 2.5% O_2 atmosphere.

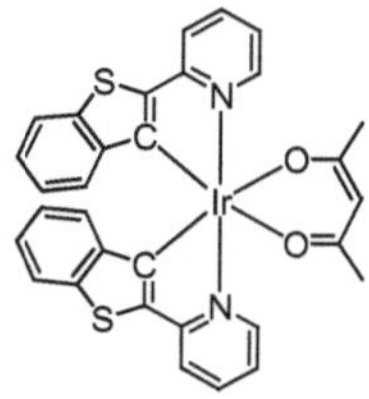

16

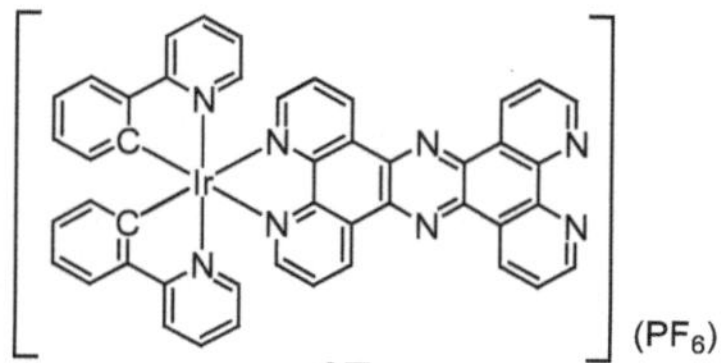

17

4. RATIOMETRIC SENSORS

Intensity-based luminescent sensors straightforwardly correlate the luminescence intensity with the concentration of analytes. However, the intensity is also determined by sensor concentration, which is usually uncertain and difficult to control or measure in an intracellular environment. Wavelength-ratiometric sensors reflect the presence and the concentration of analytes by a change in the ratio of intensities at two luminescence wavelengths. They are self-calibrating and have high accuracy and precision, especially in intracellular and in vivo sensing. The most convenient strategy to design ratiometric sensors is to tie two luminophores together, one of which displays intensity response toward analyte while the other is inert and serves as an internal standard during sensing. A ratiometric oxygen sensor **18** consisting of a fluorescent coumarin unit and a phosphorescent iridium(III) complex displayed both blue fluorescence and red phosphorescence.[43] Since only the phosphorescence was quenched by oxygen, the intensity ratio was used to determine intracellular oxygen concentration.

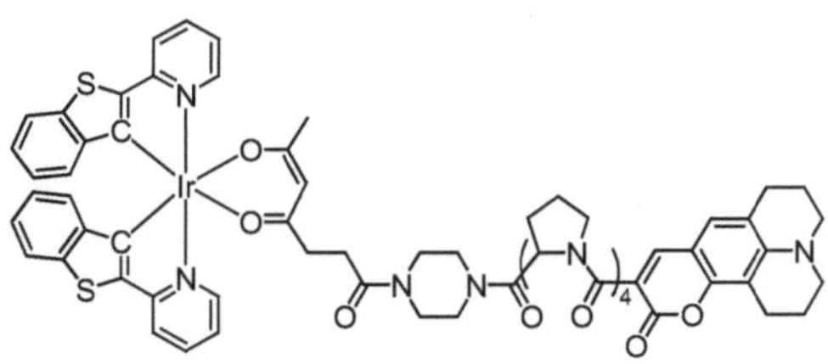

18

Core–shell structured silica nanoparticles are good candidates to construct ratiometric sensors for intracellular applications because they exhibit excellent hydrophilicity, biocompatibility, and stability.[44] Importantly, the core and shell allow independent modification to meet their respective requirements. A phosphorescent ratiometric sensor for ClO^- has been prepared by the incorporation of two iridium(III) complexes into core–shell-structured silica nanoparticles.[45] The blue-emissive complex **19** that was insensitive to ClO^- was immobilized in the inner solid core while the nonemissive complex **20** that exhibited a specific and significant red-luminescence turn-on in response to ClO^- was adsorbed in the outer mesoporous layer (Fig. 3.9). The nanoparticles were well dispersed in solution and spherical in shape with a core diameter of about 80 nm and a shell thickness of 18 nm. Upon exposure to ClO^-, the nanoparticles displayed a sharp luminescence color change from blue to red. The nanoparticles have been used to image exogenous and endogenous ClO^- in living RAW 264.7 cells via ratiometric imaging where phosphorescence signals in the blue (480 ± 20 nm) and red (600 ± 20 nm) windows were collected for analysis. In the absence of ClO^-, the cells incubated with the nanoparticles only exhibited blue luminescence. Upon addition of exogenous ClO^- or stimulation of the cells to produce endogenous ClO^-, the blue luminescence maintained a similar intensity while the red luminescence was significantly enhanced. While both the blue and red luminescence intensities were remarkably influenced by the dose of the nanoparticles, the incubation time and temperature, and excitation laser power, their intensity ratios were almost unaffected, demonstrating the strong resistance of the ratiometric detection toward external influences.

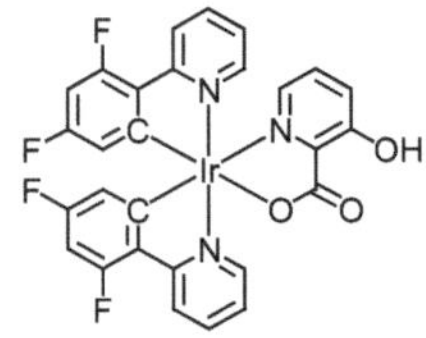

19

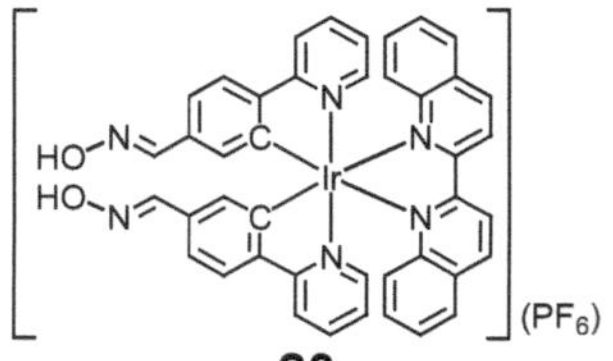

20

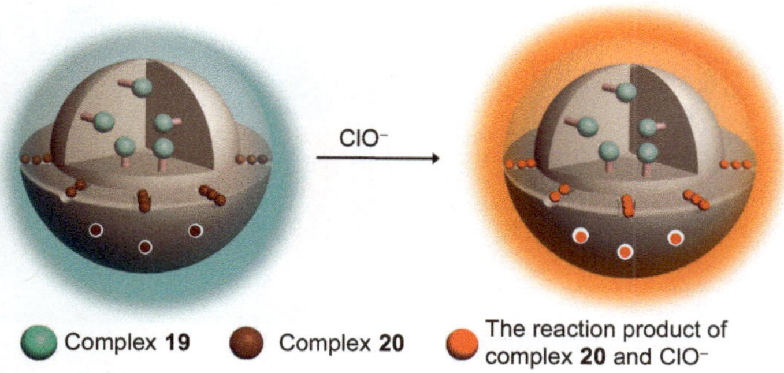

Figure 3.9 Design concept of the ratiometric sensor for ClO⁻ based on core–shell-structured nanoparticles. *Reproduced from Zhang, K.Y.; Zhang, J.; Liu, Y.; Liu, S.; Zhang, P.; Zhao, Q.; Tang, Y.; Huang, W.* Chem. Sci. ***2015***, *6, 301–307 by permission of the Royal Society of Chemistry.*

Another strategy to tie two luminophores together is to incorporate them into the same polymer chain. Two iridium(III) complexes have been incorporated into an acrylamide-based thermosensitive polymer, affording a ratiometric phosphorescent thermometer **21**.[46] Increasing temperature resulted in coil-globule transition of this polymer, which led to a decrease in microenvironment polarity of the iridium(III) complexes and hence induced luminescence enhancement of both complexes. Because of the different sensitivity of the two complexes to surrounding environment, their luminescence intensity ratio was correlated to the temperature. The luminescence color of the polymer changed from orange to sky-blue when the temperature increased from 10 to 40°C (Fig. 3.10). The polymer has been used for temperature determination in live cells and zebrafish via ratiometric luminescence imaging and PLIM.

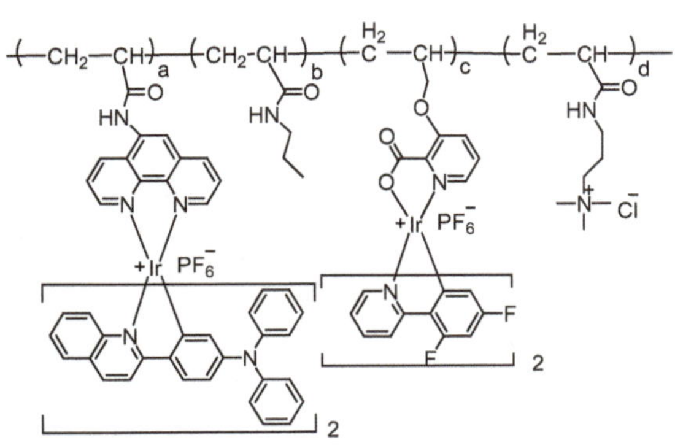

21

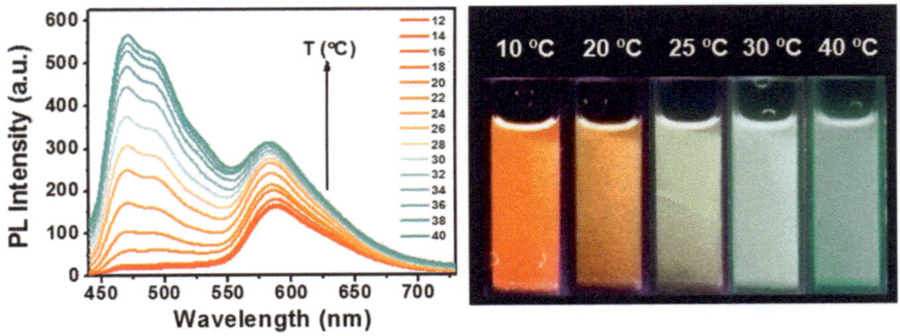

Figure 3.10 Emission spectra (*left*) and photographs (*right*) of the polymer **21** (0.01 w/v%) in phosphate buffer saline (pH = 7.4) at various temperatures. *PL*, photoluminescence. *Reproduced from Chen, Z.; Zhang, K.Y.; Tong, X.; Liu, Y.; Hu, C.; Liu, S.; Yu, Q.; Zhao, Q.; Huang, W.* Adv. Funct. Mater. ***2016***, 26, *4386–4396 by permission of Wiley-VCH.*

Luminescent ion pairs involve a luminescent cation and a luminescent anion that are bonded together via electrostatic attraction. A luminescent ion pair consisting of two iridium(III) complexes (**22**) has been developed as a ratiometric pH sensor.[47] The cationic complex containing two pyridine pendants displayed red luminescence, which was quenched upon protonation of the pyridine units. The anionic complexes showed blue luminescence that was independent of pH values. Interestingly, the ion pair exhibited blue, yellow, and red luminescence at pH 2, 5, and 8, respectively. The ion pair has been used for quantitative measurements of intracellular pH values. The red and blue luminescence of the ion pair-loaded HepG-2 cells was highly colocalized, indicating that the cationic and anionic complexes of the ion pair remained associated with each other after cellular uptake. The intensity ratio of the blue and red luminescence was linearly related to the intracellular pH value in the range of 4–8. According to this calibration curve, the averaged intracellular pH value of living HepG-2 cells was determined to be 6.80 ± 0.20. Treatment of the cells with H_2O_2 increased the pH value to 7.20 ± 0.15.

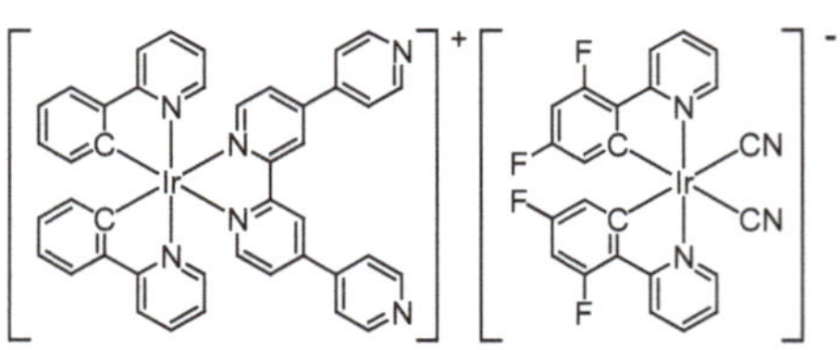

22

5. PHOTODYNAMIC THERAPY

Luminescence quenching of transition metal complexes by molecular oxygen is usually accompanied with the generation of singlet oxygen ($^1O_2$), which can cause cell death. Thus luminescent transition metal complexes have a great potential to be used as PDT reagents. A series of phosphorescent cyclometalated iridium(III) polypyridine poly(ethylene glycol) (PEG) complexes [Ir(N^C)$_2$(bpy-PEG)](PF$_6$) (**23**) has been developed for this purpose.[48] The complexes exhibited intense and long-lived green to orange–red phosphorescence, which can be quenched by O_2. The quantum yields for $^1O_2$ production of the complexes in aerated dimethyl sulfoxide (0.24–0.83) depended on the excited-state lifetimes of the complexes. Because of the protection of the complexes from interacting nonspecifically with the intracellular proteins and triggering immunogenicity and antigenicity inside the cells by the long PEG chains, these complexes exhibited exceptionally low cytotoxicity in the dark ($IC_{50} > 300\ \mu M$). However, most of them became significantly cytotoxic upon irradiation ($IC_{50} = 3.4–23.2\ \mu M$). Laser-scanning confocal microscopy images and flow cytometric analysis revealed the induction of rapid necrotic cell death upon light activation. Another iridium(III) PEG complex **24** has been modified with a nitroveratryl linker as a photolabile protecting group between the iridium(III) polypyridine core and the PEG pendant.[49] Upon photoactivation, the nitroveratryl linker was photocleaved and the complex core was released from the PEG chain, leading to significantly enhanced cytotoxicity.

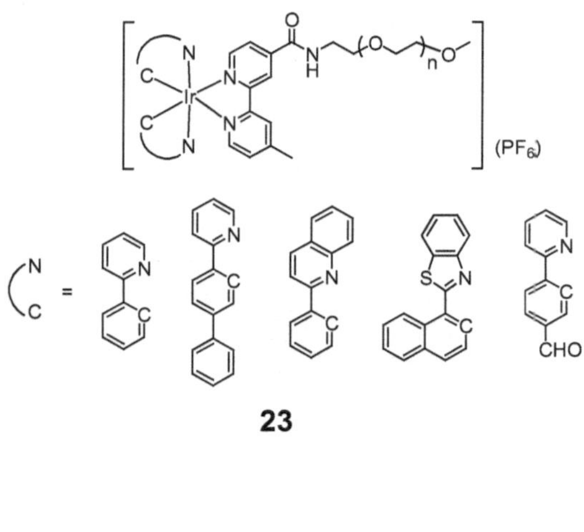

23

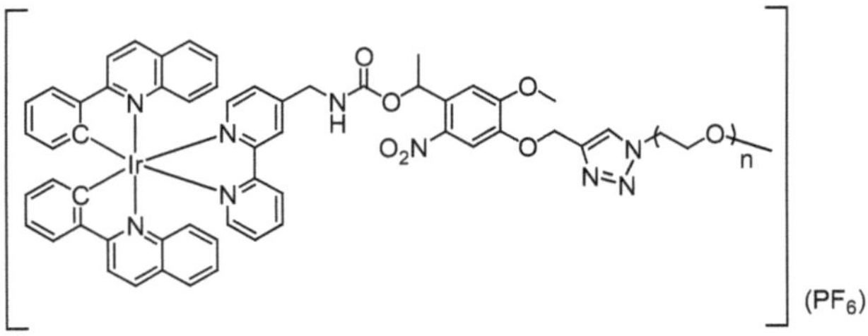

24

Two iridium(III) complexes **25** and **26**, which can stain the mitochondria and lysosomes, respectively, of living cells, have been developed to investigate the effect of organelle targeting on PDT efficiency.[50] Both complexes exhibited long-lived phosphorescence, which is sensitive to oxygen quenching. The photophysical properties and $^1O_2$ quantum yields of the complexes were similar as both complexes had the same iridium(III) polypyridine core and cyclometalating ligands. As oxygen is often inadequate in solid tumor cells, the photocytotoxicity of the complexes was evaluated under both normoxic and hypoxic conditions. The results showed that HeLa cells treated with the mitochondria-targeted complex maintained a slower respiration rate, leading to a higher intracellular oxygen level under hypoxia. As a consequence, this complex exhibited an improved PDT effect compared to the lysosome-targeted complex, especially under hypoxia conditions, suggestive of high practicable potential of mitochondria-targeted PDT agents in cancer therapy.

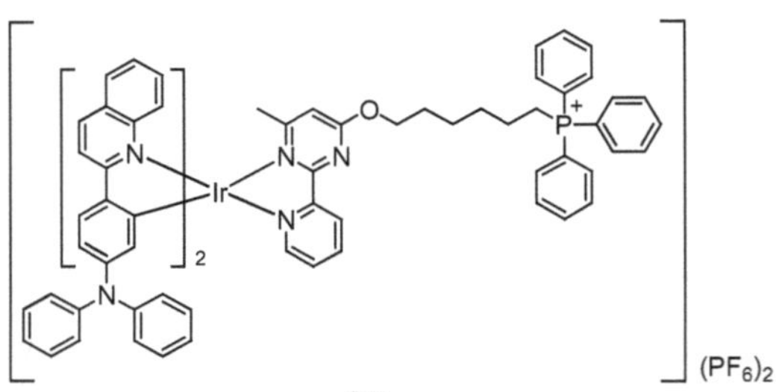

25

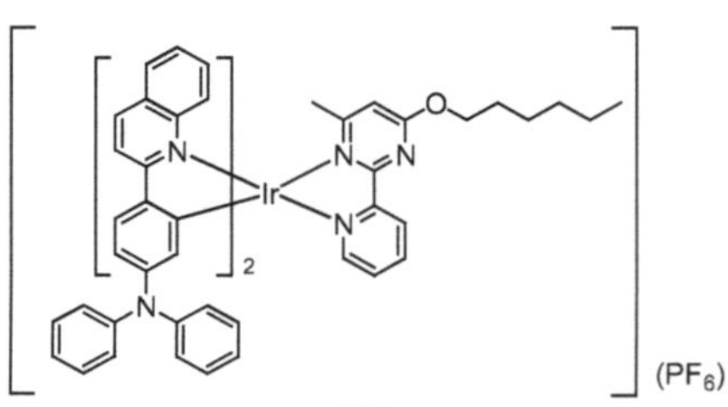

26

6. CONCLUSION

In this chapter, we have summarized the recent design of phosphorescent rhenium(I) and iridium(III) complexes as labeling reagents for intracellular proteins and structures in their native environment. The development of phosphorescent bioorthogonal probes for specific intracellular protein labeling has been emphasized. Additionally, the use of rhenium(I) and iridium(III) complexes to sense intracellular RONS, biothiols, and hypoxia has been discussed. The responsive and insensitive luminophores can be tied together via covalent bonds, hybridized nanoparticles, copolymerization, and electrostatic attraction, to afford ratiometric wavelength-ratiometric sensors. Finally, the potential applications of related complexes in photodynamic therapy have been discussed. On the basis of the studies summarized in this chapter and other related work, we believe that the favorable photophysical, photochemical, and biological properties of rhenium(I) and iridium(III) complexes will continue to contribute to the development of new theranostic reagents with a high application potential.

ACKNOWLEDGMENTS

The author acknowledges the financial support from the National Natural Science Foundation of China (21501098) and Natural Science Foundation of Jiangsu Province of China (BK20150833).

REFERENCES

1. Zhang, K.Y.; Lo, K. K. -W. Chemosensing and Diagnostics. In Reedijk, J., Poeppelmeier, K., Eds.; *Comprehensive Inorganic Chemistry II*, Vol. 8; Elsevier: Oxford, 2013; pp 657–732.
2. Zhao, Q.; Li, F.; Huang, C. *Chem. Soc. Rev.* **2010,** *39,* 3007–3030.
3. Fernández-Moreira, V.; Thorp-Greenwood, F. L.; Coogan, M. P. *Chem. Commun.* **2010,** *46,* 186–202.
4. Zhao, Q.; Huang, C.; Li, F. *Chem. Soc. Rev.* **2011,** *40,* 2508–2524.
5. Lo, K. K. -W.; Choi, A. W. -T.; Law, W. H. -T. *Dalton Trans.* **2012,** *41,* 6021–6047.
6. Coogan, M. P.; Fernández-Moreira, V. *Chem. Commun.* **2014,** *50,* 384–399.
7. Lo, K. K. -W. *Acc. Chem. Res.* **2015,** *48,* 2985–2995.
8. Lee, P.-K.; Liu, H.-W.; Yiu, S.-M.; Louie, M.-W.; Lo, K. K. -W. *Dalton Trans.* **2011,** *40,* 2180–2189.
9. Wang, B.; Liang, Y.; Dong, H.; Tan, T.; Zhan, B.; Cheng, J.; Lo, K. K. -W.; Lam, Y. W.; Cheng, S. H. *ChemBioChem* **2012,** *13,* 2729–2737.
10. Shoshan-Barmatz, V.; Golan, M. *Curr. Med. Chem.* **2012,** *19,* 714–735.
11. Park, J.; Kim, Y.; Choi, S.; Koh, H.; Lee, S.-H.; Kim, J.-M.; Chung, J. *PLoS One* **2010,** *5,* e13151.
12. Saxon, E.; Bertozzi, C. R. *Science* **2000,** *287,* 2007–2010.
13. Prescher, J. A.; Bertozzi, C. R. *Nat. Chem. Biol.* **2005,** *1,* 13–21.
14. Choi, A. W. -T.; Liu, H.-W.; Lo, K. K. -W. *J. Inorg. Biochem.* **2015,** *148,* 2–10.

15. Lo, K. K. -W.; Chan, B. T. -N.; Liu, H.-W.; Zhang, K.Y.; Li, S. P. -Y.; Tang, T. S. -M. *Chem. Commun.* **2013,** *49,* 4271–4273.
16. Lee, L. C. -C.; Lau, J. C. -W.; Liu, H.-W.; Lo, K. K. -W. *Angew. Chem. Int. Ed.* **2016,** *55,* 1046–1049.
17. Devaraj, N. K.; Weissleder, R. *Acc. Chem. Res.* **2011,** *44,* 816–827.
18. Knall, A.-C.; Slugovc, C. *Chem. Soc. Rev.* **2013,** *42,* 5131–5142.
19. Choi, A. W. -T.; Tso, K. K. -S.; Yim, V. M. -W.; Liu, H.-W.; Lo, K. K. -W. *Chem. Commun.* **2015,** *51,* 3442–3445.
20. Bredt, D. S.; Snyder, S. H. *Annu. Rev. Biochem.* **1994,** *63,* 175–195.
21. Wink, D. A.; Vodovotz, Y.; Laval, J.; Laval, F.; Dewhirst, M. W.; Mitchell, J. B. *Carcinogenesis* **1998,** *19,* 711–721.
22. Bredt, D. S.; Hwang, P. M.; Snyder, S. H. *Nature* **1990,** *347,* 768–770.
23. Choi, A. W. -T.; Poon, C.-S.; Liu, H.-W.; Cheng, H.-K.; Lo, K. K. -W. *New J. Chem.* **2013,** *37,* 1711–1719.
24. Choi, A. W. -T.; Yim, V. M. -W.; Liu, H.-W.; Lo, K. K. -W. *Chem. Eur. J.* **2014,** *20,* 9633–9642.
25. Law, W. H. -T.; Leung, K.-K.; Lee, L. C. -C.; Poon, C.-S.; Liu, H.-W.; Lo, K. K. -W. *ChemMedChem* **2014,** *9,* 1316–1329.
26. Winterbourn, C. C. *Nat. Chem. Biol.* **2008,** *4,* 278–286.
27. Prokopowicz, Z. M.; Arce, F.; Biedron, R.; Chiang, C. L.; Ciszek, M.; Katz, D. R.; Nowakowska, M.; Zapotoczny, S.; Marcinkiewicz, J.; Chain, B. M. *J. Immunol.* **2010,** *184,* 824–835.
28. Harrison, J. E.; Shultz, J. *J. Biol. Chem.* **1976,** *251,* 1371–1374.
29. Gomez-Mejiba, S. E.; Zhai, Z.; Gimenez, M. S.; Ashby, M. T.; Chilakapati, J.; Kitchin, K.; Mason, R. P.; Ramirez, D. C. *J. Biol. Chem.* **2010,** *285,* 20062–20071.
30. Roberts, R. A.; Laskin, D. L.; Smith, C. V.; Robertson, F. M.; Allen, E. M.; Doorn, J. A.; Slikker, W. *Toxicol. Sci.* **2009,** *112,* 4–16.
31. Li, G.; Lin, Q.; Ji, L.; Chao, H. *J. Mater. Chem. B* **2014,** *2,* 7918–7926.
32. Li, G.; Lin, Q.; Sun, L.; Feng, C.; Zhang, P.; Yu, B.; Chen, Y.; Wen, Y.; Wang, H.; Ji, L.; Chao, H. *Biomaterials* **2015,** *53,* 285–295.
33. Refsum, H.; Smith, A. D.; Ueland, P. M.; Nexo, E.; Clarke, R.; McPartlin, J.; Johnston, C.; Engbaek, F.; Schneede, J. *Clin. Chem.* **2004,** *50,* 3–32.
34. Janáky, R.; Varga, V.; Hermann, A.; Saransaari, P.; Oja, S. S. *Neurochem. Res.* **2000,** *25,* 1397–1405.
35. Xiong, L.; Zhao, Q.; Chen, H.; Wu, Y.; Dong, Z.; Zhou, Z.; Li, F. *Inorg. Chem.* **2010,** *49,* 6402–6408.
36. Tang, Y.; Yang, H. R.; Sun, H. B.; Liu, S. J.; Wang, J. X.; Zhao, Q.; Liu, X. M.; Xu, W. J.; Li, S. B.; Huang, W. *Chem. Eur. J.* **2013,** *19,* 1311–1319.
37. Xu, W.; Zhao, X.; Lv, W.; Yang, H.; Liu, S.; Liang, H.; Tu, Z.; Xu, H.; Qiao, W.; Zhao, Q.; Huang, W. *Adv. Healthcare Mater.* **2014,** *3,* 658–669.
38. Mao, Z.; Wang, M.; Liu, J.; Liu, L.-J.; Lee, S. M. -Y.; Leung, C.-H.; Ma, D.-L. *Chem. Commun.* **2016,** *52,* 4450–4453.
39. Acker, T.; Acker, H. *J. Exp. Biol.* **2004,** *207,* 3171–3188.
40. Harris, A. L. *Nat. Rev. Cancer* **2002,** *2,* 38–47.
41. Zhang, S.; Hosaka, M.; Yoshihara, T.; Negishi, K.; Iida, Y.; Tobita, S.; Takeuchi, T. *Cancer Res.* **2010,** *70,* 4490–4498.
42. Liu, S.; Liang, H.; Zhang, K.Y.; Zhao, Q.; Zhou, X.; Xu, W.; Huang, W. *Chem. Commun.* **2015,** *51,* 7943–7946.
43. Yoshihara, T.; Yamaguchi, Y.; Hosaka, M.; Takeuchi, T.; Tobita, S. *Angew. Chem. Int. Ed.* **2012,** *51,* 4148–4151.
44. Chaudhuri, R. G.; Paria, S. *Chem. Rev.* **2012,** *112,* 2373–2433.
45. Zhang, K.Y.; Zhang, J.; Liu, Y.; Liu, S.; Zhang, P.; Zhao, Q.; Tang, Y.; Huang, W. *Chem. Sci.* **2015,** *6,* 301–307.

46. Chen, Z.; Zhang, K.Y.; Tong, X.; Liu, Y.; Hu, C.; Liu, S.; Yu, Q.; Zhao, Q.; Huang, W. *Adv. Funct. Mater.* **2016,** *26,* 4386–4396.
47. Ma, Y.; Liang, H.; Zeng, Y.; R Yang, H.; Ho, C.-L.; Xu, W. J.; Zhao, Q.; Huang, W.; Wong, W.-Y. *Chem. Sci.* **2016,** *7,* 3338–3346.
48. Li, S. P. -Y.; Lau, C. T. -S.; Louie, M.-W.; Lam, Y.-W.; Cheng, S. H.; Lo, K. K. -W. *Biomaterials* **2013,** *34,* 7519–7532.
49. Tso, K. K. -S.; Leung, K.-K.; Liu, H.-W.; Lo, K. K. -W. *Chem. Commun.* **2016,** *53,* 4557–4560.
50. Lv, W.; Zhang, Z.; Zhang, K. Y.; Yang, H.; Liu, S.; Xu, A.; Guo, S.; Zhao, Q.; Huang, W. *Angew. Chem. Int. Ed.* **2016**. http://dx.doi.org/10.1002/ange.201604130.

CHAPTER FOUR

Organoruthenium(II)-Arene Complexes: Structural Building Blocks for Anticancer Drug Discovery

M.J. Chow[1], W.H. Ang[1,2]
[1]National University of Singapore, Singapore; [2]NUS Graduate School for Integrative Science and Engineering, Singapore

Contents

1. INTRODUCTION

Special considerations are needed in the design of biologically active metal complexes for therapy. Not all metals can be used as drug scaffolds because some of them can cause acute or chronic systemic toxicity, are

Inorganic and Organometallic Transition Metal Complexes with Biological Molecules and Living Cells
ISBN 978-0-12-803814-7
http://dx.doi.org/10.1016/B978-0-12-803814-7.00004-6

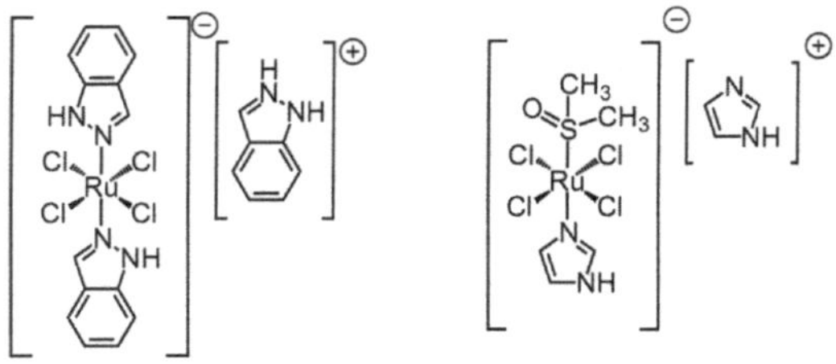

Figure 4.1 Structures of KP1019 and new antitumor metastasis inhibitor-A (NAMI-A).

unstable in an aqueous physiological environment, or there is a general lack of understanding on their chemical behavior, particularly in a biological setting. This greatly limits the range of metals that can be considered for anticancer drug development and the types of complexes that will be suitable for use as chemotherapeutic agents. Among the different transition metal centers that have been exhaustively studied for therapeutic applications, Ru is particularly suitable for the design of metal-based drugs.[1] First, Ru forms coordination bonds that are substitutionally inert,[2] and Ru complexes are less susceptible to ligand substitution in biological conditions. Second, Ru causes relatively low systemic toxicity as compared to other heavy metals such as Cd, Pb, and Os. Third, Ru has been known to be trafficked and transported via the same cellular mechanisms as Fe. Since cancer cells have a high Fe demand because of their high metabolism and rapid proliferation, there is often an upregulation in Fe transport proteins such as transferrin. This characteristic has been exploited to enhance the uptake of Ru metallodrugs in cancer cells compared to healthy tissue.[3] Lastly, the synthetic chemistry of Ru is very well established leading to the surge of stable Ru complexes being prepared and systematically investigated. These factors combine to fuel current interest in the development of Ru-based metallopharmaceuticals.[4] Research on Ru complexes as therapeutic agents has already begun to bear fruit. Ru(III) complexes such as new antitumor metastasis inhibitor-A (NAMI-A) and KP1019 are currently being investigated in clinical trials (Fig. 4.1).[5,6] It has been proposed that these compounds act as "prodrugs" that will undergo in vivo reduction to generate their active Ru(II) progeny.[3] Not surprisingly, this finding has drawn the attention of researchers to Ru(II) complexes and stimulated much research interest.

Ru(II)-arene complexes in particular have piqued much interest because of their stability, ease of synthesis, and possible tuning of pharmacologically

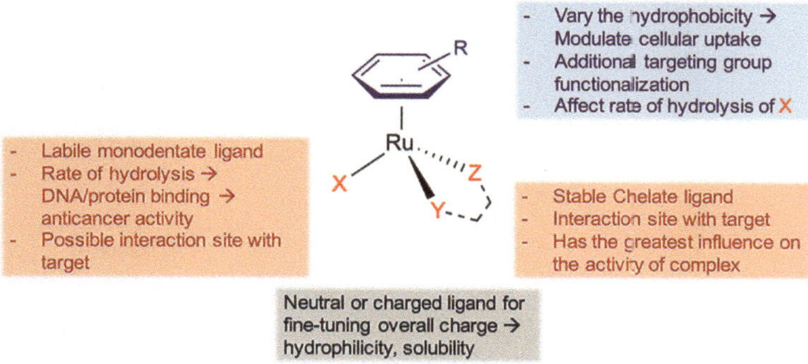

Figure 4.2 General structure of Ru(II)-arene complexes and design considerations.

relevant features such as solubility and lipophilicity. These complexes typically adopt a pseudooctahedral geometry, which resembles the shape of a three-legged piano stool (Fig. 4.2). In this chapter, the most recent synthetic strategies and applications of mononuclear Ru(II)-arene complexes in medicinal chemistry are discussed, with a special focus on their anticancer properties. Many of these studies have demonstrated the therapeutic properties or inhibitory activity toward cancer-related enzyme targets. This makes Ru(II)-arene complexes promising candidates as anticancer drugs.

2. ACCESSING THE RUTHENIUM(II)-ARENE SCAFFOLD

Comprehensive synthetic strategies for many types of Ru(II)-arene complexes have been published,[7] thus we limit our discussion to the synthesis of structures most often seen in the field of medicinal chemistry. Biologically active Ru(II)-arene complexes are usually accessed via a synthetic route involving the $[(\eta^6\text{-arene})RuCl_2]_2$ precursor (Fig. 4.3), which can be synthesized using several methods. The first method takes advantage of a labile facially bound arene ligand for direct exchange with a desired arene ligand. For example, ethylbenzoate is reduced to its cyclohexadiene product via Birch reduction (1) and reacted with hydrated $RuCl_3$ to give $[(\eta^6\text{-ethylbenzoate})RuCl_2]_2$.[8] The electron-withdrawing carboxylate group renders the arene ligand labile and subsequent exchange (2) could be performed using a wide range of electron-rich substituted aromatic ligands to afford the desired $[(\eta^6\text{-arene})RuCl_2]_2$ precursor. Alternatively, the desired ligand can be directly converted to its cyclohexadiene product by Birch reduction (3) before treatment with hydrated $RuCl_3$.[9,10] The main limitation is that many functional groups do not tolerate Birch reduction conditions. Lastly, stable $[(\eta^6\text{-cymene})RuCl_2]_2$,

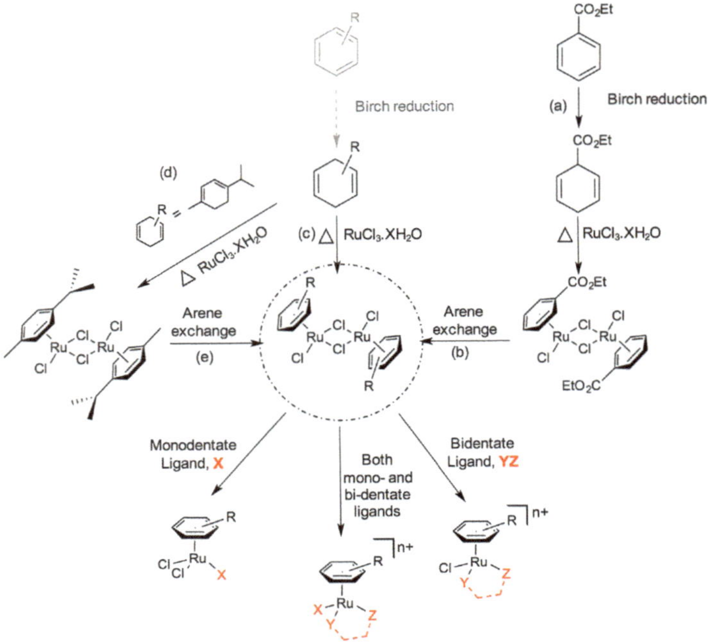

Figure 4.3 Synthetic strategies to access the Ru(II)-arene scaffold.

which can be readily prepared from commercially available cyclohexadienes (4), can undergo arene exchange (5) with highly substituted benzenes such as hexamethylbenzene and triisopropylbenzene.[11,12] $[(\eta^6\text{-Hexamethylbenzene})RuCl_2]_2$ and $[(\eta^6\text{-1,3,5-triisopropyl-benzene})RuCl_2]_2$ are typically prepared from $[(\eta^6\text{-cymene})RuCl_2]_2$ in good yields using this approach. After obtaining the desired $[(\eta^6\text{-arene})RuCl_2]_2$ precursors, the target Ru(II)-arene complexes could be synthesized by reaction with a stoichiometric amount of monodentate or bidentate chelating ligands.

A large number of studies focusing on drug targets and binding modes, drug transport, and cellular accumulation have led to deeper understanding on how the different components of an Ru(II)-arene complex contribute to its overall activity.[13–18] The arene and its substituent R, monodentate ligand **X**, and chelating ligand **YZ** can be harnessed to modulate the overall physical, chemical, and pharmacological properties of the Ru(II) complex (Fig. 4.2). The facially bound neutral arene ligand stabilizes the +2 oxidation state of the Ru metal center and modulates the hydrophobicity of the complexes depending on the degree of alkyl/aryl functionalization. This directly influences cellular accumulation of the complexes, presumably via a passive diffusion mechanism. The arene can also be functionalized with targeting groups, through direct conjugation or via reactive handles such as maleimide and benzaldehyde moieties, to achieve facilitated transport. The degree of π-donation from the arene can also influence the lability of the monodentate ligand **X**. In addition, the nature of **X** directly affects the rate of hydrolysis of the Ru(II)-arene complex. Consequently, this modulates the rate of Ru binding toward reactive biomolecules such as DNA and proteins, the presumed biological targets, and affects the cytotoxicity. The chelating ligands **YZ** are typically electron donating and stable to hydrolysis, and constitute an integral part of the intended Ru(II) complex design. They are often involved in direct interaction with the intended biological targets, e.g., ethylenediamine ligand (en) H-bonds with guanine bases on the DNA, and can greatly influence the activity and the mode-of-action. Lastly, the overall charge of the Ru(II) complexes can be tuned accordingly depending on the nature of **X** and **YZ** ligands. This intricate understanding of the Ru(II)-arene structure has led to the rational design of many potent drug candidates.

3. RU(II)-ARENE DRUG CANDIDATES UNDER INVESTIGATION

First-generation metallodrugs such as cisplatin and other Pt(II) complexes often induce cell death by targeting nuclear DNA, causing DNA damage by forming adducts and leading to p53 activation, cell cycle arrest, and apoptosis.[19] In response to this, cancer cells are able to adapt via several drug-resistance mechanisms such as increased DNA repair or p53 loss-of-function.[20] Therefore the identification of new drug candidates with varied cellular targets and modes of cell killing may be a way to circumvent drug resistance. One approach to the identification of new metallodrugs is a "top-down" approach, driven by the nature of the coordination chemistry or ligand composition. This approach necessitates the screening of biological activity, followed by exhaustive investigation, to establish the biological targets of the candidates. Many Ru(II)-arene complexes with novel structures and modes-of-action have been identified via this approach. Two of the most studied Ru(II)-arene complexes are RAPTA-C[21] by Dyson's group and RM175[22] by Sadler's group (Fig. 4.4).

RAPTA-C is a member of a family of water-soluble and air-stable Ru(II)-arene complexes containing the monodentate 1,3,5-triaza-7-phosphatricyclo-[3.3.1.1]decane (pta) ligand. Earlier generations of RAPTA complexes demonstrated low cytotoxicity in vitro, structure- and pH-dependent binding to DNA (the presumptive biological target), as well as

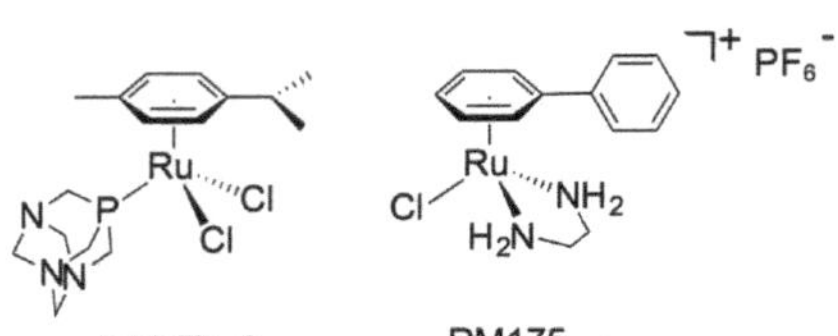

Figure 4.4 Structures of RAPTA-C and RM175.

strong antimetastatic activity in vivo while having limited effects on primary tumor.[23] Other studies have shown that RAPTA-C was also able to inhibit the growth of primary tumor.[24] It significantly reduced the growth of A2780 ovarian and LS174 colorectal carcinoma primary tumors in separate xenograph tumor models. In both models, the observed antiangiogenic effects were in agreement with previously established antiangiogenic activity of RAPTA-C.[25]

Results from detailed molecular mechanism studies suggest that RAPTA-C is multitargeting and exerts its anticancer activity through a combination of factors. RAPTA-C demonstrated pH-dependent DNA damage, and was selective toward the hypoxic environment around cancer cells.[26] It was known to inhibit enzyme targets that are implicated in cancer therapy, such as cathepsin-B and thioredoxin reductase.[27] One study involving inductively coupled plasma mass spectrometry and X-ray crystallographic analyses demonstrated that RAPTA-C bound preferentially to the histone proteins and not DNA in a nucleosome core particle.[28] Thus the activity of RAPTA-C might arise from its interactions with several cellular targets. Further studies showed that RAPTA-C induced p53-dependent G2/M phase cell cycle arrest and apoptosis.[29] Its high selectivity toward rapidly growing cancer, excellent in vivo activity against both primary tumor and metastases, and unique mode-of-action make RAPTA-C a promising drug candidate for clinical studies. This also spurs ongoing research into structural variation of the RAPTA scaffold in the hope of discovering more potent and selective anticancer agents.[30–35]

RM175 is another Ru(II)-arene complex that has undergone detailed in vitro and in vivo investigations. The metallodrug comprises a biphenyl ligand facially bound to Ru with en as the chelating **YZ** ligand. The chlorido ligand occupies the remaining coordination site, giving rise to an overall cationic charge, which is balanced with a hexafluorophosphate counteranion. RM175 demonstrates high cytotoxicity in vitro with IC_{50} in the same order of magnitude as cisplatin.[36] Structure–activity relationship studies showed that the rate of hydrolysis of RM175-type complexes is correlated to their cytotoxicity,[37] suggesting that the DNA-binding of RM175 is crucial to its activity. This is in agreement with studies that show that RM175 has a strong binding preference to DNA (specifically at guanine residues) but not other sulfur- or nitrogen-containing biomolecules.[28,38] Although RM175 shares the same molecular target as cisplatin, it forms a distinct type of adduct with DNA that is more resistant to DNA repair mechanisms. In addition, RM175 does not show any in vitro or in vivo cross-resistance in cisplatin-resistant A2780cis ovarian carcinoma, indicating a different mechanism of action as compared to cisplatin.[36] RM175 induces p53-dependent cell cycle arrest and apoptosis, which were absent in p53-null subtypes.[39] Similar to RAPTA-C,[36,40] RM175 also exhibits excellent antitumor in vivo and antimetastatic activity.

4. CYTOTOXIC RU(II)-ARENE COMPLEXES VIA CLASSICAL APPROACHES

RAPTA-C and RM175 remain the most rigorously investigated among the family of anticancer Ru(II)-arene complexes and their success propelled this area of research to the forefront (Fig. 4.5). Subsequently, a wider range of ligand types, particularly those based on N-coordinating ligands, was systematically studied as the basis for drug design. While most of these complexes were evaluated in vitro for their cytotoxicity against established cancer cell lines, particularly in the context of their cross-resistance with cisplatin or oxaliplatin, a small number of complexes were further characterized in vivo. By drawing parallels with cisplatin and RM175, most studies typically presume DNA to be the biological targets, which formed the basis of the investigations.

4.1 DNA Binding as the Primary Basis for Cytotoxicity

Ru(II)-arene complexes **1–4** are cytotoxic because of their ability to target and bind DNA.[41–44] Complex **1**, synthesized using diketonate and aminophosphine ligands, was found to be highly cytotoxic toward pancreatic and breast cancer cell lines and acted by inducing apoptosis.[41] The hydrolysis of the Ru-P bond was required before binding to DNA, and structural variation of the complex also showed that its activity was correlated to the rate of hydrolysis of the Ru-P bond. Complex **2** was isolated as the aquated form ($X = H_2O$), since its chlorido analog ($X = Cl$) was not active because of its slow hydrolysis rate.[42] It exhibited moderate activity against ovarian cancer cell lines but not breast cancer cell lines. In addition, it was not

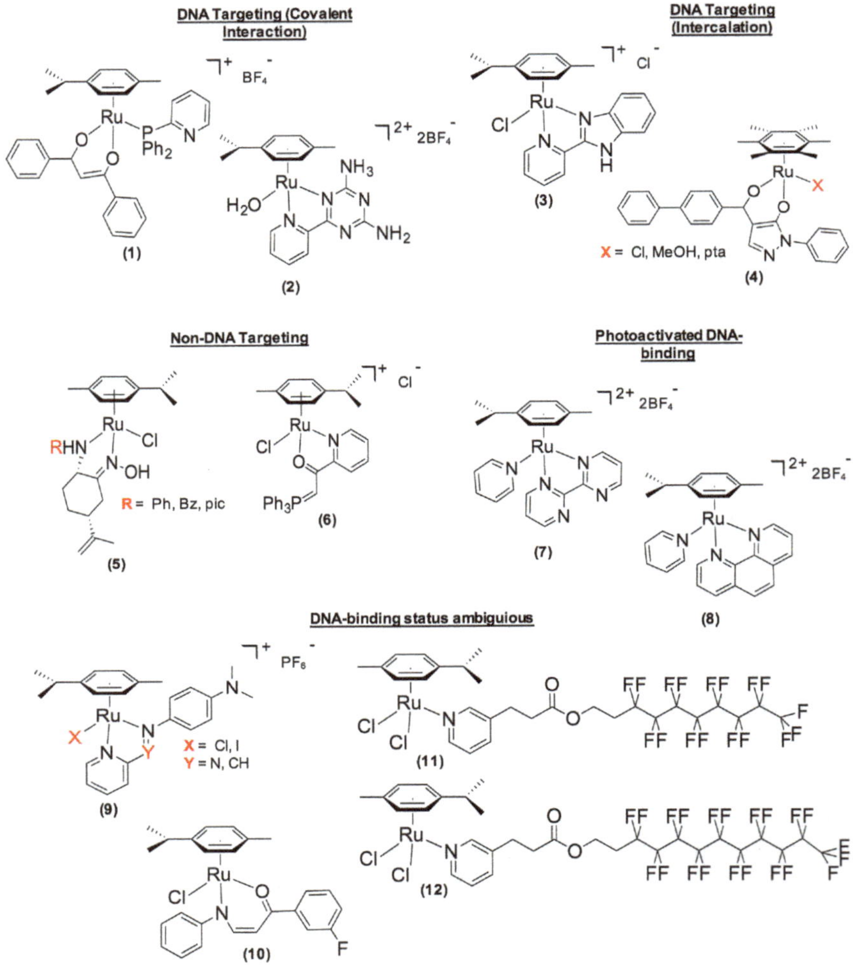

Figure 4.5 Examples of Ru(II)-arene complexes investigated.

cross-resistant with cisplatin and displayed improved activity against cisplatin-resistant phenotype compared to wild type. Complex **3** was designed to interact DNA through both covalent binding and intercalation.[43] It showed modest activities against breast cancer and ovarian cancer cells, with no cross-resistance in cisplatin-resistant ovarian cancer. The observed ability of complexes **2** and **3** to circumvent cross-resistance could be because of the formation of distinct types of Ru-DNA adducts that were more resistant to DNA repair mechanisms, similar to the case of RM175. However, further validation would be required to confirm this hypothesis. Complex

4 interacted with DNA mainly by intercalation and its activity was not dependent on its rate of aquation.[44] It showed modest efficacy in cervical, breast, kidney, and colorectal cancer cell lines and inhibited cell viability via p53-dependent cell-cycle arrest and apoptosis, consistent with the downstream effects of DNA damage. Other cytotoxic Ru(II)-arene complexes that targeted DNA have also been reported.[45–48]

4.2 Cytotoxicity Independent of DNA Binding

The activity of some other complexes did not depend on their ability to interact with DNA. Both complexes **5** and **6** were unable to interact with DNA and their activity was presumably because of interactions with other biological targets.[49,50] Complex **5** showed modest activity against cisplatin-resistant PC3 prostate cancer[49] and prevented cell migration in wound healing assay. Its inability to interact with plasmid DNA suggested that DNA was not the target, although its actual target was not elucidated. Complex **6** demonstrated excellent activity in a panel of cell lines, including MDA-MD-231 triple-negative breast cancer cells, and induced p53-independent apoptosis.[50] Moreover, it had excellent in vivo activity against MDA-DM-231 xenograft tumor. Intriguingly, it accumulated selectively in the tumor instead of the kidney or liver. This p53-independent activity and selective accumulation in tumor may show potential as an effective treatment with low systemic toxicity for resistant cancer with a loss of p53 functions. The design premise for complexes **7** and **8** was to bind DNA after the photodissociation of the pyridine ligands.[51,52] However, the complexes were markedly cytotoxic even prior to photoactivation, suggesting that other biological targets may be in play. Nevertheless, their structures serve as a starting point of the design of photoactivated Ru(II)-arene anticancer drugs.[53–55]

4.3 DNA Binding Status Not Known

The biological targets of other Ru(II)-arene complexes, e.g., **9–12**, remained unclear but they exhibited potent in vitro activity.[56–60] The mode-of-action of complex **9** was strongly structure dependent.[56] Although these derivatives were all active toward a panel of ovarian, lung, breast, and colorectal cancer cell lines through cell cycle arrest and apoptosis induction, only the iodido analogs demonstrated p53-independent activity and did not show cross-resistance with cisplatin or oxaliplatin. The chlorido analogs, in contrast, demonstrated p53-dependent activity and lower potency in cisplatin- and oxaliplatin-resistant cell lines, highlighting how minor structural changes can significantly affect the mode-of-action of these complexes. Complex **10** induced significant DNA fragmentation in vitro as evidenced by the comet assay but its direct interaction with DNA was not studied.[57] It had anticancer activity that was not cross-resistant with cisplatin, induced apoptotic cell death, and was a strong inhibitor of thioredoxin reductase. However, the direct link between the observed cytotoxicity and the inhibition of thioredoxin reductase remained to be established. Complexes **11** and **12** with perfluoroalkyl-modified ligands also exhibited structure-dependent properties. The cytotoxicity of complex **11** with a longer chain required heat activation; it was nontoxic at a physiological temperature of 37°C but highly toxic at 42°C against ovarian, breast, and lung cancer cell lines.[58] In contrast, complex **12** with a shorter chain did not exhibit a similar temperature dependency and it was equally cytotoxic at both temperatures. Further studies based on complex **11** could yield thermoactive drugs for targeted therapy. Complex **12** underwent further in vitro and in vivo investigations and was found to be more selectively toxic toward A2780 compared to MDA-MD-231.[59,60] It also prevented MDA-MD-231 cell migration in a wound healing assay and demonstrated more significant antiangiogenic activity than RAPTA-C. Its low toxicity toward healthy cells and potent antiangiogenic activity at nontoxic dose made it a promising candidate for further preclinical studies, alongside RAPTA-C.

Many more cytotoxic Ru(II)-arene complexes have been reported. They have been synthesized using a wide range of mono- and bidentate ligands that include bipyridine- and imidazole-based ligands,[61–63] bithiazole,[64] Schiff-base,[65] maleimide,[66] alkylphenylsulfide or sulfoxide,[67,68] mercaptobenzothiazole, β-ketoamine,[69,70] and others.[71–76]

5. NONCLASSICAL RU(II)-ARENE COMPLEXES BY RATIONAL DESIGN

5.1 Disrupting Cellular Redox Through Biocatalysis

One strategy of targeting cancer cells is to disrupt the intricate cellular redox balance within the cells. In a normal cell, reactive oxygen species (ROS) and reactive nitrogen species (RNS) are part of the signaling pathway of cellular processes and their levels are tightly modulated.[77] In contrast, cancer cells are constantly under oxidative stress because of the disruption of this regulation,[78] making them more susceptible to perturbation caused by redox-active therapeutic agents. This kink-in-the-chain can be exploited for selective targeting of cancer cells. In this aspect, metal complexes are advantageous over organic compounds because of their high redox mobility, particularly catalytic complexes that can sustain redox chemistry within a complex intracellular environment.[79,80] Ru(II)-arene complexes have been reported to catalyze reactions implicated in cellular redox modulation ($NAD^+ \leftrightarrow NADH$; GSH ↔ GS-SG).[81] However, only a limited number of such studies have been able to map the catalytic functions of the complexes to their anticancer activity. Complexes **13** and **14** are representative examples (Fig. 4.6).[82,83]

Complex **13** incurred an increase in oxidative stress by catalyzing the oxidation and deactivation of antioxidant GSH to GS-SG, which led to the increase in cellular ROS.[82] The quenching of ROS with *N*-acetylcysteine (NAC) resulted in an increase in cell viability, suggesting that the catalytic ability of complex **13** was responsible for its low micromolar IC_{50} in A2780 and A549 cancer cells. In contrast, complex **14** induced reductive stress in the presence of a hydride source via the reduction of NAD^+ to NADH.[83] The complex exhibited modest activity in A2780 cells but its cytotoxicity increased 14-fold in the presence of formate as the hydride source. Cotreatment of complex **14** and formate induced a corresponding reduction in cellular NAD^+/NADH ratio and enhancement in cytotoxicity, in a manner that is strongly dependent on the formate concentration. The levels of ROS also decreased, presumably because of the decrease in the oxidant NAD^+. In addition, complex **14** did not induce the classical hallmarks of intrinsic apoptosis. This suggests a distinct mechanism of action as compared to cisplatin or other Ru(II)-arene complexes discussed so far, which may have significant implications in the treatment of drug-resistant cancers.

5.2 Conjugation to Biologically Active or Targeting Ligands

Another approach commonly used for the design of Ru(II)-arene anticancer agents involves the conjugation of bioactive ligands to the complex scaffold. The design premise is to exploit the antioxidant, antiinflammatory, antimicrobial, antiviral, antifungal, or anticancer activity of the ligands to potentiate the activity of the Ru(II)-arene scaffold. Some of the commonly used bioactive ligands are the curcumin derivatives,[84–86] paullones,[87–93] flavonoids,[94–97] and thiosemicarbozones,[98–102] although a diverse range of other molecules have also been reported (Fig. 4.7).[103–111]

A class of Ru(II)-arene complexes based on paullones, which are potent cyclin-dependent kinase (CDK) inhibitors, were tested for anticancer activity.[87] These complexes, including complex **15**, demonstrated good in vitro

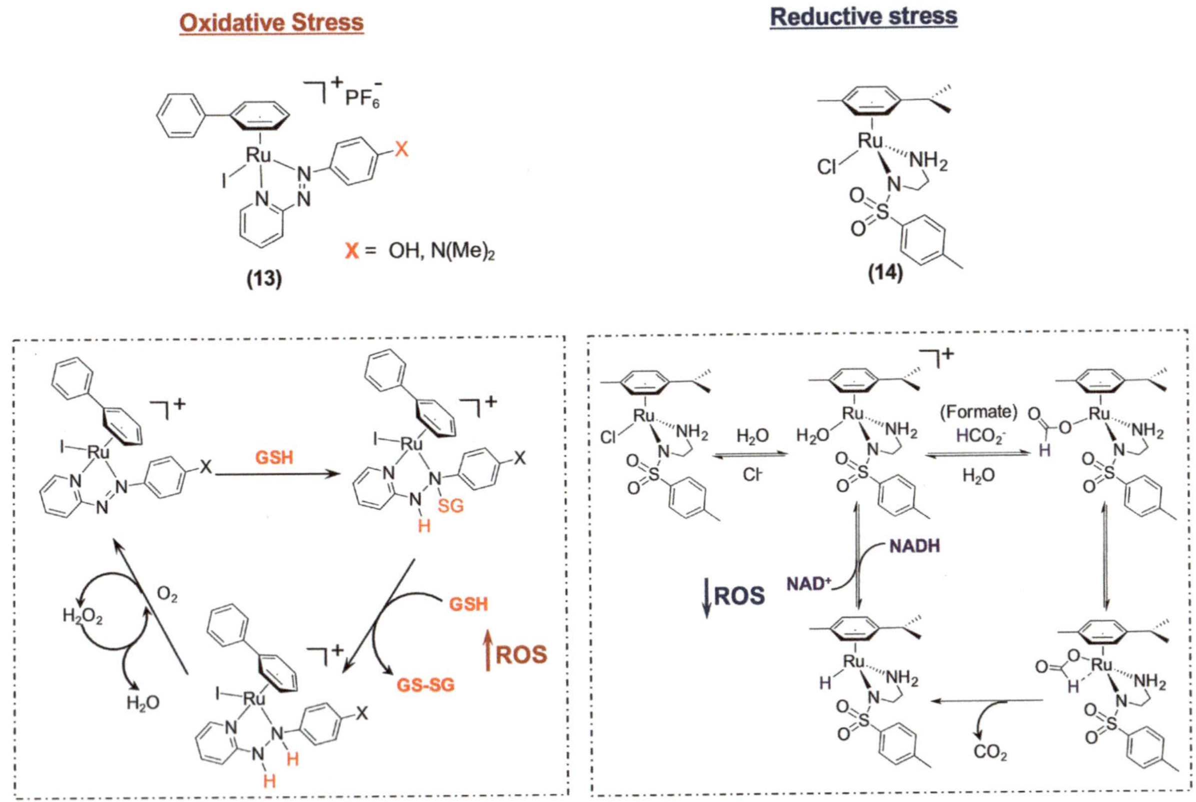

Figure 4.6 Examples of biocatalytic anticancer Ru(II)-arene complexes. *GSH*, glutathione; *GS-SG*, oxidised, disulphide-bridged glutathione; *NADH*, nicotinamide adenine dinucleotide; *NAD*$^+$, oxidised nicotinamide adenine dinucleotide; *ROS*, reactive oxygen species.

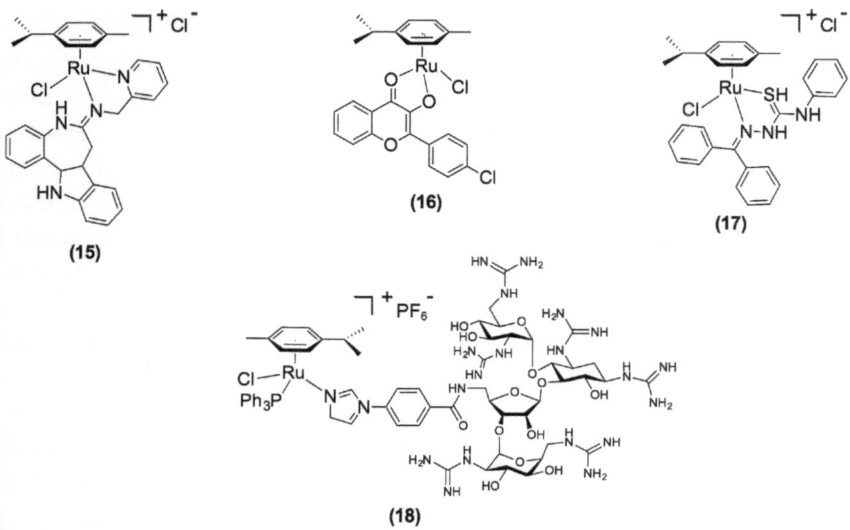

Figure 4.7 Examples of Ru(II)-arene complexes conjugated to biologically active ligands.

activity and inhibited DNA synthesis. In particular, complex **15** displayed a concentration-dependent dual mode-of-action; it induced G_0/G_1 phase cell cycle arrest at low concentration and apoptosis at high concentration. However, whether the observed effects were indeed caused by the inhibition CDKs remains unknown. Another class of Ru(II)-arene complexes, including complex **16**, contained the flavonoid 3-hydroxyflavone within its structure.[94,95] The Ru(II)-arene conjugates displayed similar activity in vitro compared with the unconjugated 3-hydroxyflavones but the complexes exhibited strong topoisomerase IIα inhibitory activity. By virtue of introducing the Ru(II)-arene moiety, the solubility was significantly enhanced while an additional site is available for DNA interaction. In another study, an Ru-conjugated thiosemicarbozone complex **17** demonstrated stronger anticancer activity and displayed a different mode of interaction with DNA compared to the unconjugated ligand.[98] More complex bioactive ligands had also been used in this broad approach. Guanidinoneomycin, an antibiotic with high affinity and selectivity toward RNA, was conjugated to the Ru(II)-arene scaffold and tested for anticancer activity.[103] The conjugated complex **18** was more than 20 times more cytotoxic than the unconjugated guanidinoneomycin. However, complex **18** showed no selectivity toward cancer cells, and had similar cytotoxicity in breast, prostate cancer cells, and primary lung fibroblast cells, which limits its potential applications.

5.3 Targeting Enzymatic Pathways Through Rational Design

Organometallic Ru(II)-arene inhibitors of therapeutic enzyme targets can be rationally designed specifically for biological targets implicated in the progression of diseases. The major advantage of this targeted approach would be a significant reduction in the physiological side effects that hamper nontargeted therapy. To achieve this, therapeutic enzyme inhibitors need to exhibit a high degree of selectivity and inhibitory activity toward their intended targets. There are several reasons why it would be useful to design an inhibitor scaffold around an Ru(II) metal center compared to a classical organic structure. First, the higher valency of the d^6 Ru(II) metal center prefers an octahedral configuration, which gives rise to greater structural diversity. Second, Ru(II) complexes react favorably toward soft binding partners, e.g., S- and P-based ligands as well as chelating motifs, which makes Ru(II) coordination chemistry predictable and ideal for structural design. Meggers pioneered the development of metalloinhibitors,[112] and showed that the Ru(II) scaffold can be used to design highly potent and selective protein kinase inhibitors.[113] There have been many attempts to design enzyme inhibitors using transition metal scaffolds.[112,114,115] Several examples of Ru(II)-arene complexes were designed to inhibit specific targets using the [(η^6-arene)Ru]-[Linker]-[Substrate] approach (Fig. 4.8). Typically, small molecular organic moieties that are antagonist or substrates to targeted enzymes are separated from the Ru(II)-arene scaffold via a linker. Enhancement in inhibitory activity is achieved through a careful

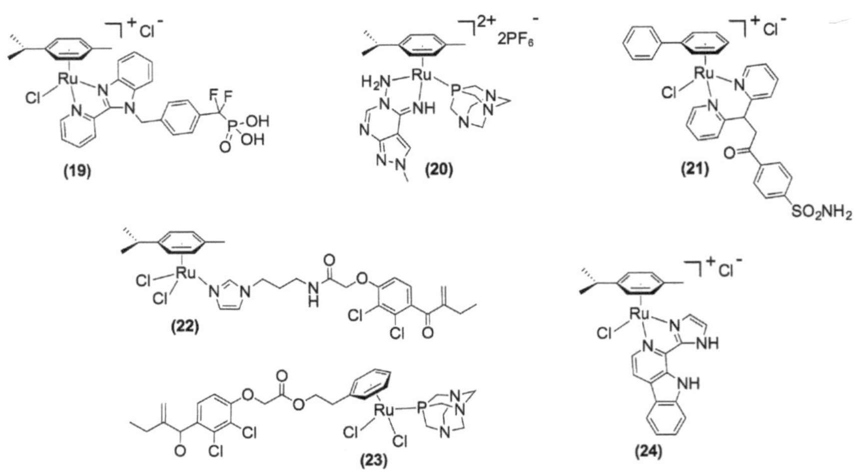

Figure 4.8 Examples of Ru(II)-arene enzyme inhibitors.

design of the whole construct, including positioning of the substrate and Ru(II)-arene fragments, as well as the nature of these components.

A series of Ru(II)-arene complexes was synthesized containing a phenyl-difluoromethylphosphonic acid (DFPA) moiety for the selective targeting of protein tyrosine phosphatase 1B (PTP1B).[116] DFPA acts as a substrate mimetic for PTP enzymes and competes for binding with phosphorylated tyrosine at the active site, thereby inhibiting PTP activity. The complex was designed such that different Ru(II)-arene fragments could be easily introduced by taking advantage of the chelating pyridyl-benzimidazole ligand that will enhance structural diversity. The main challenge in attempting to inhibit PTP1B was its high structural homology with T-cell PTP (TCPTP), an essential enzyme required for the function of immune T-cells. Among the class of candidates tested, complex **19** showed a 10-fold selectivity for PTP-1B over TCPTP. In contrast, the uncoordinated-DFPA ligand or the Ru(II)-arene fragment did not exhibit any selectivity. The proof-of-concept demonstrated that molecular recognition of PTPs can be achieved by targeting both the substrate site and its periphery using Ru(II)-arene combined with appropriate substrate groups. Similarly, complex **20** selectively bound to G-protein coupled receptor human adenosine receptor 3 (hA_3), which was implicated in a variety of diseases as an antagonist but the complex did not bind to the other subtypes hA_1, hA_{2A}, and hA_{2b}.[117] The design of the metalloantagonist was based on the pyrazolo-triazolo-pyrimidine scaffold, which previously displayed a nanomolar affinity to hA_3. By replacing the triazolo segment of the potent antagonist with the [(η^6-cymene)Ru(pta)] unit, a new class of selective Ru(II)-based antagonists were developed but with lower potency. Changing the pta ligand to a chloride increased the potency but decreased the selectivity toward hA_3, highlighting the effect of structural variations on activity and selectivity. Against other biomolecular targets, the addition of the Ru(II)-arene scaffold does not improve activity or selectivity of the inhibitors. For example, the Ru(II)-arene scaffold was attached to the phenylsulfonamide moiety via a linker for the inhibition of human carbonic anhydrase (hCA) II.[118] This attempt was met with limited success, as the uncoordinated ligand group was more active than the coordinated complex **21**. In other studies, ethacrynic acid was conjugated to the Ru(II)-arene scaffold at two different positions to develop metalloinhibitors of glutathione-*S*-transferase P1-1 (GSTP1-1),[119,120] either via an imidazole linker as in complex **22** or directly coupled to the arene motif as in complex **23**. In both cases, only marginal improvement to the inhibitory activity was observed compared to the unconjugated ligand. However, the Ru(II)-arene fragment was designed such that it would be strategically positioned toward the solvent accessible region when bound to the GSTP1-1 H-site. This allowed the Ru(II)-arene fragment to react covalently to Cys residues present at the protein dimer interface, changing the mode of inhibition from a reversible to an irreversible one. This led to five-fold improvement in cytotoxicity observed for complex **23** compared to the uncoordinated ligand.[119]

There have also been examples demonstrating that selective inhibition by rationally designed metallocomplexes translated to downstream modulation of specific biochemical pathways, which can be harnessed for therapeutic application. A Ru(II)-arene β-carboline complex **24** was synthesized as a potent inhibitor of CDK1, which was implicated in cell-cycle regulation.[121] The inhibition of CDK1 triggered both growth arrest at the G2/M phase and downregulation of CDK1 and cyclin B1. The complex was found to display a more potent and selective antitumor activity compared to the uncoordinated ligand. Other Ru(II)-arene complexes have been synthesized as enzyme inhibitors with varying levels of success.[122–126]

6. NONCLASSICAL RU(II)-ARENE COMPLEXES FROM CLASSICAL APPROACHES

Combinatorial coordination chemistry and multicomponent reactions can also be used in tandem with the Ru(II)-arene scaffold for library synthesis, screening, and identification of anticancer leads. Although they have seldom been applied to coordination-based metallodrug discovery, they can potentially accelerate the classical discovery process by increasing the probability of identifying active compounds. For instance, large libraries of structurally and functionally diverse Ru(II)-arene complexes were assembled to screen for cytotoxic activities distinct from classical anticancer drugs.

6.1 Coordination-Directed Three-Component Assembly

One of the few such examples use a coordination-directed three-component assembly (C3A) methodology to synthesize a library of 450 stable Ru(II)-arene Schiff-base (RAS) complexes for screening (Fig. 4.9a).[127] The reaction involves an Ru-arene precursor, a picolinaldehyde, and an aniline. Imine condensation between the picolinaldehyde and aniline forms a Schiff-base ligand that is suitable for chelation to the metal center. Without metal chelation, such ligands are usually unstable because of hydrolysis of the imine bond in an aqueous environment. The incorporation of the

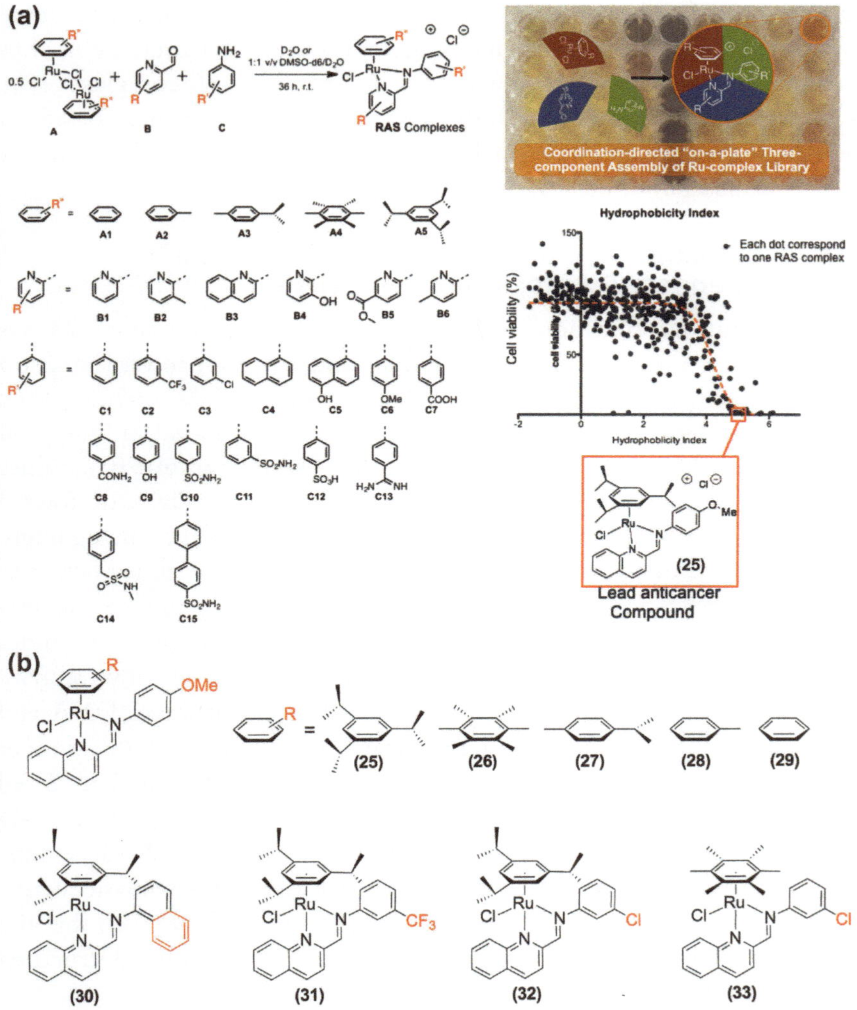

Figure 4.9 (a) Coordination-directed three-component assembly (C3A) library synthesis, screening, and lead discovery. (b) Compounds included in further structure–activity relationship (SAR) studies. *RAS*, Ru(II)-arene Schiff-base.

Ru-arene precursor yields an organoruthenium complex that is stabilized by the electron-donating Schiff-base chelate, pushing the assembly reaction favorably to completion. Reaction mixtures can be analyzed by mass spectrometry and NMR, and failed reaction assembly is indicated by the formation of insoluble aggregates. Even without purification C3A yields RAS complexes with 70–90% purity in aqueous solvents, which is ideal for screening. Furthermore, organic components with different functional groups are widely available from commercial sources and could be used in a combinatorial, one-pot fashion to give a library of RAS complexes (Fig. 4.9a) assembled on 96-well plates in a high-throughput manner. Based on an initial screening against a panel of different cell lines, several possible hits were obtained out of the 442 RAS complexes assembled by C3A.

6.2 Identification of Lead RAS Complexes With p53-Independent Cytotoxic Modes-of-Action

Following further validation of several initial hits, a lead compound **25** was uncovered. Complex **25** had good solubility and was stable toward aquation and reaction with other "N"- or "S"-bearing nucleophiles such as DNA model bases, deoxyguanosine monophosphate, and the abundant biological reductant, glutathione. The complex exhibited low micromolar IC_{50} values in a panel of cancer cell lines, and was found to be equally toxic toward both cisplatin-sensitive and -resistant ovarian cancer lineages. Intriguingly, complex **25** demonstrated p53-independent activity, with similar activities against p53 wild type and p53-null cell types, and only induced significant cell-cycle arrest and apoptosis at extreme concentrations (10 times IC_{50}) that did not correspond with its observed antiproliferative activity. These observations suggested that complex **25** did not target DNA and had a distinct mode of cell killing compared to cisplatin and many other Ru(II)-arene anticancer complexes. The identification of complex **25** led to more detailed SAR and mode-of-action studies of several structurally related compounds (Fig. 4.9b).[128] All nine RAS complexes **25–33** exhibited a p53-independent mode-of-action, regardless of their physiochemical properties such as stability and hydrophobicity. This suggested that the quinoline moiety, common to all nine complexes, was a required structural motif for the observed p53 independence.

6.3 Induction of Nonapoptotic Programmed Cell Death via ER Stress Pathways

A closer inspection of the properties of complexes **25** and **26** revealed that both induced a form of nonapoptotic cell death via the activation of endoplasmic reticulum (ER) stress pathways distinct for each compound (Fig. 4.10).[129] Both pathways and types of cell death induced also differed greatly from clinical drugs like cisplatin, which acted via apoptosis. Treatment of gastric cancer cells with complexes **25** and **26** failed to induce apoptotic biomarkers regardless of concentration or duration of treatment. In addition, the inhibition of proapoptotic

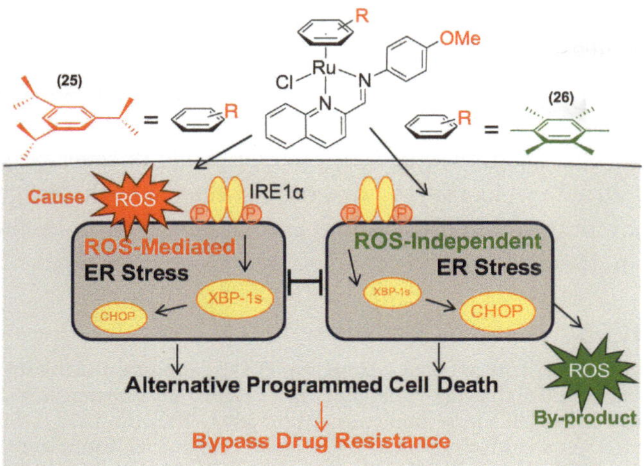

Figure 4.10 Ru(II)-arene complexes with unique modes-of-action identified using the three-component assembly (C3A) approach. *CHOP*, CCAAT-enhancer-binding protein homologous protein; *ER*, endoplasmic reticulum; *ROS*, reactive oxygen species; *XBP-1s*, spliced X-box binding protein 1.

caspases did not decrease the efficacy of both compounds. Detailed mode-of-action studies revealed that complex **25** acted via ROS-mediated ER stress while complex **26** induced cell death via an ER stress pathway independent of ROS production. This study further demonstrated how subtle structural changes could significantly perturb the modes-of-action of these Ru(II)-arene complexes, and that structural tuning could be a way to modulate not only their cytotoxicity but also the molecular pathways involved. It was further established that the nonapoptotic cell death induced by both complexes **25** and **26** could be exploited for the treatment of apoptosis-resistant cancer phenotypes, since they have greater activity in drug-resistant colorectal cancer TC7 compared to many other clinical drugs including etoposide and 5-fluorouracil. The discovery of strong anticancer candidates like complexes **25** and **26** is an affirmation of the efficiency and robustness of the C3A methodology in discovering drug candidates with a unique mode-of-action. The development of similar methodology in the future could potentially accelerate the drug discovery process.

7. CONCLUSION

In this chapter, we discussed the use of the Ru(II)-arene scaffold as a structural building block for the design of anticancer compounds. Their ease of preparation, unique chemical–structural properties and interaction with biomolecules make them useful structural scaffolds for anticancer drug design. The different approaches and strategies that have been employed in their design have led to the discovery of promising anticancer drug candidates, with modes-of-action distinct from those of the current clinical drugs. There is no doubt that the future studies could uncover potent cancer drugs with varied mechanisms of action and circumvent the drug resistance phenomenon observed in classical cancer treatment. This holds promise for cancer chemotherapy, especially for drug-resistant cancer types.

REFERENCES

1. Meggers, E.; Atilla-Gokcumen, G.; Bregman, H.; Maksimoska, J.; Mulcahy, S.; Pagano, N.; Williams, D. Exploring Chemical Space with Organometallics: Ruthenium Complexes as Protein Kinase Inhibitors. *Synlett* **2007,** *2007* (8), 1177–1189.
2. Taube, H. Rates and Mechanisms of Substitution in Inorganic Complexes in Solution. *Chem. Rev.* **1952,** *50* (1), 69–119.
3. Antonarakis, E. S.; Emadi, A. Ruthenium-Based Chemotherapeutics: Are They Ready for Prime Time? *Cancer Chemother. Pharmacol.* **2010,** *66,* 1–9.
4. Smith, G. S.; Therrien, B. Targeted and Multifunctional Arene Ruthenium Chemotherapeutics. *Dalton Trans.* **2011,** *40* (41), 10793–10800.
5. Rademaker-Lakhai, J. M.; van den Bongard, D.; Pluim, D.; Beijnen, J. H.; Schellens, J. H. M. Phase I and Pharmacological Study with Imidazolium- Trans-DMSO-imidazole-tetrachlororuthenate, a Novel Ruthenium Anticancer Agent. *Clin. Cancer Res.* **2004,** *10,* 3717–3727.
6. Hartinger, C. G.; Jakupec, M. A.; Zorbas-Seifried, S.; Groessl, M.; Egger, A.; Berger, W.; Zorbas, H.; Dyson, P. J.; Keppler, B. K. KP1019, a New Redox-active Anticancer Agent – Preclinical Development and Results of a Clinical Phase I Study in Tumor Patients. *Chem. Biodiversity* **2008,** *5,* 2140–2155.
7. Bozec, H. L.; Touchard, D.; Dixneuf, P. H. Organometallic Chemistry of Arene Ruthenium and Osmium Complexes. **1989,** *29,* 163–247.
8. Geldbach, T. J.; Brown, M. R. H.; Scopelliti, R.; Dyson, P. J. Ruthenium–benzo-crownether Complexes: Synthesis, Structures, Catalysis and Immobilisation in Ionic Liquids. *J. Organomet. Chem.* **2005,** *690,* 5055–5065.
9. Zelonka, R. A.; Baird, M. C. Benzene Complexes of Ruthenium(II). *Can. J. Chem.* **1972,** *50,* 3063–3072.
10. Bennett, M. A.; Smith, A. K. Arene Ruthenium(II) Complexes Formed by Dehydrogenation of Cyclo-hexadienes with Ruthenium(III) Trichloride. *J. Chem. Soc. Dalton Trans.* **1974,** 223–241.
11. Bennett, M. A.; Huang, T. N.; Matheson, T. W.; Smith, A. K.; Ittel, S.; Nickerson, W. (η6-Hexamethylbenzene)Ruthenium Complexes. *Inorg. Synth.* **1982,** *21,* 74–78.
12. Hull, J. W. J.; Gladfelter, W. L. η4-Bonding in (Arene)ruthenium Complexes of Octamethylnapthalene. *Organometallics* **1982,** *3,* 605–613.
13. Peacock, A. F.; Sadler, P. J. Medicinal Organometallic Chemistry: Designing Metal Arene Complexes as Anticancer Agents. *Chem. Asian J.* **2008,** *11,* 1890–1899.
14. Pizarro, A. M.; Habtemariam, A.; Sadler, P. J. Activation Mechanisms for Organometallic Anticancer Complexes. *Top. Organomet. Chem.* **2010,** *32,* 21–56.
15. Patra, M.; Joshi, T.; Pierroz, V.; Ingram, K.; Kaiser, M.; Ferrari, S.; Spingler, B.; Keiser, J.; Gasser, G. DMSO-mediated Ligand Dissociation: Renaissance for Biological Activity of N-heterocyclic-[Ru(η6-arene)Cl2] Drug Candidates. *Chem. Eur. J.* **2013,** *19,* 14768–14772.

16. Serli, B.; Zangrando, E.; Gianferrara, T.; Scolaro, C.; Dyson, P. J.; Bergamo, A.; Alessio, E. Is the Aromatic Fragment of Piano-stool Ruthenium Compounds an Essential Feature for Anticancer Activity? The Development of New Ru(II)-[9]aneS_3 Analogues. *Eur. J. Inorg. Chem.* **2005,** *2005,* 3423–3434.
17. Fernandez, R.; Melchart, M.; Habtemariam, A.; Parsons, S.; Sadler, P. J. Use of Chelating Ligands to Tune the Reactive Site of Half-sandwich Ruthenium(II)-arene Anticancer Complexes. *Chem. Eur. J.* **2004,** *10,* 5173–5179.
18. Habtemariam, A.; Melchart, M.; Fernandez, R.; Parsons, S.; Oswald, I. D.; Parkin, A.; Fabbiani, F. P.; Davidson, J. E.; Dawson, A.; Aird, R. E.; Jodrell, D. I.; Sadler, P. J. Structure-activity Relationships for Cytotoxic Ruthenium(II) Arene Complexes Containing N,N-, N,O-, and O,O-Chelating Ligands. *J. Med. Chem.* **2006,** *49,* 6858–6868.
19. Siddik, Z. H. Cisplatin: Mode of Cytotoxic Action and Molecular Basis of Resistance. *Oncogene* **2003,** *22,* 7265–7279.
20. Brabec, V.; Kasparkova, J. Molecular Aspects of Resistance to Antitumor Platinum Drugs. *Drug Resist. Updates* **2002,** *5,* 147–161.
21. Suss-Fink, G. Arene Ruthenium Complexes as Anticancer Agents. *Dalton Trans.* **2010,** *39,* 1673–1688.
22. Yan, Y. K.; Melchart, M.; Habtemariam, A.; Sadler, P. J. Organometallic Chemistry, Biology and Medicine: Ruthenium Arene Anticancer Complexes. *Chem. Commun.* **2005,** *38,* 4764–4776.
23. Scolaro, C.; Bergamo, A.; Brescacin, L.; Delfino, R.; Cocchietto, M.; Laurenczy, G.; Geldbach, T. J.; Sava, G.; Dyson, P. J. In Vitro and In Vivo Evaluation of Ruthenium(II)-Arene PTA Complexes. *J. Med. Chem.* **2005,** *48,* 4161–4171.
24. Weiss, A.; Berndsen, R. H.; Dubois, M.; Müller, C.; Schibli, R.; Griffioen, A. W.; Dyson, P. J.; Nowak-Sliwinska, P. In Vivo Anti-tumor Activity of the Organometallic Ruthenium(II)-Arene Complex [Ru(η^6-*p*-cymene)- Cl_2(pta)] (RAPTA-c) in Human Ovarian and Colorectal Carcinomas. *Chem. Sci.* **2014,** *5,* 4742–4748.
25. Nowak-Sliwinska, P.; van Beijnum, J. R.; Casini, A.; Nazarov, A. A.; Wagnieres, G.; van den Bergh, H.; Dyson, P. J.; Griffioen, A. W. Organometallic Ruthenium(II) Arene Compounds with Antiangiogenic Activity. *J. Med. Chem.* **2011,** *54,* 3895–3902.
26. Allardyce, C. S.; Dyson, P. J.; Ellis, D. J.; Heath, S. L. [Ru(η^6-*p*-cymene)Cl_2(pta)] (Pta = 1,3,5-triaza-7-phosphatricyclo[3.3.1.1]decane): A Water Soluble Compound that Exhibits PH Dependent DNA Binding Providing Selectivity for Diseased Cells. *Chem. Commun.* **2001,** *15,* 1396–1397.
27. Casini, A.; Gabbiani, C.; Sorrentino, F.; Rigobello, M. P.; Bindoli, A.; Geldbach, T. J.; Marrone, A.; Re, N.; Hartinger, C. G.; Dyson, P. J.; Messori, L. Emerging Protein Targets for Anticancer Metallodrugs: Inhibition of Thioredoxin Reductase and Cathepsin B by Antitumor Ruthenium(II)-Arene Compounds. *J. Med. Chem.* **2008,** *51,* 6773–6781.
28. Adhireksan, Z.; Davey, G. E.; Campomanes, P.; Groessl, M.; Clavel, C. M.; Yu, H.; Nazarov, A. A.; Yeo, C. H.; Ang, W. H.; Droge, P.; Rothlisberger, U.; Dyson, P. J.; Davey, C. A. Ligand Substitutions Between Ruthenium-Cymene Compounds Can Control Protein versus DNA Targeting and Anticancer Activity. *Nat. Commun.* **2014,** *5,* 3462.
29. Chatterjee, S.; Kundu, S.; Bhattacharyya, A.; Hartinger, C. G.; Dyson, P. J. The Ruthenium(II)-Arene Compound RAPTA-c Induces Apoptosis in EAC Cells Through Mitochondrial and p53-JNK Pathways. *J. Biol. Inorg. Chem.* **2008,** *13,* 1149–1155.
30. Tan, Y. Q.; Dyson, P. J.; Ang, W. H. Acetal-Functionalized RAPTA Complexes for Conjugation and Labeling. *Organometallics* **2011,** *30,* 5965–5971.
31. Renfrew, A. K.; Juillerat-Jeanneret, L.; Dyson, P. J. Adding Diversity to Ruthenium(II)–Arene Anticancer (RAPTA) Compounds via Click Chemistry: The Influence of Hydrophobic Chains. *J. Organomet. Chem.* **2011,** *696,* 772–779.
32. Murray, B. S.; Menin, L.; Scopelliti, R.; Dyson, P. J. Conformational Control of Anticancer Activity: The Application of Arene-linked Dinuclear Ruthenium(II) Organometallics. *Chem. Sci.* **2014,** *5,* 2536–2545.

33. Pettinari, R.; Marchetti, F.; Condello, F.; Pettinari, C.; Lupidi, G.; Scopelliti, R.; Mukhopadhyay, S.; Riedel, T.; Dyson, P. J. Ruthenium(II)–Arene RAPTA Type Complexes Containing Curcumin and Bisdemethoxycurcumin Display Potent and Selective Anticancer Activity. *Organometallics* **2014,** *33,* 3709–3715.
34. Sersen, S.; Kljun, J.; Kryeziu, K.; Panchuk, R.; Alte, B.; Korner, W.; Heffeter, P.; Berger, W.; Turel, I. Structure-Related Mode-of-Action Differences of Anticancer Organoruthenium Complexes with β-Diketonates. *J. Med. Chem.* **2015,** *58,* 3984–3996.
35. Scolaro, C.; Chaplin, A. B.; Hartinger, C. G.; Bergamo, A.; Cocchietto, M.; Keppler, B. K.; Sava, G.; Dyson, P. J. Tuning the Hydrophobicity of Ruthenium(II)-Arene (RAPTA) Drugs to Modify Uptake, Biomolecular Interactions and Efficacy. *Dalton Trans.* **2007,** *43,* 5065–5072.
36. Aird, R. E.; Cummings, J.; Ritchie, A. A.; Muir, M.; Morris, R. E.; Chen, H.; Sadler, P. J.; Jodrell, D. I. In Vitro and In Vivo and Cross Resistance Profile of Novel Ruthenium (II) Organometallic Arene Complexes in Human Ovarian Cancer. *Br. J. Cancer* **2002,** *86,* 1652–1657.
37. Wang, F.; Habtemariam, A.; van der Geer, E. P.; Fernandez, R.; Melchart, M.; Deeth, R. J.; Aird, R.; Guichard, S.; Fabbiani, F. P.; Lozano-Casal, P.; Oswald, I. D.; Jodrell, D. I.; Parsons, S.; Sadler, P. J. Controlling Ligand Substitution Reactions of Organometallic Complexes: Tuning Cancer Cell Cytotoxicity. *Proc. Natl. Acad. Sci. USA* **2005,** *102* (51), 18269–18274.
38. Chen, H.; Parkinson, J. A.; Morris, R. E.; Sadler, P. J. Highly Selective Binding of Organometallic Ruthenium Ethylenediamine Complexes to Nucleic Acids- Novel Recognition Mechanisms. *J. Am. Chem. Soc.* **2003,** *125,* 173–186.
39. Hayward, R. L.; Schornagel, Q. C.; Tente, R.; Macpherson, J. S.; Aird, R. E.; Guichard, S.; Habtemariam, A.; Sadler, P.; Jodrell, D. I. Investigation of the Role of Bax, p21/Waf1 and p53 as Determinants of Cellular Responses in HCT116 Colorectal Cancer Cells Exposed to the Novel Cytotoxic Ruthenium(II) Organometallic Agent, RM175. *Cancer Chemother. Pharmacol.* **2005,** *55,* 577–583.
40. Bergamo, A.; Masi, A.; Peacock, A. F.; Habtemariam, A.; Sadler, P. J.; Sava, G. In Vivo Tumour and Metastasis Reduction and In Vitro Effects on Invasion Assays of the Ruthenium RM175 and Osmium AFAP51 Organometallics in the Mammary Cancer Model. *J. Inorg. Biochem.* **2010,** *104,* 79–86.
41. Aliende, C.; Perez-Manrique, M.; Jalon, F. A.; Manzano, B. R.; Rodriguez, A. M.; Cuevas, J. V.; Espino, G.; Martinez, M. A.; Massaguèr, A.; Gonzalez-Bartulos, M.; de Llorens, R.; Moreno, V. Preparation of New Half Sandwich Ruthenium Arene Complexes with Aminophosphines as Potential Chemotherapeutics. *J. Inorg. Biochem.* **2012,** *117,* 171–188.
42. Busto, N.; Valladolid, J.; Martinez-Alonso, M.; Lozano, H. J.; Jalon, F. A.; Manzano, B. R.; Rodriguez, A. M.; Carrion, M. C.; Biver, T.; Leal, J. M.; Espino, G.; Garcia, B. Anticancer Activity and DNA Binding of a Bifunctional Ru(II) Arene Aqua-Complex with the 2,4-diamino-6-(2-pyridyl)-1,3,5-triazine Ligand. *Inorg. Chem.* **2013,** *52,* 9962–9974.
43. Martinez-Alonso, M.; Busto, N.; Jalon, F. A.; Manzano, B. R.; Leal, J. M.; Rodriguez, A. M.; Garcia, B.; Espino, G. Derivation of Structure-Activity Relationships from the Anticancer Properties of Ruthenium(II) Arene Complexes with 2-aryldiazole Ligands. *Inorg. Chem.* **2014,** *53,* 11274–11288.
44. Pettinari, R.; Pettinari, C.; Marchetti, F.; Skelton, B. W.; White, A. H.; Bonfili, L.; Cuccioloni, M.; Mozzicafreddo, M.; Cecarini, V.; Angeletti, M.; Nabissi, M.; Eleuteri, A. M. Arene-Ruthenium(II) Acylpyrazolonato Complexes: Apoptosis-Promoting Effects on Human Cancer Cells. *J. Med. Chem.* **2014,** *57,* 4532–4542.
45. Wu, Q.; Fan, C.; Chen, T.; Liu, C.; Mei, W.; Chen, S.; Wang, B.; Chen, Y.; Zheng, W. Microwave-assisted Synthesis of Arene Ruthenium(II) Complexes that Induce S-phase Arrest in Cancer Cells by DNA Damage-Mediated p53 Phosphorylation. *Eur. J. Med. Chem.* **2013,** *63,* 57–63.

46. Betanzos-Lara, S.; Novakova, O.; Deeth, R. J.; Pizarro, A. M.; Clarkson, G. J.; Liskova, B.; Brabec, V.; Sadler, P. J.; Habtemariam, A. Bipyrimidine Ruthenium(II) Arene Complexes: Structure, Reactivity and Cytotoxicity. *J. Biol. Inorg. Chem.* **2012,** *17,* 1033–1051.
47. Gligorijevic, N.; Arandelovic, S.; Filipovic, L.; Jakovljevic, K.; Jankovic, R.; Grguric-Sipka, S.; Ivanovic, I.; Radulovic, S.; Tesic, Z. Picolinate Ruthenium(II)-Arene Complex with In Vitro Antiproliferative and Antimetastatic Properties: Comparison to a Series of Ruthenium(II)-Arene Complexes with Similar Structure. *J. Inorg. Biochem.* **2012,** *108,* 53–61.
48. Caruso, F.; Monti, E.; Matthews, J.; Rossi, M.; Gariboldi, M. B.; Pettinari, C.; Pettinari, R.; Marchetti, F. Synthesis, Characterization, and Antitumor Activity of Water-Soluble (Arene) ruthenium(II) Derivatives of 1,3-dimethyl-4-acylpyrazolon-5-ato Ligands. First Example of Ru(arene)(ligand) Antitumor Species Involving Simultaneous Ru-N7(guanine) Bonding and Ligand Intercalation to DNA. *Inorg. Chem.* **2014,** *53,* 3668–3677.
49. Benabdelouahab, Y.; Muñoz-Moreno, L.; Frik, M.; de la Cueva-Alique, I.; El Amrani, M. A.; Contel, M.; Bajo, A. M.; Cuenca, T.; Royo, E. Hydrogen Bonding and Anticancer Properties of Water-Soluble Chiral *p*-Cymene Ru(II) Compounds with Amino-Oxime Ligands. *Eur. J. Inorg. Chem.* **2015,** *2015,* 2295–2307.
50. Frik, M.; Martinez, A.; Elie, B. T.; Gonzalo, O.; Ramirez de Mingo, D.; Sanau, M.; Sanchez-Delgado, R.; Sadhukha, T.; Prabha, S.; Ramos, J. W.; Marzo, I.; Contel, M. In Vitro and In Vivo Evaluation of Water-soluble Iminophosphorane Ruthenium(II) Compounds. A Potential Chemotherapeutic Agent for Triple Negative Breast Cancer. *J. Med. Chem.* **2014,** *57,* 9995–10012.
51. Betanzos-Lara, S.; Salassa, L.; Habtemariam, A.; Sadler, P. J. Photocontrolled Nucleobase Binding to an Organometallic Ru(II) Arene Complex. *Chem. Commun.* **2009,** *43,* 6622–6624.
52. Betanzos-Lara, S.; Salassa, L.; Habtemariam, A.; Novakova, O.; Pizarro, A. M.; Clarkson, G. J.; Liskova, B.; Brabec, V.; Sadler, P. J. Photoactivatable Organometallic Pyridyl Ruthenium(II) Arene Complexes. *Organometallics* **2012** *31,* 3466–3479.
53. Zhou, Q. X.; Lei, W. H.; Hou, Y. J.; Chen, Y. J.; Li, C.; Zhang, B. W.; Wang, X. S. Bodipy-Modified Ru(II) Arene Complex-A New Ligand Dissociation Mechanism and a Novel Strategy to Red Shift the Photoactivation Wavelength of Anticancer Metallodrugs. *Dalton Trans.* **2013,** *42,* 2786–2791.
54. Chen, Y.; Lei, W.; Jiang, G.; Hou, Y.; Li, C.; Zhang, B.; Zhou, Q.; Wang, X. Fusion of Photodynamic Therapy and Photoactivated Chemotherapy: A Novel Ru(II) Arene Complex with Dual Activities of Photobinding and Photocleavage Toward DNA. *Dalton Trans.* **2014,** *43,* 15375–15384.
55. Barragan, F.; Lopez-Senin, P.; Salassa, L.; Betanzos-Lara, S.; Habtemariam, A.; Moreno, V.; Sadler, P. J.; Marchan, V. Photocontrolled DNA Binding of a Receptor-Targeted Organometallic Ruthenium(II) Complex. *J. Am. Chem. Soc.* **2011,** *133,* 14098–14108.
56. Romero-Canelon, I.; Salassa, L.; Sadler, P. J. The Contrasting Activity of Iodido versus Chlorido Ruthenium and Osmium Arene Azo- and Imino-Pyridine Anticancer Complexes: Control of Cell Selectivity, Cross-Resistance, p53 Dependence, and Apoptosis Pathway. *J. Med. Chem.* **2013,** *56,* 1291–1300.
57. Lord, R. M.; Hebden, A. J.; Pask, C. M.; Henderson, I. R.; Allison, S. J.; Shepherd, S. L.; Phillips, R. M.; McGowan, P. C. Hypoxia-Sensitive Metal β-Ketoiminato Complexes Showing Induced Single-Strand DNA Breaks and Cancer Cell Death by Apoptosis. *J. Med. Chem.* **2015,** *58,* 4940–4953.
58. Clavel, C. M.; Păunescu, E.; Nowak-Sliwinska, P.; Dyson, P. J. Thermoresponsive Organometallic Arene Ruthenium Complexes for Tumour Targeting. *Chem. Sci.* **2014,** *5,* 1097–1101.
59. Clavel, C. M.; Paunescu, E.; Nowak-Sliwinska, P.; Griffioen, A. W.; Scopelliti, R.; Dyson, P. J. Discovery of a Highly Tumor-Selective Organometallic Ruthenium(II)-Arene Complex. *J. Med. Chem.* **2014,** *57,* 3546–3558.

60. Clavel, C. M.; Paunescu, E.; Nowak-Sliwinska, P.; Griffioen, A. W.; Scopelliti, R.; Dyson, P. J. Modulating the Anticancer Activity of Ruthenium(II)-Arene Complexes. *J. Med. Chem.* **2015,** *58,* 3356–3365.
61. Purkait, K.; Karmakar, S.; Bhattacharyya, S.; Chatterjee, S.; Dey, S. K.; Mukherjee, A. A Hypoxia Efficient Imidazole-Based Ru(II) Arene Anticancer Agent Resistant to Deactivation by Glutathione. *Dalton Trans.* **2015,** *44,* 5969–5973.
62. Wu, Q.; He, J.; Mei, W.; Zhang, Z.; Wu, X.; Sun, F. Arene Ruthenium(II) Complex, a Potent Inhibitor Against Proliferation, Migration and Invasion of Breast Cancer Cells, Reduces Stress Fibers, Focal Adhesions and Invadopodia. *Metallomics* **2014,** *6,* 2204–2212.
63. Tsolis, T.; Manos, M. J.; Karkabounas, S.; Zelovitis, I.; Garoufis, A. Synthesis, X-ray Structure Determination, Cytotoxicity and Interactions with 9-methylguanine, of Ruthenium(II) η6-Arene Complexes. *J. Organomet. Chem.* **2014,** *768,* 1–9.
64. Kalidasan, M.; Forbes, S.; Mozharivskyj, Y.; Ahmadi, M.; Ahmadihosseini, Z.; Phillips, R. M.; Kollipara, M. R. Mononuclear Half-Sandwich Cyclic-π-Perimeter Platinum Group Metal Complexes Having Bithiazole Ligands: Synthesis, Molecular and Anti-Cancer Studies. *Inorg. Chim. Acta* **2014,** *421,* 349–358.
65. Palepu, N. R.; Nongbri, S. L.; Premkumar, J. R.; Verma, A. K.; Bhattacharjee, K.; Joshi, S. R.; Forbes, S.; Mozharivskyj, Y.; Thounaojam, R.; Aguan, K.; Kollipara, M. R. Synthesis and Evaluation of New Salicylaldehyde-2-picolinylhydrazone Schiff Base Compounds of Ru(II), Rh(III) and Ir(III) as In Vitro Antitumor, Antibacterial and Fluorescence Imaging Agents. *J. Biol. Inorg. Chem.* **2015,** *20,* 619–638.
66. Moon, S.; Hanif, M.; Kubanik, M.; Holtkamp, H.; Söhnel, T.; Jamieson, S. M. F.; Hartinger, C. G. Organoruthenium and Osmium Anticancer Complexes Bearing a Maleimide Functional Group: Reactivity to Cysteine, Stability, and Cytotoxicity. *ChemPlusChem* **2015,** *80,* 231–236.
67. Ludwig, G.; Kaluderovic, G. N.; Ruffer, T.; Bette, M.; Korb, M.; Block, M.; Paschke, R.; Lang, H.; Steinborn, D. Cationic Arene Ruthenium(II) Complexes with Chelating P-Functionalized Alkyl Phenyl Sulfide and Sulfoxide Ligands as Potent Anticancer Agents. *Dalton Trans.* **2013,** *42,* 3771–3774.
68. Ludwig, G.; Kaluderovic, G. N.; Bette, M.; Block, M.; Paschke, R.; Steinborn, D. Highly Active Neutral Ruthenium(II) Arene Complexes: Synthesis, Characterization, and Investigation of Their Anticancer Properties. *J. Inorg. Biochem.* **2012,** *113,* 77–82.
69. Pettinari, R.; Pettinari, C.; Marchetti, F.; Clavel, C. M.; Scopelliti, R.; Dyson, P. J. Cytotoxicity of Ruthenium–Arene Complexes Containing β-Ketoamine Ligands. *Organometallics* **2013,** *32,* 309–316.
70. Pettinari, R.; Marchetti, F.; Pettinari, C.; Petrini, A.; Scopelliti, R.; Clavel, C. M.; Dyson, P. J. Synthesis, Structure, and Antiproliferative Activity of Ruthenium(II) Arene Complexes with N,O-chelating Pyrazolone-Based β-Ketoamine Ligands. *Inorg. Chem.* **2014,** *53* (24), 13105–13111.
71. Sommer, M. G.; Kureljak, P.; Urankar, D.; Schweinfurth, D.; Stojanovic, N.; Bubrin, M.; Gazvoda, M.; Osmak, M.; Sarkar, B.; Kosmrlj, J. Combining [Arene-Ru] with Azocarboxamide to Generate a Complex with Cytotoxic Properties. *Chem. Eur. J.* **2014,** *20,* 17296–17299.
72. Hanif, M.; Meier, S. M.; Kandioller, W.; Bytzek, A.; Hejl, M.; Hartinger, C. G.; Nazarov, A. A.; Arion, V. B.; Jakupec, M. A.; Dyson, P. J.; Keppler, B. K. From Hydrolytically Labile to Hydrolytically Stable Ru(II)-Arene Anticancer Complexes with Carbohydrate-Derived Co-Ligands. *J. Inorg. Biochem.* **2011,** *105,* 224–231.
73. Meier, S. M.; Hanif, M.; Adhireksan, Z.; Pichler, V.; Novak, M.; Jirkovsky, E.; Jakupec, M. A.; Arion, V. B.; Davey, C. A.; Keppler, B. K.; Hartinger, C. G. Novel Metal(II) Arene 2-pyridinecarbothioamides: A Rationale to Orally Active Organometallic Anticancer Agents. *Chem. Sci.* **2013,** *4,* 1837–1846.

74. Stepanenko, I. N.; Novak, M. S.; Muhlgassner, G.; Roller, A.; Hejl, M.; Arion, V. B.; Jakupec, M. A.; Keppler, B. K. Organometallic 3-(1H-benzimidazol-2-yl)-1H-pyrazolo[3,4-b] pyridines as Potential Anticancer Agents. *Inorg. Chem.* **2011,** *50,* 11715–11728.
75. Mitra, R.; Samuelson, A. G. Substitution-Modulated Anticancer Activity of Half-Sandwich Ruthenium(II) Complexes with Heterocyclic Ancillary Ligands. *Eur. J. Inorg. Chem.* **2014,** *2014,* 3536–3546.
76. Egbewande, F. A.; Paul, L. E. H.; Therrien, B.; Furrer, J. Synthesis, Characterization and Cytotoxicity of (η^6-*p*-cymene)ruthenium(II) Complexes of α-Amino Acids. *Eur. J. Inorg. Chem.* **2014,** 1174–1184.
77. Trachootham, D.; Lu, W.; Ogasawara, M. A.; Nilsa, R. D.; Huang, P. Redox Regulation of Cell Survival. *Antioxid. Redox Signal.* **2008,** *10,* 1343–1374.
78. Jungwirth, U.; Kowol, C. R.; Keppler, B. K.; Hartinger, C. G.; Berger, W.; Heffeter, P. Anticancer Activity of Metal Complexes: Involvement of Redox Processes. *Antioxid. Redox Signal.* **2011,** *15,* 1085–1127.
79. Romero-Canelon, I.; Sadler, P. J. Next-Generation Metal Anticancer Complexes: Multitargeting via Redox Modulation. *Inorg. Chem.* **2013,** *52,* 12276–12291.
80. Trachootham, D.; Alexandre, J.; Huang, P. Targeting Cancer Cells by ROS-Mediated Mechanisms: A Radical Therapeutic Approach? *Nat. Rev. Drug Discov.* **2009,** *8,* 579–591.
81. Soldevila-Barreda, J. J.; Sadler, P. J. Approaches to the Design of Catalytic Metallodrugs. *Curr. Opin. Chem. Biol.* **2015,** *25,* 172–183.
82. Dougan, S. J.; Habtemariam, A.; McHale, S. E.; Parsons, S.; Sadler, P. J. Catalytic Organometallic Anticancer Complexes. *Proc. Natl. Acad. Sci. USA* **2008,** *105,* 11628–11633.
83. Soldevila-Barreda, J. J.; Romero-Canelon, I.; Habtemariam, A.; Sadler, P. J. Transfer Hydrogenation Catalysis in Cells as a New Approach to Anticancer Drug Design. *Nat. Commun.* **2015,** *6,* 6582.
84. Bonfili, L.; Pettinari, R.; Cuccioloni, M.; Cecarini, V.; Mozzicafreddo, M.; Angeletti, M.; Lupidi, G.; Marchetti, F.; Pettinari, C.; Eleuteri, A. M. Arene-Ru(II) Complexes of Curcumin Exert Antitumor Activity via Proteasome Inhibition and Apoptosis Induction. *ChemMedChem* **2012,** *7,* 2010–2020.
85. Lei, X.; Su, W.; Li, P.; Xiao, Q.; Huang, S.; Qian, Q.; Huang, C.; Qin, D.; Lan, H. Ruthenium(II) Arene Complexes of Curcuminoids: Synthesis, X-ray Diffraction Structure and Cytotoxicity. *Polyhedron* **2014,** *81,* 614–618.
86. Caruso, F.; Rossi, M.; Benson, A.; Opazo, C.; Freedman, D.; Monti, E.; Gariboldi, M. B.; Shaulky, J.; Marchetti, F.; Pettinari, R.; Pettinari, C. Ruthenium-Arene Complexes of Curcumin: X-ray and Density Functional Theory Structure, Synthesis, and Spectroscopic Characterization, In Vitro Antitumor Activity, and DNA Docking Studies of (p-cymene)Ru(curcuminato)chloro. *J. Med. Chem.* **2012,** *55,* 1072–1081.
87. Schmid, W. F.; John, R. O.; Muhlgassner, G.; Heffeter, P.; Jakupec, M. A.; Galanski, M.; Berger, W.; Arion, V. B.; Keppler, B. K. Metal-Based Paullones as Putative CDK Inhibitors for Antitumor Chemotherapy. *J. Med. Chem.* **2007,** *50,* 6343–6355.
88. Muhlgassner, G.; Bartel, C.; Schmid, W. F.; Jakupec, M. A.; Arion, V. B.; Keppler, B. K. Biological Activity of Ruthenium and Osmium Arene Complexes with Modified Paullones in Human Cancer Cells. *J. Inorg. Biochem.* **2012,** *116,* 180–187.
89. Stepanenko, I. N.; Casini, A.; Edafe, F.; Novak, M. S.; Arion, V. B.; Dyson, P. J.; Jakupec, M. A.; Keppler, B. K. Conjugation of Organoruthenium(II) 3-(1H-benzimidazol-2-yl) pyrazolo[3,4-b]pyridines and Indolo[3,2-d]benzazepines to Recombinant Human Serum Albumin: A Strategy to Enhance Cytotoxicity in Cancer Cells. *Inorg. Chem.* **2011,** *50,* 12669–12679.
90. Filak, L. K.; Kalinowski, D. S.; Bauer, T. J.; Richardson, D. R.; Arion, V. B. Effect of the Piperazine Unit and Metal-Binding Site Position on the Solubility and Anti-proliferative Activity of Ruthenium (II)- and Osmium (II)- Arene Complexes of Isomeric Indolo[3,2-c]quinoline-piperazine Hybrids. *Inorg. Chem.* **2014,** *53,* 6934–6943.

91. Filak, L. K.; Goschl, S.; Heffeter, P.; Ghannadzadeh Samper, K.; Egger, A. E.; Jakupec, M. A.; Keppler, B. K.; Berger, W.; Arion, V. B. Metal-Arene Complexes with Indolo[3,2-c]-quinolines: Effects of Ruthenium vs Osmium and Modifications of the Lactam Unit on Intermolecular Interactions, Anticancer Activity, Cell Cycle, and Cellular Accumulation. *Organometallics* **2013,** *32,* 903–914.
92. Filak, L. K.; Muhlgassner, G.; Bacher, F.; Roller, A.; Galanski, M.; Jakupec, M. A.; Keppler, B. K.; Arion, V. B. Ruthenium- and Osmium-Arene Complexes of 2-Substituted Indolo[3,2-c]quinolines: Synthesis, Structure, Spectroscopic Properties, and Antiproliferative Activity. *Organometallics* **2011,** *30,* 273–283.
93. Arion, V. B.; Dobrov, A.; Goschl, S.; Jakupec, M. A.; Keppler, B. K.; Rapta, P. Ruthenium- and Osmium-Arene-based Paullones Bearing a TEMPO Free-radical Unit as Potential Anticancer Drugs. *Chem. Commun.* **2012,** *48,* 8559–8561.
94. Kurzwernhart, A.; Kandioller, W.; Bachler, S.; Bartel, C.; Martic, S.; Buczkowska, M.; Muhlgassner, G.; Jakupec, M. A.; Kraatz, H. B.; Bednarski, P. J.; Arion, V. B.; Marko, D.; Keppler, B. K.; Hartinger, C. G. Structure-activity Relationships of Targeted Ru(II)(η^6-*p*-cymene) Anticancer Complexes with Flavonol-Derived Ligands. *J. Med. Chem.* **2012,** *55,* 10512–10522.
95. Kurzwernhart, A.; Kandioller, W.; Bartel, C.; Bachler, S.; Trondl, R.; Muhlgassner, G.; Jakupec, M. A.; Arion, V. B.; Marko, D.; Keppler, B. K.; Hartinger, C. G. Targeting the DNA-topoisomerase Complex in a Double-Strike Approach with a Topoisomerase Inhibiting Moiety and Covalent DNA. *Binder. Chem. Commun.* **2012,** *48,* 4839–4841.
96. Pastuszko, A.; Niewinna, K.; Czyz, M.; Jóźwiak, A.; Małecka, M.; Budzisz, E. Synthesis, X-ray Structure, Electrochemical Properties and Cytotoxic Effects of New Arene Ruthenium(II) Complexes. *J. Organomet. Chem.* **2013,** *745–746,* 64–70.
97. Kurzwernhart, A.; Kandioller, W.; Enyedy, E. A.; Novak, M.; Jakupec, M. A.; Keppler, B. K.; Hartinger, C. G. 3-Hydroxyflavones vs. 3-Hydroxyquinolinones: Structure-activity Relationships and Stability Studies on Ru(II)(arene) Anticancer Complexes with Biologically Active Ligands. *Dalton Trans.* **2013,** *42,* 6193–6202.
98. Su, W.; Qian, Q.; Li, P.; Lei, X.; Xiao, Q.; Huang, S.; Huang, C.; Cui, J. Synthesis, Characterization, and Anticancer Activity of a Series of Ketone-*N*(4)-Substituted Thiosemicarbazones and Their Ruthenium(II) Arene Complexes. *Inorg. Chem.* **2013,** *52,* 12440–12449.
99. Beckford, F.; Dourth, D.; Shaloski, M., Jr.; Didion, J.; Thessing, J.; Woods, J.; Crowell, V.; Gerasimchuk, N.; Gonzalez-Sarrias, A.; Seeram, N. P. Half-Sandwich Ruthenium-Arene Complexes with Thiosemicarbazones: Synthesis and Biological Evaluation of [(η6-p-cymene)Ru(piperonal thiosemicarbazones)Cl]Cl Complexes. *J. Inorg. Biochem.* **2011,** *105,* 1019–1029.
100. Ivanović, I.; Gligorijević, N.; Aranđelović, S.; Radulović, S.; Roller, A.; Keppler, B. K.; Tešić, Ž. L.; Grgurić-Šipka, S. New Ruthenium(II)-Arene Complexes Bearing Hydrazides and the Corresponding (Thio)semicarbazones of 3- and 4-Acetylpyridine: Synthesis, Characterization, Crystal Structure Determination and Antiproliferative Activity. *Polyhedron* **2013,** *61,* 112–118.
101. Su, W.; Zhou, Q.; Huang, Y.; Huang, Q.; Huo, L.; Xiao, Q.; Huang, S.; Huang, C.; Chen, R.; Qian, Q.; Liu, L.; Li, P. Synthesis, Crystal and Electronic Structure, Anticancer Activity of Ruthenium(II) Arene Complexes with Thiosemicarbazones. *Appl. Organomet. Chem.* **2013,** *27,* 307–312.
102. Su, W.; Tang, Z.; Xiao, Q.; Li, P.; Qian, Q.; Lei, X.; Huang, S.; Peng, B.; Cui, J.; Huang, C. Synthesis, Structures, Antiproliferative Activity of a Series of Ruthenium(II) Arene Derivatives of Thiosemicarbazones Ligands. *J. Organomet. Chem.* **2015,** *783,* 10–16.
103. Grau-Campistany, A.; Massaguer, A.; Carrion-Salip, D.; Barragan, F.; Artigas, G.; Lopez-Senin, P.; Moreno, V.; Marchan, V. Conjugation of a Ru(II) Arene Complex to Neomycin or to Guanidinoneomycin Leads to Compounds with Differential Cytotoxicities and Accumulation Between Cancer and Normal Cells. *Mol. Pharmacol.* **2013,** *10,* 1964–1976.

104. Aman, F.; Hanif, M.; Siddiqui, W. A.; Ashraf, A.; Filak, L. K.; Reynisson, J.; Söhnel, T.; Jamieson, S. M. F.; Hartinger, C. G. Anticancer Ruthenium(η^6-*p*-cymene) Complexes of Nonsteroidal Anti-inflammatory Drug Derivatives. *Organometallics* **2014,** *33,* 5546–5553.
105. Khan, F.-A.; Therrien, B.; Süss-Fink, G.; Zava, O.; Dyson, P. J. Arene Ruthenium Dichloro Complexes Containing Isonicotinic Ester Ligands: Synthesis, Molecular Structure and Cytotoxicity. *J. Organomet. Chem.* **2013,** *730,* 49–56.
106. Vock, C. A.; Ang, W. H.; Scolaro, C.; Phillips, A. D.; Lagopoulos, L.; Juillerat-Jeanneret, L.; Sava, G.; Scopelliti, R.; Dyson, P. J. Development of Ruthenium Antitumor Drugs that Overcome Multidrug Resistance Mechanisms. *J. Med. Chem.* **2007,** *50,* 2166–2175.
107. Babak, M.V.; Plazuk, D.; Meier, S. M.; Arabshahi, H. J.; Reynisson, J.; Rychlik, B.; Blauz, A.; Szulc, K.; Hanif, M.; Strobl, S.; Roller, A.; Keppler, B. K.; Hartinger, C. G. Half-sandwich Ruthenium(II) Biotin Conjugates as Biological Vectors to Cancer Cells. *Chem. Eur. J.* **2015,** *21,* 5110–5117.
108. Wang, Z.; Qian, H.; Yiu, S. M.; Sun, J.; Zhu, G. Multi-targeted Organometallic Ruthenium(II)-arene Anticancer Complexes Bearing Inhibitors of Poly(ADP-ribose) Polymerase-1: A Strategy to Improve Cytotoxicity. *J. Inorg. Biochem.* **2014,** *131,* 47–55.
109. Castonguay, A.; Doucet, C.; Juhas, M.; Maysinger, D. New Ruthenium(II)-Letrozole Complexes as Anticancer Therapeutics. *J. Med. Chem.* **2012,** *55,* 8799–8806.
110. Hudej, R.; Kljun, J.; Kandioller, W.; Repnik, U.; Turk, B.; Hartinger, C. G.; Keppler, B. K.; Miklavčič, D.; Turel, I. Synthesis and Biological Evaluation of the Thionated Antibacterial Agent Nalidixic Acid and Its Organoruthenium(II) Complex. *Organometallics* **2012,** *31,* 5867–5874.
111. Delgado, R. A.; Galdámez, A.; Villena, J.; Reveco, P. G.; Thomet, F. A. Synthesis, Characterization and In Vitro Biological Evaluation of [Ru(η^6-arene)(*N,N*)Cl]PF_6 Compounds Using the Natural Products Arenes Methylisoeugenol and Anethole. *J. Organomet. Chem.* **2015,** *782,* 131–137.
112. Dorr, M.; Meggers, E. Metal Complexes as Structural Templates for Targeting Proteins. *Curr. Opin. Chem. Biol.* **2014,** *19,* 76–81.
113. Bregman, H.; Meggers, E. Ruthenium Half-Sandwich Complexes as Protein Kinase Inhibitors- An N-Succinimidyl Ester for Rapid Derivatizations of the Cyclopentadienyl Moiety. *Org. Lett.* **2006,** *8* (24), 5465–5468.
114. Kilpin, K. J.; Dyson, P. J. Enzyme Inhibition by Metal Complexes: Concepts, Strategies and Applications. *Chem. Sci.* **2013,** *4,* 1410–1419.
115. Lu, L.; Zhu, M. Protein Tyrosine Phosphatase Inhibition by Metals and Metal Complexes. *Antioxid. Redox Signal.* **2014,** *20,* 2210–2224.
116. Ong, J. X.; Yap, C. W.; Ang, W. H. Rational Design of Selective Organoruthenium Inhibitors of Protein Tyrosine Phosphatase 1B. *Inorg. Chem.* **2012,** *51,* 12483–12492.
117. Paira, P.; Chow, M. J.; Venkatesan, G.; Kosaraju, V. K.; Cheong, S. L.; Klotz, K. N.; Ang, W. H.; Pastorin, G. Organoruthenium Antagonists of Human A_3 Adenosine Receptors. *Chem. Eur. J.* **2013,** *19,* 8321–8330.
118. Monnard, F. W.; Heinisch, T.; Nogueira, E. S.; Schirmer, T.; Ward, T. R. Human Carbonic Anhydrase II as a Host for Piano-stool Complexes Bearing a Sulfonamide Anchor. *Chem. Commun.* **2011,** *47,* 8238–8240.
119. Ang, W. H.; De Luca, A.; Chapuis-Bernasconi, C.; Juillerat-Jeanneret, L.; Bello, M. L.; Dyson, P. J. Organometallic Ruthenium Inhibitors of Glutathione-S-Transferase P1-1 as Anticancer Drugs. *ChemMedChem* **2007,** *2,* 1799–1806.
120. Ang, W. H.; Parker, L. J.; De Luca, A.; Juillerat-Jeanneret, L.; Morton, C. J.; Lo Bello, M.; Parker, M. W.; Dyson, P. J. Rational Design of an Organometallic Glutathione Transferase Inhibitor. *Angew. Chem. Int. Ed. Engl.* **2009,** *48,* 3854–3857.
121. He, L.; Liao, S. Y.; Tan, C. P.; Ye, R. R.; Xu, Y. W.; Zhao, M.; Ji, L. N.; Mao, Z. W. Ruthenium-Arene β-Carboline Complexes as Potent Inhibitors of Cyclin-Dependent Kinase 1: Synthesis, Characterization and Anticancer Mechanism Studies. *Chem. Eur. J.* **2013,** *19,* 12152–12160.

122. Carcelli, M.; Bacchi, A.; Pelagatti, P.; Rispoli, G.; Rogolino, D.; Sanchez, T. W.; Sechi, M.; Neamati, N. Ruthenium Arene Complexes as HIV-1 Integrase Strand Transfer Inhibitors. *J. Inorg. Biochem.* **2013,** *118,* 74–82.
123. Antonyan, A.; De, A.; Vitali, L. A.; Pettinari, R.; Marchetti, F.; Gigliobianco, M. R.; Pettinari, C.; Camaioni, E.; Lupidi, G. Evaluation of (Arene)Ru(II) Complexes of Curcumin as Inhibitors of Dipeptidyl Peptidase IV. *Biochimie* **2014,** *99,* 146–152.
124. Khan, R. A.; Arjmand, F.; Tabassum, S.; Monari, M.; Marchetti, F.; Pettinari, C. Organometallic Ruthenium(II) Scorpionate as Topo IIα Inhibitor; In Vitro Binding Studies with DNA, HPLC Analysis and Its Anticancer Activity. *J. Organomet. Chem.* **2014,** *771,* 47–58.
125. Traven, K.; Sinreih, M.; Stojan, J.; Sersen, S.; Kljun, J.; Bezensek, J.; Stanovnik, B.; Turel, I.; Rizner, T. L. Ruthenium Complexes as Inhibitors of the Aldo-Keto Reductases AKR1C1-1C3. *Chem. Biol. Interact.* **2015,** *234,* 349–359.
126. Hanif, M.; Nawaz, M. A.; Babak, M. V.; Iqbal, J.; Roller, A.; Keppler, B. K.; Hartinger, C. G. Ruthenium(II)(η6-arene) Complexes of Thiourea Derivatives: Synthesis, Characterization and Urease Inhibition. *Molecules* **2014,** *19,* 8080–8092.
127. Chow, M. J.; Licona, C.; Wong, D.Y. Q.; Pastorin, G.; Gaiddon, C.; Ang, W. H. Discovery and Investigation of Anticancer Ruthenium-Arene Schiff-Base Complexes via Water-Promoted Combinatorial Three-Component Assembly. *J. Med. Chem.* **2014,** *57* (14), 6043–6059.
128. Chow, M. J.; Babak, M.V.; Wong, D.Y. Q.; Pastorin, G.; Gaiddon, C.; Ang, W. H. Structural Determinants of p53-Independence in Anticancer Ruthenium Arene Schiff- Base Complexes, *Mol. Pharmacol.* **2016,** *13,* 2543–2554.
129. Chow, M. J.; Licona, C.; Pastorin, G.; Mellitzer, G.; Ang, W. H.; Gaiddon, C. Structural Tuning of Organoruthenium Compounds Allows Oxidative Switch to Control ER Stress Pathways and Bypass Multidrug Resistance. *Chem. Sci.* **2016,** 7, 4117–4124.

CHAPTER FIVE

Medicinal Chemistry of Metal *N*-Heterocyclic Carbene (NHC) Complexes

I. Ott
Technische Universität Braunschweig, Braunschweig, Germany

Contents

1. INTRODUCTION

N-Heterocyclic carbenes (NHCs) have attracted increasing attention in chemistry because of their strong σ-donor capacity that makes them very useful ligands for various transition metals. The knowledge in this field goes back to early works on carbenes by Wanzlick and Öfele in the 1960s followed by the isolation of stable carbenes by Arduengo in 1991.[1–4]

Inorganic and Organometallic Transition Metal Complexes with Biological Molecules and Living Cells
ISBN 978-0-12-803814-7
http://dx.doi.org/10.1016/B978-0-12-803814-7.00005-8

A carbene is commonly a two-coordinate form of a carbon containing two unpaired electrons, and in NHCs the carbene is additionally stabilized by two neighboring nitrogens. Regarding their binding to transition metals, NHC-to-metal σ→d donation is the driving force; however, metal-to-NHC d→π* and NHC-to-metal π→d donations are also important contributors to the coordinative bond stabilities. These properties facilitate the formation of stable complexes with various transition metals and have shaped the astonishing developments in the field of chemical catalysis over many years. More recently metal NHC complexes have found their way into modern medicinal inorganic chemistry and bioorganometallic chemistry. This is mainly driven by the comparably high stability of this type of complex as well as the structural versatility of the coordinated NHC ligands. This enables the relatively easy access to a broad chemical space that can provide the basis for structure–activity relationship studies that are required for efficient drug design and development.

The most likely first reports on biological activities triggered by metal NHC complexes appeared between 1996 and 1999 when antibacterial and antifungal activities were described for several ruthenium(II) and rhodium(I) complexes.[5–7] Some examples of these first biologically active metal NHC complexes are depicted in Fig. 5.1. These structures already contain key elements that can still be found in more recently studied complexes. The following discoveries on the exciting properties of antibacterial silver NHC complexes and gold NHC anticancer agents have further shaped the field, which nowadays belongs to one of the most active areas within inorganic medicinal chemistry.

In this chapter the most important findings on biological functions of metal NHC complexes are summarized. Complexes with the classical N–C–N NHC structure incorporated into (benz)imidazole derivatives are

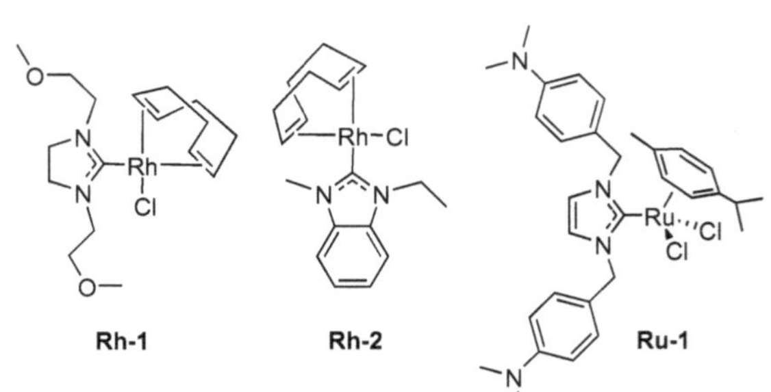

Figure 5.1 Examples of first biologically studied metal *N*-heterocyclic carbene (NHC) complexes.

ordered along the elements of the periodic table. Complexes that contain additional bioactive or biogenic ligands, additional metal centers, or "non-classical" NHC ligands are discussed in separate subchapters.

2. METAL NHC COMPLEXES WITH CLASSICAL N–C–N CARBENE LIGANDS

2.1 Copper NHC Complexes

Copper complexes can be developed as Fenton-type reagents that lead to oxidative DNA damage ultimately leading to apoptosis. The chloride complex **Cu-1** (Fig. 5.2) was more cytotoxic than the reference drug cisplatin in several cancer cell lines and induced apoptosis. In contrast to cisplatin, **Cu-1** arrested the cell cycle progression in the G1 phase.[8] Among a series of related copper NHC bromide complexes, **Cu-2** was the most cytotoxic derivative.[9] Further in vitro and in vivo investigations into this promising compound showed that it was active against a broad panel of different cancer cell lines with IC_{50} values in the nanomolar range, and had good activity in a mouse xenograft model with tolerable toxicity.[10]

2.2 Ruthenium NHC Complexes

Organometallic ruthenium complexes have been playing an enormous role in catalytic chemistry. In particular, their impact in the field of metathesis may be best exemplified by the fact that the Nobel Prize in Chemistry was awarded in 2005 to Grubbs, Schrock, and Chauvin "for the development of the metathesis method in organic synthesis." Although their applicability in biological media was found to be hampered by insufficient solubility in an aqueous environment, Grubbs-type catalysts are able to trigger effects as enzyme inhibitors, cytotoxic agents, and, at higher dosage, as modulators of cell metabolism. For example, the second-generation Hoveyda–Grubbs

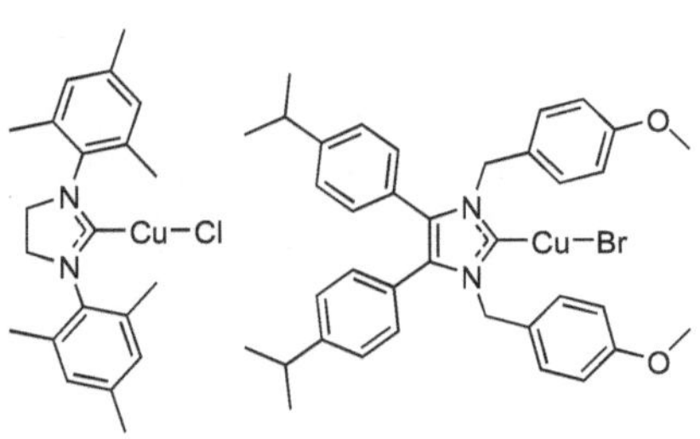

Cu-1 **Cu-2**

Figure 5.2 Examples of copper *N*-heterocyclic carbene (NHC) complexes.

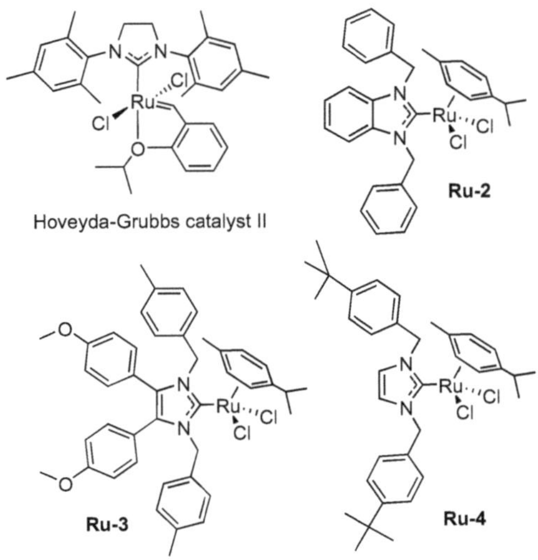

Figure 5.3 Examples of ruthenium *N*-heterocyclic carbene (NHC) complexes.

catalyst depicted in Fig. 5.3 triggered cytotoxic effects with an IC_{50} value of approximately 10 μM in cancer cells and inhibited the enzymes TrxR and cathepsin B with IC_{50} values of 2.5 and 8.0 μM, respectively.[11]

Ruthenium compounds probably represent the most intensively studied group of new metallodrugs after the class of platinum anticancer complexes and several examples have reached clinical trial stages of drug development.[12–14] Also taking into account that ruthenium complexes were among the very first biologically studied complexes, the literature has been surprisingly silent regarding the biological properties of ruthenium NHC complexes.

Complex **Ru-2** was the only good cytotoxic representative in a series of complexes differing in the substituents at the NHC nitrogens.[15] All complexes were enzyme inhibitors with activity against selected thiol and selenol containing enzymes (e.g., thioredoxin reductase, cathepsin B) but only the most lipophilic complex **Ru-2** was taken up into cells and hence triggered cytotoxic effects. The antiproliferative effects were accompanied by a decrease in cell impedance and cell respiration rate. Reactivity toward biological thiols was further confirmed by several mechanistic experiments. For example, upon incubation with acetylcysteine the NHC ligand was lost over time. Toxicity experiments in zebrafish embryos indicated that there might be a useful therapeutic window for ruthenium NHC complexes as toxic effects occurred in concentrations much higher than those applied in the cell culture. Strong cytotoxic effects in cancer cells were also noted for complex **Ru-3**, which showed particular good results in MCF-7 breast cancer cells.[16] Complex **Ru-4** was the most lipophilic and most active derivative in a series of ruthenium NHC complexes, thus further confirming the previous noted importance of lipophilicity for ruthenium NHC complexes to trigger cytotoxicity.[17] Further studies showed that **Ru-4** induced cell cycle arrest at the G2/M phase and bound better to the DNA than the less active ruthenium NHC complexes.

2.3 Rhodium NHC Complexes

Rhodium(I) is isoelectronic with platinum(II) and its complexes usually show square planar geometry. With these properties it can be speculated that rhodium(I) complexes might trigger similar pharmacological properties as observed with the DNA-targeting platinum(II) anticancer chemotherapeutics like cisplatin. In fact, for complex **Rh-3** (Fig. 5.4), cytotoxic effects, the alteration of cell migration, the induction of DNA condensation, and the inhibition of DNA replication were observed.[18]

The benzimidazole derived rhodium(I) complex **Rh-4** triggered strong cytotoxic effects in several cancer cell lines, inhibited thioredoxin reductase moderately and showed efficient binding to both albumin and DNA.[19]

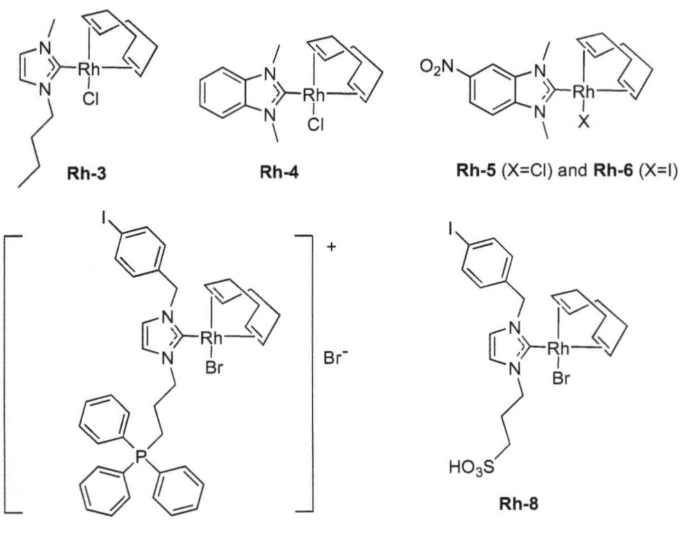

Figure 5.4 Examples of rhodium *N*-heterocyclic carbene (NHC) complexes.

Studies on the biodistribution of this complex using atomic absorption spectroscopy confirmed the transport of the metal into the nuclei, which further supported the relevance of an interaction with DNA for this type of organometallics, and demonstrated that the presence of albumin in the cell culture media had a negative effect on the cellular rhodium uptake. Of particular interest were results obtained in drug resistant cell lines, which indicated that **Rh-4** could be used for the treatment of drug resistant cancerous malignancies. In studies on the cytotoxicity of structural analogs of **Rh-4**, in which the benzimidazole NHC ligand was replaced with larger NHCs and/or the 1,5-cyclooctadiene ligand was replaced with CO ligands, problems regarding solubility and stability were evident. Physicochemical properties have definitely to be taken into account in the further development of this type of bioorganometallics.[19]

Further studies on analogs of **Rh-4**, more precisely with substituted benzimidazole NHC ligands and with iodido instead of chlorido ligands, revealed substantial effects on cellular signaling.[20] Phosphorylation of heat shock protein 27 (HSP27), p38, the extracellular signal-regulated kinases 1 and 2 (ERK-1, ERK-2), the focal adhesion kinase, and p70S6K was induced at relevant concentration levels by the nitro-NHC derivatives **Rh-5** and **Rh-6**. The mitogen-activated protein kinases (MAPKs) ERK-1, ERK-2, and p38 are known to be activated in response to oxidative stress and have implications on cell survival. This pattern of biological activity differs significantly from that of the DNA binding cytostatic drug cisplatin and, therefore, it can be concluded that platinum(II) anticancer drugs and rhodium(I) NHC species trigger different cellular pharmacodynamic effects despite the fact that both types of metallodrug target the DNA. Solution stability studies on **Rh-5** and **Rh-6** showed that the compounds were stable over approximately 3 h. However, after longer exposure the NHC ligand of the complexes was lost. Decomposition was found to be initiated by hydrolysis of the halido ligands.

The rhodium(I) NHC complex with a positively charged phosphonium side chain **Rh-7** triggered antibacterial effects with a clear preference for Gram-positive strains over Gram-negative strains.[21] This was explained with the overall negative charge of the bacterial cell wall of Gram-positive bacteria, which facilitates an interaction with positively charged drugs. In good agreement with this, a negatively charged derivative with a sulfonate group, **Rh-8**, was inactive. Furthermore, **Rh-7** also showed antileishmanial and antitrypanosomal activities against *Leishmania major* and *Trypanosoma brucei*, respectively, and triggered cytotoxic effects in cancer cells. Again, **Rh-8** was inactive in these assays. Cellular uptake studies by atomic absorption spectroscopy confirmed that **Rh-7** reached higher cellular concentrations than **Rh-8**.

2.4 Palladium NHC Complexes

Based on several similarities with platinum, palladium complexes are generally considered as possible anticancer agents and are known to trigger potent cytotoxic effects.

The *trans*-configured bis-NHC complex **Pd-1** (Fig. 5.5) triggered higher antiproliferative effects in cancer cell lines than the analogous mono-NHC complex **Pd-2**.[22] This was ascribed to the fact that the palladium center in **Pd-1** is more electron rich than that in **Pd-2**. Studies on the mode of action of **Pd-1** showed that it overexpressed cyclin B1, which is correlated with arrest in the G2/M cell cycle phase and apoptosis induction, and that the entry of the cells into the mitosis phase was prevented by inactivation of cdc2. Further studies suggested that the cell death mechanism was p53 dependent. Potent cytotoxic effects were also noted for **Pd-3**.[23] Antibacterial and anticancer properties were reported for **Pd-4**.[24] The tridentate complex **Pd-5** was the most anticancer active compound out of a series of related palladium NHC complexes.[25]

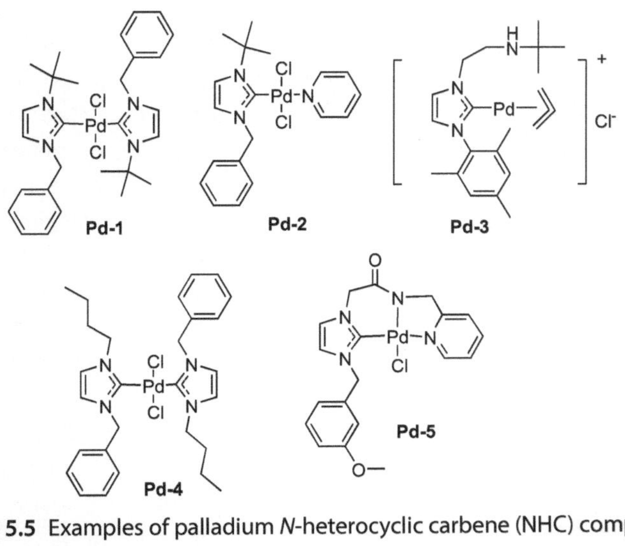

Figure 5.5 Examples of palladium *N*-heterocyclic carbene (NHC) complexes.

2.5 Silver NHC Complexes

As mentioned earlier, the first report on the biological activity of ruthenium and rhodium NHC complexes dealt with their antibacterial properties.[5] However, going back to reports by Youngs et al. there has also been an increased interest in the development of silver metal NHC complexes as antibacterials.[26,27]

Silver has been used in medicine since ancient times and the efficient antibacterial properties of silver ions are well known and widely applied (e.g., for medicinal and disinfection purposes). For example, silver nitrate and silver sulfadiazine (Fig. 5.6) are frequently used for treatment of skin ulcers, topic burns or for supporting wound healing.[28] The mechanisms of the antimicrobial action of silver are not yet clear and most likely related to multiple effects. For example, morphological changes, the interaction with thiol-containing proteins, or the inhibition of the respiratory chain are of relevance.[29] The soft Lewis acid silver can interact with, sulfur- and nitrogen-containing biomolecules, and this offers vast options of biological interactions. The high chloride levels in biological media will lead to precipitation of silver chloride unless it is bound to biomolecules.[30]

The water soluble pincer silver(I) NHC complex **Ag-1** (Fig. 5.7) showed improved antibacterial activities over that of silver nitrate, thus confirming for the first time the potential of NHCs as suitable ligands for silver antibacterials.[26] However, the low solubility and stability in chloride-containing solutions were identified as limiting factors regarding the possible in vivo applications. The complex **Ag-2** was less effective as an antibacterial agent than silver nitrate but its activity could be enhanced by encapsulation into Tecophilic, which is a polymer that can be electrospun from ethanol and has very good hydrophilic properties.[31] Upon exposure to water, the encapsulated **Ag-2** released silver nanoparticles and the system was effective against *Escherichia coli*, *Pseudomonas aeruginosa*, *Staphylococcus aureus*, *Candida albicans*, *Aspergillus niger*, and *Saccharomyces cerevisiae*. The authors concluded that the antimicrobial activity of encapsulated **Ag-2** "can be considered to be a combination of active silver species, which may include $AgCl_2^-$ ions, clusters of Ag^+ ions, AgCl, and free Ag^+ ions."[31]

The concept of improving the solubility and bioavailability of silver NHC complexes by encapsulation in suitable carriers has been evaluated successfully in several cases. Thus **Ag-3** loaded into degradable acetylated dextran nanoparticles provided an effective form of pulmonary drug delivery.[32] Complexes **Ag-4** and **Ag-5** incorporated into degradable polymeric

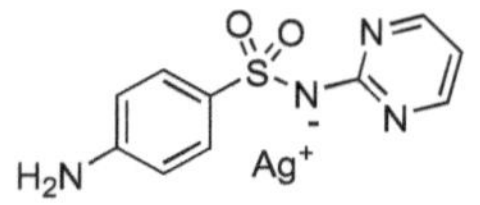

silver sulfadiazine

Figure 5.6 Silver sulfadiazine.

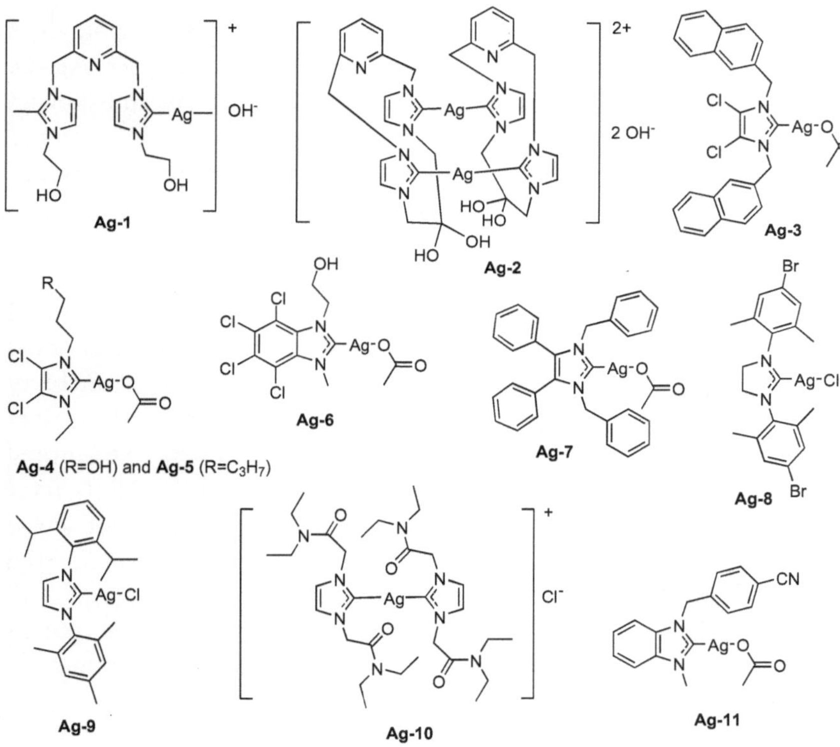

Figure 5.7 Examples of silver *N*-heterocyclic carbene (NHC) complexes.

nanoparticles showed promising results regarding possible direct epithelial treatment or the prevention of ubiquitous bacterial infections.[33]

Silver NHC acetato complexes that contain a hydroxyethyl side chain such as **Ag-6** have higher water solubility and were effective antibacterial agents.[34] Complex **Ag-7** was an effective antibacterial agent and also inhibited in vivo larvae of *Galleria mellonella*.[35]

Complexes with lipophilic side chains such as **Ag-8** were identified in a screening as strong inhibitors of biofilm formation in low concentrations in several pathogenic bacteria.[36]

Besides their promising antibacterial properties, silver NHC complexes have also demonstrated potential as anticancer drugs. Complexes such as **Ag-9** showed efficacy against several cancer cell lines, lowered the mitochondrial membrane potential and induced apoptosis, which was triggered by the translocation of the apoptosis inducing factor (AIF) and caspase-12 into the nucleus.[37]

The bis-NHC complex with hydrophilic side chains **Ag-10** was the most anticancer active derivative out of a series of related compounds.[38] Moreover, **Ag-10** was a strong TrxR inhibitor. The complex **Ag-11** triggered strong cytotoxic effects in several cancer cell lines and showed no cross-resistance with platinum anticancer drugs. Strong binding to the DNA was confirmed for **Ag-11**. However, no tumor growth reduction could be noted in tumor-bearing mice.[39]

2.6 Iridium NHC Complexes

The iridium(I) analog of **Rh-5** (see earlier), complex **Ir-1** (Fig. 5.8), caused antibacterial, antiprotozoal, and anticancer effects in a very similar manner as **Rh-5**.[21] Complex **Ir-2** was reported to trigger cytotoxic effects.[40] Stability studies in solution showed a progressive oxidation to the iridium(III) species and the mass spectrometric evaluation of the interaction with cytochrome *c* as a model protein afforded adducts containing iridium(III) NHC fragments. The complex thus might trigger an oxidative mechanism of action that is accompanied by a loss of the coordinated secondary ligands and the formation of kinetically inert protein-iridium(III) adducts.

Cyclometalated iridium(III) complexes such as **Ir-3** were potent mitochondria-targeting compounds and could be further activated by

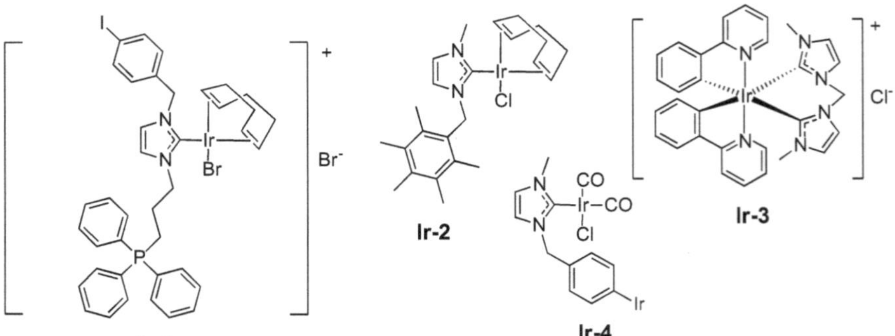

Ir-1 Ir-2 Ir-3 Ir-4

Figure 5.8 Examples of iridium *N*-heterocyclic carbene (NHC) complexes.

photoirradiation.[41] Their cellular uptake followed the order of lipophilicity and was determined by fluorescence microscopy and by metal quantification using inductively coupled plasma mass spectrometry. The mitochondrial localization was confirmed by colocalization studies with fluorescence microscopy.

Complex **Ir-4** represents an example for a possible future direction in the field of metal NHC complexes, because for this derivative its CO-releasing properties could be confirmed.[42] The CO release was fast, with a quantity correlated to approx. 0.4 equivalents. This effect could not be enhanced by irradiation, which was attributed to the lack of proper metal-to-ligand charge-transfer transitions that would weaken the metal–carbonyl bond.

2.7 Platinum NHC Complexes

The anticancer activity of the platinum(II) species cisplatin was discovered by Rosenberg and colleagues in 1965 and led to the development of the class of platinum cytostatics. It has been administered within many chemotherapeutic regimens since the 1970s and still plays a major role in current tumor treatment.[43–45] Cisplatin and other platinum(II) complexes bind to the DNA and thereby disturb cell proliferation finally leading to strong cytotoxic effects. Cisplatin is administered by intravenous injection. Triggered by the lower intracellular chloride levels compared to the extracellular environment, cisplatin hydrolyzes after cellular uptake with the formation of positively charged aqua complexes that are transported into the nucleus where they bind to the DNA. The majority of cisplatin-based platinum-DNA adducts consist of intrastrand crosslinks of two neighboring guanine bases in the same DNA strand. With this property, platinum-DNA adducts differ significantly from that caused by alkylating agents (e.g., melphalan) that lead to interstrand crosslinks between the two opposing strands of the DNA. Despite their high clinical efficacy, platinum anticancer drugs face severe problems related to strong toxic side effects (e.g., nephrotoxicity) and resistance development. Accordingly, the design and development of novel platinum complexes with fewer side effects and more promising resistance profile would be of high interest. Not surprisingly, the concept of creating novel bioactive metal complexes has also been applied to NHC complexes related to their favorable kinetic stability and structural versatility.

Square planar platinum(II) (NHC)PtI_2(amine) complexes such as complexes **Pt-1** and **Pt-2** (Fig. 5.9) were reported as efficient anticancer agents featuring a *trans* geometry.[46] This concept was further extended to analogs that feature two *trans* configured platinum(II) NHC centers. Complex

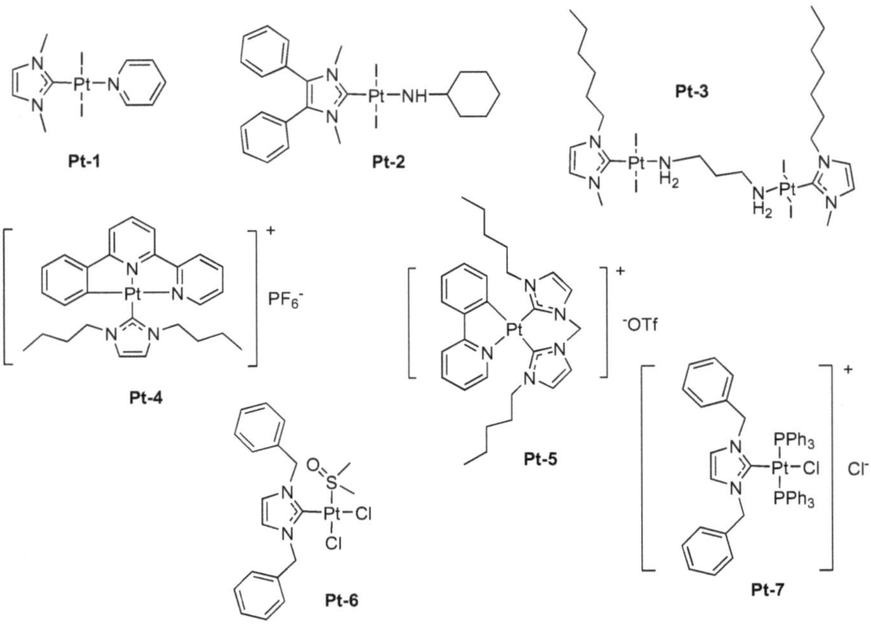

Figure 5.9 Examples of platinum *N*-heterocyclic carbene (NHC) complexes.

Pt-3 and a mononuclear analog showed more effective cellular accumulation than cisplatin; however, in relation to their higher cellular levels, DNA platination was lower than expected.[47] The DNA platination occurred at a relatively slow rate that is common for *trans*-platinum species. Interestingly the bimetallic complex **Pt-3** did not affect the cell cycle but it induced apoptosis, which was at least in part related to the translocation of AIF and caspase 12 to the nucleus.

The highly emissive complex **Pt-4** with lipophilic side chains was highly cytotoxic in vitro and was also effective in the nude mice model in vivo. The complex preferentially accumulated in cytoplasmic structures, activated poly(ADP-ribose) polymerase (PARP), and acted as a surviving suppressant finally leading to cell apoptosis.[48]

Strongly luminescent platinum(II) complexes with bidentate NHC ligands such as complex **Pt-5** were selectively localized in the endoplasmic reticulum (ER) domain, triggered ER stress and apoptosis in cells, and were phototoxic.[49] These complexes feature a planar coordination environment at the platinum(II) center and contain a chelating ligand making them useful scaffolds for noncovalent binding interactions with biological targets.

The platinum(II) NHC complexes **Pt-6** and **Pt-7**, which differ in their coordinated spectator ligands (dimethyl sulfoxide + 2 Cl vs Cl + 2 triphenylphosphine) and overall charge (neutral vs cationic) were investigated for their cytotoxicity, DNA interaction, and other biological effects.[50] Cytotoxicity was found to increase with increasing steric shielding of the platinum center. Notably, complex **Pt-7** was much more active in this aspect than complex **Pt-6**. Regarding DNA interaction, complex **Pt-6** showed a pattern that is comparable to cisplatin while complex **Pt-7** caused DNA aggregation. Complex **Pt-7** also caused vascular disruptive effects in the chorioallantoic membrane of fertilized chicken eggs.

2.8 Gold NHC Complexes

The applications of gold and its complexes in medicine date back thousands of years and gold played a central role in alchemy and in the Renaissance period. The first scientific reports on the therapeutic use of gold complexes by the famous bacteriologist Robert Koch describe their antibacterial effects. Later, gold complexes found their way into the current drug market as disease-modifying drugs for the treatment of rheumatoid arthritis.[51,52] Strong inhibition of tumor cell proliferation and other effects relevant for cancer chemotherapy therapy have been observed with many gold complexes, including the therapeutic lead compound auranofin, GoPI, or gold porphyrin 1a (Fig. 5.10), and accordingly gold complexes are nowadays also considered as potential new anticancer drug candidates.[53–56] The precise mechanisms of action of gold metallodrugs are still under discussion; however, there is strong evidence that the inhibition of the enzyme thioredoxin reductase (TrxR) and the interaction with mitochondria are key players among other targets and pathways.

Because of the insufficient physiological stability of the currently used gold drugs, there is a considerable interest in the development of

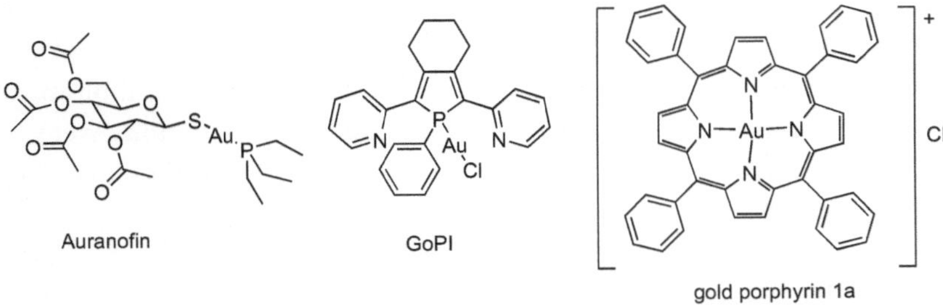

Figure 5.10 Examples of therapeutic and experimental gold metallodrugs.

improved, more stable, gold-based agents. Gold in its oxidation state I and III can be stably coordinated to NHC ligands and this fact provides one strong rationale for the development of gold NHC complexes as antitumor agents.

The pioneering works by Berners-Price and colleagues in this area have shaped the current understanding of the anticancer properties of gold NHC complexes and have motivated an increasing number of researchers to contribute to this particular topic.[57–59] Importantly, key properties such as the inhibition of thioredoxin reductase or antimitochondrial properties have been identified in these reports and make gold NHC complexes such as complexes **Au-1** and **Au-2** (Fig. 5.11) an attractive option for anticancer drug design with a mode of action very different from cisplatin.

For the cationic bis-NHC complex **Au-2** the inhibition of thioredoxin reductase in cells was confirmed for the first time and NMR-based experiments on **Au-2** and derivatives with cysteine or selenocysteine suggested that their mechanism of action involves ligand exchange processes and the formation of $[Au(Cys)_2]^-$ and $[Au(SeCys)_2]^-$ species, respectively.[57]

Gold(I) NHC chlorido complexes with an NHC ligand derived from benzimidazole, e.g., complex **Au-3**, were designed taking the structure of the highly potent gold metallodrug GoPI (Fig. 5.10) into account.[60–62] The binding of GoPI to glutathione reductase, which is structurally closely related to TrxR, has been determined by crystallography. Computational chemistry suggested that lipophilic interactions by the side chains of the inhibitor could be optimized to increase binding efficiency. Accordingly, a series of benzimidazolylidene NHC complexes with different residues at the nitrogen and a chlorido secondary ligand was prepared. Importantly, the chlorido ligand as a likely leaving group would enable efficient binding of the complexes in the TrxR active site. These compounds thus combined the structural motifs and chemistry of both GoPI and biologically active NHC complexes such as **Au-2**. The new gold NHC complexes were strong inhibitors of purified TrxR and the inhibition was selective if compared to the related enzyme glutathione reductase. The metallodrugs also induced apoptosis, increased reactive oxygen species (ROS) formation, affected the cellular metabolism, and triggered antimitochondrial effects. Interestingly, the compounds were also active in drug resistant leukemia cells and this indicates that they might find some application in the treatment of resistant cancers.

Starting from complex **Au-3**, two further complexes were investigated, in which the chlorido ligand was replaced with a phosphane (**Au-4**) or another NHC ligand (**Au-5**). The exchange of the anionic chloride

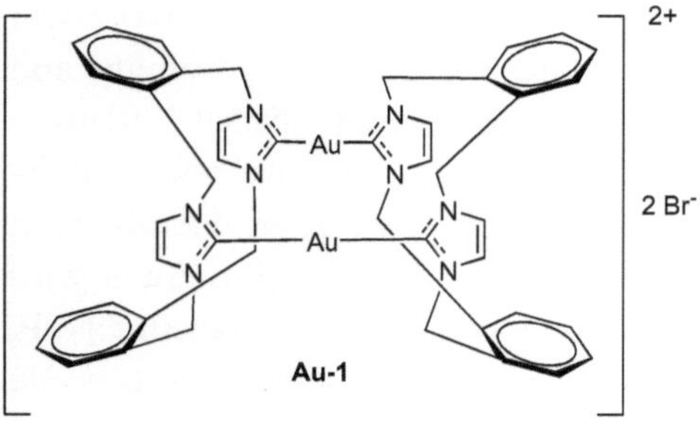

Au-1

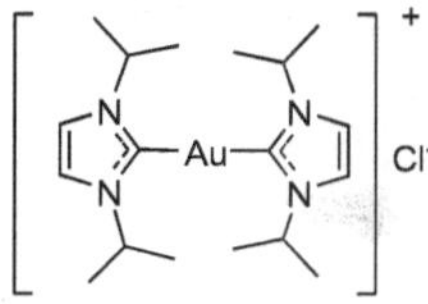

Au-2

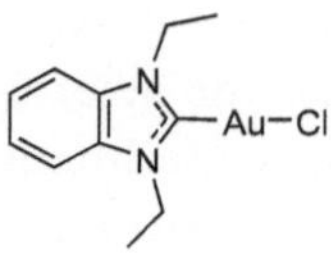

Au-3

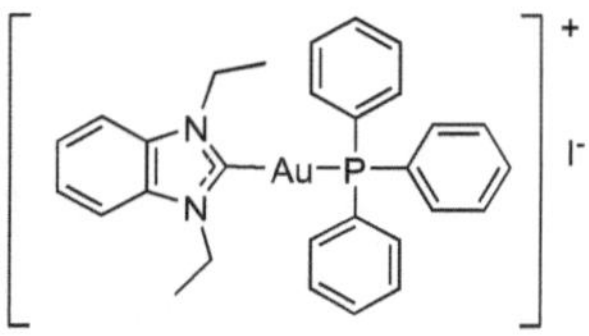

Au-4

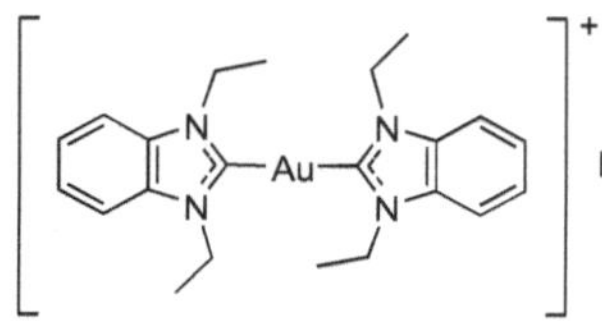

Au-5

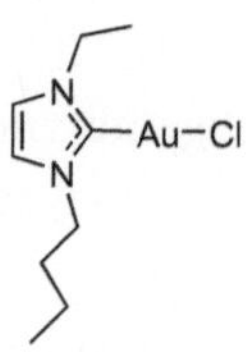

Au-6

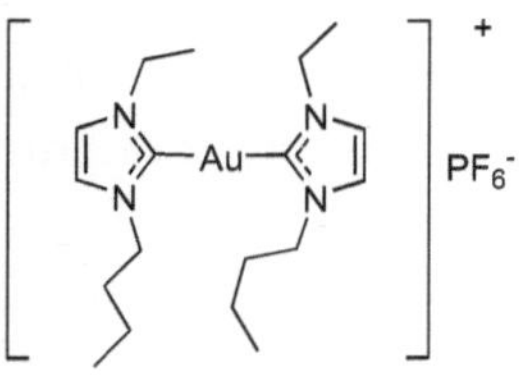

Au-7

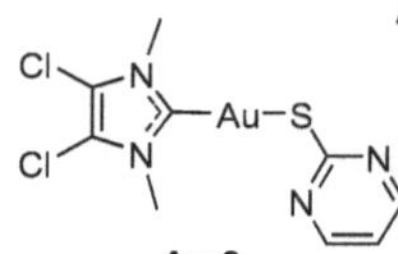

Au-8

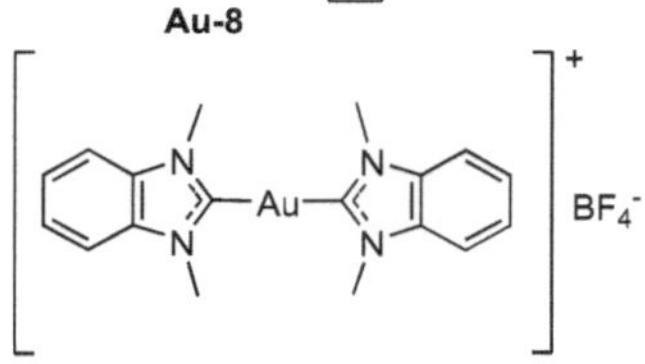

Au-9

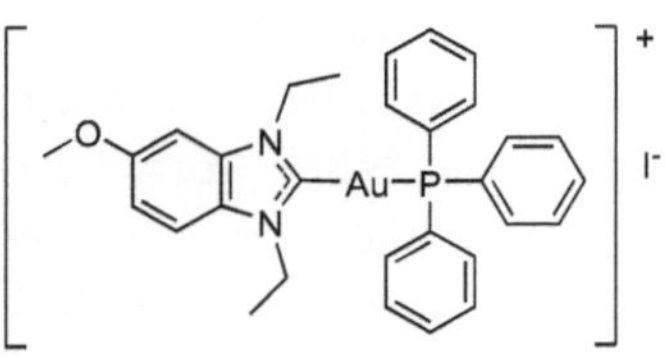

Au-10

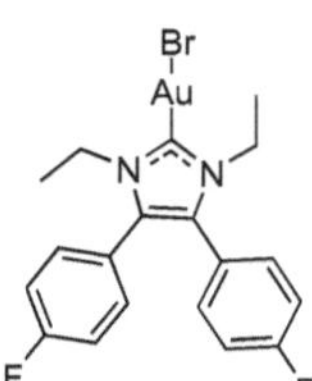

Au-12

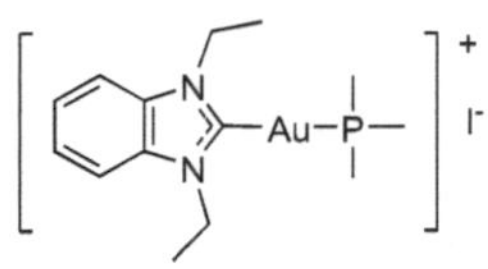

Au-11

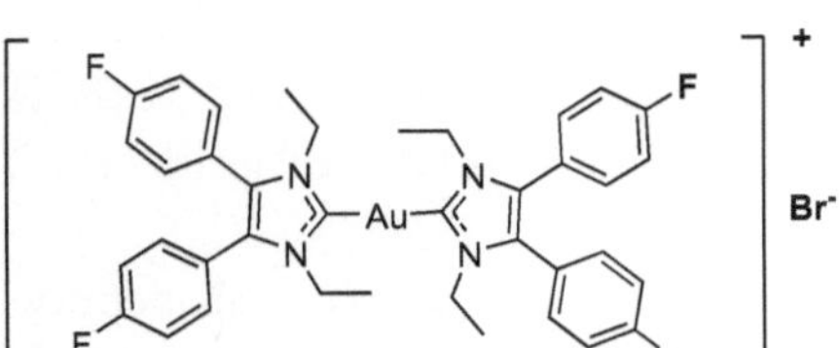

Au-13

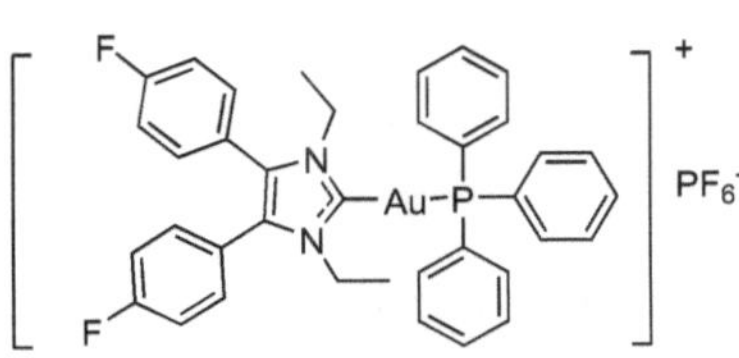

Au-14

Figure 5.11 Examples of gold *N*-heterocyclic carbene (NHC) complexes as possible anticancer drugs.

with neutral ligands led to cationic complexes and enabled a comparative study of their in vitro properties related to charge, kinetic stability, and lipophilicity. Calculation of the bond dissociation energies by DFT (density functional theory) methods indicated increasing coordinative bond stabilities with the order **Au-3** < **Au-4** < **Au-5**. The reversed order was observed regarding the inhibition of TrxR and binding to albumin. Thus **Au-3** with the most labile leaving group was the most active compound against TrxR in this series. Cellular biodistribution studies showed that the two cationic species caused higher cellular gold levels as well as higher accumulation in mitochondria, which was translated into strong antimitochondrial effects, in particular for complex **Au-5**. This is certainly related to their cationic and lipophilic character as similar properties are well known for delocalized lipophilic cations. The results from the biodistribution experiments also explained the enhanced biological activity of complexes **Au-4** and **Au-5** regarding cell proliferation inhibition, apoptosis induction, ROS formation, and general effects on tumor cell metabolism. These studies altogether confirmed that the chemical–biological properties of gold(I) NHC complexes could be modulated by a choice of the coordinated ligands and suggested that TrxR inhibition and antimitochondrial effects are key players in the pharmacology of gold complexes. The increase of cytotoxic activity in going from a halide ligand to a phosphane or a second NHC ligand was also confirmed for NHC-Au-Br complexes.[63]

The impact of TrxR inhibition on the biological profile of gold complexes has been suggested in many studies. To clarify possible correlations between the cytotoxicity and TrxR inhibition by gold NHC complexes, a series of 20 structurally diverse gold(I/III) complexes was screened for cytotoxicity and TrxR inhibition.[64] For most examples, both IC_{50} values for TrxR inhibition and cytotoxicity were low, indicating that there is a connection between TrxR and cell growth inhibition. However, a concrete (mathematical) correlation could not be identified. This may be the consequence of different additional properties such as bioavailability. Taken together, this screening study suggested again that TrxR is an important contributing factor in the biological profile of gold NHC complexes.

The synthetic linear selenocysteine containing C-terminal peptide hTrxR(488–499) was used for mechanistic studies on complexes **Au-3**, **Au-4**, and **Au-5** by electrospray ionization mass spectrometry. These experiments confirmed the selenocysteine as the primary site of coordination and showed that all complexes gave adducts that contained the same

[NHC-Au]$^+$ fragment. However, the complexes displayed different extents of protein metalation, and this was largest with complex **Au-3**.[65]

Complexes **Au-6** and **Au-7**, both of which were cytotoxic, stable in solution, and resistant to the biological reductants ascorbic acid and glutathione, were investigated for their interaction with model proteins.[66] No binding was observed with cytochrome *c* and lysozyme, whereas both complexes bound tightly to the cytoplasmic copper chaperone Atox-1.

Gold(I) NHC complexes with pyrimidinethiolate complexes such as **Au-8** were better cytotoxic agents than the respective chlorido derivatives.[67] Complex **Au-8** strongly inhibited both cytosolic (TrxR1) and mitochondrial (TrxR2) TrxRs with higher activity against TrxR1. Both thioredoxin proteins 1 and 2 were oxidized in cells and this effect could be related to the TrxR inhibition and/or the massive increase in ROS formation.

More detailed analysis of the effects of complex **Au-5** on cellular signaling showed that, as a consequence of TrxR inhibition, the complex led to an activation of the ASK1-p38-MAPK cascade and triggered apoptosis in gemcitabine resistant pancreatic cancer cells. Reduced Trx directly inhibits ASK1 and thus terminates ASK1-p38-MAPK-dependent apoptosis. Complex **Au-5** reduced the binding of Trx to ASK1. Importantly, in ASK1-deficient cells p38-MAPK was mainly in its inactive form.[68] The impact of different additional pathways on the pharmacological pattern of gold NHC complexes is underlined by the fact that a close structural analog of complex **Au-5**, more precisely its methyl analog **Au-9**, stabilized G-quadruplexes, although with a low selectivity.[69] The issue of low selectivity was solved by using caffeine-derived NHC ligands (see later).

The methoxy derivative of complex **Au-4**, namely, complex **Au-10**, caused strong and irreversible effects on the cellular metabolism indicating mitochondria as clear targets with cell death via apoptosis as the final consequence.[70] Analysis of important cell signaling pathways revealed an activation of all important MAPKs. Thus the levels of phospho-ERK1, phospho-ERK2, phospho-p38a, and phospho-JNK (total) were found to be increased upon treatment with complex **Au-10**. Moreover, phosphorylation of the chaperone HSP27 as well as the DNA damage-associated kinase Chk2 was elevated and other important signaling molecules such as the transcription factor cAMP response element binding protein showed a transient activation profile. Importantly, the specific gold-based TrxR inhibitor auranofin (Fig. 5.10) induced very similar changes in cell signaling. Integrating the results, a model was suggested, which explained that the mode of action of complex **Au-10** involved the direct inhibition of

TrxR, increasing ROS levels leading to severe cellular redox perturbations, and activation of the ERK1/2 pathway and other relevant signaling pathways. At the end, the cellular balance between cell survival and apoptosis is shifted toward the latter direction. Similar modes of action are most likely of general importance for gold NHC complexes. However, it should be noted that additional important targets have been under investigation. For example, gold NHC complexes were found to be inhibitors of protein tyrosine phosphatases.[71]

Modification of the phosphane ligands of complex **Au-4** afforded complexes with a strongly increased activity and selectivity against TrxR.[72] The trimethylphosphane analog complex **Au-11** inhibited TrxR in the low nanomolar range with an approximately 100-fold selectivity over glutathione reductase. Computational chemistry showed that the gold phosphane bond stability in complex **Au-11** and other trialkylphosphane derivatives was increased compared to that of the triphenylphosphane **Au-4**. However, this was also accompanied by a decrease in cellular accumulation. The complexes, in particular **Au-4**, were also shown to be strong inhibitors of the enzyme poly(ADP-ribose) polymerase I, which is a zinc-finger protein (PARP-1) that is important for DNA repair.

Gold(I) complexes with NHC ligands derived from 4,5-diarylimidazole such as complex **Au-12** inhibited TrxR, although with a much lower potency than auranofin, and showed a low nuclear accumulation.[73] Related to the 4,5-diarylimidazole NHC ligand, binding of the complex to estrogen receptors and/or inhibition of cyclooxygenase enzymes had been considered as possible modes of action. However, there was no evidence of estrogen receptor binding and the inhibition of cyclooxygenase enzymes did not correlate with cytotoxicity. The authors therefore concluded that the mentioned pathways and targets could be excluded. However, the selective inhibition of cyclooxygenase 1 by complex **Au-12** was of special interest. Further derivatives included complexes **Au-13** and **Au-14**, in which the bromido ligand was replaced with a second NHC ligand and a phosphane, respectively.[63] Similar to the findings on complexes **Au-3**, **Au-4**, and **Au-5**, complexes **Au-13** and **Au-14** displayed strongly enhanced cytotoxicity compared to complex **Au-12**. Derivatives containing a gold(III) ion instead of gold(I) showed similar activity.

Cationic gold(III) complexes with an NHC ligand as well as a strongly emissive secondary (N^N^N) ligand (e.g., **Au-15** and **Au-16**, Fig. 5.12) provided a unique opportunity to study the reactivity of the complexes under in vitro (cellular) conditions.[74] The gold(III) complexes could act as fluorescent thiol sensitive "switch-on" probes, meaning the fluorescence

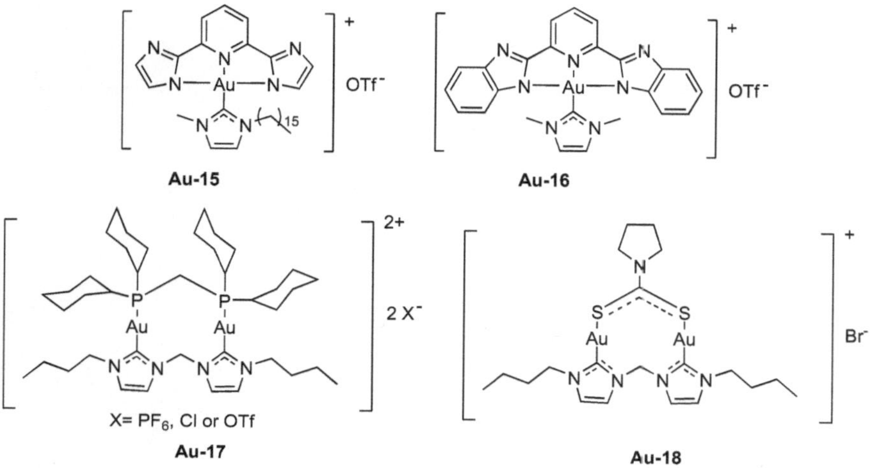

Figure 5.12 Examples of gold *N*-heterocyclic carbene (NHC) complexes as anticancer drug candidates.

emission arose from a released N^N^N ligand upon reduction of gold(III) to gold(I). In these complexes the NHC ligand is used to stabilize the supposedly anticancer-active gold(I) species, which is formed upon reduction of the nonemissive gold(III) complex by physiological thiols. The complexes were stable under aqueous conditions but reacted quickly with glutathione leading to the formation of gold(I)-NHC adduct, oxidized glutathione, and the release of the fluorescent ligand (Fig. 5.13). This enabled studying the reaction in cells by means of fluorescence microscopy.

Cells treated with complex **Au-16** showed fluorescence in the cytoplasm and colocalization experiments with organelle-specific stains indicated that a significant portion of the fluorescence was located in the mitochondria. Complex **Au-15** showed promising in vivo activity in a mouse xenograft model.

The biscationic dinuclear complex **Au-17** contains bridging bis-NHC and diphosphane ligands.[75] The complex was moderately stable toward glutathione, inhibited TrxR, and showed a good antitumor effect in mice bearing xenografts with no observable side effects. Importantly, the complex was able to inhibit angiogenesis in the animal model.

The dinuclear gold(I) pyrrolidinedithiocarbamato complex **Au-18,** which features intramolecular Au–Au interaction, triggered strong cytotoxic effects and was able to overcome cisplatin resistance in ovarian cancer cells. The complex was incorporated into a biodegradable blue-fluorescent zinc-based metal–organic framework (MOF) as a carrier to increase

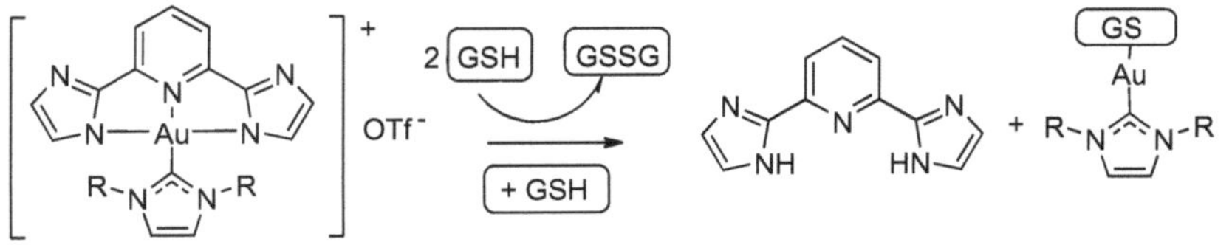

Figure 5.13 Reduction of a gold(III) complex by glutathione with ligand release. *GSH*, reduced glutathione; *GSSG*, oxidized glutathione.

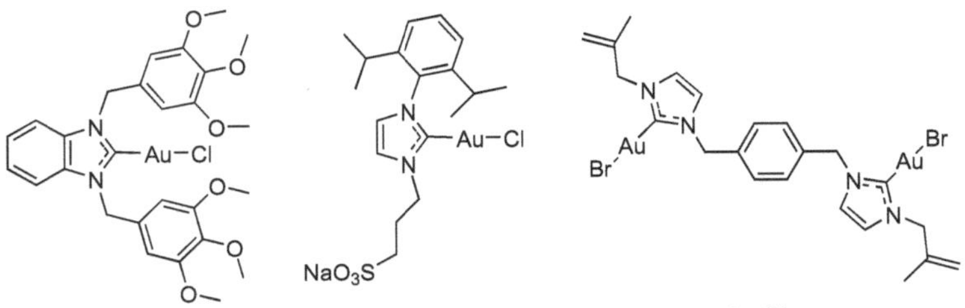

Au-19 **Au-20** **Au-21**

Figure 5.14 Examples of gold *N*-heterocyclic carbene (NHC) complexes as antibacterial agents.

its bioavailability, which resulted in a sustained release of complex **Au-18** in trans-well cell culture experiments. For both free **Au-18** and MOF-incorporated **Au-18**, possible antimigratory potential was confirmed in a wound-healing migration assay. In this established assay, the cell migration in response to an artificial wound produced on a cell monolayer is examined.[76]

Besides their excellent anticancer activities, gold NHC compounds have also demonstrated potent antiinfective properties against several pathogenic organisms.[22,36,77–80] Some examples are depicted in Fig. 5.14.

3. METAL NHC COMPLEXES WITH "NONCLASSICAL" NHC LIGANDS AND HETEROBIMETALLIC METAL NHC COMPLEXES

With many fundamental aspects of the biological properties of metal NHC complexes being established, further development in terms of drug design and development would now benefit from broader structural variations that could widen the knowledge on structure–activity relationships of this type of metallodrugs. Besides the use of different metals, this can be achieved by variations in the NHC structure in close proximity to the coordinated metal, thus affecting the chemical properties and kinetics of the complexes. Most studies reporting on biologically active metal NHC complexes have been using imidazole derivatives. Biological data on structures with NHC ligands different from imidazoles ("nonclassical" NHC ligands) are less frequent. On the other hand, the structural diversity and biological activity can be strongly modified by the incorporation of additional metal centers leading to heterobimetallic or even polymetallic metal NHC complexes. Some recent examples where those two concepts have been applied are highlighted in this section.

3.1 Metal NHC Complexes With Nonclassical Ligands

Some ruthenium complexes including the pyrazole derivative complex **Ru-5** (Fig. 5.15) were studied for toxic effects (e.g., embryotoxicity or teratogenic effects) in zebrafish embryos.[81] The ruthenium complexes were generally well tolerated and their toxicity profile appeared to correlate with oxidant properties, which were studied in vitro.

Other pyrazole-derived NHC complexes such as the cationic complexes **Au-22** and **Au-23** showed cytotoxic effects.[82] Complex **Au-22** also represents a heteroleptic (two different NHC ligands) complex and **Au-23** is most likely the first gold NHC complex containing the $Ph(OPh)_3$ ligand.

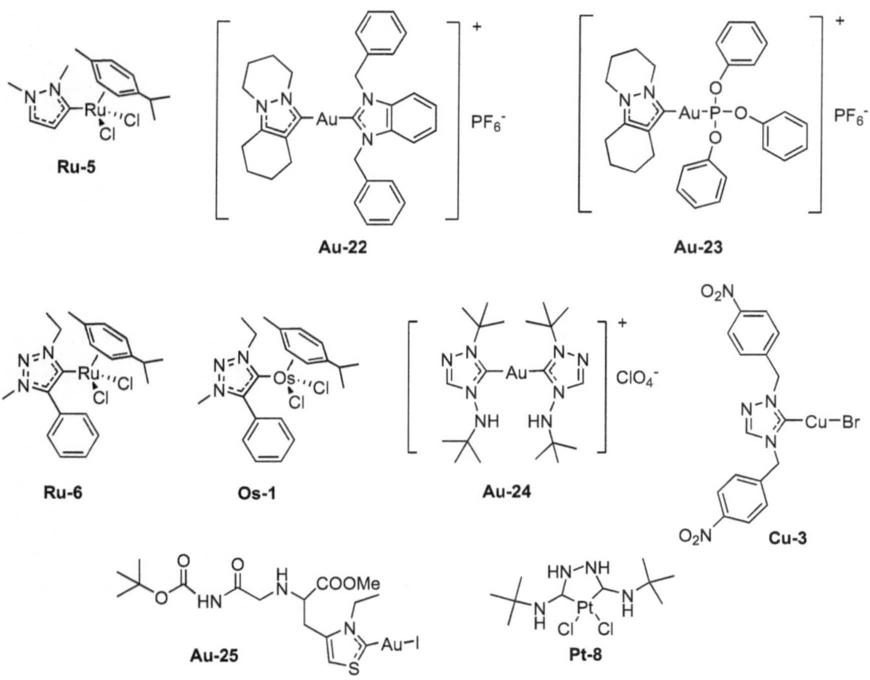

Figure 5.15 Examples of complexes with nonclassical *N*-heterocyclic carbene (NHC) ligands.

Ruthenium(II) and osmium(II) 1,2,3-triazolylidene NHC complexes triggered cytotoxic effects and the NHC ligands were found to remain coordinated to the metal center under physiological conditions.[83] Also gold complexes with 1,2,3-triazole NHC ligands attached to peptides triggered interesting biological profile (see Section 4 on bioconjugates).[84]

1,2,4-Triazole NHC ligands were reported as another promising scaffold for novel anticancer bioorganometallics.[85] Thus the 4-amino-1,2,4-triazole-derived cationic bis-NHC gold complex **Au-24** triggered cytotoxic effects, was a moderate inhibitor of TrxR, and showed an efficient cellular uptake.

A triazolium-based copper NHC complex, **Cu-3**, was reported as inactive against TrxR. However, this compound affected impairment of the ubiquitin–proteasome degradation pathway, thus inducing a nonapoptotic cell death pathway. Analogous gold and silver NHC complexes inhibited TrxR but were inactive as proteasome inhibitors. These results impressively underline the importance of the nature of the metal and demonstrate that copper NHC complexes target different cellular pathways than other NHC complexes, such as those of gold and silver.[86]

The *N,S*-heterocyclic carbene gold(I) complex **Au-25** triggered efficient cytotoxic effects in cultured tumor cells.[87]

The platinum complex **Pt-8** represents an interesting example of a Chugaev carbene and was found to be cytotoxic and interact with DNA and thiols.[88]

3.2 Heterobimetallic Metal NHC Complexes

The cationic bis-NHC gold(I) complex with two ferrocene units **Au-Fe-1** triggered cytotoxic effects with bioselectivity for tumor cells (Fig. 5.16).[89] Another antiproliferative gold NHC complex containing ferrocene units is complex **Au-Fe-2**.[90] This complex, which contains four ferrocene units, was targeting antioxidant pathways by regulating ROS via multiple mechanisms including TrxR inhibition. Cellular uptake studies afforded a 4:1 ratio for Fe:Au, which indicates that the complex entered the cells intactly.

Heterobimetallic ruthenium arene gold(I) NHC complexes such as complex **Au-Ru-a** triggered stronger cytotoxic effects in several cancer cell lines than in the human embryonic kidney cell line (HEK-293T). The complexes did not interact with the DNA but inhibited the activity of thioredoxin reductase.[91]

Other heterobimetallic complexes, e.g., complex **Au-Ru-2**, were designed to combine the biological effects of gold and the photoactivation properties of ruthenium polypyridyl units.[92] The complexes showed moderate in vitro cytotoxic, antileishmanial, and antimalarial activities.

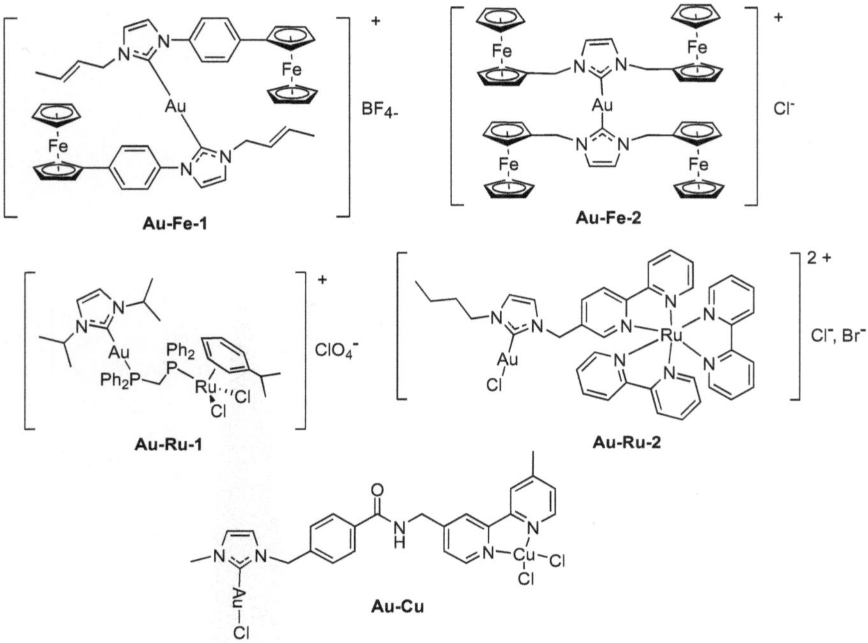

Figure 5.16 Examples of heterobimetallic *N*-heterocyclic carbene (NHC) complexes.

The Cu-Au heterodinuclear complex **Au-Cu** and derivatives triggered cytotoxic effects and were powerful inhibitors of TrxR1 and TrxR2. However, the compounds were toxic against liver tissues as determined in precision cut tissue slices.[93]

4. METAL NHC COMPLEXES WITH BIOLOGICALLY ACTIVE, BIOGENIC, OR TARGETING LIGANDS

For several reasons peptides represent ideal structures for bioconjugate formation. They can provide unique biological/pharmacological properties in terms of specific bioactivity and/or targeting abilities on the cellular or tissue level. Moreover, peptide structures are conveniently accessible via solid-phase peptide synthesis.

Amino acid and peptide conjugates such as the gold-cysteine complex **Au-26** (Fig. 5.17) showed cytotoxicity against several cancer cell lines.[94] Gold-peptide bioconjugates, such as complex **Au-27**, with a mitochondria-targeting sequence were obtained via a noncatalyzed [3 + 2] cycloaddition

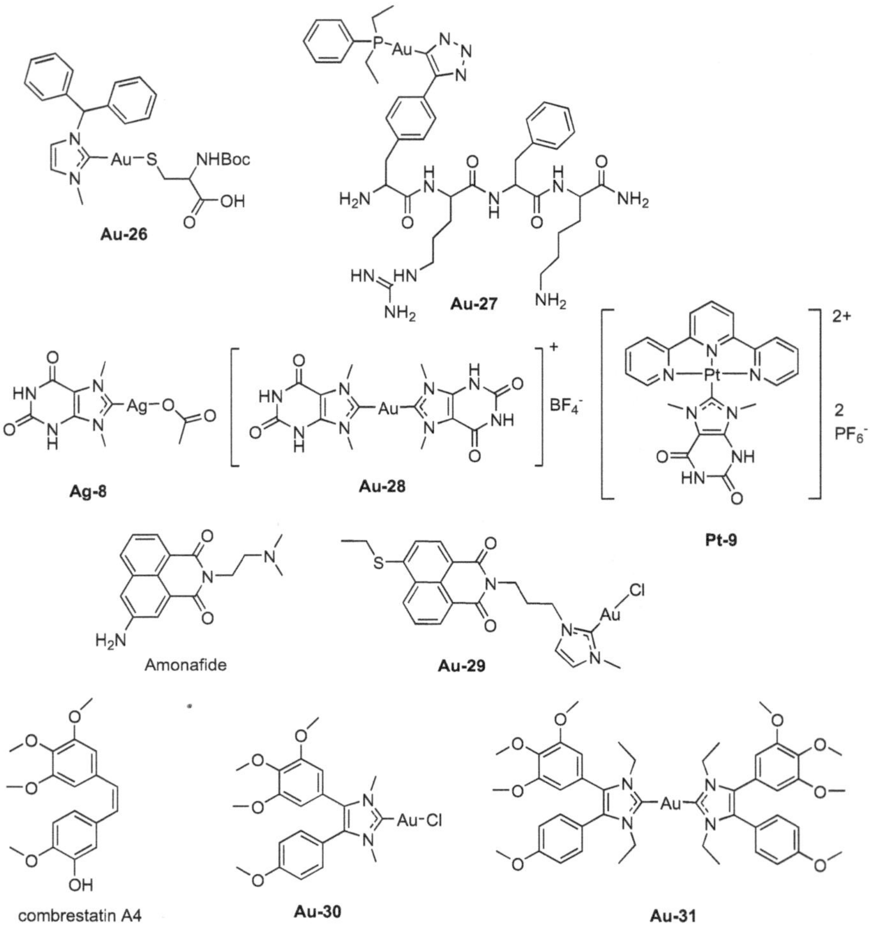

Figure 5.17 Examples of metal complexes with biogenic or bioactive *N*-heterocyclic carbene (NHC) ligands.

reaction of gold azides with alkynyl peptides.[84] In these complexes, the gold atoms were bound to a triazole NHC ligand. The bioconjugates showed prolonged stability in thiol-containing media and their cytotoxicity correlated with their lipophilicity. They were strong and selective inhibitors of TrxR and triggered strong effects in a number of biochemical and cell biological assays. Thus cellular respiration and acidification rates changed fast upon exposure to the complexes. Additionally, the ROS levels were increased, strong effects on mitochondrial respiration were observed, and apoptosis was induced. Importantly, the cisplatin resistance in p53-mutant breast cancer cells could be overcome.

Besides its well-known stimulating effects, the xanthine caffeine triggers a number of pharmacological effects and represents a nontoxic moiety. Together with the imidazole substructure, this makes them ideal candidates for coordination as biogenic NHC ligands.

The silver acetate complex **Ag-8** was evaluated as an antiinfective agent.[95] The complex showed antimicrobial activity against numerous resistant respiratory pathogens from the lungs of cystic fibrosis patients and was also found active against fungi. Notably, silver-free methylated caffeine also displayed some antibacterial and antifungal effects. Preliminary animal studies did not show toxic properties. Complex **Ag-8** was further studied in an aerosolized form in an airway infection model in mice with promising results, which suggested potential applications in the therapy of bacterial respiratory tract infections.[96]

The gold biscarbene caffeine derivative complex **Au-28** triggered cytotoxic effects and was found to be an efficient and selective quadruplex stabilizing agent.[69] G-quadruplexes are guanine-rich DNA or RNA sequences with importance in cellular signaling. Their stabilization by small molecules has been studied with the aim of controlling cellular events. As a selective agent complex **Au-28** addresses an important issue in the development of G-quadruplex stabilizing compounds. Small molecule quadruplex stabilizers often lack selectivity upon incubation with duplex DNA. Moreover, complex **Au-28** was only mildly toxic in healthy tissues and was a modest PARP-1 inhibitor.

Caffeine coordinated as an NHC ligand to a platinum terpyridine fragment afforded (bis)cationic complexes such as complex **Pt-9** that were highly cytotoxic, showed differentiation-like morphological changes in MCF-7 cells, and in vitro antiangiogenic properties in the tube formation assay.[97] Interestingly, a structural analog in which the xanthine was coordinated via the nitrogen (leading to a monocationic nonorganometallic theobromine derivative) was inactive or substantially less active than complex **Pt-9** in all assays.

Naphthalimides represent an interesting class of DNA intercalators. Some representatives such as mitonafide or amonafide had reached the clinical trial stage but finally did not progress to therapeutic applications.[98,99] Their mechanism of action is based on intercalation between the base pairs of DNA, and with this the compounds are able to cause substantial antiproliferative effects. Besides this, naphthalimides open interesting options for imaging because of their fascinating photophysical properties that enable the design of highly emissive ligands. Conjugates between a naphthalimide and a metal NHC fragment might thus lead to complexes with multitargeting abilities, combining DNA intercalation (because of the naphthalimide partial structure) with the biological effects triggered by the metal NHC fragment. In the ideal case, this would be accompanied by strong emissive properties.

In fact gold(I) NHC complexes containing naphthalimides (e.g., complex **Au-29**) turned out to be combined intercalators and thioredoxin reductase inhibiting agents with strong cytotoxic effects.[100] In complexes such as **Au-29** the gold(I) NHC center is positioned in the side chain of the naphthalimide, where usually an essential amino nitrogen is found (see amonafide, Fig. 5.17). Under physiological conditions, this nitrogen can be protonated with the formation of a cationic species, which can interact with the negatively charged phosphate backbone of DNA. This interaction facilitates the intercalation of the flat aromatic system of the naphthalimide into the DNA. Supposedly, this mechanism was still possible with **Au-29** and so it can be concluded that the crucial structure–activity relationships of the naphthalimides were maintained in the organometallic conjugate.

The *cis*-stilbene combretastatin A-4 (Fig. 5.17) is a plant metabolite that induces a rapid vascular shutdown in solid tumors. Its anticancer effects originate from several mechanisms (e.g., binding to the colchicine binding site of tubulin, interference with the vascular endothelial-cadherin/b-catenin/Akt signaling pathway in endothelial cells). Gold NHC analogs of this bioactive compound, such as complexes **Au-30** and **Au-31**, showed promising anticancer activity and very strong antivascular effects.[101,102] Interestingly, the antivascular effects observed with complexes **Au-30** and **Au-31** were not related to tubulin binding, as is the case with combretastatin A4. More detailed studies on complex **Au-31** and it derivatives showed that the antivascular effects were related to interference with the actin cytoskeleton of endothelial and cancer cells. Complex **Au-31** also showed promising effects in vivo in a xenograft mouse model.

5. CONCLUSIONS

Metal complexes with NHC ligands have experienced a fast, and still rapidly ongoing, development in the fields of medicinal inorganic chemistry and bioorganometallics. This is related to certain advantages of these systems.

The generation of metal complexes that are stable under biological/physiological conditions is possible, and the imidazole core of the NHC ligand allows many structural variations, leading to a high diversity of compounds. Regarding possible applications as anticancer agents, certainly gold NHC complexes have shaped the field. Mechanistic studies are already available demonstrating that many examples follow a biological pattern commonly triggered by cytotoxic gold complexes. Also platinum, rhodium, and other metals coordinated to NHC ligands have shown highly promising results as anticancer drug candidates. Regarding the use as antibacterial agents, silver NHC complexes have been most widely studied, also with highly encouraging results.

Research on the biological applications of metal NHC complexes is also progressing to the use of nonclassical NHC ligands, which differ from the common imidazole core (e.g., triazole-derived NHC ligands) and an increasing number of studies report on the incorporation of metal NHC fragments into heterobimetallic complexes or conjugates with biologically active or naturally occurring compounds. This clearly shows that metal NHC complexes have evolved as frequently used organometallic pharmacophores that trigger distinct pharmacological effects.

REFERENCES

1. Wanzlick, H.-W.; Schönherr, H.-J. Direct Synthesis of a Mercury Salt-Carbene Complex. *Angew. Chem. Int. Ed. Engl.* **1968,** *7,* 141–142.
2. Wanzlick, H. W. Aspects of Nucleophilic Carbene Chemistry. *Angew. Chem. Int. Ed. Engl.* **1962,** *1,* 75–80.
3. Öfele, K. 1,3-Dimethyl-4-imidazolinyliden-(2)-pentacarbonylchrom ein neuer übergangsmetall-carben-komplex. *J. Organomet. Chem.* **1968,** *12,* 42–43.
4. Arduengo, A. J.; Harlow, R. L.; Kline, M. A Stable Crystalline Carbene. *J. Am. Chem. Soc.* **1991,** *113,* 361–363.
5. Cetinkaya, B.; Cetinkaya, E.; Kucubay, H.; Durmaz, R. Antimicrobial Activity of Carbene Complexes of Rhodium(I) and Ruthenium(II). *Arzneim. Forsch. – Drug Res.* **1996,** *46,* 821–823.
6. Durmaz, R.; Kücükbay, H.; Cetinkaya, E.; Cetinkaya, B. Antimicrobial Activity of Rhodium (I) and Ruthenium (II) Carbene Complexes Derived from Benzimidazole Against *Staphylococcus aureus* Isolates. *Turk. J. Med. Sci.* **1997,** *27,* 59–61.
7. Cetinkaya, B.; Özdemir, I.; Binbasioglu, B.; Durmaz, R.; Günal, S. Antibacterial and Antifungal Activities of Complexes of Ruthenium(II). *Arzneim. Forsch. – Drug Res.* **1999,** *49,* 538–540.
8. Teyssot, M.; Jarrousse, A.; Chevry, A.; Haze, A. D.; Beaudoin, C.; Manin, M.; Nolan, S. P.; Diez-Gonzalez, S.; Morel, L.; Gautier, A. Toxicity of Copper(I)-NHC Complexes Against Human Tumor Cells: Induction of Cell Cycle Arrest, Apoptosis, and DNA Cleavage. *Chem. Eur. J.* **2009,** *15,* 314–318.
9. Streciwilk, W.; Hackenberg, F.; Müller-Bunz, H.; Tacke, M. Synthesis and Cytotoxicity Studies of *p*-Benzyl Substituted NHC–Copper(I) Bromide Derivatives. *Polyhedron* **2014,** *80,* 3–9.

10. Walther, W.; Fichtner, I.; Hackenberg, F.; Streciwilk, W.; Tacke, M. In Vitro and In Vivo Investigations into the Carbene Copper Bromide Anticancer Drug Candidate WBC4. *Lett. Drug Des. Drug Disc.* **2014,** *11,* 825–832.
11. Oehninger, L.; Alborzinia, H.; Ludewig, S.; Baumann, K.; Wölfl, S.; Ott, I. From Catalysts to Bioactive Organometallics: Do Grubbs Catalysts Trigger Biological Effects? *ChemMedChem* **2011,** *6,* 2142–2145.
12. Hartinger, C. G.; Jakupec, M. A.; Zorbas-Seifried, S.; Groessl, M.; Egger, A.; Berger, W.; Zorbas, H.; Dyson, P. J.; Keppler, B. K. KP1019, a New Redox-Active Anticancer Agent – Preclinical Development and Results of a Clinical Phase I Study in Tumor Patients. *Chem. Biodivers.* **2008,** *5,* 2140–2155.
13. Romero-Canelon, I.; Sadler, P. J. Next-Generation Metal Anticancer Complexes: Multitargeting via Redox Modulation. *Inorg. Chem.* **2013,** *52,* 12276–12291.
14. Bergamo, A.; Sava, G. Ruthenium Anticancer Compounds: Myths and Realities of the Emerging Metal-Based Drugs. *Dalton Trans.* **2011,** *40,* 7817–7823.
15. Oehninger, L.; Stefanopoulou, M.; Alborzinia, H.; Schur, J.; Ludewig, S.; Namikawa, K.; Muñoz-Castro, A.; Köster, R. W.; Baumann, K.; Wölfl, S.; Sheldrick, W. S.; Ott, I. Evaluation of Arene Ruthenium(II) N-Heterocyclic Carbene Complexes as Organometallics Interacting With Thiol and Selenol Containing Biomolecules. *Dalton Trans.* **2013,** *42,* 1657–1666.
16. Hackenberg, F.; Müller-Bunz, H.; Smith, R.; Streciwilk, W.; Zhu, X.; Tacke, M. Novel Ruthenium(II) and Gold(I) NHC Complexes: Synthesis, Characterization, and Evaluation of Their Anticancer Properties. *Organometallics* **2013,** 5551–5560.
17. Lv, G.; Guo, L.; Qiu, L.; Yang, H.; Wang, T.; Liu, H.; Lin, J. Lipophilicity-Dependent Ruthenium N-Heterocyclic Carbene Complexes as Potential Anticancer Agents. *Dalton Trans.* **2015,** *44,* 7324–7331.
18. McConnell, J. R.; Rananaware, D. P.; Ramsey, D. M.; Buys, K. N.; Cole, M. L.; McAlpine, S. R. A Potential Rhodium Cancer Therapy: Studies of a Cytotoxic Organorhodium(I) Complex that Binds DNA. *Bioorg. Med. Chem. Lett.* **2013,** *23,* 2527–2531.
19. Oehninger, L.; Kuster, L. N.; Schmidt, C.; Munoz-Castro, A.; Prokop, A.; Ott, I. A Chemical-Biological Evaluation of Rhodium(I) N-Heterocyclic Carbene Complexes as Prospective Anticancer Drugs. *Chem. Eur. J.* **2013,** *19,* 17871–17880.
20. Oehninger, L.; Spreckelmeyer, S.; Holenya, P.; Meier, S. M.; Can, S.; Alborzinia, H.; Schur, J.; Keppler, B. K.; Wolfl, S.; Ott, I. Rhodium(I) N-Heterocyclic Carbene Bioorganometallics as In Vitro Antiproliferative Agents with Distinct Effects on Cellular Signaling. *J. Med. Chem.* **2015,** *58,* 9591–9600.
21. Simpson, P. V.; Schmidt, C.; Ott, I.; Bruhn, H.; Schatzschneider, U. Synthesis, Cellular Uptake and Biological Activity Against Pathogenic Microorganisms and Cancer Cells of Rhodium and Iridium N-Heterocyclic Carbene Complexes Bearing Charged Substituents. *Eur. J. Inorg. Chem.* **2013,** 5547–5554.
22. Ray, S.; Mohan, R.; Singh, J. K.; Samantaray, M. K.; Shaikh, M. M.; Panda, D.; Ghosh, P. Anticancer and Antimicrobial Metallopharmaceutical Agents Based on Palladium, Gold, and Silver N-Heterocyclic Carbene Complexes. *J. Am. Chem. Soc.* **2007,** *129,* 15042–15053.
23. Wang, C.-H.; Shih, W.-C.; Chang, H. C.; Kuo, Y.-Y.; Hung, W.-C.; Ong, T.-G.; Li, W.-S. Preparation and Characterization of Amino-Linked Heterocyclic Carbene Palladium, Gold, and Silver Complexes and Their Use as Anticancer Agents that Act by Triggering Apoptotic Cell Death. *J. Med. Chem.* **2011,** *54,* 5245–5249.
24. Haque, R. A.; Salman, A. W.; Budagumpi, S.; Abdullah, A. A.; Majid, A. M. Sterically Tuned Ag(I)- and Pd(II)-N-Heterocyclic Carbene Complexes of Imidazol-2-ylidenes: Synthesis, Crystal Structures, and In Vitro Antibacterial and Anticancer Studies. *Metallomics* **2013,** *5,* 760–769.

25. Lee, J.-Y.; Lee, J.-Y.; Chang, Y.-Y.; Hu, C.-H.; Wang, N. M.; Lee, H. M. Palladium Complexes with Tridentate N-Heterocyclic Carbene Ligands: Selective "Normal" and "Abnormal" Bindings and Their Anticancer Activities. *Organometallics* **2015,** *34,* 4359–4368.
26. Melaiye, A.; Simons, R. S.; Milsted, A.; Pingitore, F.; Wesdemiotis, C.; Tessier, C. A.; Youngs, W. J. Formation of Water-Soluble Pincer Silver(I)−Carbene Complexes: A Novel Antimicrobial Agent. *J. Med. Chem.* **2004,** *47,* 973–977.
27. Kascatan-Nebioglu, A.; Panzner, M. J.; Tessier, C. A.; Cannon, C. L.; Youngs, W. J. N-Heterocyclic Carbene–Silver Complexes: A New Class of Antibiotics. *Coord. Chem. Rev.* **2007,** *251,* 884–895.
28. Konop, M.; Tatsiana, D.; Misicka, A.; Rudnicka, L. Certain Aspects of Silver and Silver Nanoparticles in Wound Care: A Minireview. *J. Nanomat.* **2016,** *2016.* ID 7614753.
29. Hindi, K. M.; Panzner, M. J.; Tessier, C. A.; Cannon, C. L.; Youngs, W. J. The Medicinal Applications of Imidazolium Carbene-Metal Complexes. *Chem. Rev.* **2009,** *109,* 3859–3884.
30. Chernousova, S.; Epple, M. Silver as Antibacterial Agent: Ion, Nanoparticle, and Metal. *Angew. Chem. Int. Ed. Engl.* **2013,** *52,* 1636–1653.
31. Melaiye, A.; Sun, Z.; Hindi, K.; Milsted, A.; Ely, D.; Reneker, D. H.; Tessier, C. A.; Youngs, W. J. Silver(I)−Imidazole Cyclophane Gem-Diol Complexes Encapsulated by Electrospun Tecophilic Nanofibers: Formation of Nanosilver Particles and Antimicrobial Activity. *J. Am. Chem. Soc.* **2005,** *127,* 2285–2291.
32. Ornelas-Megiatto, C.; Shah, P. N.; Wich, P. R.; Cohen, J. L.; Tagaev, J. A.; Smolen, J. A.; Wright, B. D.; Panzner, M. J.; Youngs, W. J.; Frechet, J. M.; Cannon, C. L. Aerosolized Antimicrobial Agents Based on Degradable Dextran Nanoparticles Loaded with Silver Carbene Complexes. *Mol. Pharm.* **2012,** *9,* 3012–3022.
33. Lim, Y. H.; Tiemann, K. M.; Heo, G. S.; Wagers, P. O.; Rezenom, Y. H.; Zhang, S.; Zhang, F.; Youngs, W. J.; Hunstad, D. A.; Wooley, K. L. Preparation and In Vitro Antimicrobial Activity of Silver-Bearing Degradable Polymeric Nanoparticles of Polyphosphoester-block-Poly(L-Lactide). *ACS Nano* **2015,** *9,* 1995–2008.
34. Wright, B. D.; Shah, P. N.; McDonald, L. J.; Shaeffer, M. L.; Wagers, P. O.; Panzner, M. J.; Smolen, J.; Tagaev, J.; Tessier, C. A.; Cannon, C. L.; Youngs, W. J. Synthesis, Characterization, and Antimicrobial Activity of Silver Carbene Complexes Derived from 4,5,6,7-Tetrachlorobenzimidazole Against Antibiotic Resistant Bacteria. *Dalton Trans.* **2012,** *41,* 6500–6506.
35. Browne, N.; Hackenberg, F.; Streciwilk, W.; Tacke, M.; Kavanagh, K. Assessment of In Vivo Antimicrobial Activity of the Carbene Silver(I) Acetate Derivative SBC3 Using *Galleria mellonella* Larvae. *Biometals* **2014,** *27,* 745–752.
36. Bernardi, T.; Badel, S.; Mayer, P.; Groelly, J.; de Fremont, P.; Jacques, B.; Braunstein, P.; Teyssot, M. L.; Gaulier, C.; Cisnetti, F.; Gautier, A.; Roland, S. High-Throughput Screening of Metal-N-Heterocyclic Carbene Complexes Against Biofilm Formation by Pathogenic Bacteria. *ChemMedChem* **2014,** *9,* 1140–1144.
37. Eloy, L.; Jarrousse, A.-S.; Teyssot, M.-L.; Gautier, A.; Morel, L.; Jolivalt, C.; Cresteil, T.; Roland, S. Anticancer Activity of Silver–N-Heterocyclic Carbene Complexes: Caspase-Independent Induction of Apoptosis via Mitochondrial Apoptosis-Inducing Factor (AIF). *ChemMedChem* **2012,** *7,* 805–814.
38. Pellei, M.; Gandin, V.; Marinelli, M.; Marzano, C.; Yousufuddin, M.; Dias, H. V. R.; Santini, C. Synthesis and Biological Activity of Ester- and Amide-Functionalized Imidazolium Salts and Related Water-Soluble Coinage Metal N-Heterocyclic Carbene Complexes. *Inorg. Chem.* **2012,** *51,* 9873–9882.
39. Fichtner, I.; Cinatl, J.; Michaelis, M.; Sanders, L. C.; Hilger, R.; Kennedy, B. N.; Reynolds, A. L.; Hackenberg, F.; Lally, G.; Quinn, S. J.; McRae, I.; Tacke, M. In Vitro and In Vivo Investigations into the Carbene Silver Acetate Anticancer Drug Candidate SBC1. *Lett. Drug. Des. Drug Disc.* **2012,** *9,* 815–822.

40. Gothe, Y.; Marzo, T.; Messori, L.; Metzler-Nolte, N. Cytotoxic Activity and Protein Binding Through an Unusual Oxidative Mechanism by an Iridium(I)-NHC Complex. *Chem. Commun.* **2015,** *51,* 3151–3153.
41. Li, Y.; Tan, C.-P.; Zhang, W.; He, L.; Ji, L.-N.; Mao, Z.-W. Phosphorescent Iridium(III)-bis-N-Heterocyclic Carbene Complexes as Mitochondria-Targeted Theranostic and Photodynamic Anticancer Agents. *Biomaterials* **2015,** *39,* 95–104.
42. Simpson, P. V.; Radacki, K.; Braunschweig, H.; Schatzschneider, U. An Iridium N-Heterocyclic Carbene Complex $IrCl(CO)_2$(NHC) as a Carbon Monoxide-Releasing Molecule (CORM). *J. Organomet. Chem.* **2015,** *782,* 116–123.
43. Rosenberg, B.; L, V.; Krigas, T. Inhibition of Cell Division in *Escherichia coli* by Electrolysis Products From a Platinum Electrode. *Nature* **1965,** *205,* 698–699.
44. Barnes, K. R.; Lippard, S. J. Cisplatin and Related Anticancer Drugs: Recent Advances and Insights. *Met. Ions Biol. Syst.* **2004,** *42,* 143–177.
45. Wang, D.; Lippard, S. J. Cellular Processing of Platinum Anticancer Drugs. *Nat. Rev. Drug Disc.* **2005,** *4,* 307–320.
46. Skander, M.; Retailleau, P.; Bourrie, B.; Schio, L.; Mailliet, P.; Marinetti, A. N-Heterocyclic Carbene-Amine Pt(II) Complexes, a New Chemical Space for the Development of Platinum-Based Anticancer Drugs. *J. Med. Chem.* **2010,** *53,* 2146–2154.
47. Chtchigrovsky, M.; Eloy, L.; Jullien, H.; Saker, L.; Segal-Bendirdjian, E.; Poupon, J.; Bombard, S.; Cresteil, T.; Retailleau, P.; Marinetti, A. Antitumor Trans-N-Heterocyclic Carbene-Amine-Pt(II) Complexes: Synthesis of Dinuclear Species and Exploratory Investigations of DNA Binding and Cytotoxicity Mechanisms. *J. Med. Chem.* **2013,** *56,* 2074–2086.
48. Sun, R. W. -Y.; Chow, A. L. -F.; Li, X.-H.; Yan, J. J.; Chui, S. S. -Y.; Che, C.-M. Luminescent Cyclometalated Platinum(II) Complexes Containing N-Heterocyclic Carbene Ligands With Potent In Vitro and In Vivo Anti-cancer Properties Accumulate in Cytoplasmic Structures of Cancer Cells. *Chem. Sci.* **2011,** *2,* 728–736.
49. Zou, T.; Lok, C.-N.; Fung, Y. M. E.; Che, C.-M. Luminescent Organoplatinum(II) Complexes Containing Bis(N-Heterocyclic Carbene) Ligands Selectively Target the Endoplasmic Reticulum and Induce Potent Photo-toxicity. *Chem. Commun.* **2013,** *49,* 5423–5425.
50. Muenzner, J. K.; Rehm, T.; Biersack, B.; Casini, A.; de Graaf, I. A.; Worawutputtapong, P.; Noor, A.; Kempe, R.; Brabec, V.; Kasparkova, J.; Schobert, R. Adjusting the DNA Interaction and Anticancer Activity of Pt(II) N-Heterocyclic Carbene Complexes by Steric Shielding of the Trans Leaving Group. *J. Med. Chem.* **2015,** *58,* 6283–6292.
51. Koch, R. Über bakteriologische Forschung. In *Verhandlungen des X. Internationalen Medizinischen Kongresses, Berlin 1890. 1891 Bd.I;* Verlag von August Hirschwald: Berlin, 1890.
52. Kean, W. F.; Kean, I. R. Clinical Pharmacology of Gold. *Inflammopharmacology* **2008,** *16,* 112–125.
53. Madeira, J. M.; Gibson, D. L.; Kean, W. F.; Klegeris, A. The Biological Activity of Auranofin: Implications for Novel Treatment of Diseases. *Inflammopharmacology* **2012,** *20,* 297–306.
54. Ott, I. On the Medicinal Chemistry of Gold Complexes as Anticancer Drugs. *Coord. Chem. Rev.* **2009,** *253,* 1670–1681.
55. Bertrand, B.; Casini, A. A Golden Future in Medicinal Inorganic Chemistry: The Promise of Anticancer Gold Organometallic Compounds. *Dalton Trans.* **2014,** *43,* 4209–4219.
56. Zou, T.; Lum, C. T.; Lok, C. N.; Zhang, J. J.; Che, C. M. Chemical Biology of Anticancer Gold(III) and Gold(I) Complexes. *Chem. Soc. Rev.* **2015**.
57. Hickey, J. L.; Ruhayel, R. A.; Barnard, P. J.; Baker, M. V.; Berners-Price, S. J.; Filipovska, A. Mitochondria-Targeted Chemotherapeutics: The Rational Design of Gold(I) N-Heterocyclic Carbene Complexes That Are Selectively Toxic to Cancer Cells and Target Protein Selenols in Preference to Thiols. *J. Am. Chem. Soc.* **2008,** *130,* 12570–12571.

58. Baker, M.; Barnard, P. J.; Berners-Price, S. J.; Brayshaw, S. K.; Hickey, J. L.; Skelton, B. W.; White, A. H. Cationic, Linear Au(I) N-Heterocyclic Carbene Complexes: Synthesis, Structure and Anti-mitochondrial Activity. *Dalton Trans.* **2006,** 3708–3715.
59. Barnard, P. J.; Baker, M.V.; Berners-Price, S. J.; Day, D. A. Mitochondrial Permeability Transition Induced by Dinuclear Gold(I)–Carbene Complexes: Potential New Antimitochondrial Antitumor Agents. *J. Inorg. Biochem.* **2004,** *98,* 1642–1647.
60. Rubbiani, R.; Kitanovic, I.; Alborzinia, H.; Can, S.; Kitanovic, A.; Onambele, L. A.; Stefanopoulou, M.; Geldmacher, Y.; Sheldrick, W. S.; Wolber, G.; Prokop, A.; Wölfl, S.; Ott, I. Benzimidazol-2-ylidene Gold(I) Complexes Are Thioredoxin Reductase Inhibitors with Multiple Antitumor Properties. *J. Med. Chem.* **2010,** *53,* 8608–8618.
61. Urig, S.; Fritz-Wolf, K.; Reau, R.; Herold-Mende, C.; Toth, K.; Davioud-Charvet, E.; Becker, K. Undressing of Phosphine Gold(I) Complexes as Irreversible Inhibitors of Human Disulfide Reductases. *Angew. Chem. Int. Ed.* **2006,** *45,* 1881–1886.
62. Deponte, M.; Urig, S.; Arscott, L. D.; Fritz-Wolf, K.; Réau, R.; Herold-Mende, C.; Koncarevic, S.; Meyer, M.; Davoiud-Charvet, E.; Ballou, D. P.; Williams, C. H.; Becker, K. Mechanistic Studies on a Novel, Highly Potent Gold-Phosphole Inhibitor of Human Glutathione Reductase. *J. Biol. Chem.* **2005,** *280,* 20628–20637.
63. Liu, W.; Bensdorf, K.; Proetto, M.; Hagenbach, A.; Abram, U.; Gust, R. Synthesis, Characterization, and In Vitro Studies of Bis[1,3-diethyl-4,5-diarylimidazol-2-ylidene] Gold(I/III) Complexes. *J. Med. Chem.* **2012,** *55,* 3713–3724.
64. Rubbiani, R.; Schuh, E.; Meyer, A.; Lemke, J.; Wimberg, J.; Metzler-Nolte, N.; Meyer, F.; Mohr, F.; Ott, I. TrxR Inhibition and Antiproliferative Activities of Structurally Diverse Gold N-Heterocyclic Carbene Complexes. *Med. Chem. Commun.* **2013,** *4,* 942–948.
65. Pratesi, A.; Gabbiani, C.; Michelucci, E.; Ginanneschi, M.; Papini, A. M.; Rubbiani, R.; Ott, I.; Messori, L. Insights on the Mechanism of Thioredoxin Reductase Inhibition by Gold N-Heterocyclic Carbene Compounds Using a Synthetic Selenododecapeptide: An ESI-MS Investigation. *J. Inorg. Biochem.* **2014**. accepted.
66. Messori, L.; Marchetti, L.; Massai, L.; Scaletti, F.; Guerri, A.; Landini, I.; Nobili, S.; Perrone, G.; Mini, E.; Leoni, P.; Pasquali, M.; Gabbiani, C. Chemistry and Biology of Two Novel Gold(I) Carbene Complexes as Prospective Anticancer Agents. *Inorg. Chem.* **2014,** *53,* 2396–2403.
67. Schuh, E.; Pflüger, C.; Citta, A.; Folda, A.; Rigobello, M. P.; Bindoli, A.; Casini, A.; Mohr, F. Gold(I) Carbene Complexes Causing Thioredoxin 1 and Thioredoxin 2 Oxidation as Potential Anticancer Agents. *J. Med. Chem.* **2012,** *55,* 5518–5528.
68. Cheng, X.; Holenya, P.; Can, S.; Alborzinia, H.; Rubbiani, R.; Ott, I.; Wolfl, S. A TrxR Inhibiting Gold(I) NHC Complex Induces Apoptosis Through ASK1-p38-MAPK Signaling in Pancreatic Cancer Cells. *Mol. Cancer* **2014,** *13,* 221.
69. Bertrand, B.; Stefan, L.; Pirrotta, M.; Monchaud, D.; Bodio, E.; Richard, P.; Le Gendre, P.; Warmerdam, E.; de Jager, M. H.; Groothuis, G. M.; Picquet, M.; Casini, A. Caffeine-Based Gold(I) N-Heterocyclic Carbenes as Possible Anticancer Agents: Synthesis and Biological Properties. *Inorg. Chem.* **2014,** *53,* 2296–2303.
70. Holenya, P.; Can, S.; Rubbiani, R.; Alborzinia, H.; Junger, A.; Cheng, X.; Ott, I.; Wolfl, S. Detailed Analysis of Pro-Apoptotic Signaling and Metabolic Adaptation Triggered by a N-Heterocyclic Carbene-Gold(I) Complex. *Metallomics* **2014,** *6,* 1591–1601.
71. Krishnamurthy, D.; Karver, M. R.; Fiorillo, E.; Orru, V.; Stanford, S. M.; Bottini, N.; Barrios, A. M. Gold(I)-Mediated Inhibition of Protein Tyrosine Phosphatases: A Detailed In Vitro and Cellular Study. *J. Med. Chem.* **2008,** *51,* 4790–4795.
72. Rubbiani, R.; Salassa, L.; de Almeida, A.; Casini, A.; Ott, I. Cytotoxic Gold(I) N-Heterocyclic Carbene Complexes With Phosphane Ligands as Potent Enzyme Inhibitors. *ChemMedChem* **2014,** *9,* 1205–1210.
73. Liu, W.; Bensdorf, K.; Proetto, M.; Abram, U.; Hagenbach, A.; Gust, R. NHC Gold Halide Complexes Derived from 4,5-Diarylimidazoles: Synthesis, Structural Analysis, and Pharmacological Investigations as Potential Antitumor Agents. *J. Med. Chem.* **2011,** *54,* 8605–8615.

74. Zou, T.; Lum, C. T.; Chui, S. S.-Y.; Che, C.-M. Gold(III) Complexes Containing N-Heterocyclic Carbene Ligands: Thiol "Switch-on" Fluorescent Probes and Anti-cancer Agents. *Angew. Chem. Int. Ed. Engl.* **2013,** *52,* 1–5.
75. Zou, T.; Lum, C. T.; Lok, C.-N.; To, W.-P.; Low, K.-H.; Che, C.-M. A Binuclear Gold(I) Complex with Mixed Bridging Diphosphine and Bis(N-Heterocyclic Carbene) Ligands Shows Favorable Thiol Reactivity and Inhibits Tumor Growth and Angiogenesis In Vivo. *Angew. Chem. Int. Ed.* **2014,** *53,* 5810–5814.
76. Sun, R. W. -Y.; Zhang, M.; Li, D.; Zhang, Z.-F.; Cai, H.; Li, M.; Xian, Y.-J.; Ng, S. W.; Wong, A. S. -T. Dinuclear Gold(I) Pyrrolidinedithiocarbamato Complex: Cytotoxic and Antimigratory Activities on Cancer Cells and the Use of Metal–Organic Framework. *Chem. Eur. J.* **2015,** *21,* 18534–18538.
77. Fernandez, G. A.; Vela Gurovic, M. S.; Olivera, N. L.; Chopa, A. B.; Silbestri, G. F. Antibacterial Properties ofWater-Soluble Gold(I) N-Heterocyclic Carbene Complexes. *J. Inorg. Biochem.* **2014,** *135,* 54–57.
78. Samanta, T.; Roymahapatra, G.; Porto, W. F.; Seth, S.; Ghorai, S.; Saha, S.; Sengupta, J.; Franco, O. L.; Dinda, J.; Mandal, S. M. N, N′-Olefin Functionalized Bis-imidazolium Gold(I) Salt Is an Efficient Candidate to Control Keratitis-Associated Eye Infection. *PLoS One* **2013,** *8,* e58346.
79. Hemmert, C.; Fabie, A.; Fabre, A.; Benoit-Vical, F.; Gornitzka, H. Synthesis, Structures, and Antimalarial Activities of Some Silver(I), Gold(I) and Gold(III) Complexes Involving N-Heterocyclic Carbene Ligands. *Eur. J. Med. Chem.* **2013,** *60,* 64–75.
80. Özdemir, I.; Temelli, N.; Günal, S.; Demir, S. Gold(I) Complexes of N-Heterocyclic Carbene Ligands Containing Benzimidazole: Synthesis and Antimicrobial Activity. *Molecules* **2010,** *15,* 2203–2210.
81. Alfaro, J. M.; Prades, A.; del Carmen Ramos, M.; Peris, E.; Ripoll-Gomez, J.; Poyatos, M.; Burgos, J. S. Biomedical Properties of a Series of Ruthenium-N-Heterocyclic Carbene Complexes Based on Oxidant Activity In Vitro and Assessment In Vivo of Biosafety in Zebrafish Embryos. *Zebrafish* **2010,** *7,* 13–21.
82. Sivaram, H.; Tan, J.; Huynh, H. V. Cationic Gold(I) Heteroleptic Complexes Bearing a Pyrazole-Derived N-Heterocyclic Carbene: Syntheses, Characterizations, and Cytotoxic Activities. *Dalton Trans.* **2013,** *42,* 12421–12428.
83. Kilpin, K. J.; Crot, S.; Riedel, T.; Kitchen, J. A.; Dyson, P. J. Ruthenium(II) and Osmium(II) 1,2,3-Triazolylidene Organometallics: A Preliminary Investigation into the Biological Activity of 'Click' Carbene Complexes. *Dalton Trans.* **2014,** *43,* 1443–1448.
84. Köster, S. D.; Alborzinia, H.; Can, S.; Kitanovic, I.; Wölfl, S.; Rubbiani, R.; Ott, I.; Riesterer, P.; Prokop, A.; Merz, K.; Metzler-Nolte, N. A Spontaneous Gold(I)-Azide Alkyne Cycloaddition Reaction Yields Gold-Peptide Bioconjugates Which Overcome Cisplatin Resistance in a p53-Mutant Cancer Cell Line. *Chem. Sci.* **2012,** *3,* 2062–2072.
85. Serebryanskaya, T. V.; Zolotarev, A. A.; Ott, I. A Novel Aminotriazole-Based NHC Complex for the Design of Gold(I) Anti-cancer Agents: Synthesis and Biological Evaluation. *Med. Chem. Commun.* **2015,** *6,* 1186–1189.
86. Pellei, M.; Gandin, V.; Marinelli, M.; Orsetti, A.; Del Bello, F.; Santini, C.; Marzano, C. Novel Triazolium Based 11th Group NHCs: Synthesis, Characterization and Cellular Response Mechanisms. *Dalton Trans.* **2015,** *44,* 21041–21052.
87. Gutiérrez, A.; Gimeno, M. C.; Marzo, I.; Metzler-Nolte, N. Synthesis, Characterization, and Cytotoxic Activity of AuI N,S-Heterocyclic Carbenes Derived from Peptides Containing L-Thiazolylalanine. *Eur. J. Inorg. Chem.* **2014,** 2512–2519.
88. Alves, G.; Morel, L.; El-Ghozzi, M.; Avignant, D.; Legeret, B.; Nauton, L.; Cisnetti, F.; Gautier, A. A Platinum Chugaev Carbene Complex as a Potent Anticancer Agent. *Chem. Commun.* **2011,** *47,* 7830–7832.
89. Horvath, U. E. I.; Bentivoglio, G.; Hummel, M.; Schottenberger, H.; Wurst, K.; Nell, M. J.; van Rensburg, C. E. J.; Cronje, S.; Raubenheimer, H. G. A Cytotoxic Bis(Carbene) Gold(I) Complex of Ferrocenyl Complexes: Synthesis and Structural Characterisation. *New J. Chem.* **2008,** *32,* 533–539.

90. Arambula, J. F.; McCall, R.; Sidoran, K. J.; Magda, D.; Mitchell, N. A.; Bielawski, C. W.; Lynch, V. M.; Sessler, J. L.; Arumugam, K. Targeting Antioxidant Pathways with Ferrocenylated N-Heterocyclic Carbene Supported Gold(I) Complexes in A549 Lung Cancer Cells. *Chem. Sci.* **2016,** *7,* 1245–1256.
91. Fernandez-Gallardo, J.; Elie, B. T.; Sanau, M.; Contel, M. Versatile Synthesis of Cationic N-Heterocyclic Carbene-Gold(I) Complexes Containing a Second Ancillary Ligand. Design of Heterobimetallic Ruthenium-Gold Anticancer Agents. *Chem. Commun.* **2016,** *52,* 3155–3158.
92. Boselli, L.; Carraz, M.; Mazères, S.; Paloque, L.; González, G.; Benoit-Vical, F.; Valentin, A.; Hemmert, C.; Gornitzka, H. Synthesis, Structures, and Biological Studies of Heterobimetallic Au(I)−Ru(II) Complexes Involving N-Heterocyclic Carbene-Based Multidentate Ligands. *Organometallics* **2015,** *34,* 1046–1055.
93. Bertrand, B.; Citta, A.; Franken, I. L.; Picquet, M.; Folda, A.; Scalcon, V.; Rigobello, M. P.; Le Gendre, P.; Casini, A.; Bodio, E. Gold(I) NHC-Based Homo- and Heterobimetallic Complexes: Synthesis, Characterization and Evaluation as Potential Anticancer Agents. *J. Biol. Inorg. Chem.* **2015,** *20,* 1005–1020.
94. Lemke, J.; Pinto, A.; Niehoff, P.; Vasylyeva, V.; Metzler-Nolte, N. Synthesis, Structural Characterisation and Anti-proliferative Activity of NHC Gold Amino Acid and Peptide Conjugates. *Dalton Trans.* **2009,** 7063–7070.
95. Kascatan-Nebioglu, A.; Melaiye, A.; Hindi, K.; Durmus, S.; Panzner, M. J.; Hogue, L. A.; Mallett, R. J.; Hovis, C. E.; Coughenour, M.; Crosby, S. D.; Milsted, A.; Ely, D. L.; Tessier, C. A.; Cannon, C. L.; Youngs, W. J. Synthesis from Caffeine of a Mixed N-Heterocyclic Carbene−Silver Acetate Complex Active Against Resistant Respiratory Pathogens. *J. Med. Chem.* **2006,** *49,* 6811–6818.
96. Cannon, C. L.; Hogue, L. A.; Vajravelu, R. K.; Capps, G. H.; Ibricevic, A.; Hindi, K. M.; Kascatan-Nebioglu, A.; Walter, M. J.; Brody, S. L.; Youngs, W. J. In Vitro and Murine Efficacy and Toxicity Studies of Nebulized SCC1, a Methylated Caffeine-Silver(I) Complex, for Treatment of Pulmonary Infections. *Antimicrob. Agents Chemother.* **2009,** *53,* 3285–3293.
97. Zhang, J.-J.; Che, C.-M.; Ott, I. Caffeine Derived Platinum(II) N-Heterocyclic Carbene Complexes with Multiple Anti-cancer Activities. *J. Organomet. Chem.* **2015,** *782,* 37–41.
98. Brana, M. F.; Ramos, A. Naphthalimides as Anticancer Agents: Synthesis and Biological Activity. *Curr. Med. Chem. Anti-cancer Agents* **2001,** *1,* 237–255.
99. Banerjee, S.; Veale, E. B.; Phelan, C. M.; Murphy, S. A.; Tocci, G. M.; Gillespie, L. J.; Frimannsson, D. O.; Kelly, J. M.; Gunnlaugsson, T. Recent Advances in the Development of 1,8-Naphthalimide Based DNA Targeting Binders, Anticancer and Fluorescent Cellular Imaging Agents. *Chem. Soc. Rev.* **2013,** *42,* 1601–1618.
100. Meyer, A.; Oehninger, L.; Geldmacher, Y.; Alborzinia, H.; Wolfl, S.; Sheldrick, W. S.; Ott, I. Gold(I) N-Heterocyclic Carbene Complexes with Naphthalimide Ligands as Combined Thioredoxin Reductase Inhibitors and DNA Intercalators. *ChemMedChem* **2014,** *9,* 1794–1800.
101. Kaps, L.; Biersack, B.; Müller-Bunz, H.; Mahal, K.; Münzner, J.; Tacke, M.; Müller, T.; Schobert, R. Gold(I)–NHC Complexes of Antitumoral Diarylimidazoles: Structures, Cellular Uptake Routes and Anticancer Activities. *J. Inorg. Biochem.* **2012,** *106,* 52–58.
102. Muenzner, J. K.; Biersack, B.; Kalie, H.; Andronache, I. C.; Kaps, L.; Schuppan, D.; Sasse, F.; Schobert, R. Gold(I) Biscarbene Complexes Derived From Vascular-Disrupting Combretastatin A-4 Address Different Targets and Show Antimetastatic Potential. *ChemMedChem* **2014,** *9,* 1195–1204.

CHAPTER SIX

Metal Complexes as Delivery Systems for CO, NO, and H_2S to Explore the Signaling Network of Small-Molecule Messengers

U. Schatzschneider
Julius-Maximilians-Universität Würzburg, Würzburg, Germany

Contents

1. INTRODUCTION

Although mostly known for their toxicity, carbon monoxide (CO), nitric oxide (NO), and hydrogen sulfide (H_2S) are possibly the smallest natural products. They are endogenously produced by well-controlled enzymatic

Inorganic and Organometallic Transition Metal Complexes with Biological Molecules and Living Cells
ISBN 978-0-12-803814-7
http://dx.doi.org/10.1016/B978-0-12-803814-7.00006-X

processes in higher organisms, including humans, and serve a vital signaling function at low physiological concentration, in particular in response to cellular stress conditions. In the latter context, these small molecules are increasingly explored for their therapeutic potential in human medicine.[1–3] In addition to stimulation of their enzymatic generation in cells, artificial delivery systems are also evaluated for CO, NO, and H_2S. While most prodrugs for hydrogen sulfide are based on organic sulfur compounds, transition metal carbonyl and nitrosyl complexes are particularly relevant as carriers for CO and NO. In the present chapter, we will first briefly cover the endogenous pathways of enzymatic CO, NO, and H_2S production and then contrast these with the most relevant artificial delivery systems. Next, molecular probes for in vitro detection of these small-molecule messengers will be discussed. Finally, the chapter concludes with a critical discussion of what is currently known about the downstream signaling pathways triggered by CO, NO, and H_2S.

2. ENZYMATIC GENERATION OF CO, NO, AND H_2S

The enzymatic pathways of CO, NO, and H_2S generation are generally well understood these days. NO and H_2S originate from the amino acids L-arginine and L-methionine, respectively, while CO is generated by degradation of heme. Only one endogenous pathway exists for CO and NO production, albeit involving different isoforms of heme oxygenase (HO) and nitric oxide synthase (NOS) enzymes, while there are multiple alternative processes for H_2S generation, involving either cystathionine γ-lyase (CSE), cystathionine β-synthase (CBS), 3-mercaptopyruvate sulfurtransferase (3-MST), or cysteine lyase (CL) in the key step.

2.1 Nitric Oxide Synthase

NO is endogenously produced by the enzymatic oxidation of L-arginine via intermediate N^{ω}-hydroxy-L-arginine to L-citrulline. Both reactions are catalyzed by NOS, with the NO released during the second step (Fig. 6.1).

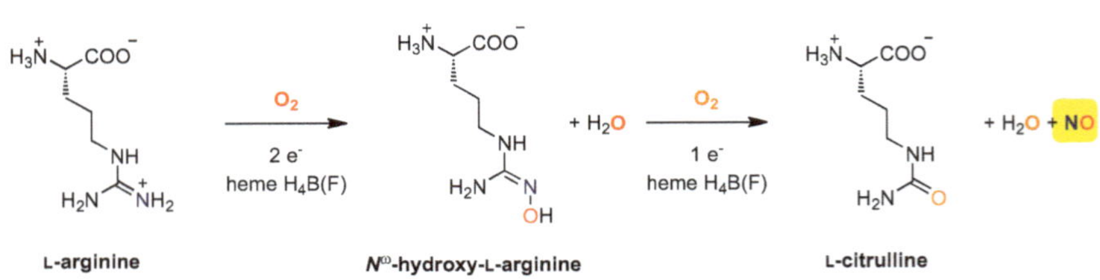

Figure 6.1 Nitric oxide synthase (NOS) oxidizes L-arginine to L-citrulline and nitric oxide (NO).

In this process, two equivalents of dioxygen as well as three electrons delivered by tetrahydrobiopterin or tetrahydrofolate are consumed. Three different isoforms of the enzyme have been identified in different tissues, with endothelial NOS active in the cardiovascular system, neuronal NOS operating in the nervous system, and inducible NOS involved in immune response.[4–6] Because of the highly reactive nature of NO, its concentration is difficult to determine in biological systems, but is generally assumed to be in the nanomolar range, and thus much lower than that of CO and H_2S.[7]

2.2 Hydrogen Sulfide Generation

The endogenous generation of H_2S is characterized by multiple alternative pathways, which, however, all share L-cysteine or L-homocysteine as the common precursor.[8–13] L-Cysteine can either undergo α,β-elimination to pyruvate and ammonia catalyzed by CSE or participate in group exchange reactions with water or sulfite mediated by CBS or CL, leading to L-serine or L-cysteate, respectively. All three processes result in the release of one equivalent of H_2S per mole of L-cysteine (Fig. 6.2). CSE can also utilize L-homocysteine instead of L-cysteine in an α,γ-elimination, leading to formation of α-ketobutyrate and ammonia together with H_2S. An alternative reaction pathway is also observed for CBS. Using L-homocysteine instead of water as the substrate for the reaction with L-cysteine, the product is the mixed thioether L-cystathionine together with H_2S. Alternatively, CBS can also metabolize two equivalents of either L-cysteine or L-homocysteine, leading to L-lanthionine or L-homolanthionine, respectively, all accompanied by generation of H_2S. An additional pathway starts from oxidation of L-cysteine to disulfide L-cysteine, which also acts as a substrate for CSE, leading to pyruvate, ammonia, and L-thiocysteine, incorporating a hydropersulfide (R-S-SH) group, which can be reduced to H_2S and L-cysteine.

Finally, 3-mercaptopyruvate, resulting from a transamination process of L-cysteine with α-ketoglutarate, is reduced by 3-MST to pyruvate and H_2S. The latter process is also distinct from the others since it occurs in both mitochondria and cytosol, while the other reactions are exclusively cytosolic. It might also be of interest to the bioinorganic chemist that CBS, CSE, and CL are pyridoxal-5′-phosphate-dependent enzymes while 3-MST harbors a zinc in the active center. The physiological concentration of H_2S generated by these systems has been determined in the 10–100-μM range for mammals, although an accurate determination is particularly difficult in this case because of the different and interconvertible low-molecular weight sulfur species in cells.[7,12,14]

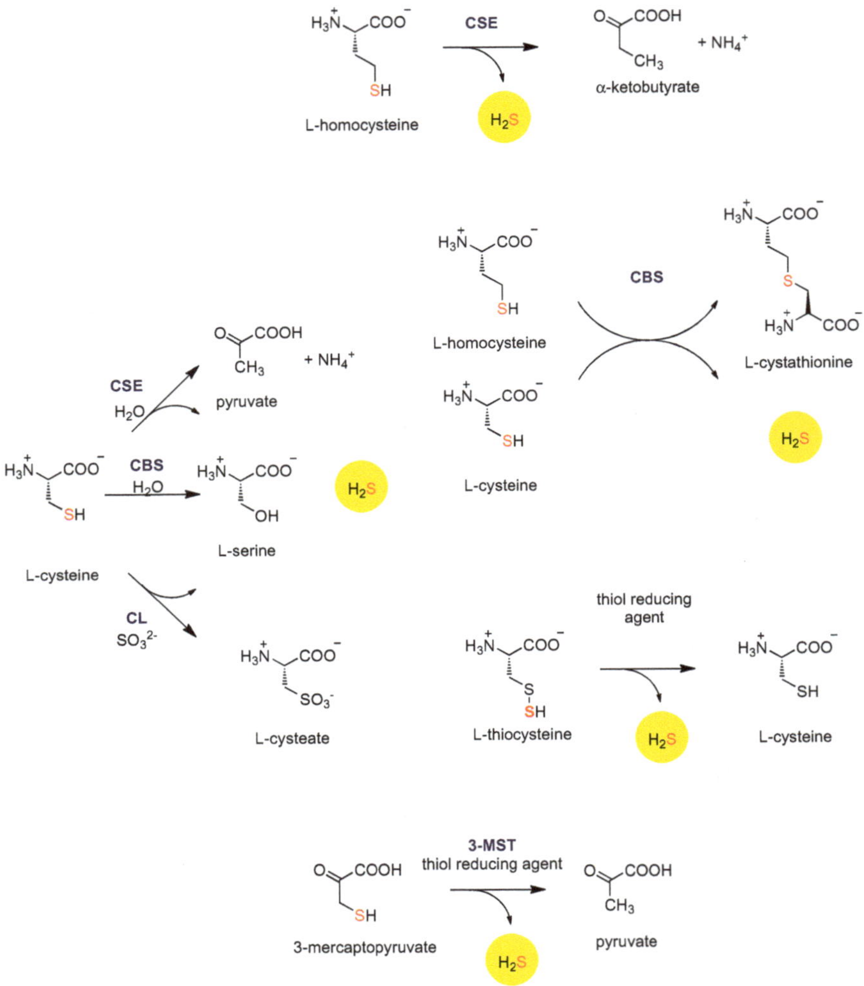

Figure 6.2 Different pathways for H_2S generation from sulfur-containing amino acids and their derivatives. *3-MST*, 3-mercaptopyruvate sulfurtransferase; *CBS*, cystathionine β-synthese; *CL*, cysteine lyase; *CSE*, cystathionine γ-lyase.

2.3 Heme Oxygenases

CO is generated by HOs, which are unique since the heme prosthetic group has a dual role as active center and substrate.[15–18] In the first step, iron protoporphyrin IX binds dioxygen, forming an iron(III) hydroperoxo intermediate, which regiospecifically hydroxylates the α-meso carbon atom of the heme (Fig. 6.3). In a spontaneous autoxidation step using a second

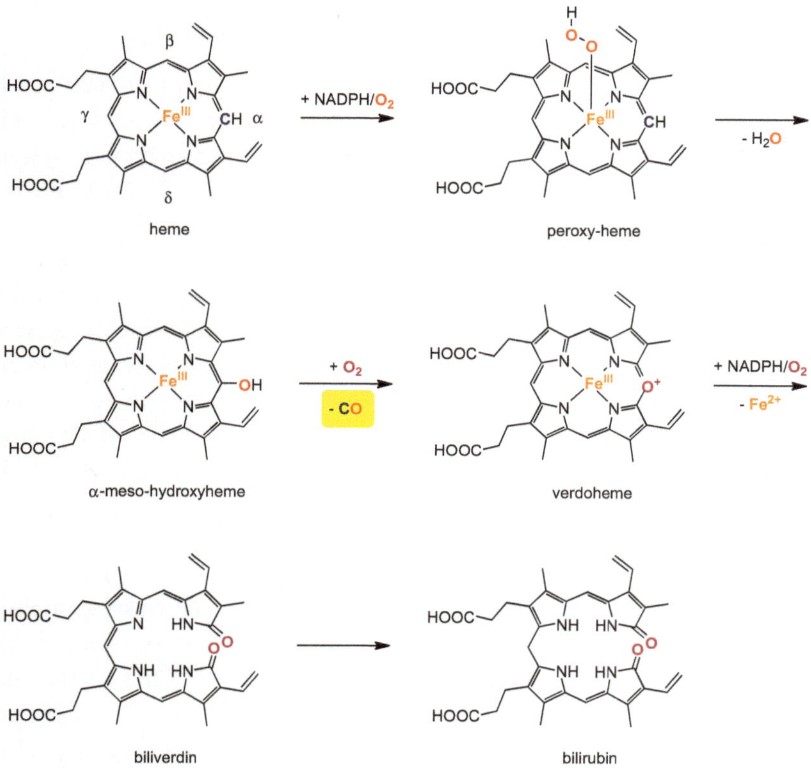

Figure 6.3 Heme oxygenase (HO) regiospecifically oxidizes the α-meso-position of heme to carbon monoxide and verdoheme, which is then further cleaved to biliverdin, the latter finally converted to bilirubin by biliverdin reductase.

equivalent of dioxygen, verdoheme is formed while the carbon atom is released as CO. Finally, the cyclic tetrapyrrole is cleaved under consumption of a third molecule of O_2, resulting in the formation of biliverdin with concomitant release of ferrous iron. In addition to one molecule of heme and three equivalents of dioxygen, this process also consumes two equivalents of nicotinamide adenine dinucleotide phosphate. Finally, in a separate step, linear tetrapyrrole biliverdin is converted to bilirubin by biliverdin reductase. Two different isoforms of HO have been identified. The baseline production of CO is caused by constitutively expressed HO-2 while inducible HO-1 is upregulated under stress conditions. Although variable between different tissues, it is generally assumed that CO concentrations in cells are in the low micromolar range.[7]

3. DELIVERY SYSTEMS FOR CO, NO, AND H_2S

To increase tissue levels of the three small signaling molecules CO, NO, and H_2S for potential therapeutic applications, two distinct strategies are explored. On the one hand, the naturally occurring enzymatic systems can be stimulated by exogenous drugs, or alternatively, CO, NO, and H_2S directly or indirectly delivered to target sites in the organism. Although there is some work on devices for the direct application of CO gas to patients,[2] this method is rather difficult to control and thus poses an inherent danger to both the patient and the medical personnel involved. More importantly, however, the inhalative uptake of CO via the lungs does not allow much control of its tissue distribution beyond some simple and fixed differences in accumulation because of variable lipophilicity of the target organs. Therefore small molecules as delivery vectors for CO, NO, and H_2S are in the focus of attention.

3.1 Nitric Oxide Delivery

In the case of NO, organic NO donors as well as metal-nitrosyl complexes have received considerable attention for more than 150 years, although NO was only identified much later as the active agent.[19–21] One of the most prominent examples of such use is possibly the prescription of nitroglycerin (glyceryl trinitrate) to Alfred Nobel for treatment of his angina pectoris (Fig. 6.4). Another prominent example of a metal-nitrosyl complex in clinical use to control hypertensive emergencies is sodium nitroprusside, $Na_2[Fe(CN)_5(NO)]$, which undergoes both enzymatic as well as non-enzymatic degradation to NO. A host of other compounds has also been explored over the last couple of decades. Among the organic NO donors, organic nitrites and nitrates as well as *N*-nitrosamines, *N*-nitrosimines, nitrosothiols, and *C*-nitroso compounds feature prominently, in addition to heterocycles and other more exotic species.[19] While sodium nitroprusside is probably the most prominent transition metal-based NO donor, nitrosyl complexes based on other metals, in particular ruthenium and manganese, have also been studied. On the other hand, iron-nitrosyl complexes are still a matter of much active research, with a particular focus on dinitrosyl-iron complexes (DNICs).[22–24]

3.2 Hydrogen Sulfide Delivery

There has been quite some controversy in the community over the use of proper H_2S donors.[12,25,26] The most simple pathway of application is

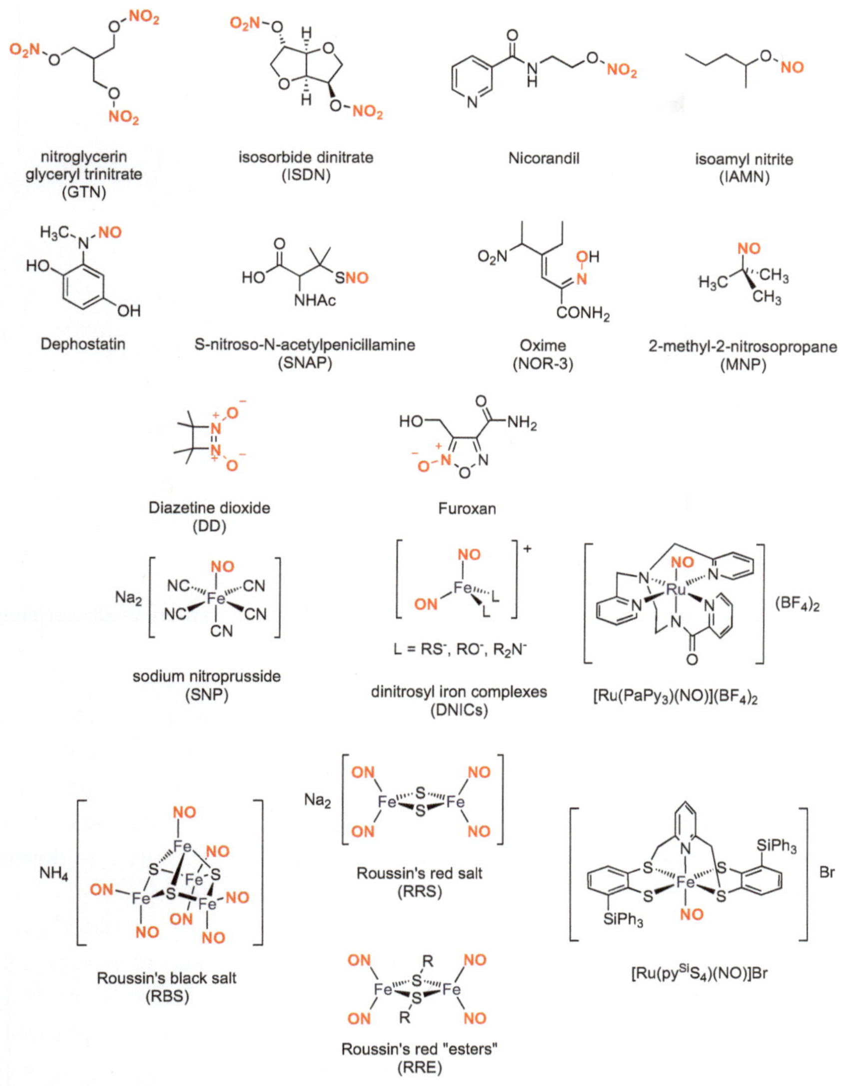

Figure 6.4 Organic nitric oxide (NO) donors (top) and metal-nitrosyl complexes (bottom) as artificial delivery systems for NO.

possibly H_2S gas inhalation, but this is severely limited by issues with concentration control and prevention of overdosing.

Many studies use simple commercially available salts such as sodium hydrosulfide or sodium sulfide. However, their dissolution in aqueous medium results in a very fast increase in H_2S concentration within just a few seconds, which is

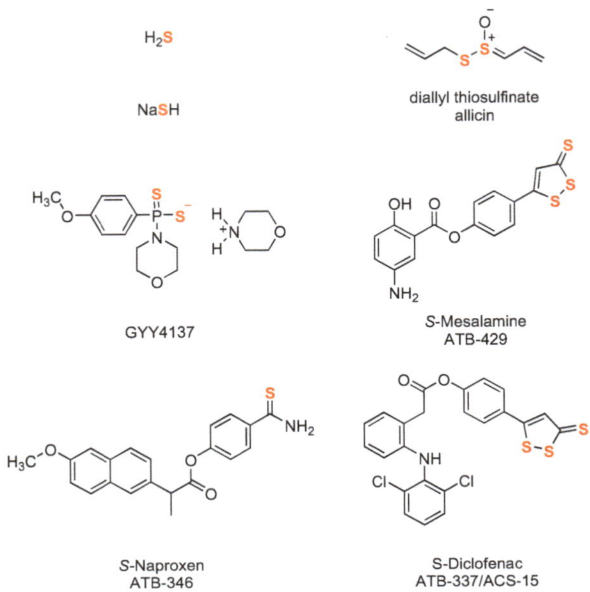

Figure 6.5 A selection of hydrogen sulfide delivery systems based on small inorganic compounds (top left) and organic sulfur-containing molecules.

nonphysiological compared to slow enzymatic H_2S generation. Furthermore, H_2S is lost to the atmosphere in open containers within just a few minutes.[27] An additional problem is contamination of commercial samples, in particular of sodium hydrosulfide, and formation of polysulfides.[28] Therefore, use of these compounds is discouraged and research has now shifted to slow-release donors that liberate H_2S on a physiologically more relevant time profile.

For a long time, the positive effects of garlic consumption on blood pressure and heart protection have been noticed, and are linked to natural sulfur compounds such as diallyl thiosulfinate (allicin, Fig. 6.5). Novel recently published synthetic compounds for slow H_2S release include the phosphinodithioate GYY4137, which enables H_2S delivery over several days, as well as several hybrid compounds combining an H_2S-releasing moiety based on carbothioamides or 1,2-dithiole-3-thiones with established nonsteroidal anti-inflammatory drugs such as mesalamine, naproxen, or diclofenac (Fig. 6.5).[12,25]

3.3 Carbon Monoxide Delivery

In the case of CO, some devices have been developed for the controlled quantitative delivery of CO by inhalation to patients through a nasal cannula.[2] However, the major issue with this form of delivery is its low target specificity because of the CO accumulating in the body based on the distribution coefficient between blood and tissue, which is a fixed value that cannot be modified by molecular design in this diatomic compound. Therefore small molecules are the focus of research as *CO-releasing molecules* (CORMs), which allow full control over both release kinetics and biodistribution.[29,30] The type of trigger mechanism to initiate the CO release from a carrier molecule as well as the timescale on which this process takes place is controlled by proper choice of the metal–(co)ligand combination, commonly referred to as the "CORM sphere," while tissue specificity and targeting can be modulated by variation of the outer ligand periphery, called the "drug sphere" in this context (Fig. 6.6a).

So far, three major types of trigger mechanism have been explored to initiate CO release from a carrier molecule acting as a prodrug for CO. Most CORMs currently used in biological studies are based on ligand-triggered CO release, in which a solvent molecule from the medium displaces one or more CO ligands from a metal–carbonyl CORM. A wide range of half-lives for CO release have been reported for this methodology from minutes to hours. However, a major disadvantage of this strategy is related to the fact that the decomposition of the CO prodrug starts as soon as the solid material is dissolved in proper medium and therefore a controlled starting point for the study of biological effects is rather difficult to achieve; tissue-specific activation is also not possible because of high abundance of potential donor group capable of CO replacement in biological systems.[31]

To address this problem, *enzyme-triggered CORMs* were developed by the group of Schmalz, in which an enzymatic transformation of a peripheral group in a η^4-cyclohexadiene ligand triggers a hapticity change leading to increased susceptibility to decomposition, resulting in release of iron(III), CO, and the organic ligand.[32–36]

An alternative way to achieve tight spatial and temporal control of the CO release was inspired by organic "caged" compounds, in which a photolabile protective group is attached to a bioactive molecule rendered inactive by this modification. In these *photoactivatable CO-releasing molecules* (photoCORMs), a metal–coligand fragment acts as the protective group to the carbonyl ligands attached, which are released from the metal coordination sphere upon illumination.[30,37–41] A major issue with photoCORMs is the inverse correlation between the tissue penetration depth of light and its wavelength, which requires careful tuning of the photochemical properties of the compounds to shift the activation wavelength toward the red part of the spectrum.

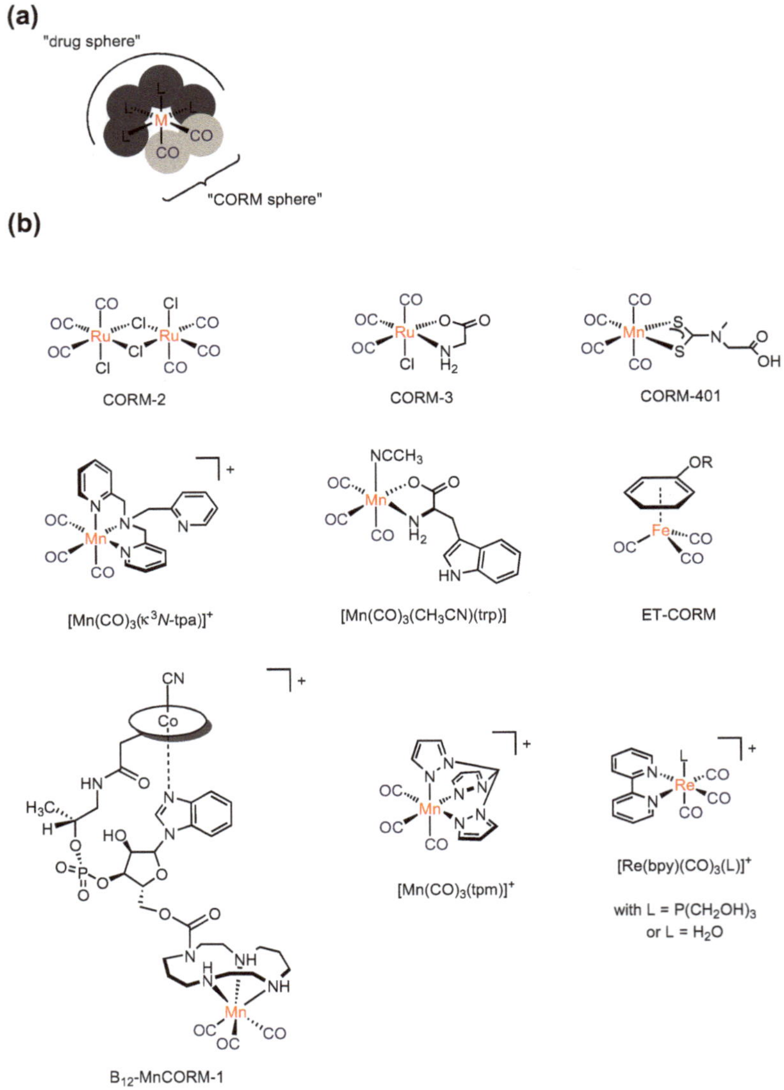

Figure 6.6 (a) The metal–(co)ligand combination, referred to as the "CORM sphere," controls the trigger mechanism as well as timescale of the carbon monoxide release process while the outer "drug sphere" determines the tissue specificity of a *CO-releasing molecule* (CORM); (b) selection of CORM lead structures explored over the last couple of years. *ET-CORM*, enzyme-triggered CO-releasing molecule.

An alternative way of systematizing CORMs is by the type of central element to which the carbonyl groups are bound. So far, most CORMs are based on transition metal-carbonyl complexes, since they offer most flexibility in molecular design, to tune both the CO release kinetics and the bioavailability. Group VI–IX elements have been utilized as the metal center and there are now CORMs based on chromium, molybdenum, tungsten, manganese, rhenium, iron, ruthenium, cobalt, rhodium, and iridium. However, a major focus has been on carbonyl complexes of essential bioelements such as Mn, Mo, and Fe as well as Ru because of its close analogy with iron, possibly to alleviate potential issues with metal-based toxicity.[30] In addition, a main group carbonyl compound, sodium boranocarbonate, $Na_2[H_3BCO_2]$ (CORM-A1), has been introduced by Alberto and coworkers.[42–44] Also some initial examples of purely organic photoCORMs have appeared in the literature, which, however, are outside the scope of this chapter, which is focused on metal-based systems.[45–47]

Some selected compounds are shown in Fig. 6.6 with ruthenium-carbonyl complexes CORM-2 and -3 commercially available and most commonly utilized in biological studies. These compounds should, however, be used with great care, since CO release is always accompanied by formation of a metal–coligand fragment, called iCORM (inactivated CORM), which might have a biological activity of its own.[31,48] Therefore it is mandatory to carry out careful control experiments in which, in addition to the CORM, the iCORM is also independently prepared and assessed for biological activity. Preferentially, solutions saturated with CO gas should also be used as an additional control to unequivocally demonstrate that any biological effect observed is caused by the CO only and not the iCORM. Since this is often difficult to achieve experimentally, an alternative is the addition of hemoglobin to the assay solution as a CO scavenger. If this abolishes the biological activity, it is a strong but still indirect indicator of key involvement of diffusible CO in the effects observed.

Other issues still actively debated in CORM research are focused on whether slow or fast CO release is preferred, how many equivalents of CO should be delivered per mole of carrier molecule, and whether the CORM should be able to penetrate cell membranes, leading to intracellular accumulation of the iCORM fragment, or rather an inability to internalize in cells, with the CO "payload" released outside and then diffused into the cells and the iCORM quickly washed away by the circulation. Very likely, CORMs need to be carefully tuned to each particular biomedical application that is envisioned for them and much more focus is needed for proper design of the "drug sphere" than an ever-increasing number of novel metal–carbonyl fragments explored in this context, which often lack the required "drug-likeness" for further development.

4. MOLECULAR PROBES FOR SMALL-MOLECULE MESSENGERS

In this section, only methods applicable to in vitro detection of small signaling molecules are covered, which essentially restricts the treatment to fluorescent probes.

4.1 Fluorescent Probes for Nitric Oxide

One of the oldest and simplest methods for the determination of NO is the Griess assay,[49] which is based on the reaction of nitrite produced by NO autoxidation with sulfanilamide producing a diazonium salt, which further reacts with *N*-1-naphthylethylenediamine to an azo dye that is quantified by UV/Vis spectroscopy (Fig. 6.7a). However, this method is not applicable to in vitro determination.[32,50]

Most fluorescent probes for NO or derived species are based on a 1,2-diamino-substituted phenyl ring attached to a fluorophore (Fig. 6.7b). These react with NO in the presence of dioxygen to a triazole ring, which results in a highly fluorescent product.[51,52] The reaction with NO is indirect,

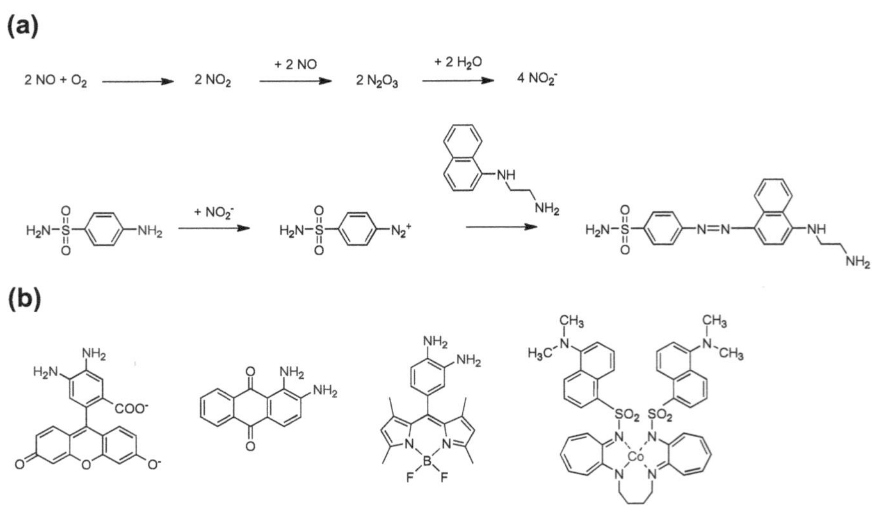

Figure 6.7 (a) Reaction sequence of the Griess assay leading to formation of a diazo dye upon autoxidation of nitric oxide and reaction with sulfanilamide and *N*-1-naphthylethylenediamine and (b) some examples of fluorescent probes for nitric oxide.

however, and it is assumed that N_2O_3 is in fact the active species. In addition to fluorescein, anthraquinone and borondipyrromethane difluoride (BODIPY) chromophores have also been utilized in this context. An interesting metal-based system, on the other hand, was reported by Franz and Lippard, in which two aminotroponiminate (ATI) moieties functionalized with a dansyl fluorophore are each linked together by a butylene chain to chelate a cobalt(II) center as a tetradentate ligand.[53] This compound readily reacts with nitric oxide as followed by IR spectroscopy, in which bands indicative of a cobalt dinitrosyl moiety are detected, resulting in detachment of one ATI group from the metal center. This leads to restoration of the dansyl fluorescence with a detection limit for NO of about 50–100 μM.

4.2 Fluorescent Probes for Hydrogen Sulfide[54,55]

Three different general strategies have been utilized for H_2S detection in biological systems: the reduction of organoazides to amines, thiolysis of dinitrophenyl ether protective groups, and copper sulfide precipitation.[55] One of the advantages of organoazides is their straightforward synthesis and compatibility with biological systems. A wide range of fluorophores has been modified this way, with two examples of rhodamine- and dansyl-based probes shown in Fig. 6.8a.[56,57] A related approach is based on nitro instead

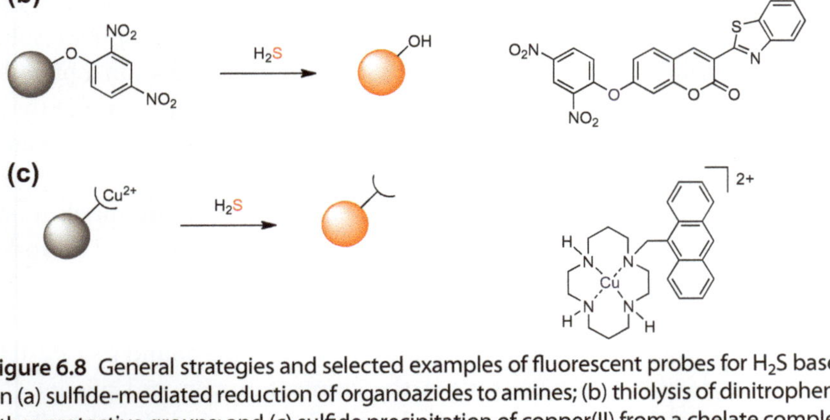

Figure 6.8 General strategies and selected examples of fluorescent probes for H_2S based on (a) sulfide-mediated reduction of organoazides to amines; (b) thiolysis of dinitrophenyl ether protective groups; and (c) sulfide precipitation of copper(II) from a chelate complex.

of azide group reduction (not shown). Since H_2S is an excellent nucleophile, additional probes have been developed based on HS^--mediated cleavage of protective groups from fluorophores, which become emissive upon bond scission. Both the thiolysis of dinitrophenyl ether protective groups as well as disruption of conjugated π-systems in coumarin dyes have been reported. As an illustrative example, Fig. 6.8b shows a benzothiophene-functionalized coumarin fluorophore inactivated by attachment of a 2,4-dinitrophenyl group.[58] Metal-based probes for H_2S have also been reported, which are based on H_2S-mediated precipitation of transition metals very well established in quantitative analysis. For example, the copper(II) cyclam complex with a pendant anthracene moiety in Fig. 6.8c is nonemissive because of quenching of the fluorescence by the paramagnetic copper ion.[59] However, precipitation of the metal as copper sulfide by H_2S turns the probe on.

4.3 Fluorescent Probes for Carbon Monoxide[60,61]

The first fluorescent probe for detection of CO in living cells was reported by the group of He in 2012.[62] This system, called "COser" (CO sensor), is genetically encoded on a plasmid and based on a fusion protein of circularly permuted yellow fluorescent protein inserted between two monomers of CooA, a dimeric CO-sensing heme protein from *Rhodospirillum rubrum*. Axial CO binding to the heme group of the CooA building block leads to conformational changes in the overall protein structure, which results in a switch-on of emission with a maximum at 528 nm. Cross-sensitivity of the system with imidazole, cyanide, NO, dioxygen, sodium sulfide, and glutathione was examined and found to be negligible at up to 100 μM. The response of COSer to incubation of HeLa cells with CORM-2 was investigated with fluorescence microscopy and allowed in vitro detection of the fluorescence increase at 1–5 μM CORM-2. Besides these promising results, however, a major drawback of this system is that it is encoded on a plasmid vector and thus cannot easily be used in a cell-free system. Because of these difficulties, COSer apparently has not been used in biological experiments beyond the initial publication so far.

In contrast, small molecule fluorescent probes for CO are much more universally applicable and easier to handle for chemists. Chang and coworkers reported on "CO probe 1" (COP-1), which is a BODIPY dye functionalized with a *para*-(dimethylaminomethyl)phenyl moiety on the central carbon atom (Fig. 6.9a). Two of these BODIPY units are bound to a Pd_2Cl_2 diamond core as a C,N-chelator via the amino nitrogen and phenyl C3 carbon atoms.[63] Insertion of CO in the palladium–carbon bond leads to

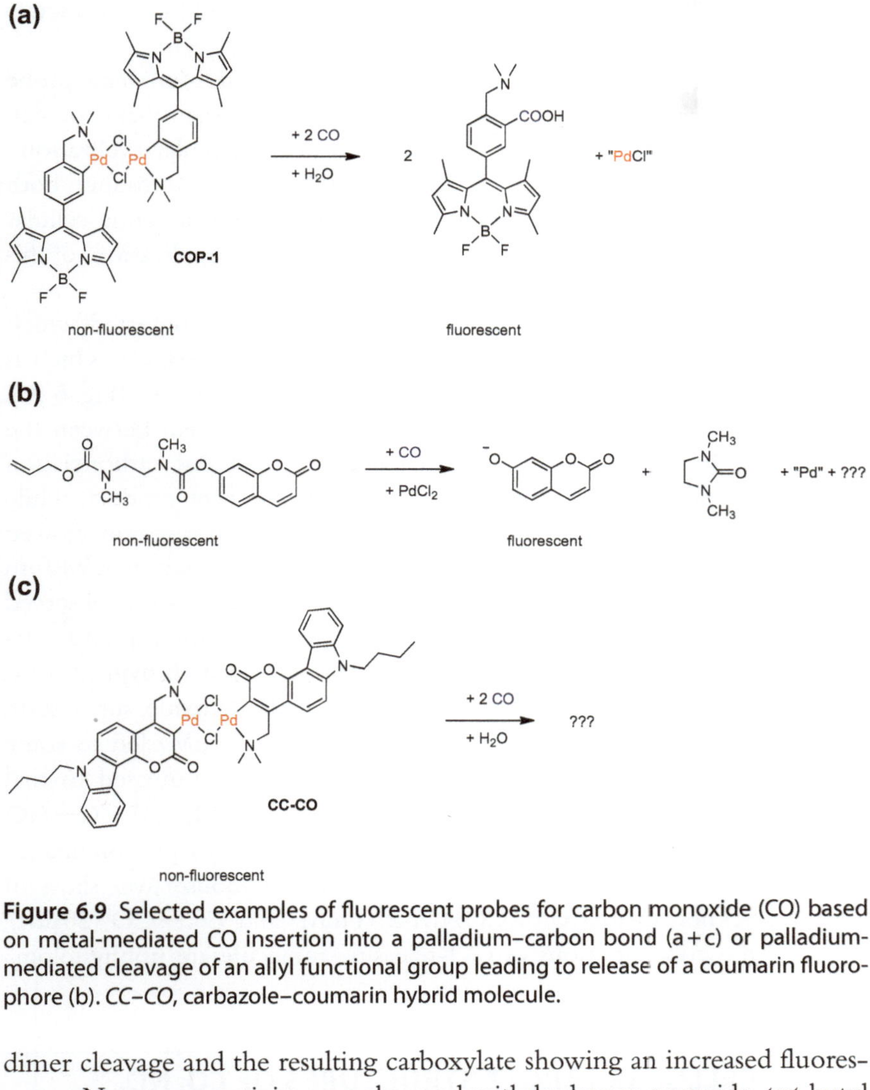

Figure 6.9 Selected examples of fluorescent probes for carbon monoxide (CO) based on metal-mediated CO insertion into a palladium–carbon bond (a+c) or palladium-mediated cleavage of an allyl functional group leading to release of a coumarin fluorophore (b). *CC–CO*, carbazole–coumarin hybrid molecule.

dimer cleavage and the resulting carboxylate showing an increased fluorescence. No cross-reactivity was observed with hydrogen peroxide, *tert*-butyl hydroperoxide, sodium hypochlorite, superoxide, NO, peroxy nitrite, and H_2S. Treatment of HEK293 T-cells with CORM-3 at 5 or 5 μM and 1 μM COP-1 led to an increase in cellular fluorescence by a factor of about 2–3, as determined by confocal fluorescence microscopy with the emission centered at 507 nm. In contrast to COser, COP-1 has already been used in a number cellular imaging experiments,[64–69] although it should be kept in mind that in addition to the fluorescent BODIPY unit, also a potentially bioactive "PdCl" moiety remains in the system, which might have a reactivity with bio(macro)molecules.

Dhara and coworkers also reported a coumarin-based fluorogenic probe "PCO-1," which consists of a protected 7-hydroxycoumarin that shows an increased fluorescence upon a Pd(0)-mediated intramolecular cyclization–elimination reaction (Fig. 6.9b).[70] However, the need to introduce both the organic probe and the required palladium species in the same cellular compartment will likely prevent widespread bioimaging applications of this system.

Interestingly, also a two-photon probe for CO has been reported, which is based on a carbazole–coumarin hybrid molecule (CC—CO), which is bound to a Pd_2Cl_2 diamond core in the same way as in COP-1 (Fig. 6.9c). However, the mechanism of activation seems to be different between the two probes. In the case of COP-1, a palladium-mediated carbonylation takes place, with the CO incorporated in the fluorescent product, while CC—CO acts by protonolysis of the Pd–C bond. The latter system allowed detection of CO in living tissue for the first time, with excitation at 740 nm and emission at 477 nm.[71] An 11-fold increase in emission was observed upon treatment with CORM-2 with the detection limit for CO estimated to be 0.65 μM. No cross-reactivity was observed with hypochlorite, hydrogen peroxide, hydroxyl radicals, *tert*-butyl hydroperoxide, superoxide, and NO. Addition of high concentrations of H_2S (100 μM) led to some increase in fluorescence, which, however, was modest compared to that of CO. Treatment of HeLa cells with CORM-2 (200 μM) and CC—CO (5 μM) led to strong fluorescence in both the one- and two-photon images. Interestingly, a response could also be observed in mouse liver slices of 400 μm thickness treated with CORM-2 (1 mM) and CC—CO (20 μM), although the concentrations of CO-releasing molecule are unphysiologically high.

5. CELLULAR TARGET STRUCTURES OF CO, NO, AND H_2S

While the mechanisms for enzymatic generation of CO, NO, and H_2S are now well understood at the molecular level, elucidation of the cellular target structures of these small signaling molecules has been hampered by their highly reactive nature, in particular in the case of NO, the complex interplay of their signaling pathway, and the rather nonspecific way these molecules, which are among the smallest natural products, interact with proteins. Major targets are assumed to be accessible axial sites in heme proteins and possibly also other nonheme iron proteins. In addition, NO and H_2S, but not CO, can modify amino acid side chains in proteins, leading to *S*-nitrosylated and *S*-sulfhydrated thiols.[72] Furthermore, much research in the field has been devoted to systemic studies of the biological action of small molecule messengers so far, while molecular biology investigations aimed at an understanding of the modulation of enzyme function at the atomic level significantly lag behind. Furthermore, H_2S is able to rapidly migrate from its site of generation and might excert activity at rather distant locations, with its "sphere of action" estimated to be as large as 200 neighboring cells.[73,74] Hopefully, comparable data on the migration of CO and NO in cells and tissues will also become available in the near future.

5.1 Cellular Targets of Nitric Oxide

It is generally assumed that soluble guanylate cyclase (sGC), which converts GTP to 3′,5′-cyclic GMP (cGMP) and pyrophosphate, is the major cellular target of NO. cGMP then triggers a wide range of cell-specific downstream signaling pathways such as protein kinases, ion-gated channels, and several families of phosphodiesterases. This NO signaling is particularly prominent in the vasculature, regulating blood flow.[75] Studies of sGC have been hampered by the fact that this enzyme is rather difficult to purify and no crystal structure of the full protein has been obtained so far. sGC is an $\alpha\beta$ heterodimer, with tissue-dependent expression of the different subunits $\alpha1$, $\alpha2$, $\beta1$, and $\beta2$.[76–78] Both subunits consist of an *N*-terminal heme-NO/O_2 binding domain (H-NOX), central Per/ARNT/Sim and coiled-coil domains, and a *C*-terminal catalytic domain, where GTP binding and conversion to cGMP takes place. In the rat $\alpha1\beta1$ enzyme, the two subunits are 690 and 619 amino acids in length, respectively. The H-NOX domain contains a heme group with a five-coordinated iron center anchored to the protein by a histidine ligand, which can bind CO and NO in the free axial position, while dioxygen binding varies from enzyme to enzyme.[79] Small molecules can modulate sGC activity upon NO binding by a factor of 200–800, which indicates additional regulatory binding sites in the enzyme.[76,80] Direct NO signaling via modification of iron-sulfur centers has also moved in the focus of research.[23] For example, SoxR, a redox-sensitive transcription factor incorporating a [2Fe-2S] center is activated upon exposure to NO because of conversion to a protein-bound DNIC-dithiol adduct.[81] NO-mediated transformation has also been reported for cubane-type [4Fe-4S] clusters, e.g., in cytosolic aconitase.[82]

5.2 Cellular Targets of Hydrogen Sulfide

H_2S can undergo both specific and nonspecific interactions with other constituents of cells and tissues. As a strong reducing agent, H_2S reacts with reactive oxygen species such as peroxynitrite, superoxide, and hydrogen peroxide.[26,83] The reaction with *S*-nitrosothiols results in formation of thionitrous acid, which is further metabolized to nitrosyl species. Possibly, the most important specific reaction is protein *S*-sulfhydration of cysteines, leading to R-S-SH species. It has been estimated that about 10–25% of many liver proteins are sulfhydrated under physiological conditions.[84] Since this modification is relatively stable, at least compared to *S*-nitrosylation, it should be amendable to mass spectrometric identification. H_2S also reacts with heme iron,[85] e.g., resulting in formation of sulfhemoglobin from methemoglobin. A distinct feature in which H_2S reactivity differs from that of CO and to some degree also of NO is because while the latter two bind low oxidation state metals, H_2S has a preference for oxidized metal centers and as such prefers Fe(III) over Fe(II), for example. While the toxicity of high levels of H_2S are thought to be caused by heme binding, in particular in cytochrome *c* oxidase, with concomitant disturbance of the mitochondrial electron transfer chain, little seems to be known about the role of such modifications in normal cellular signaling pathways.

5.3 Cellular Targets of Carbon Monoxide

While the cellular production of CO by HOs is well understood at the molecular level, this is not the case for the downstream signaling pathways triggered by CO. This is pretty surprising, since in contrast to NO and H_2S, CO is much less reactive and very likely requires a transition metal center for binding to a protein target. For some time sGC has been implicated in CO signaling, but the activation is much lower compared to NO in the absence of additional mediators (fivefold by CO compared to 200-fold by NO) and thus likely not of physiological relevance. Activation by CO can be increased by exogenous application of heme-dependent sGC stimulators (YC-1, riociguat) or heme-independent sGC activators (cinaciguat) and "...the existence of an endogenous YC-1-like substance would provide the molecular basis for a physiological role in the regulation of sGC," but such a discovery has remained elusive so far.[86] The activity of some ion channels such as large conductance Ca^{2+} or voltage-activated K^+ channels (BK_{Ca}, MaxiK, Slo1) is also modulated by CO or CORMs, but their mechanism remains unclear to date, possibly because of difficulties in the biophysical characterization of such large membrane proteins.[87–89]

6. SUMMARY

CO, NO, and H_2S are endogenously produced by well-controlled enzymatic processes in higher organisms, including humans, and serve a vital signaling function at low, physiological concentration, in particular in response to cellular stress conditions. NO and H_2S are synthesized from amino acid sources, while CO is produced from heme by HO enzymes. Physiologically relevant concentrations of H_2S and CO are in the micromolar range, while the NO concentration in the body is nanomolar. To exploit the signaling pathways triggered by these small-molecule messengers, in addition to stimulation of the endogenous enzymatic pathways, a wide range of artificially exogenous delivery systems has also been developed. Both organic as well as transition metal-based systems for CO, NO, and H_2S delivery to tissues and cells are available now and used both in fundamental studies as well as in experimental therapeutic approaches. A number of fluorescent probes to study small-molecule speciation in biological systems are now available for all three small messenger molecules, but generally the "sphere of action," that is the distance CO, NO, and H_2S can travel before being captured by a cellular target molecule, is mostly unknown so far. While the mechanisms of endogenous production of these small messengers is generally well known to date, this is not the case with their downstream target structures, in particular in the case of CO.

Major directions of research in the field should therefore in the future be directed at the interaction of CO with metalloproteins at the atomic level, CO and H_2S carrier systems with improved drug-like properties tuned for a specific therapeutic application and adjustable release kinetics, and next-generation fluorescent probes for highly sensitive studies of the speciation of these messenger molecules in cells and tissues at physiological concentrations. Almost 20 years after the Nobel Prize in Physiology or Medicine was awarded to Furchgott, Ignarro, and Murad "for their discoveries concerning nitric oxide as a signaling molecule in the cardiovascular system" there is certainly still much to discover on the biological activity and therapeutic potential of some of the smallest natural products around.

REFERENCES

1. Szabo, C. Gasotransmitters: New Frontiers for Translational Science. *Sci. Transl. Med.* **2010,** *2,* 59ps54.
2. Motterlini, R.; Otterbein, L. E. The Therapeutic Potential of Carbon Monoxide. *Nat. Rev. Drug Discov.* **2010,** *9,* 728–743.

3. Szabo, C. Hydrogen Sulphide and Its Therapeutic Potential. *Nat. Rev. Drug Discov.* **2007,** *6,* 917–935.
4. Crane, B. R.; Sudhamsu, J.; Patel, B. A. Bacterial Nitric Oxide Synthases. *Annu. Rev. Biochem.* **2010,** *79,* 445–470.
5. Feng, C. Mechanism of Nitric Oxide Synthase Regulation: Electron Transfer and Interdomain Interactions. *Coord. Chem. Rev.* **2012,** *256,* 393–411.
6. Poulos, T. L. Heme Enzyme Structure and Function. *Chem. Rev.* **2014,** *114,* 3919–3962.
7. Kajimura, M.; Fukuda, R.; Bateman, R. M.; Yamamoto, T.; Suematsu, M. Interactions of Multiple Gas-Transducting Systems: Hallmarks and Uncertainties of CO, NO, and H_2S Gas Biology. *Antioxid. Redox Signal.* **2010,** *13,* 157–192.
8. Thomas, D.; Surdin-Kerjan, Y. Metabolism of Sulfur Amino Acids in *Saccharomyces cerevisiae. Microbiol. Mol. Biol. Rev.* **1997,** *61,* 503–532.
9. Chiku, T.; Padovani, D.; Zhu, W.; Singh, S.; Vitvitsky, V.; Banerjee, R. H_2S Biogenesis by Human Cystathionine γ-Lyase Leads to the Novel Sulfur Metabolites Lanthionine and Homolanthionine and Is Responsive to the Grade of Hyperhomocysteinemia. *J. Biol. Chem.* **2009,** *284,* 11601–11612.
10. Li, L.; Rose, P.; Moore, P. K. Hydrogen Sulfide and Cell Signaling. *Annu. Rev. Pharmacol. Toxicol.* **2011,** *51,* 169–187.
11. Singh, S.; Banerjee, R. PLP-Dependent H_2S Biogenesis. *Biochim. Biophys. Acta* **2011,** *1814,* 1518–1527.
12. Wang, R. Physiological Implications of Hydrogen Sulfide: A Whiff Exploration That Blossomed. *Physiol. Rev.* **2012,** *92,* 791–896.
13. Yadav, P. K.; Martinov, M.; Vitvitsky, V.; Seravalli, J.; Wedmann, R.; Filipovic, M.; Banerjee, R. Biosynthesis and Reactivity of Cysteine Persulfides in Signaling. *J. Am. Chem. Soc.* **2016,** *138,* 289–299.
14. Frune, J.; Saeed, A.; Levitt, M. D. Whole Tissue Hydrogen Sulfide Concentrations Are Orders of Magnitude Lower than Presently Accepted Values. *Am. J. Physiol. Regul. Integr. Comp. Physiol.* **2008,** *295,* 1479–1485.
15. Maines, M. D. The Heme Oxygenase System: A Regulator of Second Messenger Gases. *Annu. Rev. Pharmacol. Toxicol.* **1997,** *37,* 517–554.
16. Matsui, T.; Iwasaki, M.; Sugiyama, R.; Unno, M.; Ikeda-Saito, M. Dioxygen Activation for the Self-degradation of Heme: Reaction Mechanism and Regulation of Heme Oxygenase. *Inorg. Chem.* **2010,** *49,* 3602–3609.
17. Matsui, T.; Unno, M.; Ikeda-Saito, M. Heme Oxygenase Reveals Its Strategy for Catalyzing Three Successive Oxygenation Reactions. *Acc. Chem. Res.* **2010,** *43,* 240–247.
18. Wilks, A.; Heinzl, G. Heme Oxygenation and the Widening Paradigm of Heme Degradation. *Arch. Biochem. Biophys.* **2014,** *544,* 87–95.
19. Wang, P. G.; Xian, M.; Tang, X.; Xu, X.; Wen, Z.; Cai, T.; Janczuk, A. J. Nitric Oxide Donors: Chemical Activities and Biological Applications. *Chem. Rev.* **2002,** *102,* 1091–1134.
20. Carpenter, A. W.; Schoenfisch, M. H. Nitric Oxide Release: Part II. Therapeutic Applications. *Chem. Soc. Rev.* **2012,** *41,* 3742–3752.
21. Riccio, D. A.; Schoenfisch, M. H. Nitric Oxide Release: Part I. Macromolecular Scaffolds. *Chem. Soc. Rev.* **2012,** *41,* 3731–3741.
22. Lewandowska, H.; Kalinowska, M.; Brzoska, K.; Wojciuk, K.; Wojciuk, G.; Kruszewski, M. Nitrosyl Iron Complexes – Synthesis, Structure and Biology. *Dalton Trans.* **2011,** *40,* 8273–8289.
23. Fitzpatrick, J.; Kim, E. Synthetic Modeling Chemistry of Iron-Sulfur Clusters in Nitric Oxide Signaling. *Acc. Chem. Res.* **2015,** *48,* 2453–2461.
24. Tsai, M.-L.; Tsou, C.-C.; Liaw, W.-F. Dinitrosyl Iron Complexes (DNICs): From Biomimetic Synthesis and Spectroscopic Characterization toward Unveiling the Biological and Catalytic Roles of DNICs. *Acc. Chem. Res.* **2015,** *48,* 1184–1193.

25. Whiteman, M.; Le Trionnair, S.; Chopra, M.; Fox, B.; Whatmore, J. Emerging Role of Hydrogen Sulfide in Health and Disease: Critical Appraisal of Biomarkers and Pharmacological Tools. *Clin. Sci.* **2011,** *121,* 459–488.
26. Zhao, Y.; Biggs, T. D.; Xian, M. Hydrogen Sulfide (H_2S) Releasing Agents: Chemistry and Biological Applications. *Chem. Commun.* **2014,** *50,* 11788–11805.
27. DeLeon, E. R.; Stoy, G. F.; Olson, K. R. Passive Loss of Hydrogen Sulfide in Biological Experiments. *Anal. Biochem.* **2012,** *421,* 203–207.
28. Greiner, R.; Palinkas, Z.; Bäsell, K.; Becher, D.; Antelmann, H.; Nagy, P.; Dick, T. P. Polysulfides Link H_2S to Protein Thiol Oxidation. *Antioxid. Redox Signal.* **2013,** *19,* 1749–1765.
29. Romao, C. C.; Blättler, W. A.; Seixas, J. D.; Bernardes, G. J. L. Developing Drug Molecules for Therapy with Carbon Monoxide. *Chem. Soc. Rev.* **2012,** *41,* 3571–3583.
30. Schatzschneider, U. Novel Lead Structures and Activation Mechanisms for CO-Releasing Molecules (CORMs). *Br. J. Pharmacol.* **2015,** *172,* 1638–1650.
31. Seixas, J. D.; Santos, M. F. A.; Mukhopadhyay, A.; Coelho, A. C.; Reis, P. M.; Veiros, L. F.; Marques, A. R.; Penacho, N.; Goncalves, A. M. L.; Romao, M. J.; Bernardes, G. J. L.; Santos-Silva, T.; Romao, C. C. A Contribution to the Rational Design of $Ru(CO)_3Cl_2L$ Complexes for In Vivo Delivery of CO. *Dalton Trans.* **2015,** *44,* 5058–5075.
32. Romanski, S.; Kraus, B.; Schatzschneider, U.; Neudörfl, J.; Amslinger, S.; Schmalz, H.-G. Acyloxybutadiene Iron Tricarbonyl Complexes as Enzyme-Triggered CO-Releasing Molecules (ET-CORMs). *Angew. Chem. Int. Ed.* **2011,** *50,* 2392–2396.
33. Romanski, S.; Kraus, B.; Guttentag, M.; Schlundt, W.; Rücker, H.; Adler, A.; Neudörfl, J.-M.; Alberto, R.; Amslinger, S.; Schmalz, H.-G. Acyloxybutadiene Tricarbonyl Iron Complexes as Enzyme-Triggered CO-Releasing Molecules (ET-CORMs): A Structure-Activity Relationship Study. *Dalton Trans.* **2012,** *41,* 13862–13875.
34. Romanski, S.; Rücker, H.; Stamellou, E.; Guttentag, M.; Neudörfl, J.; Alberto, R.; Amslinger, S.; Yard, B.; Schmalz, H.-G. Iron Dienylphosphate Tricarbonyl Complexes as Water-Soluble Enzyme-Triggered CO-Releasing Molecules (ET-CORMs). *Organometallics* **2012,** *31,* 5800–5809.
35. Botov, S.; Stamellou, E.; Romanski, S.; Guttentag, M.; Alberto, R.; Neudörfl, J.; Yard, B.; Schmalz, H.-G. Synthesis and Performance of Acyloxy-diene-$Fe(CO)_3$ Complexes with Variable Chain Lengths as Enzyme-Triggered Carbon Monoxide-Releasing Molecules. *Organometallics* **2013,** *32,* 3587–3594.
36. Sitnikov, N. S.; Li, Y.; Zhang, D.; Yard, B.; Schmalz, H.-G. Design, Synthesis, and Functional Evaluation of CO-Releasing Molecules Triggered by Penicillin G Amidase as a Model Protease. *Angew. Chem. Int. Ed.* **2015,** *54,* 12314–12318.
37. Schatzschneider, U. Photoactivated Biological Activity of Transition Metal Complexes. *Eur. J. Inorg. Chem.* **2010,** 1451–1467.
38. Schatzschneider, U. PhotoCORMs: Light-Triggered Release of Carbon Monoxide from the Coordination Sphere of Transition Metal Complexes for Biological Applications. *Inorg. Chim. Acta* **2011,** *374,* 19–23.
39. Rimmer, R. D.; Pierri, A. E.; Ford, P. C. Photochemically Activated Carbon Monoxide Release for Biological Targets. Toward Developing Air-Stable PhotoCORMs Labilized by Visible Light. *Coord. Chem. Rev.* **2012,** *256,* 1509–1519.
40. Chakraborty, I.; Carrington, S. J.; Mascharak, P. K. Design Strategies to Improve the Sensitivity of Photoactive Metal Carbonyl Complexes (PhotoCORMs) to Visible Light and Their Potential as CO-Donors to Biological Targets. *Acc. Chem. Res.* **2014,** *47,* 2603–2611.
41. Wright, M. A.; Wright, J. A. PhotoCORMs: CO Release Moves into the Visible. *Dalton Trans.* **2016,** *45,* 6801–6811.
42. Motterlini, R.; Sawle, P.; Bains, S.; Hammad, J.; Alberto, R.; Foresti, R.; Green, C. J. CORM-A1: A New Pharmacologically Active Carbon Monoxide-Releasing Molecule. *FASEB J.* **2004,** *18,* 284–286.

43. Pitchumony, T. S.; Spingler, B.; Motterlini, R.; Alberto, R. Derivatives of Sodium Boranocarbonate as Novel CO-Releasing Molecules (CO-RMs). *Chimia* **2008,** *62,* 277–279.
44. Pitchumony, T. S.; Spingler, B.; Motterlini, R.; Alberto, R. Syntheses, Structural Characterization and CO Releasing Properties of Boranocarbonate $[H_3BCO_2H]^-$ Derivatives. *Org. Biomol. Chem.* **2010,** *8,* 4849–4954.
45. Antony, L. A. P.; Slanina, T.; Sebej, P.; Solomek, T.; Klan, P. Fluorescein Analogue Xanthene-9-carboxylic Acid: A Transition-Metal-Free CO Releasing Molecule Activated by Green Light. *Org. Lett.* **2013,** *15,* 4552–4555.
46. Peng, P.; Wang, C.; Shi, Z.; Johns, V. K.; Ma, Y.; Oyer, J.; Copik, A.; Igarashi, R.; Liao, Y. Visible-light Activatable Organic CO-Releasing Molecules (PhotoCORMs) that Simultaneously Generate Fluorophores. *Org. Biomol. Chem.* **2013,** *11,* 6671–6674.
47. Palao, E.; Slanina, T.; Muchova, L.; Solomek, T.; Vitek, L.; Klan, P. Transition-Metal-Free CO-Releasing BODIPY Derivatives Activatable by Visible to NIR Light as Promising Bioactive Molecules. *J. Am. Chem. Soc.* **2016,** *138,* 126–133.
48. Johnson, T. R.; Mann, B. E.; Teasdale, I. P.; Adams, H.; Foresti, R.; Green, C. J.; Motterlini, R. Metal Carbonyls as Pharmaceuticals? $[Ru(CO)_3Cl(glycinate)]$, a CO-Releasing Molecule with an Extensive Aqueous Solution Chemistry. *Dalton Trans.* **2007,** 1500–1508.
49. Tsikas, D. Analysis of Nitrite and Nitrate in Biological Fluids by Assays Based on the Griess Reaction: Appraisal of the Griess Reaction in the L-arginine/nitric Oxide Area of Research. *J. Chromatogr. B* **2007,** *851,* 51–70.
50. Cnoeski, P. N.; Schoenfisch, M. H. Nitric Oxide Release: Part III. Measurement and Reporting. *Chem. Soc. Rev.* **2012,** *41,* 3753–3758.
51. Hetrick, E. M.; Schoenfisch, M. H. Analytical Chemistry of Nitric Oxide. *Annu. Rev. Anal. Chem.* **2009,** *2,* 409–433.
52. Gomes, A.; Fernandes, E.; Lima, J. L. F.C. Use of Fluorescent Probes for Detection of Reactive Nitrogen Species: A Review. *J. Fluoresc.* **2006,** *16,* 119–139.
53. Franz, K. J.; Singh, N.; Lippard, S. J. Metal-Based NO Sensing by Selective Ligand Dissociation. *Angew. Chem. Int. Ed.* **2000,** *39,* 2120–2122.
54. Lin, V. S.; Chang, C. J. Fluorescent Probes for Sensing and Imaging Biological Hydrogen Sulfide. *Curr. Opin. Chem. Biol.* **2012,** *16,* 595–601.
55. Lin, V. S.; Chen, W.; Xian, M.; Chang, C. J. Chemical Probes for Molecular Imaging and Detection of Hydrogen Sulfide and Reactive Sulfur Species in Biological Systems. *Chem. Soc. Rev.* **2015,** *44,* 4596–4618.
56. Lippert, A. R.; New, N. J.; Chang, C. J. Reaction-Based Fluorescent Probes for Selective Imaging of Hydrogen Sulfide in Living Cells. *J. Am. Chem. Soc.* **2011,** *133,* 10078–10080.
57. Peng, H.-I.; Cheng, Y.; Dai, C.; King, A. L.; Predmore, B. L.; Lefer, D. J.; Wang, B. A Fluorescent Probe for Fast and Quantitative Detection of Hydrogen Sulfide in Blood. *Angew. Chem. Int. Ed.* **2011,** *50,* 9672–9675.
58. Yuan, L.; Zuo, Q.-P. Reaction-Based Fluorescent Probe for Hydrogen Sulfide with Large Signal-to-Noise Ratio in Living Cells and Tissues. *Sens. Actuators B* **2014,** *196,* 151–155.
59. Sasakura, K.; Hanaoka, K.; Shibuya, N.; Mikami, Y.; Kimura, Y.; Komatsu, T.; Ueno, T.; Terai, T.; Kimura, H.; Nagano, T. Development of a Highly Selective Fluorescent Probe for Hydrogen Sulfide. *J. Am. Chem. Soc.* **2011,** *133,* 18003–18005.
60. Yuan, L.; Lin, W.; Tan, L.; Zheng, K.; Huang, W. Lighting Up Carbon Monoxide: Fluorescent Probes for Monitoring CO in Living Cells. *Angew. Chem. Int. Ed.* **2013,** *52,* 1628–1630.
61. Kumar, N.; Bhalla, V.; Kumar, M. Recent Developments of Fluorescent Probes for the Detection of Gasotransmitters (NO, CO and H_2S). *Coord. Chem. Rev.* **2013,** *257,* 2335–2347.

62. Wang, J.; Karpus, J.; Zhao, B. S.; Luo, Z.; Chen, P. R.; He, C. A Selective Fluorescent Probe for Carbon Monoxide Imaging in Living Cells. *Angew. Chem. Int. Ed.* **2012,** *51,* 9652–9656.
63. Michel, B. W.; Lippert, A. R.; Chang, C. J. A Reaction-Based Fluorescent Probe for Selective Imaging of Carbon Monoxide in Living Cells Using a Palladium-Mediated Carbonylation. *J. Am. Chem. Soc.* **2012,** *134,* 15668–15671.
64. Pai, S.; Hafftlang, M.; Atongo, G.; Nagel, C.; Niesel, J.; Botov, S.; Schmalz, H.-G.; Yard, B.; Schatzschneider, U. New Modular Manganese(I) Tricarbonyl Complexes as PhotoCORMs: In Vitro Detection of Photoinduced Carbon Monoxide Release Using COP-1 as a Fluorogenic Switch-on Probe. *Dalton Trans.* **2014,** *43,* 8664–8678.
65. Wilson, J. L.; Kobeissi, S. F.; Oudir, S.; Haas, B.; Michel, B.; Rande, J.-L.; Ollivier, A.; Martens, T.; Rivard, M.; Motterlini, R.; Foresti, R. Design and Synthesis of New Hybrid Molecules that Activate the Transcription Factor Nrf2 and Simultaneously Release Carbon Monoxide. *Chem. Eur. J.* **2014,** *20,* 14698–14704.
66. Stamellou, E.; Storz, D.; Botov, S.; Ntasis, E.; Wedel, J.; Sollazzo, S.; Krämer, B. K.; van Son, W.; Seelen, M.; Schmalz, H.-G.; Schmidt, A.; Hafner, M.; Yard, B. Different Design of Enzyme-Triggered CO-Releasing Molecules (ET-CORMs) Reveals Quantitative Differences in Biological Activities in Terms of Toxicity and Inflammation. *Redox Biol.* **2014,** *2,* 739–748.
67. Chaves-Ferreira, M.; Albuquerque, I. S.; Matak-Vinkovic, D.; Coelho, A. C.; Carvalho, S. M.; Saraiva, L. M.; Romao, C. C.; Bernardes, G. J. L. Spontaneous CO Release from $Ru^{II}(CO)_2$-Protein Complexes in Aqueous Solution, Cells, and Mice. *Angew. Chem. Int. Ed.* **2014,** *54,* 1172–1175.
68. Nobre, L. S.; Jeremias, H.; Romao, C. C.; Saraiva, L. M. Examining the Antimicrobial Activity and Toxicity to Animal Cells of Different Types of CO-Releasing Molecules. *Dalton Trans.* **2016,** *45,* 1455–1466.
69. Nikam, A.; Ollivier, A.; Rivard, M.; Wilson, J. L.; Mebarki, K.; Martens, T.; Dubois-Randé, J.-L.; Motterlini, R.; Foresti, R. Diverse Nrf2 Activators Coordinated to Cobalt Carbonyls Induce Heme Oxygenase-1 and Release Carbon Monoxide In Vitro and In Vivo. *J. Med. Chem.* **2016,** *59,* 756–762.
70. Pal, S.; Mukherjee, M.; Sen, B.; Mandal, S. K.; Lohar, S.; Chattopadhyay, P.; Dhara, K. A New Fluorogenic Probe for the Selective Detection of Carbon Monoxide in Aqueous Medium Based on Pd(0) Mediated Reaction. *Chem. Commun.* **2015,** *51,* 4410–4413.
71. Zheng, K.; Lin, W.; Tan, L.; Chen, H.; Cui, H. A Unique Carbazole–Coumarin Fused Two-Photo Platform: Development of a Robust Two-Photo Fluorescent Probe for Imaging Carbon Monoxide in Living Tissue. *Chem. Sci.* **2014,** *5,* 3439–3448.
72. Kajimura, M.; Nakanishi, T.; Takenouchi, T.; Morikawa, T.; Hishiki, T.; Yukutake, Y.; Suematsu, M. Gas Biology: Tiny Molecules Controlling Metabolic Systems. *Respir. Physiol. Neurobiol.* **2012,** *184,* 139–148.
73. Riahi, S.; Rowley, C. N. Why Can Hydrogen Sulfide Permeate Cell Membranes? *J. Am. Chem. Soc.* **2014,** *136,* 15111–15113.
74. Cuevasanta, E.; Denicola, A.; Alvarez, B.; Möller, M. N. Solubility and Permeability of Hydrogen Sulfide in Lipid Membranes. *PLoS One* **2011,** *7,* e34562.
75. Cary, S. P. L.; Winger, J. A.; Derbyshire, E. R.; Marletta, M. A. Nitric Oxide Signaling: No Longer Simply On or Off. *Trends Biochem. Sci.* **2006,** *31,* 231–239.
76. Derbyshire, E. R.; Marletta, M. A. Structure and Regulation of Soluble Guanylate Cyclase. *Annu. Rev. Biochem.* **2012,** *81,* 533–559.
77. Gileadi, O. Structures of Soluble Guanylate Cyclase: Implications for Regulatory Mechanisms and Drug Development. *Biochem. Soc. Trans.* **2014,** *42,* 108–113.
78. Lee, Y.-C.; Martin, E.; Murad, F. Human Recombinant Soluble Guanylyl Cyclase: Expression, Purification, and Regulation. *Proc. Natl. Acad. Sci. U.S.A.* **2000,** *97,* 10763–10768.

79. Boon, E. M.; Marletta, M. A. Ligand Discrimination in Soluble Guanylate Cyclase and the H-NOX Family of Heme Sensor Proteins. *Curr. Opin. Chem. Biol.* **2005,** *9,* 441–446.
80. Evgenov, O.V.; Pacher, P.; Schmidt, P. M.; Hasko, G.; Schmidt, H. H. H.W.; Stasch, J.-P. NO-Independent Stimulators and Activators of Soluble Guanylate Cyclase: Discovery and Therapeutic Potential. *Nat. Rev. Drug Discov.* **2006,** *5,* 755–768.
81. Ding, H.; Demple, B. Direct Nitric Oxide Signal Transduction via Nitrosylation of Iron-Sulfur Centers in the SoxR Transcription Activator. *Proc. Natl. Acad. Sci. U.S.A.* **2000,** *97,* 5146–5150.
82. Victor, E.; Lippard, S. J. A Tetranitrosyl [4Fe–4S] Cluster Forms En Route to Roussin's Black Anion: Nitric Oxide Reactivity of $[Fe_4S_4(LS_3)L']^{2-}$. *Inorg. Chem.* **2014,** *53,* 5311–5320.
83. Fukuto, J. M.; Carrington, S. J.; Tantillo, D. J.; Harrison, J. G.; Ignarro, L. J.; Freeman, B. A.; Chen, A.; Wink, D. A. Small Molecule Signaling Agents: The Integrated Chemistry and Biochemistry of Nitrogen Oxides, Oxides of Carbon, Dioxygen, Hydrogen Sulfide, and Their Derived Species. *Chem. Res. Toxicol.* **2012,** *25,* 769–793.
84. Mustafa, A. K.; Gadalla, M. M.; Sen, N.; Kim, S.; Mu, W.; Gazi, S. K.; Barrow, R. K.; Yang, G.; Wang, R.; Snyder, S. H. H_2S Signals Through Protein S-sulfhydration. *Sci. Signal.* **2009,** *2,* ra72 (71–78).
85. Pavlik, J. W.; Noll, B. C.; Oliver, A. G.; Schulz, C. E.; Scheidt, W. R. Hydrosulfide (HS^-) Coordination in Iron Porphyrinates. *Inorg. Chem.* **2010,** *49,* 1017–1026.
86. Friebe, A.; Schultz, G.; Koesling, D. Sensitizing Soluble Guanylyl Cyclase to Become a Highly CO-Sensitive Enzyme. *EMBO J.* **1996,** *15,* 6863–6868.
87. Hou, S.; Heinemann, S. H.; Hoshi, T. Modulation of BKCa Channel Gating by Endogenous Signaling Molecules. *Physiology* **2009,** *24,* 26–35.
88. Wilkinson, W. J.; Gadeberg, H. C.; Harrison, A. W. J.; Allen, N. D.; Riccardi, D.; Kemp, P. J. Carbon Monoxide Is a Rapid Modulator of Recombinant and Native P2X2 Ligand-Gated Ion Channels. *Br. J. Pharmacol.* **2009,** *158,* 862–871.
89. Wilkinson, W. J.; Kemp, P. J. Carbon Monoxide: An Emerging Regulator of Ion Channels. *J. Physiol.* **2011,** *589,* 3055–3062.

CHAPTER SEVEN

Antimicrobial Metallodrugs

D.H. Simpson, P. Scott
University of Warwick, Coventry, United Kingdom

Contents

1. INTRODUCTION

The development of antibiotics in the early 20th century revolutionized the field of medicine, preventing countless deaths from diseases caused by microbial pathogens. In the early 21st century, issues such as immunodeficiency, which was increasingly common in the age of HIV/AIDS and for patients undergoing cancer chemotherapy, have boosted the demand for agents to combat opportunistic infections.[1,2] Our ability to respond to such challenges is, however, threatened by the rise of antimicrobial resistance,

Inorganic and Organometallic Transition Metal Complexes with Biological Molecules and Living Cells
ISBN 978-0-12-803814-7
http://dx.doi.org/10.1016/B978-0-12-803814-7.00007-1

which has created an extensive problem for the medical establishment.[3–7] Infection by drug-resistant pathogens has become a prominent and growing cause of global mortality.[8]

1.1 Antimicrobial Resistance and the Case for Metallo-Antimicrobials

Whereas the genetic changes required for bacterial resistance are substantial, rapid cell division at human body temperature (approximately 20 min for *Escherichia coli*) and the vast number of bacteria involved in infections means that these changes occur often. DNA replication produces on average one error in every 10^7 bases,[9,10] and in a typical bacterial genome of about 3×10^6 base pairs, this is equivalent to approximately 0.3 errors per generation.[10] Although only a few resistant mutations are produced, those mutants survive and flourish where other drug-susceptible bacteria die. This rapid adaptation ensures that antibiotic-resistant strains often appear soon after clinical use of a drug.[10–12] Methicillin-resistant *Staphylococcus aureus* (MRSA), multidrug-resistant *Mycobacterium tuberculosis*, and fluoroquinolone-resistant *Pseudomonas aeruginosa* are much-publicized examples.[8,13,14]

Although some of the mechanisms by which microbes resist drug action are more generalized, many depend on specific molecular recognition of the active antimicrobial[15–19] (Fig. 7.1). For example, drug-deactivating enzymes produced by microbes (e.g., β-lactamases) commonly provide resistance to natural product or closely related, "semisynthetic" agents that contain structures and chemical motifs familiar to biology.[20] In contrast, such enzymes are not evolved and have not been found for "purely synthetic" antimicrobials, which has led to the argument that the ideal compound might be produced by rational design rather than by the modification of a natural product,[21,22] or indeed by exploratory phenotypic screening (see subsequent discussion).

Transition metal complexes, in many cases representing completely anthropogenic chemistry, thus emerge as an alternative source of chemotherapeutic antimicrobial agents.[23,24] They offer structural diversity not available to organic chemistry alone, alongside a range of ligand exchange kinetics, redox activities, and tunable electrostatic topologies.[25] However, synthetic transitional metal complexes, e.g., simple *tris*-chelate complexes used by Dwyer (see subsequent discussion), have traditionally offered these features at the cost of a lack of functional complexity compared with biomolecules and synthetic organic compounds. Also, whereas the unique properties of metal complexes create new opportunities for interactions with biological machinery, possibly leading to new modes of action, they also present new questions with respect to biological compatibility and

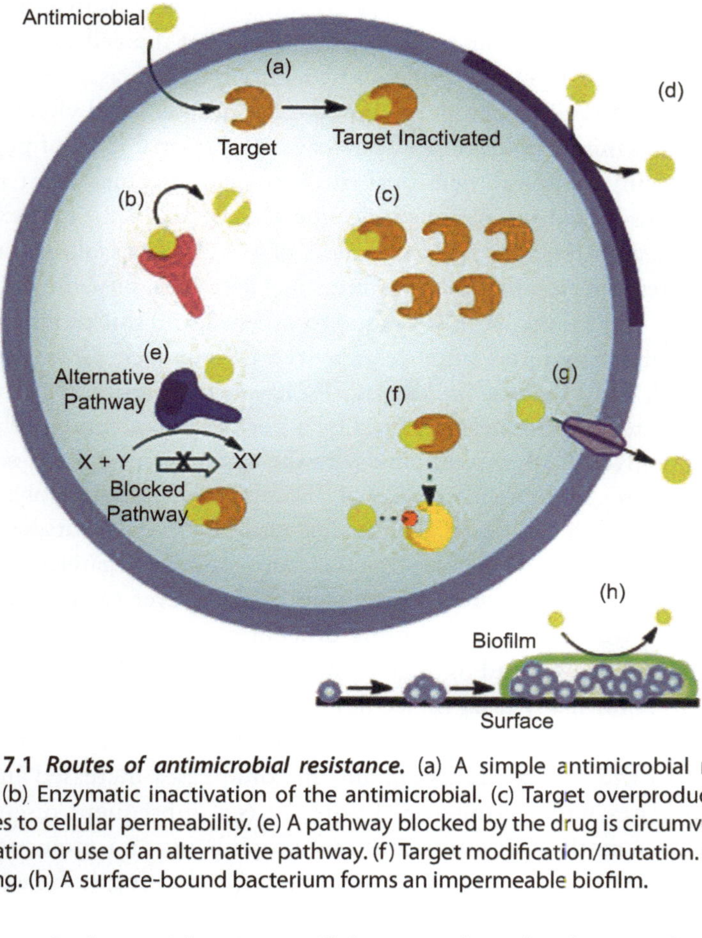

Figure 7.1 ***Routes of antimicrobial resistance.*** (a) A simple antimicrobial mode of action. (b) Enzymatic inactivation of the antimicrobial. (c) Target overproduction. (d) Changes to cellular permeability. (e) A pathway blocked by the drug is circumvented by the creation or use of an alternative pathway. (f) Target modification/mutation. (g) Efflux pumping. (h) A surface-bound bacterium forms an impermeable biofilm.

transport. Such considerations will be crucial to the design of synthetic metallodrugs, and are briefly addressed here.

1.2 Metallodrugs and the Bacterial Cell Envelope

Before a complex may access an endocellular target, it is required to cross the bacterial cell envelope somehow, which provides defense against chemical and physical stresses from the environment. This barrier may prove to be a considerable challenge to synthetic chemotherapeutics.[26] This brief account of the issues facing metallodrugs with respect to cellular entry follows the prevailing assumption that passive diffusion is the principal route of ingress, although there is a highly credible argument for a carrier-mediated view of drug uptake, i.e., that drugs predominantly enter cells via promiscuous proteinaceous carriers.[27–29]

Pathogenic bacteria are commonly classified as gram-positive or -negative according to the response of the outer surface of the cell to the Gram test, developed in 1884.[30] Gram-positive bacteria such as *Staphylococcus*, *Streptococcus*, and *Bacillus* stain purple upon exposure to crystal violet (hexamethyl pararosaniline chloride).[31] They have a substantial layer of peptidoglycan (murein), a three-dimensional lattice structure formed from a sugar polymer, cross-linked via oligopeptide side chains (Fig. 7.2). The exterior is decorated with teichoic acids, polymers of, e.g., glycerol phosphate and carbohydrates, whose function is to bind metal cations, providing rigidity. Biosynthesis of peptidoglycan is unique to bacteria, and this process is thus an important antibacterial target (e.g., by the β-lactams).[32] Gram-negative bacteria such as *Escherichia*, *Yersinia*, and *Pseudomonas* are enveloped by both inner and outer membranes separated by a periplasm containing relatively little peptidoglycan[31]; they stain pink with the application of Gram-staining assays. The outer membrane consists of phospholipids and lipopolysaccharides (LPSs), which are found on the cell surface, directed outward. LPSs are a unique feature to these bacteria; their amphiphilic, negatively charged nature allows them to confer a negatively charged layer to the outer cell membrane,[33] which is stabilized using divalent cations. The double plasma membrane studded with LPSs in gram-negative species makes them a problematic target for many drug molecules.

The lipophilicity of a complex is taken as a good measure of its ability to pass into the cell by diffusion, and in some cases increased lipophilicity correlates with antimicrobial potency (see subsequent discussion). The standard method of evaluating the lipophilicity of a molecule is to measure the partition between octanol and water, i.e., log P, for example, by using the "shake-flask" method.[34] This criterion is part of "Lipinski's rule of five" (see the following text). Ideal lipophilicity for antimicrobial activity lies within a target, organism-dependent range: high enough to be able to cross the cell membrane but low enough to be sufficiently soluble in an aqueous medium (i.e., amphiphilic).[35] Advantageously for coordination complexes, the combination of metal centers that are usually charged (hydrophilic) and lipophilic organic ligands may result in systems of amphipathic topography. This would imply that their hydrophobicity could be optimized in this respect. Relating changes in structure, geometry, and physical properties of a complex to their influence on biological activity may lead to the development of structure–activity relationships or quantitative structure–activity relationships, a common exercise in drug design and development.[36]

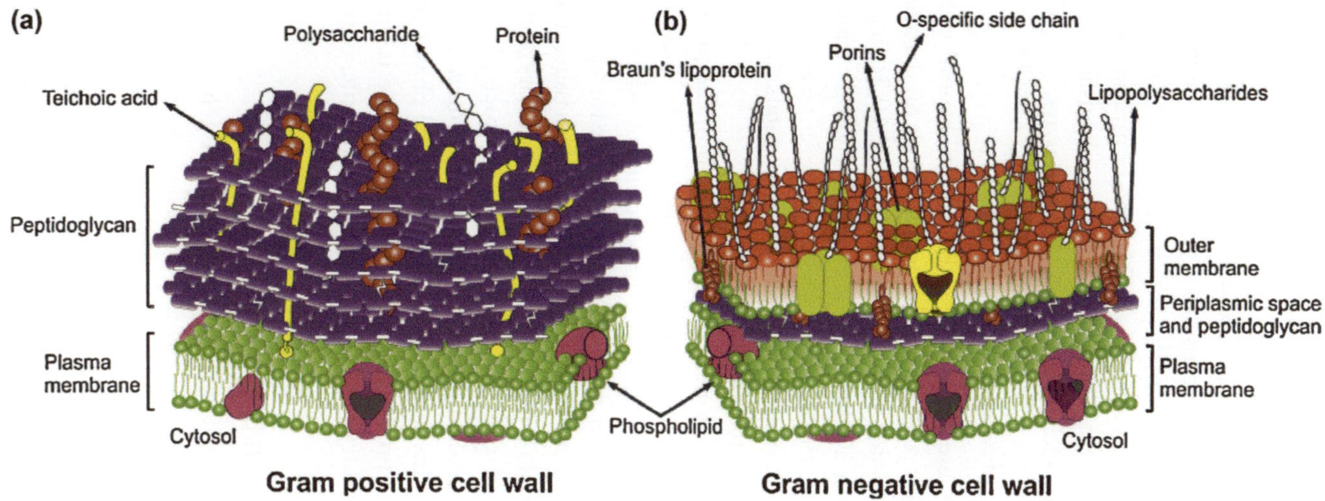

Figure 7.2 ***The bacterial cell wall.*** (a) Gram-positive envelope. (b) Gram-negative envelope. In gram-positive bacteria, the lipidic plasma membrane with embedded proteins is covered by a multilayered peptidoglycan shell decorated with polysaccharides, teichoic acids, and proteins. In gram-negative bacteria, a thin peptidoglycan layer surrounds the plasma membrane and is covered by an asymmetrical outer membrane containing lipopolysaccharides, which lies on the peptidoglycan layer. *Reprinted from Tripathi, P.; Beaussart, A.; Andre, G.; Rolain, T.; Lebeer, S.; Vanderleyden, J.; Hols, P.; Dufrêne, Y.F. Towards a nanoscale view of lactic acid bacteria.* Micron, ***48*** *(12), 1323–1330. Copyright 2012, with permission from Elsevier.*

In more extreme instances of amphiphilicity, like detergents, metal complexes may be able to destroy the integrity of the lipid bilayer of the cell wall and thus compromise the bacterium.[37] This is not thought of as a viable antibacterial route, however, because any untargeted mechanism generally compromising the cell wall is likely also to severely damage eukaryotic cell membranes, leading to low selectivity and unacceptable toxicity toward humans. The high positive charge on some metal complexes may naturally lead to surfactancy in this context, and this mechanism needs to be excluded via standard tests such as red blood cell lysis assays.[38]

1.3 Lipinski's Rule of Five and Other Criteria

It is common practice to assess the orally active "drug-likeness" of a molecule according to Lipinski's rule of five. This places a limit on the acceptable number of hydrogen bond donors and acceptors (5 and 10, respectively), a molecular weight less than 500 Da and a log *P* value (water–octanol partition coefficient) no greater than 5, with others later refining this to a range of −0.4 to +5.6 from observations using large drug databases.[39] Suitable candidates will break one criterion at most.

These guidelines were developed from observation of the properties of successful drug molecules of the time, and so are naturally biased toward small compounds of natural or synthetic-organic origin. Although therefore these particular guidelines may not be appropriate for application to metal complex design, the underlying physicochemical issues remain. It thus makes sense for metallo-organic chemists to consider overarching criteria for properties of their chosen region of chemical space at the start of discovery programs. It is challenging, for example, to synthesize complexes that combine water solubility and stability, stereochemical purity, and ready availability. Without these properties, however, the molecule is unlikely to progress out of the chemistry laboratory and into the clinic, regardless of in vitro activity.

1.4 Molecular Targets in the Cell

Coordination and organometallic chemistry are not yet generally equipped to address specific biomolecular targets, i.e., to make a compound for a specific receptor as in the target-oriented approach to drug discovery, although elegant approaches are emerging.[40,41] Nevertheless, there are great opportunities to take part in the resurgence of phenotypic drug discovery,[42–44] in which the biological targets of a drug (see subsequent examples) are established after observation of the useful biological effect. Indeed, it has been asserted that this approach has led to a disproportionately high number of first-in-class drugs with novel mechanisms of action (1999–2008).[45,46]

DNA is fundamentally important to the machinery governing cells and their proliferation. It is involved directly or indirectly in all cellular functions through the coding of proteins via genes, which making DNA a key anti-microbial target, particular for reactive-at-metal metallodrugs.[47] The major "canonical" form observed biologically, B-DNA, has a wide, shallow major groove and a relatively deep and narrow minor groove. This anionic structure lends itself to various interactions with coordination complexes,[48,49] e.g., by dative or covalent bonds or shape selectivity or via the stacked bases[50] (Fig. 7.3). Such binding or reactions may alter its shape, reactivity, or chemistry, and ultimately the function of DNA.

Within the bacterium, the chromosomal DNA exists as a supercoiled, non–organelle bound structure: the nucleoid. Maintenance of the topologic integrity of this chromosomal DNA is enacted by two enzymes classes: topoisomerases and DNA gyrases, both of which control the extent of the

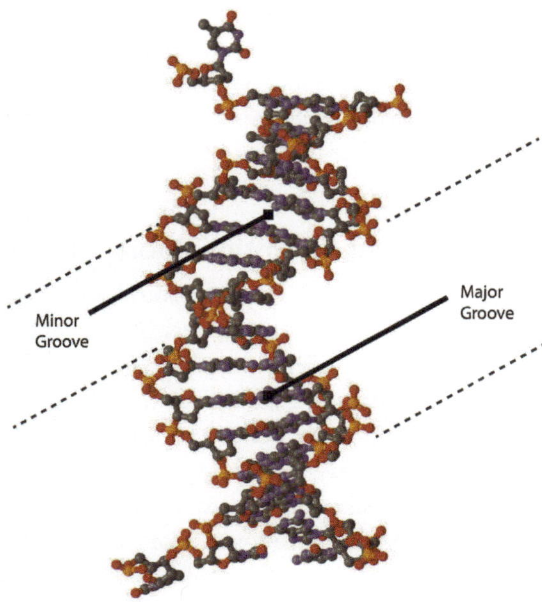

Figure 7.3 ***A short B-DNA sequence.*** *Based on coordinates obtained from the Nucleic Acid Database project (NDB), based at Rutgers University (NDB ID: NA2970 PDB ID: 4Q10). Narayanan, B.C.; Westbrook, J.; Ghosh, S.; Petrov, A.I.; Sweeney, B.; Zirbel, C.L.; Leontis, N.B.; Berman, H.M.* Nucleic Acids Res ***2014***, 42, *D114–D122.*

supercoiling,[51] which makes these enzymes useful antimicrobial targets in their own right.

Composed almost entirely of RNA, the complex machinery of the bacterial ribosome tackles the task of translating the encoded information of DNA into proteins that catalyze the cellular processes of the cell.[53] This important function makes RNA a potential target: in particular, the ribosome complex. RNA rarely takes the B-form observed in DNA[54] (Fig. 7.3), although it may be considered chemically similar and can produce analogous structures at a local level. In rapidly growing bacterial cells, most of the ribosomes appear in the aggregated form of polysomes, where several ribosomes carry out translation on a single RNA transcript (messenger RNA) of a particular gene.[55]

The demand for structural and chemical diversity required for the biocatalysis supporting life falls overwhelmingly to the polypeptides. Protein complexity is built in at the primary structural level from the large pool of chemically diverse common amino acids monomers available (more than 20), increased through the secondary structures available (α-helices and β-sheets), and increased again at the tertiary and quaternary structure levels (macrostructure and cofactor inclusion).[56] Targeting these diverse species poses a challenge to drug designers, and no antimicrobial metallodrugs have been developed to our knowledge with this specific purpose, although protein-inhibiting metallodrugs have been developed as anticancer agents,[25] notably Meggers and coworkers' staurosporine-mimetic, Ru(II) protein kinase inhibitors.[57]

1.5 Preliminary Screening for Antimicrobial Activity

Most of the studies reported here include some kind of antimicrobial screening as part of the early discovery process. The essentials are summarized here, with an excellent list of more detailed advice available.[38]

Measurement of the ability of a candidate to inhibit the growth of a bacterium [the minimum inhibitory concentration (MIC) usually reported in micrograms per milliliter] may be carried out using standardized protocols.[58] This allows rapid comparability to existing antimicrobials. MIC values may also be contrasted with the toxicity of a compound to eukaryote as a rough measure of its selectivity (for prokaryotes), using another rapid in vitro method to measure half-maximal inhibitory concentration (IC_{50}, usually in micromoles or nanomoles) against eukaryote cells. Furthermore, the minimum bactericidal concentration (MBC) (in micrograms per milliliter, the lowest concentration of an antibacterial agent required to kill rather than merely inhibit growth of a particular bacterium) can be readily determined.[58] A wide range of other studies are available, and these are mentioned in the next relevant sections.

Against the background, outlined at the start of this section, of the need for drugs that may experience less acquisition of resistance, and of the need to consider practical barriers that must be overcome, we explore here the recruitment of transition metal coordination chemistry for the development of new classes of antimicrobial chemotherapeutics. Alongside naturally occurring metallodrugs, increasingly sophisticated antimicrobial complexes are being developed and tested. These metallodrugs fall broadly into two categories: (1) those using a labile or reactive metal center as an implicit part of their mode of action, particularly natural products that are activated by metal coordination; and (2) those using metal ions as a template. The latter area is of increasing interest as we gain better control over complex stoichiometry and stereochemistry.[25]

2. REDOX-ACTIVE METAL COMPLEXES

The intrinsic antimicrobial (antibacterial) activity of certain group 11 metal ions or solubilized atoms is well-known.[59] Cu, and particularly Ag, vessels were used by early civilizations to store water because of an understanding that this would help disinfect them.[60] Since then, Cu, Ag, and other transition metal ions have been shown in water to be bactericidal, primarily via in-cell redox activity (notably cycles of Fenton/Haber–Wiess reactions[59]) and/or sporadic interference with native metal cofactors as per the Irving–Williams series[61] even at low concentration.[62,63] However, aquo or other simple complexes would not be suitable for use in antimicrobial chemotherapy because of their generally low solubility in water, poor stability, and resultant potential for toxicity in humans.[64]

In this context we note that some complexes that are unsuitable as drugs find commercial use as topical antimicrobial agents. The antifungal/antibacterial zinc pyrithione[65] (pyrithione = *N*-hydroxypyridine-2-thione) present in many shampoos is used to treat forms of dermatitis and dandruff. Although Zn(II) is not a redox active metal, it was determined that the complex acts by increasing cellular copper levels, leading to loss of activity of iron-sulfur cluster-containing proteins. It proposed that pyrithione acts as a copper ionophore, enabling copper to enter cells and distribute across intracellular membranes.[66] The topical antibacterial natural product bacitracin, a cyclic peptide, is active as a complex of a range of divalent metal ions[67] and is used in, e.g., Zn and Ag preparations.

Research has appeared on complexes that are designed to release antimicrobial metals into solution by triggers such as local redox activity. Screening of copper complexes by Wolschendorf and coworkers led to the discovery of antibacterial complexes that are implied to release copper ions.[68,69] Fromm and coworkers developed silver coordination species, some of which comprise polynuclear frameworks, with antimicrobial properties.[70] Useful approaches to work in this area would be to promote selective ingress into prokaryote cells or to create systems for organism-specific release. Such hypothetical "stealth" complexes could be similar in nature to complexes in which the metal center acts as a structural template, e.g., helicates or siderephore complexes and their analogues (see subsequent discussion).

2.1 Bleomycin

Bleomycin (BLM); (Fig. 7.4), also known as Blenoxane, was discovered in 1966 as a Cu(II)-chelating antibiotic derived from a culture of the strain *Streptomyces verticullus*,[71] although the complexes Fe and Co are the most studied.[72–74] Although BLM and its derivatives are noted anticancer and antiviral agents,[75,76] they have also shown promising prokaryote (*Bacillus subtilis*) toxicity.[77] The iron complex is considered to be the in vivo active species owing to the relative abundance of iron in living organisms relative to other transition metals. In the Fe(II)-chelate form and in the presence of dioxygen (O_2) and a reducing agent, BLM actively cleaves DNA in a directed manner.[74] It is a key example of a compound

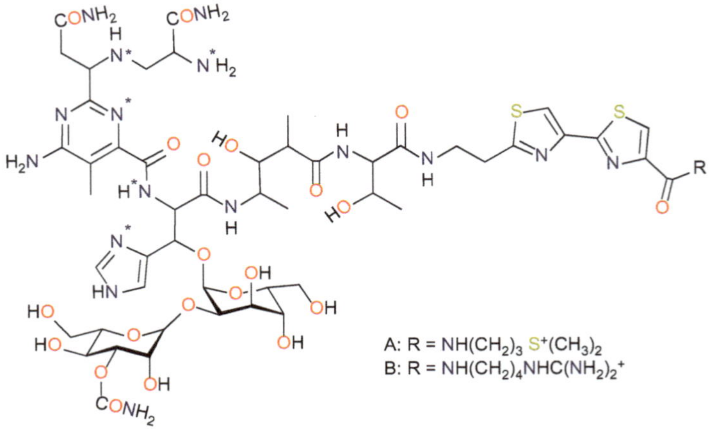

Figure 7.4 ***Structure of "free" bleomycin (A and B variants).*** The metal binding sites, proposed through spectroscopic studies, are marked with *asterisks*.[78]

in which the organic component provides an element of targeting but the principal payload is the metal ion.

The nature of the coordination geometry of the complex formed by chelation of a metal ion by BLM has been thoroughly studied. Through detection of hyperfine splitting of ^{14}N, electron spin-echo envelope spectroscopy of complexes allowed identification of the heteroatoms involved in binding[78] as well as the presence of a distorted five- or six-coordinate coordination sphere (the sixth being a possible site for a monodentate ligand). Determination of the structure of a Cu(II) complex of a BLM intermediate, lacking the sugar and peptidyl bithiazole moieties, but inclusive of all metal binding sites, has appeared.[79] Nuclear magnetic resonance (NMR) studies using diamagnetic Zn(II) species on BLM itself as well as the structurally similar tallysomycin as chelators have been used to probe the coordination of BLM.[80,81]

The proposed antimicrobial mechanism of action involves the formation of an "active form" in which the BLM-Fe(II) reacts (most likely via a superoxide intermediate) to form a peroxy species BLM-Fe(III)-OOH, with the peroxy species occupying the sixth coordination site of the distorted octahedron, as observed by Raman spectroscopy.[82] O—O bond cleavage results in the generation of O=Fe(IV)-BLM or O=Fe(V)-BLM, which oxides ribose moieties in DNA, in regions immediately preceding a 5′-GC-3′ or 5′-GT-3′ sequence, causing the ribose moiety and thus the DNA structure to break down.[74,83,84]

Despite the good early candidacy of such BLM complexes, they have largely been ruled out as a prospect for a clinical antimicrobial role. Issues include the lack of specificity for prokaryote DNA, the severity/irreversibility of DNA double-strand scission, and evidence for human toxicity (causing lung fibrosis).[85]

2.2 Streptonigrin and the Anthracyclines

Streptonigrin (SN) is a quinone-based antibiotic[86] that undergoes a dramatic structural change upon metal binding as a whole (Fig. 7.5a), toward an active form.[87] The anthracyclines (ACs) (notably daunomycin and doxorubicin) are constructed of four fused rings linked to a pyran derivative (Fig. 7.5b), which may be a pyranose sugar or group of sugars in more complex examples.[88] Although their core structures vary, these compounds share notable key features: (1) Both were originally isolated from the *Streptomyces* genus (although synthetic derivatives are known)[89–91]; (2) they contain moieties that allow specific binding/intercalation of DNA (their primary target, although ACs may also bind DNA topoisomerase enzymes)[92,93]; and (3)

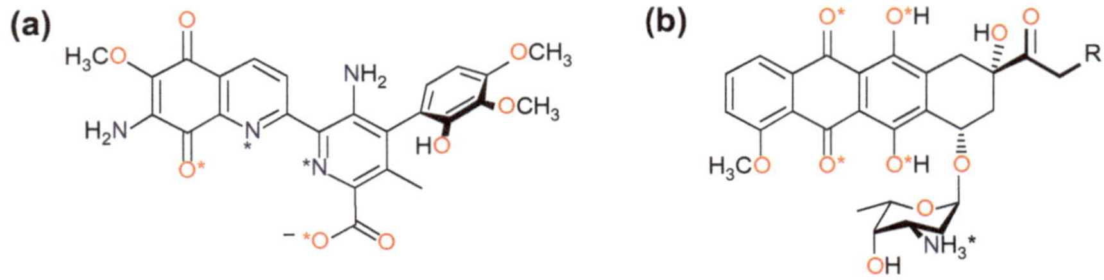

Figure 7.5 ***Structures of quinone-based antibiotics.*** (a) Streptonigrin and (b) the simple anthracyclines daunomycin (R=H) and doxorubicin (R=OH). Proposed metal binding sites are marked with *asterisks*.

although both are quinone-based antibiotics in their own right (as unbound species), they require binding of a redox-active metal to be fully active. For the latter, evidence suggests that the increase in activity is caused by improving DNA binding by modification of geometry and electrostatic topography, and the availability of localized catalysis of radical species formation.[94,95] It is suggested that ACs may be bound to one or two metal ions simultaneously, using their opposing β-ketophenolate moieties.[96]

The antimicrobial effects of SN toward *E. coli* and *B. subtilis* were described in the 1960s and 1970s.[97,98] The mechanism has been linked to its generation of O_2^- and/or peroxides, a feature closely associated with the presence of iron and copper, leading to the inhibition of DNA synthesis.[99] Similarly, metal–AC complexes [notably 1:1 complexes with Fe(III)] are implicated in the induction of DNA cleavage via the increased generation of radicals,[100,101] especially in the presence of reducing agents such as reduced nicotinamide adenine dinucleotide. Reportedly, this activity occurs through (metal-assisted) reduction to semiquinone forms, as with other quinone-based drugs,[102] a process found to be inhibited by the enzymes catalase[103] and superoxide dismutase,[104] as would be expected for a mechanism dependent on reactive superoxides and peroxides. Intercalation of the free AC in DNA, combined with limited redox activity of the organic nucleus, accounts for the mild antibacterial effect of metal-free species, particularly against *B. subtilis* strains lacking DNA-repair mechanisms.[105]

Some antifungal activity (daunomycin) has also been noted.[106]

3. COMPLEXES WITH REACTIVE METAL CENTERS

Numerous attempts have been made to screen reactive-at-metal species for antimicrobial activity, with varying results. Prominent historic examples have been reviewed in detail, notably by Ming.[107]

Although cisplatin is renowned for its anticancer activity, and accordingly is treated thoroughly in Chapter 4 alongside other platinum compounds, it was discovered serendipitously as a drug candidate by Barnett Rosenberg in the 1960s via its ability to inhibit cell division in *E. coli*.[108] Other than its completely synthetic origin, it is differentiated from the earlier antimicrobial species discussed in this section in that the metal does not contribute to the mechanism of action via redox activity, but as a reaction center in the formation of strong cross-linking bonds with the target biomolecule DNA in vivo.[49]

The focus on synthetic reactive-at-metal antimicrobials has been with the aim to control and modulate reactivity to reduce the side effects seen in chemotherapeutic use. Attempts have subsequently been made to "tame" cisplatin-like heavy transition metal complexes using strongly bound *N*-heterocyclic carbene (NHC) ancillary ligands.[109] Whereas *bis* complexes such as that in Fig. 7.6a have modest activity,[110] Panda and coworkers[111] reported a slightly more potent complex of gold chloride (Fig. 7.6b) in gram-positive *B. subtilis*, although this and similar examples tested were inactive against gram-negative *E. coli*. Gurovic and coworkers[112] synthesized much more soluble gold and silver complexes (Fig. 7.6c), but the MIC values reported were extremely high. In contrast, bimetallic compounds (Fig. 7.6d)

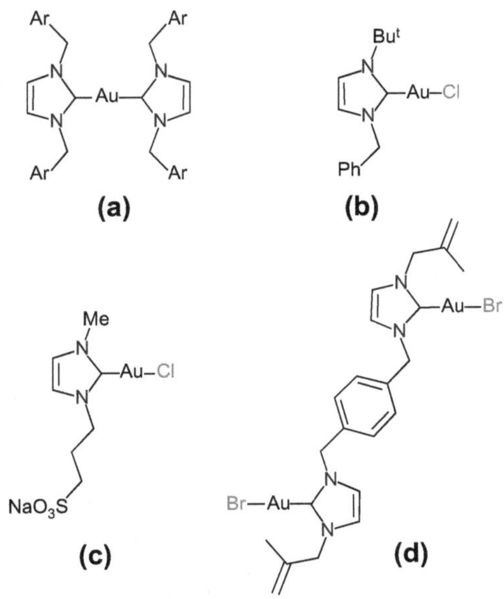

Figure 7.6 *N*-Heterocyclic carbene complexes used in antimicrobial screening.

were screened by Mandal and coworkers for their antimicrobial activities against multiple keratitis-associated human eye pathogens, and revealed some impressive MIC figures and the ability to eradicate biofilm on contact lenses.[113] An alternative strategy for controlled reactivity is the use of metal centers with "intermediate" reactivity: for example, Co(III) complexes with noninnocent (redox) ligands.[114]

4. COMPLEXES THAT RELEASE CARBON MONOXIDE

The ability of coordination complexes to act as small molecule carriers, particularly CO-releasing molecules (CORMs), has been explored as a source of chemotherapeutic agents, including antimicrobials.[115]

The efforts initially of Saraiva,[116] and later Poole in particular,[117] have driven the antimicrobial research of CORMs, and despite the analytical challenge involved demonstrating CO release under biological conditions.[118] Modest antimicrobial effects of prototypic examples have been described[116] that were markedly improved compared with the use of CO alone. This perhaps suggests selective release, or a synergistic effect of the remaining complex.[117] Interestingly, these complexes would appear to affect "nontraditional' targets" (e.g., cellular respiration).[119] Fig. 7.7 shows two notable CORMs: Ru(II)-based "CORM-3," noted for its water solubility owing to the use of glycinate and chloride ligands, and an Mn(I)-based "photoCORM," capable of releasing CO upon irradiation with near-UV light, with an initially uncoordinated pyridyl group capping the now free octahedral site.[117]

The use of electromagnetic radiation to photoactivate an inert metallodrug precursor at the site of infection, has garnered some interest.[121,122]

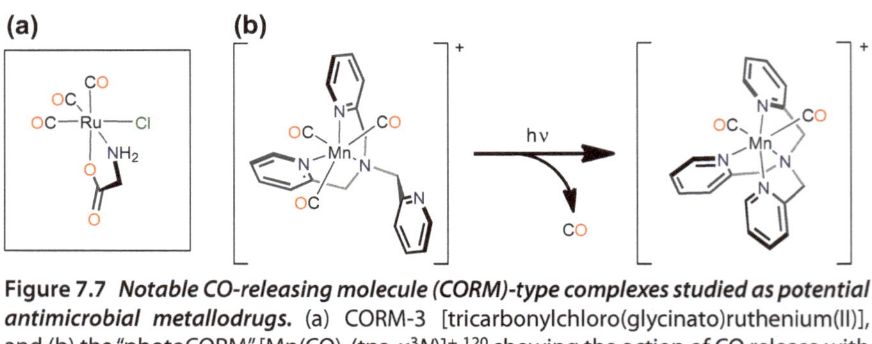

Figure 7.7 ***Notable CO-releasing molecule (CORM)-type complexes studied as potential antimicrobial metallodrugs.*** (a) CORM-3 [tricarbonylchloro(glycinato)ruthenium(II)], and (b) the "photoCORM" $[Mn(CO)_3(tpa\text{-}\kappa^3N)]^+$,[120] showing the action of CO release with UV irradiation.

However, the poor penetrability of shorter-wavelength light into the body[123] and the often-widespread nature of infection needs to be considered.

5. METAL CENTERS AS STRUCTURAL TEMPLATES

Rather than using metal-centered redox activity or substitution chemistry to damage DNA or other biomolecular target(s) permanently, more subtle inhibitory mechanisms are being addressed with complexes in which the metal is intended to be purely structure-forming. Indeed, depending on the nature of the ligands, the control of stereochemistry becomes important, and this is an area where much development is needed.[25] Nevertheless, systems have been reported with promising antimicrobial activity along with supporting studies toward an understanding of underlying mechanism(s).

5.1 Early Advances: Simple *Tris* Complexes

An early body of work that nevertheless continues to inspire new research was described by Dwyer and coworkers in the 1950s and 1960s.[124,125] These researchers tested simple complexes of transition metals with small, symmetric bidentate ligands such as the examples shown in Fig. 7.8. These complexes were found to have variable activity against strains of *Escherichia, Staphylococcus*, and *Salmonella*, with the placement of *exo*-steric groups (e.g., methyl groups), and proved to have an important influence on antibacterial activity.

5.2 DNA-Intercalating Ruthenium Complexes

Many *tris*(bipyridine) and related complexes of Ru(II) are known. These are inert and not expected to mediate any sort of metal-centered activity. However, their ability to intercalate DNA is well-studied[48,49] using, e.g.,

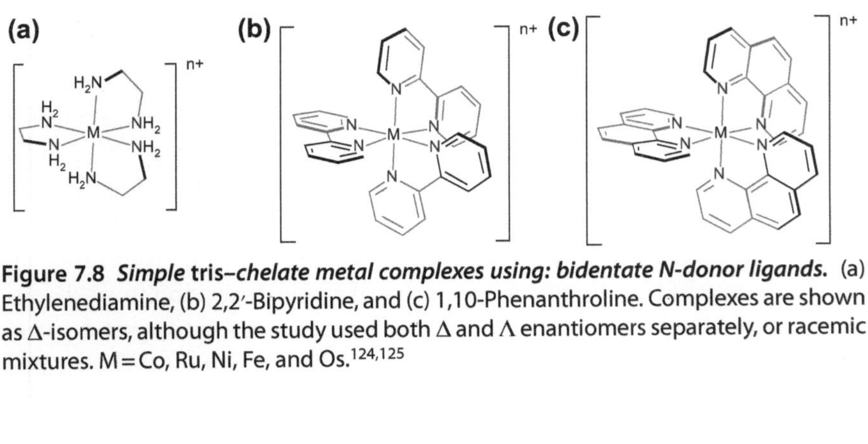

Figure 7.8 ***Simple* tris–*chelate metal complexes using: bidentate N-donor ligands.*** (a) Ethylenediamine, (b) 2,2′-Bipyridine, and (c) 1,10-Phenanthroline. Complexes are shown as Δ-isomers, although the study used both Δ and Λ enantiomers separately, or racemic mixtures. M = Co, Ru, Ni, Fe, and Os.[124,125]

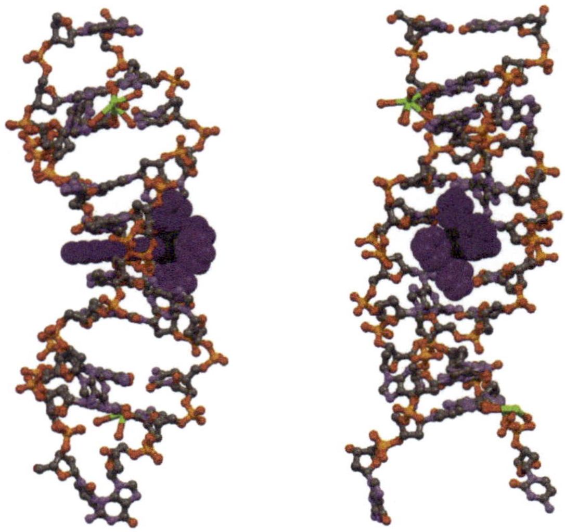

Figure 7.9 ***B-DNA complex of Δ-[Ru(bby)$_2$(dppz)]$^{2+}$.*** The complex is shown as a spacefill model in blue/gray. One dipyrido[3,2-a:2′,3′-c]phenazine (dppz) ligand intercalates two B-DNA base pairs. The structure is shown from two perspectives (90 degrees rotation about the vertical axis). Only the double helix and a single complex are shown for clarity. *Created from an RCSB Protein Data Bank structure (PDB ID:* 4E1U*). Song, H.; Kaiser, J.T.; Barton, J.K.* Nat. Chem. ***2012****, 4, 615–620.*

NMR, linear dichroism (LD) and circular dichroism.[126,127] The complex Δ-[Ru(bpy)$_2$(dppz)]$^{2+}$, for example, co-crystallizes with B-DNA containing a base pair mismatch, as shown in Fig. 7.9.[128]

Whereas much of the motivation for use of these complexes is in the arena of cancer,[129–134] in 2010 Aldrich-Wright and coworkers published an account of the antimicrobial potential of three Ru(II)-based DNA intercalators (Fig. 7.10).[135] Using standardized screening techniques,[58] MICs were determined against four gram-positive bacterial strains [*B. subtilis* 168, methicillin-sensitive *S. aureus* (MSSA) 160, and MRSA strains 41 and 252]. Results appeared to follow earlier reported trends in the ability to intercalate DNA; [Ru(phen)$_2$(dpq)]$^{2+}$ showed little or no potency, [Ru(bpy)$_2$(dpqC)]$^{2+}$ showed slight activity (MIC: 16–64 μg/mL), and [Ru(2,9-Me$_2$phen)$_2$(dppz)]$^{2+}$ displayed the highest potency, with MICs ranging from 8 μg/mL for MSSA 160 to a remarkable 2 μg/mL for MRSA 252. MBCs two to four times the corresponding MICs support a bacteriocidal rather than inhibitory (bacteriostatic) mode of action. In contrast, no activity was found against gram-negative *E. coli*, perhaps because of the low permeability of the cell envelope.

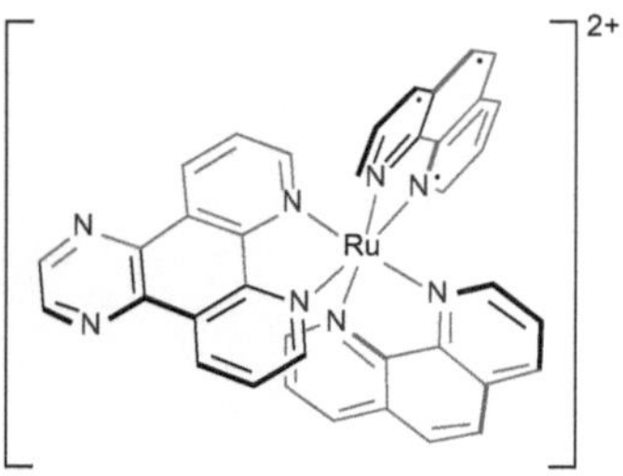

$[Ru(phen)_2(dpq)]_2^+$

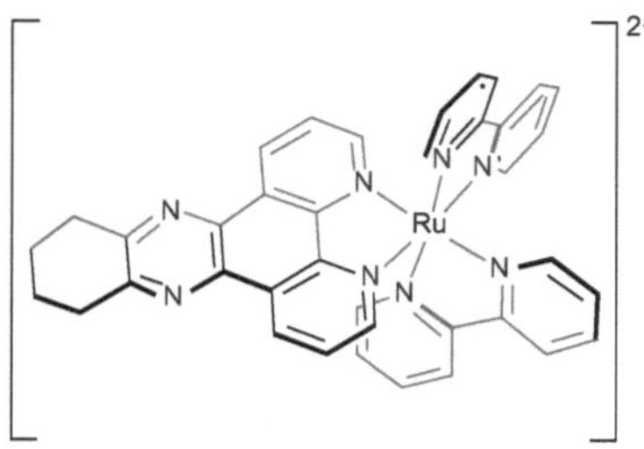

$[Ru(bpy)_2(dpqC)]_2^+$

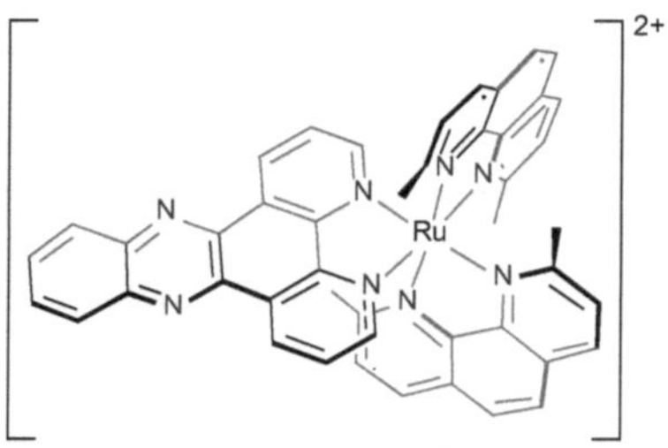

$[Ru(2,9\text{-}Me_2phen)_2(dppz)]_2^+$

Figure 7.10 ***DNA-intercalating complexes studied by Aldrich-Wright and coworkers.*** *dpq,* dipyrido[3,2-d:2′,3′-f]-quinoxaline); *dpqC,* dipyrido[3,2-a:2′,3′-c]-(6,7,8,9-tetrahydro)-phenazine); *dppz,* dipyrido[3,2-a:2′,3′-c]-phenazine. Complexes are shown as Δ-isomers.

In further work, a time-kill assay using $[Ru(2,9\text{-}Me_2phen)_2(dppz)]^{2+}$ was carried out. This established the rate with which live bacteria are killed by measuring the population growth or decline from a set initial population. In the presence of 1 μg/mL (0.5 of the MIC) or less, MRSA 252 continued to increase in population. At 8 μg/mL (the MBC) and 32 μg/mL, the population showed a sharp decline, which suggested a relatively rapid mode of action. Furthermore, the nematode worm *Caenorhabditis elegans*, a multicellular eukaryote and noted infection model,[136,137] was infected by MRSA 252, then subjected to various concentrations of $[Ru(2,9\text{-}Me_2phen)_2(dppz)]^{2+}$, i.e., an infection/recovery experiment. Over several days, the number of surviving worms was monitored (Fig. 7.11). It was observed that a relatively low concentration of $[Ru(2,9\text{-}Me_2phen)_2(dppz)]^{2+}$ (1 μg/mL) showed little improvement over the control. All worms died by day 6 in both cases, although at a slightly

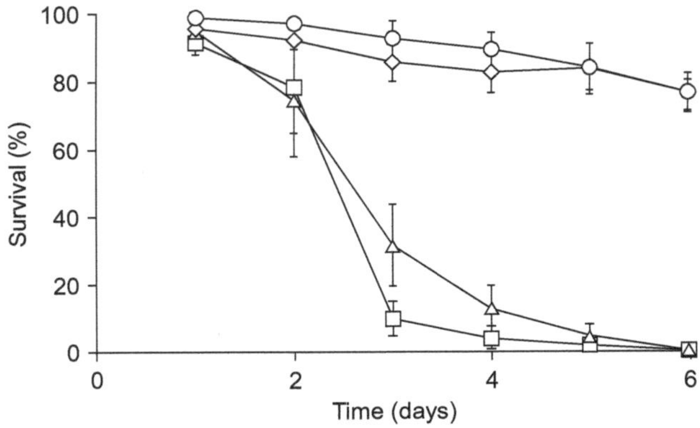

Figure 7.11 ***Survival of age-synchronized*** **C. elegans** ***in the presence of [Ru(Me₂phen)₂dppz]²⁺.*** After an initial 1-h incubation with *S. aureus* MRSA252, the number of living worms was scored daily. *Circles* indicate 32 μg/mL; *diamonds*, 8 μg/mL; *triangles*, 1 μg/mL; *squares*, no compound added (control). *Reprinted from the Bolhuis, A.; Hand, L.; Marshall, J. E.; Richards, A.D.; Rodger, A.; Aldrich-Wright, J. Antimicrobial activity of ruthenium-based intercalators.* Eur. J. Pharm. Sci. 42 *(4), 313–317. Copyright 2011, with permission from Elsevier.*

slower rate with the complex present. Concentrations of 8 and 32 μg/mL showed marked improvement over the control with about 80% nematode survival by day 6.

Overall, these experiments create one of the more advanced cases for potential use of inert metal complexes as antimicrobials: relatively low MIC, a rapid mode of action relative to infection spread, direct evidence for recovery from otherwise fatal infection in a higher eukaryote, and a credible mechanistic hypothesis. Significant questions remain, however, such as selectivity of the compounds here or in higher organisms, and the issue of enantiomeric separation.[127,138–143]

5.3 Oligonuclear Ruthenium Complexes

O'Reilly et al. found that "dimers" of $[Ru(phen)_2(Me_2bpy)]^{2+}$ units linked via an alkyl chains (Fig. 7.12) (n = 5, 7, and 10) are excellent and robust DNA-groove binders and more efficient than the mononuclear analogues. The binding strength was found to increase with the linker chain length because this allows it to "follow" the helical DNA structure.[144–146] Aldrich-Wright and coworkers also demonstrated enhanced binding, compared with its mononuclear analogue, of a dinuclear Ru(II) complex incorporating a 2-mercapto-ethyl ether linker.[147] Similarly, Lincoln and Thomas demonstrated extremely

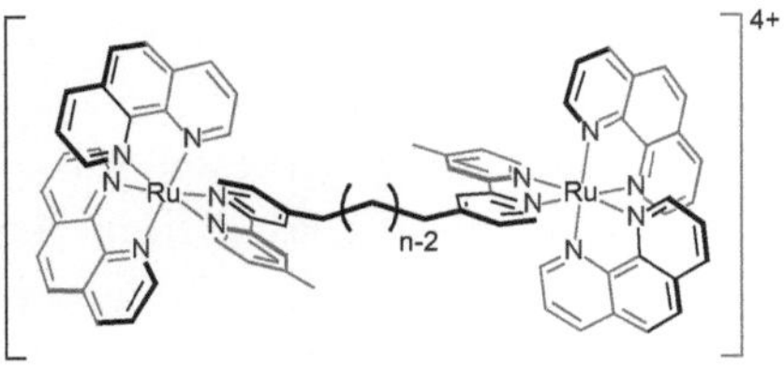

Δ,Δ-Rubb$_n$

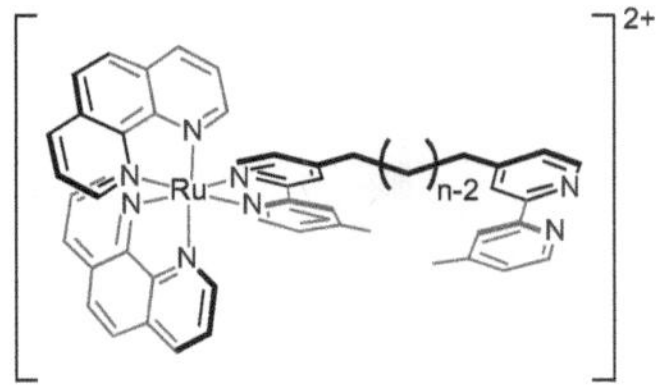

Δ,Δ-Rubb$_n$-mono

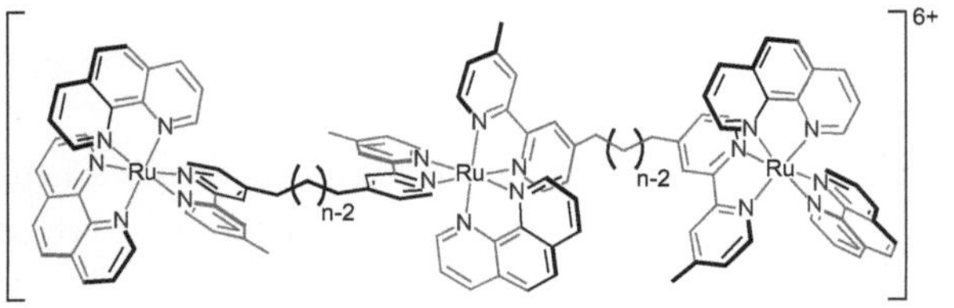

Δ,Δ,Δ-Rubb$_n$-tri

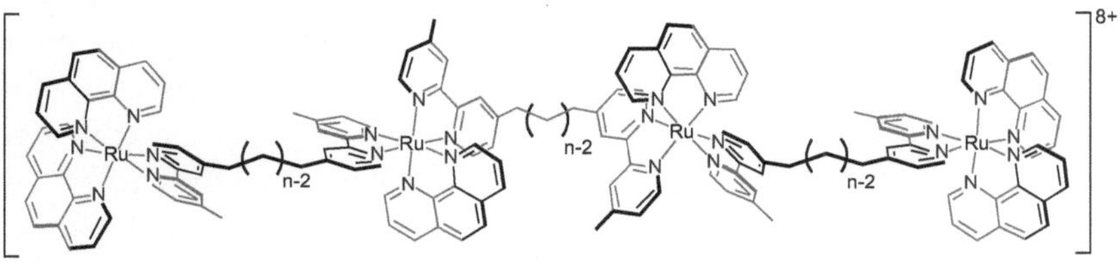

Linear Δ,Δ,Δ,Δ-Rubb$_n$-tetra

Figure 7.12 ***Oligonuclear ruthenium used by Collins et al.*** Rubb$_n$-mono, Rubb$_n$ lacking a second Ru^{II}(phen)$_2$ center (a), trinuclear (b), and linear tetranuclear (c) analogues. All metal centers are shown in the Δ configuration, and where applicable, with methylene bridges anchored *trans* across Ru(II) centers. *Rubb$_n$*, [(Ru(phen)$_2$)$_2$(μ-bb$_n$)]$^{4+}$.

high binding affinity of Ru(II) dinuclear complexes at DNA sites.[148–151] Keene and Collins presented an extensive body of work in this area, including discovery of antimicrobials and study of mechanism(s) of action.[152]

The [(Ru(phen)$_2$)$_2$(μ-bb$_n$)]$^{4+}$ (Rubb$_n$) systems in separated stereoisomeric forms (Δ,Δ or Λ,Λ, excluding the *meso*-isomer)[153] were studied alongside various controls.[154] Rubb$_{12–16}$ complexes were found to be active against a range of pathogenic bacteria, particularly gram-positive *S. aureus*, both methicillin-susceptible (ATCC 25,923) and "clinical wild-type' MRSA (MICs of 1–2 μg/mL). Activities against gram-negative *E. coli* ATCC 25,922 (2–4 μg/mL) and ATCC 27,853 *P. aeruginosa* ATCC 27,853 (8–16 μg/mL) were more modest but nevertheless promising. Against all gram-negative and gram-positive organisms tested, the Δ,Δ-isomers were marginally more active. Interestingly, relatively short (n = 2 and 5) and heteroatom-linked dinuclear complexes were shown to have little or no activity.

Selectivity for the "lead" Δ,Δ-$Rubb_{12-16}$ complexes was assessed subsequently. Human red blood cell lysis (indicating membrane disruption) and toxicity to THP-1 (monocytic leukemia) cells increased with chain length, but the most promising candidate tested (ΔΔ-$Rubb_{12}$) gave HC_{50} and IC_{50} values, respectively, over 100-fold higher than the *S. aureus* MIC.[154] Later screening by the same team showed a lower toxicity to hamster kidney (BHK), human embryonic kidney (HEK-293), and human liver carcinoma (HEP-G2) cell lines.[155]

Microbial cellular uptake for this class of compound was measured for these complexes via their inherent fluorescence.[156] Uptake fell with lower linker chain length n, reflecting the trend in log *P* values[154] $Rubb_{16} > Rubb_{12} > Rubb_7$, which suggests that the higher hydrophobicity is beneficial, although it is difficult to deconvolute other effects of the longer chain such as increased mechanical flexibility. At the same time, monometallic species $Rubb_n$-mono, which contain a pendant ligand (n = 7 and 12) and $[Ru(Me_4phen)_3]^{2+}$, did not follow the same trend, even though $[Ru(Me_4phen)_3]^{2+}$, a compound remarkably similar to those used in the early work of Dwyer,[124,125] was shown here to be a potent antimicrobial (MIC range: 0.5 μg/mL versus *S. aureus* to 32 μg/mL versus *P. aeruginosa*). This may indicate that the mechanisms of cellular uptake and/or antimicrobial action of $Rubb_n$ systems are different from mononuclear Ru(II) systems.

Uptake into gram-negative bacteria was significantly less than for gram-positive bacteria for $Rubb_n$ systems, and was shown in all cases to occur in an energy-independent manner (passive diffusion) because, for example, dead or adenosine triphosphate–inhibited cells would still accumulate the complexes.[156,157] Furthermore, the dinuclear complexes significantly depolarize and permeabilize the cell membrane of *S. aureus*,[157] although this occurred at concentrations double the corresponding MICs.[154]

Cellular localization of $Rubb_n$ in *E. coli* MG1665 was studied using wide-field fluorescence microscopy (Fig. 7.13).[155] $Rubb_{16}$, the most hydrophobic $Rubb_n$ complex reported, was found to concentrate in the vicinity of RNA (polysomes) in vivo, despite the significant membrane activity reported previously.[157] In the context of the pedigree described for $Rubb_n$ systems as DNA-groove binders, an RNA/polysome-targeting mechanism is feasible, which could halt translation of RNA to proteins, thereby preventing protein synthesis. A preference for the RNA-rich nucleolus over

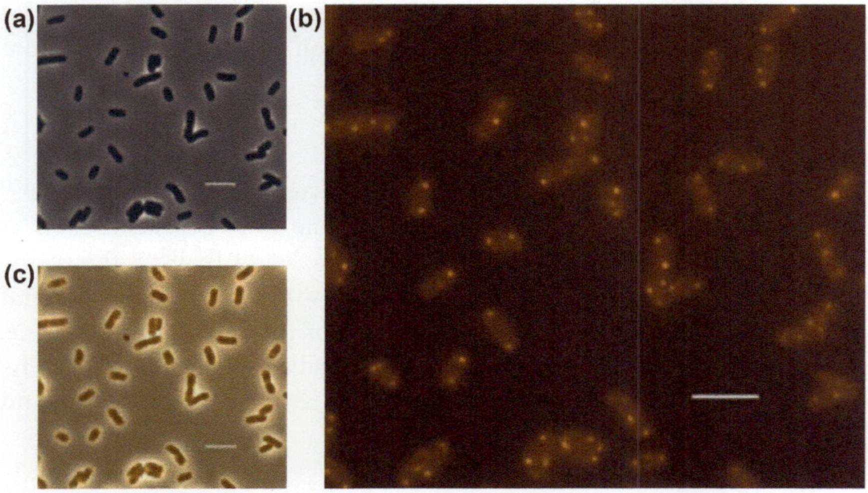

Figure 7.13 ***Rubb$_{16}$* *localization in* E. coli *MG1665 cells at an MIC of 4 mg/mL.*** Fluorescence microscopy images show: (a) phase-contrast, (b) Rubb$_{16}$ phosphorescence, and (c) merged. Scale bar = 5 μm. This image and others from the same work illustrate the concentration of Rubb$_{16}$ at the RNA-based polysomes.[155] *Published by the Royal Society of Chemistry.*

chromosomal DNA for all linear Rubb$_{12}$ systems in eukaryotic cell lines was observed via colocalization studies using confocal microscopy.[158]

These concepts were extended through the study of oligonuclear Rubb$_n$ complexes (Fig. 7.12) built through modular syntheses.[159] The application of similar methodology to earlier work[154] showed that these complexes have strong antimicrobial activity; several MICs against *S. aureus* were below 1 μM, with low MICs against gram-negatives *E. coli* (1–6 μM) and to a lesser extent, *P. aeruginosa* (9–50 μM). Linear species with chain lengths of n = 12 and 16 were found to be more active, and in particular the linear tetranuclear complexes were more active than the branched species. This is consistent with the hypothetical target as RNA with the compounds acting as polynucleotide groove binders.[155] Furthermore, although the trinuclear species were the most lipophilic (log *P*) and showed relatively high uptake, they were not as active as their linear-tetranuclear counterparts. Unlike their dinuclear counterparts, cellular uptake of Rubb$_n$-tri and linear Rubb$_n$-tetra (Fig. 7.12) in the gram-negative species was greater than or equal to that observed in the gram-positive series, and yet the MICs for gram-positive species were significantly lower. This may suggest in this instance that the resistance of gram-negative species to these compounds is not the result of the protective membrane but that for some reason intracellular processes are less susceptible.[156,157]

5.4 Siderophore Analogues

Siderophores are molecules produced by bacteria to sequester iron[160] in competition with strategies evolved in the host organism such as nutritional immunity.[161,162] Enterobactin (Fig. 7.14a)[163,164] is the strongest Fe(III)-binding siderophore known ($K = 10^{52}\,M^{-1}$).[165] The chiral backbone, biosynthesized from a trilactone of the amino acid L-serine, preorganizes for Δ configuration at the metal[166]; the structure of Fig. 7.14b is based on the analogue Δ-$[V(ent)]^{2-}$, reported by Karpishin and Raymond.[167]

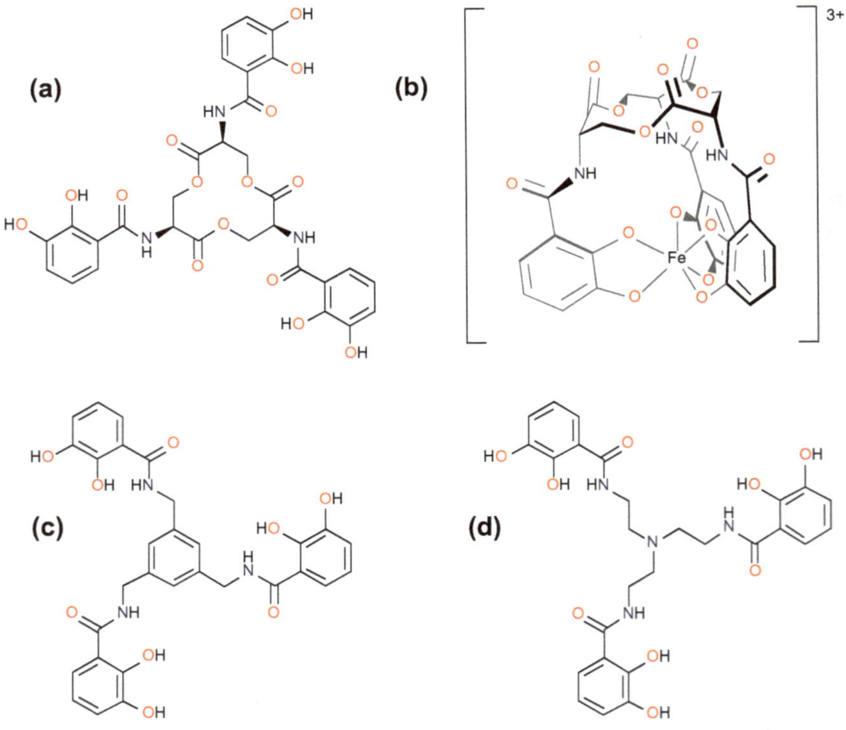

Figure 7.14 ***Siderophore and analogue chemistry of Raymond et al.*** (a) The catechol ligand enterobactin (ent) from *E. coli*. (b) Structure of $[Fe^{III}(ent)]^{3-}$ based on a crystallographic study of $[V(ent)_3]^{2-}$. (c) 1,3,5-*tris*[(2,3-dihyroxybenzoyl)aminomethyl]benzene. (d) 1,3,5-*N,N′,N′*-*tris*(2,3-dihydroxybenzoylamide)tri(aminoethyl)amine.

Raymond and co-workers conducted a detailed study of the uptake of siderophore complexes in *E. coli*, underpinning our understanding of bacterial uptake of ferric iron. Of particular interest to this chapter was the ability of an Rh(III) complex (Rh{1,3,5-*tris*[(2,3-dihyroxybenzoyl)aminomethyl]benzene (MECAM)}$^{3-}$) (ligand shown in Fig. 7.14c) to inhibit bacterial uptake of the $[Fe(ent)]^{3-}$ complex[163] nominally via the *E. coli* FepA receptor protein, a critical pathway for essential iron uptake in *E. coli*.[168,169] Fe(1,3,5-*N,N′,N′*-*tris*(2,3-dihydroxybenzoylamide)tri(aminoethyl)amine) (TRENCAM)$^{3-}$ [170] (ligand shown in Fig. 7.14d), and in particular $[Fe(MECAM)]^{3-}$, were later found to be taken up by the iron-starved gram-negative, enterobactin-dependent species *Bordetella bronchiseptica* using the same BfeA outer membrane receptor as enterobactin.[171] Other $[Fe(ent)]^{3-}$ mimics have shown similar uptake in *Salmonella typhimurium* through enterobactin receptors IroN and FepA.[172] Work on the TRENCAM ligand and its complexes has led to an intriguing set of supramolecular structures that are bicapped rather than monocapped: Two *tris*(2-aminoethyl)amine moieties link the three catechol bidentate units to two axial tertiary amine "hubs" at either pole of the complex.[173]

Through this work we can see that a siderophore analogue might be used to block the uptake of iron or to ferry more toxic ions into the prokaryote cell in "Trojan horse" strategy.

5.5 Metallohelices

Myriad coordination architectures formed through the application of ligands designed to span multiple metal centers have been reported, including helicates,[174–177] knots,[178] grids,[179,180] catenanes,[181] rotaxanes,[182] boxes,[183] and cages[184] among others.[185] Few of these have been shown to have potential use as drug molecules, probably because of the difficulty in making water-compatible compounds in significant quantities. This practical barrier arises because inert metal ions such as Ru(II), which are expected to give water-stable compounds, give mixtures of kinetic products on reaction with polydentate ligands, leading to low yields. On the other hand, the cleaner thermodynamic products produced by relatively labile metals are almost invariably subject to hydrolysis in water or biological media.

Nevertheless, Lehn's prototypical helicate chemistry[177] (Fig. 7.15) has inspired a number of workers to explore the idea that such compounds might provide entry to a new area of biomimetic chemical space unavailable to organic chemistry.[186,187]

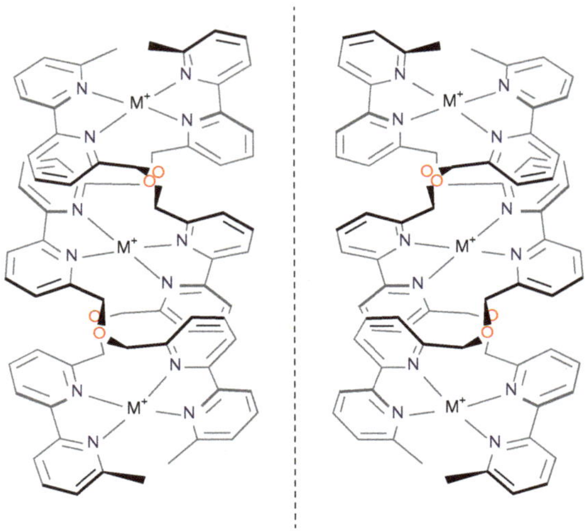

Figure 7.15 Lehn's prototypic helicate structure showing the two optical isomers (M = Cu or Ag).[177,188]

In 1997, Hannon and coworkers reported the synthesis of a helicate structure $[Fe_2L^{H}_3]^{4+}$ from simple starting materials[189]: an aromatic diamine, 2-picolinaldehyde, and an Fe(II) source that provides the anchoring octahedral metal centers (Fig. 7.16). The rigid ligand system mechanically couples the helical coordination environments, requiring them to adopt the same stereochemistry (Δ,Δ or Λ,Λ). The ensuing triple-helix structure has a well-defined pitch. The reported syntheses of this compound involve the use of weakly coordinating anions (PF_6^-) for ease of isolation, followed by exchange with chloride to provide water solubility.

Related examples include the results of simple modifications to the original Fe(II) helicate: replacing the bridging methylene with O; several peripheral modifications[190]; and Ru(II) analogue(s).[191,192] The latter, along with the original bis-Fe(II) helicate, is reported to be resolvable into enantiomers using a brine elution on a cellulose column or even paper,[190,193] as supported by circular dichroism, although without quantification of the optical purity (e.g., through the use of an NMR shift reagent such as TRISPHAT).[194] Helicates with arginine[195] and short peptide fragments[196] (Fig. 7.16) were subsequently reported via a multiple-step route. Helicity was seen to be controlled to within the limit of signal-to-noise for the NMR spectra observed.[195] Great attention has been given to the ability of

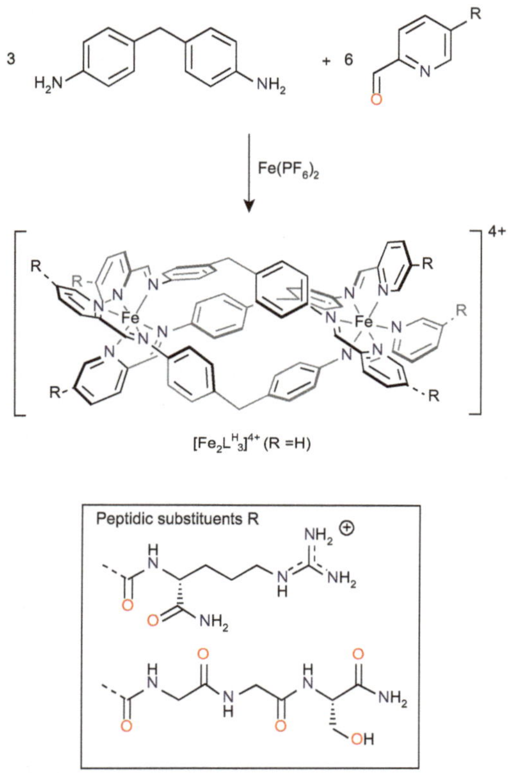

Figure 7.16 Hannon's Fe(II) helicate, shown as the Λ,Λ isomer.

these helicates to bind DNA motifs in vitro, especially the B-DNA major groove[192,197,198] and three-way junctions (3WJ)[195,199–201] inducing conformational changes.[192]

The helicate $[Fe_2L^H{}_3]^{4+}$ of Fig. 7.16 as the chloride salt was subject to in vivo screening as an antimicrobial agent.[202] MICs of 32 μg/mL were found against gram-positive *B. subtilis* strain 168, and 64 μg/mL against gram-negative *E. coli* strain GM2163, respectively. Both MBCs were less than four times the MIC, and although the potencies are low this was taken to confirm a bacteriocidal mode of action.[38] Unfortunately the compound was found to be incompatible with standard broth,[58] so a special in-house medium was devised. This makes comparison with other compounds difficult. Spectroscopic analysis of the DNA extracted from *B. subtilis* after exposure to the helicate at 100 μM indicated an approximately 3.7-fold rise in the absorbance at 575 nm compared with that of a control sample, which presumably had a low absorbance in this region. Further work is needed to establish a link between the ability of this helicate to bind DNA and the antimicrobial or other activity. A time-kill assay supports a rapid cell death for *B. subtilis*, and sodium dodecyl sulfate–polyacrylamide gel electrophoresis (protein electrophoresis) assays suggest a loss in overall protein production and an upregulation of stress proteins in that bacterium, although this may be due to membrane damage.

The low stability of $[Fe_2L^{H}{}_3]^{4+}$ to hydrolysis in water has been noted.[41] As mentioned earlier, such matters must always be a consideration for labile metal assemblies, and we recommend that simple kinetic stability studies be performed. If the helicate architecture is merely a vehicle for delivery of organic components, then in the case of $[Fe_2L^{H}{}_3]^{4+}$ that is a mixture of a relatively benign aldehyde and a carcinogenic and toxic[203,204] diamine component. Despite these limitations, this work is tantalizing and has inspired others to consider how much safer, more tractable, and more drug-like metallohelices can be synthesized. In moving this work forward, a number of criteria need to be considered if any helicate or related compound can find its way to the clinic. Specifically, the complexes need to be readily available on a reasonable scale in optically pure form, have good solubility and high stability in water, and be capable of derivatization so that structures can be optimized. Although these criteria are challenging, and made more so by the need as alluded to previously to use inherently labile metals, we are unlikely to see practical progress if we do not address them.[187]

Scott and coworkers' approach to this problem was to avoid relying on the helicate paradigm of mechanical coupling for stereochemical control at the metal, and instead to develop a diastereoselective synthesis of monometallic complexes that might be used to form the end units of a helicate-like structure. After the development of several prototype ligand designs for this purpose,[205–208] a simple but highly effective synthesis of highly stereochemically pure complexes of various first row transition metal was developed (Fig. 7.17).[209–211] For diamagnetic Fe(II) the selectivity for a single *fac* diastereoisomer (~200:1) is delivered in this α-phenyliminopyridine system by combining optimal stereodirecting steric effects and a series of hydrophobic arene π-interactions. The latter also provides the complexes with remarkable stability to hydrolysis (see subsequent discussion).

The next step to linking two such monometallic units together proceeded smoothly, and the subsequent so-called flexicates[212] were composed

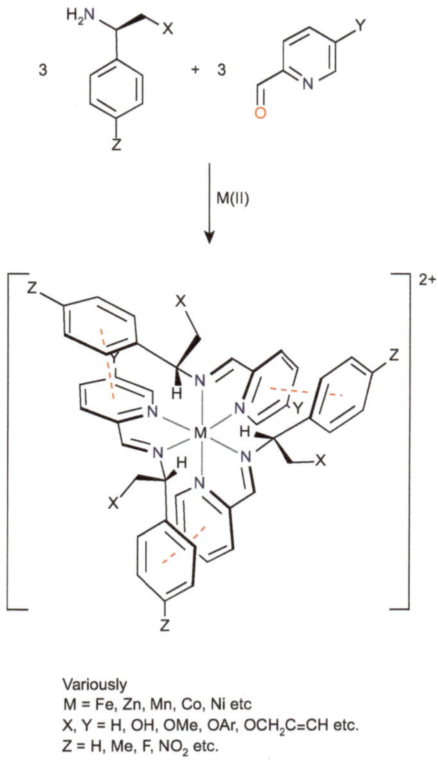

Figure 7.17 ***Highly diastereoselective self-assembly of M(II) complexes.*** **α-Phenyliminopyridine ligands from optically pure amines and pyridine-2-aldehydes engage in synergistic π-stacking arrangements as part of the mechanism of stereoselection.[209–211]**

of two diastereomerically pure tris(chelate)Fe(II) units linked at either side by relatively flexible chains. Two structural classes are readily available: a diamine linker leading to *exo*-pyridine and a dialdehyde leading to *exo*-imine architectures (Fig. 7.18).

MICs were determined in standard media against MRSA strain 252 and *E. coli* strain MC4100.[212] It was found that the more flexible architectures of the *exo*-imine flexicates (Fig. 7.18) showed little or no antibiotic activity. In stark contrast, the more rigid system of the *exo*-pyridine class of flexicate had much higher antimicrobial activity, with MICs of 8 μg/mL against gram-positive MRSA for both isomers and MICs of 8 and 4 μg/mL for Δ and Λ isomers, respectively, against gram-negative *E. coli*. Similar MICs were observed in pathogenic *MRSA* USA.300, and *E. coli* TOP10.[41] MBCs

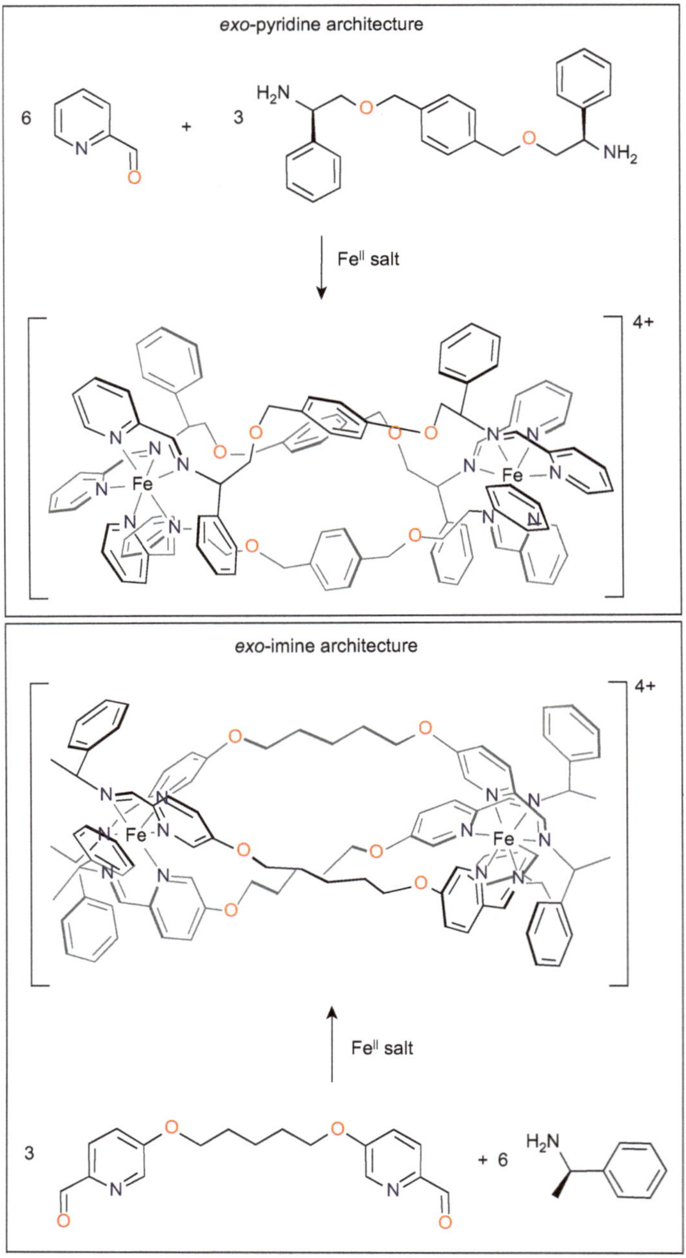

Figure 7.18 Subcomponent self-assemblies of two flexicate architectures.[212]

in these organisms of about 2 × MIC suggest a bacteriocidal mode of action. Promisingly, the same flexicates were found to have low toxicity (LC_{50} of 400–500 μg/mL) toward the nematode *C. elegans*.[212]

As with the Hannon helicate, there is evidence for binding to DNA architectures in some of these compounds. Binding affinities were generally stronger for Λ complexes and markedly stronger for the more rigid, antimicrobial *exo*-imine architectures (Fig. 7.18), and evidence was presented from LD spectra for binding in the major groove.[212] A more detailed analysis of the biophysical interactions with different DNA motifs was presented.[213] Atomic force microscopy revealed that the enantiomers caused supercoiling of circulating tumor DNA, and DNase I footprinting supported B-DNA major groove binding for this complex and suggested a sequence selectivity for 5′-CACATA and 5′-CACTAT sequences. The interaction of this flexicate with DNA motifs that are potential chemotherapeutic targets was demonstrated: four-way (Holliday) junctions (4WJ) (implemented in homologous genetic recombination),[214] 3WJs (both T-shaped and Y-shaped variations), and double-stranded DNA with bulges of different sizes. In contrast, in all such studies the *exo*-imine architecture gave essentially no DNA binding.[213]

At this stage, although the correlation between the DNA interactions seen in vitro and the antibiotic activity may be coincidental, a wide selection of further *exo*-imine flexicates[215] with little or no binding affinity for cell-free DNA were also inactive as antimicrobials. At the same time, observations of low toxicity to the amoeba, *Acanthamoeba polyphaga*, and an apparent probiotic effect on the moth larvae, *Manduca sexta*, in which the systemic distribution of the complex was observed, are highly encouraging.

Inspired by the structures of small antimicrobial peptides such as human cathelicidin LL-37,[216,217] Scott used a new set of supramolecular design principles to develop a synthesis of a range of functionalized, optically pure amphipathic metallohelices.[41] These highly water-stable "triplex" metallohelices (Fig. 7.19) are composed of three directional strands arranged head-to-tail, leading, via the helix fold, to structures with a polar functionalized face opposite a hydrophobic ridge. Perhaps surprisingly, given their high toxicity to cancer cell lines, antimicrobial activity for at least the parent complexes against MRSA and *E. coli* was negligible (MICs all >128 μg/mL).

Crowley and coworkers reported a simple synthesis of a dinuclear Ru(II) triple helicates using a Huisgen alkyne-azide cycloaddition ("click") reaction to furnish a pyridine-triazole bidentate unit at either end of a

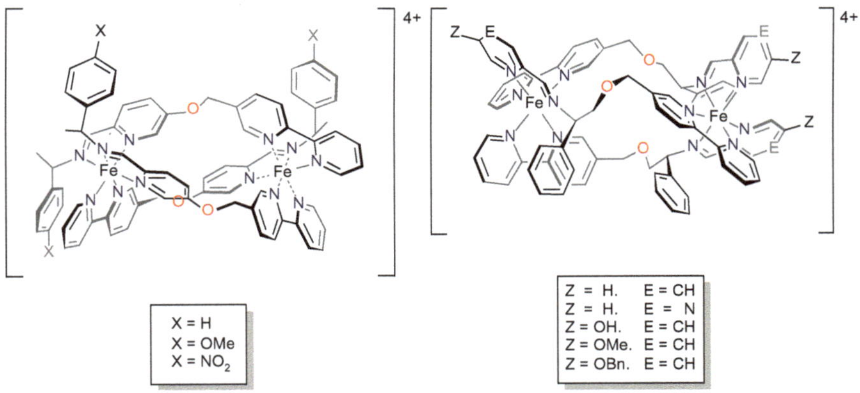

Figure 7.19 Cationic units of optically pure triplex metallohelices.[41]

symmetric AB–BA ligand.[218] Here also, negligible antimicrobial activity was observed (MICs >256 μg/mL) for this, and indeed the Hannon helicate, which was used as a control.

6. OUTLOOK

Antimicrobial transition metal complexes, which rely on metal-centered redox or reactivity, generally cause lesions to a DNA target by producing highly reactive species such as peroxides in the vicinity, or by forming new covalent bonds (cross-linking). Despite the use in many cases of ligands designed to confer selectivity of metallo-antimicrobials toward DNA, the drugs are likely to be promiscuous as a natural consequence of overt reactivity at the metal center. Poor selectivity toward the microbial target and depletion by unspecific binding are inevitable.[219] Many reactive-at-metal metallodrugs thus have found greater clinical acceptance as anti-cancer agents or as topical treatments of infection. However, overall, the feasibility of use of new systems such as NHC complexes of gold and silver in antimicrobial treatments must be low given the affinity of these ions to ubiquitous thiol groups in vivo.[112]

More promising reactive-at-metal antimicrobials display improved activity by making use of the way coordination chemistry may be used to define the geometry of their complexes, allowing engineered targeting of the endocellular target; there may be potential for development of the targeted delivery of reactive metal species using complexes engineered to contain reactivity carefully until required, a "Trojan horse" strategy.[220]

Since the late 1980s, major developments in supramolecular-coordination chemistry have allowed the design of myriad transition metal complexes with a plethora of well-defined geometries with no overtly reactive metal center. These architectures have been discussed for some time as being useful in the design of synthetic antimicrobials because they could be engineered to have a geometry ideal for inhibiting a particular cellular process while evading the resistance mechanisms experienced by natural products, because their chemistry is unfamiliar to nature. Despite this attention, it has only been since the mid-1990s that such complexes have been reported with appropriate properties for use in medicinal research, and since 2010 that substantial works on screening for antimicrobial potential and mechanisms of action have appeared.

The most promising examples thus far of antimicrobial complexes of this type have been the ruthenium-based intercalators of Aldrich-Wright, the oligonuclear ruthenium systems of Collins and Keene, and dinuclear iron metallohelix systems. As with reactive-at-metal complexes, much attention has been given to their interaction with DNA in vitro.[49,221] Little attention has been given thus far to interactions with other prokaryote cellular structures such as membrane chemistry or the possibility of protein–protein interactions (PPIs). The latter are crucial to the vast majority of cellular processes and offer high specificity. Meggers and coworkers[57] show that there is a precedent for targeting PPIs with coordination complexes through precise architectural design. In this context, self-assembling bimetallic helicate-based systems show great promise because of their structural similarity to α-helices, but control over surface topology that is so readily provided in nature is in its infancy.[41]

If inert template complexes are versatile enough to be tailored to provide potent and selective antibiotic agents, it is argued that their synthetic design ought to be driven by a better and fuller understanding of their molecular interactions within (and at the surface of) the prokaryote pathogen, i.e., understanding by what mechanism(s) it has an antibacterial effect. Advances in high-throughput assays in molecular biology and computation methods, for example, are making such awareness easier to achieve. As well as a scarcity of mechanistic understanding, it is also noted that to date the few inert template coordination complexes that have been reported to have biological activity have been tested against only a few genera (*Bacillus*, *Staphylococcus*, and *Escherichia*), and the selectivity for prokaryotes over eukaryotes in vivo has rarely been studied.

The demonstrated usefulness of CORM-type complexes as CO release vectors adds another potential weapon in the fight against microbial infection, and the idea of small molecule-releasing complexes may well be expanded toward other useful small molecules that may benefit from targeted delivery (nitrous oxide, hydrogen sulfide, etc.).

There is a growing interest in the field toward synergistic combinations of drugs, hitting different prokaryote biomolecular targets in complementary fashion and, it is hoped, \ reducing resistance and toxic side effects, allowing the "repurposing" of existing chemic space.[222–225] This has not been widely studied for coordination complexes; the report of Wolschendorf and coworkers of copper ion–releasing complexes acting synergistically with disulfiram is an exception.[69] Some of the potential of coordination complexes may be realized here as they may act on pathways (protein synthesis, DNA replication, iron intake, etc.) that could sensitize the pathogen to traditional chemotherapeutics.

The chemistry necessary to develop the well-defined structures with tailored properties is rapidly advancing. It is proposed that to guide this research effectively to reach its full potential in antimicrobial chemotherapy, we must employ a rigorous approach to understanding fully the potency, selectivity, cell wall activity, and mechanisms of action of each complex type, using a wide range of organisms. We need to conduct this interdisciplinary work with a view to advancing toward in vivo trials. Realistically, to make such an investment, coordination and organometallic chemists should be asking themselves hard questions about their chosen system. Can it ever be produced and used as a drug in humans? What are the barriers to acceptance by regulators and industry? Given the increasingly strict limitations on heavy metal impurities in drugs,[226] the deliberate use of such metals outside existing cancer treatments is perhaps unlikely. In the fight against antimicrobial resistance, there is nevertheless a need to explore chemical space outside that of natural antibiotics, and so long as we address practical criteria of safety and utility, the unique structures of metallo-organic chemistry provide a promising avenue.

ACKNOWLEDGMENTS

We thank EPSRC and MOAC doctoral training center for a studentship to DHS, and the University of Warwick for support.

REFERENCES

1. Brooks, J. T.; Kaplan, J. E.; Holmes, K. K.; Benson, C.; Pau, A.; Masur, H. *Clin. Infect. Dis.* **2009,** *48,* 609–611.
2. Freifeld, A. G.; Bow, E. J.; Sepkowitz, K. A.; Boeckh, M. J.; Ito, J. I.; Mullen, C. A.; Raad, II; Rolston, K. V.; Young, J. A. H.; Wingard, J. R. *Clin. Infect. Dis.* **2011,** *52,* E56–E93.

3. Gaynes, R. *Infecti. Dis. Clin. North Am.* **1997,** *11,* 757.
4. Gaynes, R.; Monnet, D. In *Antibiotic Resistance: Origins, Evolution, Selection and Spread, 207*; 1997; pp 47–56.
5. Knobler, S.L.; Lemon, S.M.; Najafi, M.; Burroughs, T. *The Resistance Phenomenon in Microbes and Infectious Disease Vectors: Implications for Human Health and Strategies for Containment - Workshop Summary*; National Academies Press: Washington, **2003**; pp 44–106.
6. Fischbach, M. A.; Walsh, C. T. *Science* **2009,** *325,* 1089–1093.
7. Livermore, D. M. *Int. J. Antimicrob. Agents* **2012,** *39,* 283–294.
8. Boucher, H. W.; Talbot, G. H.; Bradley, J. S.; Edwards, J. E.; Gilbert, D.; Rice, L. B.; Scheld, M.; Spellberg, B.; Bartlett, J. *Clin. Infect. Dis.* **2009,** *48,* 1–12.
9. Cairns, J.; Overbaugh, J.; Miller, S. *Nature* **1988,** *335,* 142–145.
10. Walsh, C. *Antibiotics: Actions, Origins, Resistance;* ASM Press: Washington, DC, 2003.
11. Palumbi, S. R. *Science* **2001,** *293,* 1786–1790.
12. Perron, G. G.; Lee, A. E. G.; Wang, Y.; Huang, W. E.; Barraclough, T. G. *Proc. R. Soc. B-Biol. Sci.* **2012,** *279,* 1477–1484.
13. Gill, S. R.; Fouts, D. E.; Archer, G. L.; Mongodin, E. F.; DeBoy, R. T.; Ravel, J.; Paulsen, I. T.; Kolonay, J. F.; Brinkac, L.; Beanan, M.; Dodson, R. J.; Daugherty, S. C.; Madupu, R.; Angiuoli, S.V.; Durkin, A. S.; Haft, D. H.; Vamathevan, J.; Khouri, H.; Utterback, T.; Lee, C.; Dimitrov, G.; Jiang, L. X.; Qin, H. Y.; Weidman, J.; Tran, K.; Kang, K.; Hance, I. R.; Nelson, K. E.; Fraser, C. M. *J. Bacteriol.* **2005,** *187,* 2426–2438.
14. Zignol, M.; Hosseini, M. S.; Wright, A.; Lambregts-van Weezenbeek, C.; Nunn, P.; Watt, C. J.; Williams, B. G.; Dye, C. *J. Infect. Dis.* **2006,** *194,* 479–485.
15. Donlan, R. M.; Costerton, J. W. *Clin. Microbiol. Rev* **2002,** *15,* 167.
16. Nikaido, H. *Science* **1994,** *264,* 382–388.
17. McDermott, P. F.; Walker, R. D.; White, D. G. *Int. J. Toxicol* **2003,** *22,* 135–143.
18. Poole, K. *J. Antimicrob. Chemother.* **2005,** *56,* 20–51.
19. Jorgensen, J. H.; Pfaller, M. A.; Carroll, K. C.; Funke, G.; Landry, M. L.; Richter, S. S.; Warnock, D. W. *Manual of Clinical Microbiology*, 11th ed.; American Society of Microbiology, 2015.
20. Wright, G. D. *Adv. Drug Del. Rev.* **2005,** *57,* 1451–1470.
21. Spratt, B. G. *Science* **1994,** *264,* 388–393.
22. Chopra, I. *J. Antimicrob. Chemother.* **2013,** *68,* 496–505.
23. Mandal, S. M.; Roy, A.; Ghosh, A. K.; Hazra, T. K.; Basak, A.; Franco, O. L. *Front. Pharmacol* **2014,** *5,* 12.
24. Hobman, J. L.; Crossman, L. C. *J. Med. Microbiol..* **2015,** *64,* 471–497.
25. Meggers, E. *Chemical Communications* **2009,** 1001–1010.
26. Hancock, R. E. W. *Trends Microbiol.* **1997,** *5,* 37–42.
27. Kell, D. B. *Nat. Rev. Drug Discov.* **2016,** *15,* 143.
28. Kell, D. B.; Oliver, S. G. *Front. Pharmacol.* **2014,** 5.
29. Dobson, P. D.; Kell, D. B. *Nat. Rev. Drug Discov.* **2008,** *7,* 205–220.
30. Mann, J.; Crabbe, M. J. C. *Bacteria and Antibacterial Agents;* Spektrum University Science Books: Oxford, 1996.
31. de Tejada, G. M.; Sanchez-Gomez, S.; Razquin-Olazaran, I.; Kowalski, I.; Kaconis, Y.; Heinbockel, L.; Andra, J.; Schurholz, T.; Hornef, M.; Dupont, A.; Garidel, P.; Lohner, K.; Gutsmann, T.; David, S. A.; Brandenburg, K. *Curr. Drug Targets* **2012,** *13,* 1121–1130.
32. Teo, A.; Roper, D. *Antibiotics* **2015,** *4,* 495.
33. Erridge, C.; Bennett-Guerrero, E.; Poxton, I. R. *Microb. Infect.* **2002,** *4,* 837–851.
34. Paschke, A.; Neitzel, P. L.; Walther, W.; Schuurmann, G. *J. Chem. Eng. Data* **2004,** *49,* 1639–1642.
35. Pliska, V.; Testa, B.; Waterbeemd, H.; Mannhold, R.; Kubinyi, H.; Timmerman, H. *Lipophilicity in Drug Action and Toxicology;* Wiley, 2008. Published Online.

36. Arnott, J. A.; Planey, S. L. *Expert Opin. Drug Discovery* **2012,** *7,* 863–875.
37. Schuck, S.; Honsho, M.; Ekroos, K.; Shevchenko, A.; Simons, K. *Proc. Natl. Acad. Sci. USA* **2003,** *100,* 5795–5800.
38. O'Neill, A. J.; Chopra, I. *Expert Opin. Invest. Drugs* **2004,** *13,* 1045–1063.
39. Ghose, A. K.; Viswanadhan, V. N.; Wendoloski, J. J. *J. Comb. Chem.* **1999,** *1,* 55–68.
40. Blanck, S.; Maksimoska, J.; Baumeister, J.; Harms, K.; Marmorstein, R.; Meggers, E. *Angew. Chem. Int. Ed.* **2012,** *51,* 5244–5246.
41. Faulkner, A. D.; Kaner, R. A.; Abdallah, Q. M. A.; Clarkson, G.; Fox, D. J.; Gurnani, P.; Howson, S. E.; Phillips, R. M.; Roper, D. I.; Simpson, D. H.; Scott, P. *Nat. Chem.* **2014,** *6,* 797–803.
42. Eggert, U. S. *Nat. Chem. Biol.* **2013,** *9,* 206–209.
43. Lee, J. A.; Uhlik, M.T.; Moxham, C. M.; Tomandl, D.; Sall, D. J. *J. Med. Chem.* **2012,** *55,* 4527–4538.
44. Sams-Dodd, F. *Drug Discovery Today* **2013,** *18,* 211–217.
45. Swinney, D. C.; Anthony, J. *Nat. Rev. Drug Discovery* **2011,** *10,* 507–519.
46. Moffat, J. G.; Rudolph, J.; Bailey, D. *Nat Rev Drug Discovery* **2014,** *13,* 588–602.
47. Bolhuis, A.; Aldrich-Wright, J. R. *Bioorg. Chem.* **2014,** *55,* 51–59.
48. Zeglis, B. M.; Pierre, V. C.; Barton, J. K. *Chem. Commun.* **2007,** 4565–4579.
49. Pages, B. J.; Ang, D. L.; Wright, E. P.; Aldrich-Wright, J. R. *Dalton Trans.* **2015,** *44,* 3505–3526.
50. Oguey, C.; Foloppe, N.; Hartmann, B. *PLoS One* **2010,** 5.
51. Wang, J. C. *Annu. Rev. Biochem.* **1996,** *65,* 635–692.
52. Narayanan, B. C.; Westbrook, J.; Ghosh, S.; Petrov, A. I.; Sweeney, B.; Zirbel, C. L.; Leontis, N. B.; Berman, H. M. *Nucleic Acids Res.* **2014,** *42,* D114–D122.
53. Nissen, P.; Hansen, J.; Ban, N.; Moore, P. B.; Steitz, T. A. *Science* **2000,** *289,* 920–930.
54. Arias-Gonzalez, J. R. *Integr. Biol.* **2014,** *6,* 904–925.
55. Warner, J. R.; Knopf, P. M.; Rich, A. *Proc. Natl. Acad. Sci. USA* **1963,** *49,* 122–129.
56. Alberts, B.; Johnson, A.; Lewis, J.; Raff, M.; Roberts, K.; Walter, P. *Molecular Biology of the Cell;* Oxford: Garland, 2002.
57. Feng, L.; Geisselbrecht, Y.; Blanck, S.; Wilbuer, A.; Atilla-Gokcumen, G. E.; Filippakopoulos, P.; Kraling, K.; Celik, M. A.; Harms, K.; Maksimoska, J.; Marmorstein, R.; Frenking, G.; Knapp, S.; Essen, L. O.; Meggers, E. *J. Am. Chem. Soc.* **2011,** *133,* 5976–5986.
58. *J. Antimicrob. Chemother.* **1991,** *27* (Suppl. D), 1–50.
59. Lemire, J. A.; Harrison, J. J.; Turner, R. J. *Nat. Rev. Microbiol.* **2013,** *11,* 371–384.
60. Alexander, J.W. *Surg. Infect. (Larchmt.)* **2009,** *10,* 289–292.
61. Irving, H.; Williams, R. J. P. *J. Chem. Soc. (Resumed)* **1953,** 3192–3210.
62. Macomber, L.; Imlay, J. A. *Proc. Natl. Acad. Sci. USA* **2009,** *106,* 8344–8349.
63. Harrison, J. J.; Ceri, H.; Stremick, C. A.; Turner, R. J. *Environ. Microbiol.* **2004,** *6,* 1220–1227.
64. Gaetke, L. M.; Chow, C. K. *Toxicology* **2003,** *189,* 147–163.
65. Wisher, D. *J. Med. Libr. Assoc. JMLA* **2012,** *100,* 75–76.
66. Reeder, N. L.; Kaplan, J.; Xu, J.; Youngquist, R. S.; Wallace, J.; Hu, P.; Juhlin, K. D.; Schwartz, J. R.; Grant, R. A.; Fieno, A.; Nemeth, S.; Reichling, T.; Tiesman, J. P.; Mills, T.; Steinke, M.; Wang, S. L.; Saunders, C. W. *Antimicrob. Agents Chemother.* **2011,** *55,* 5753–5760.
67. Stone, K. J.; Stroming, J. *Proc. Natl. Acad. Sci. USA* **1971,** *68,* 3223.
68. Haeili, M.; Moore, C.; Davis, C. J. C.; Cochran, J. B.; Shah, S.; Shrestha, T. B.; Zhang, Y.; Bossmann, S. H.; Benjamin, W. H.; Kutsch, O.; Wolschendorf, F. *Antimicrob. Agents Chemother.* **2014,** *58,* 3727–3736.
69. Dalecki, A. G.; Haeili, M.; Shah, S.; Speer, A.; Niederweis, M.; Kutsch, O.; Wolschendorf, F. *Antimicrob. Agents Chemother.* **2015,** *59,* 4835–4844.

70. Chevrier, I.; Sague, J. L.; Brunetto, P. S.; Khanna, N.; Rajacic, Z.; Fromm, K. M. *Dalton Trans.* **2013,** *42,* 217–231.
71. Umezawa, H.; Maeda, K.; Takeuchi, T.; Okami, Y. *J. Antibiot.* **1966,** *19,* 200–209.
72. Petering, D. H.; Byrnes, R. W.; Antholine, W. E. *Chem.-Biol. Interact.* **1990,** *73,* 133–182.
73. Xu, R. X.; Nettesheim, D.; Otvos, J. D.; Petering, D. H. *Biochemistry* **1994,** *33,* 907–916.
74. Burger, R. M. *Chem. Rev.* **1998,** *98,* 1153–1169.
75. Takeshit, M.; Horwitz, S. B.; Grollman, A. P. *Virology* **1974,** *60,* 455–465.
76. Gothelf, A.; Mir, L. M.; Gehl, J. *Cancer Treat. Rev.* **2003,** *29,* 371–387.
77. Sikic, B. I.; Rozencweig, M.; Carter, S. K. *Bleomycin Chemotherapy;* Elsevier Science, 2013.
78. Burger, R. M.; Adler, A. D.; Horwitz, S. B.; Mims, W. B.; Peisach, J. *Biochemistry* **1981,** *20,* 1701–1704.
79. Itaka, Y.; Nakamura, H.; Nakatani, T.; Muraoka, Y.; Fujii, A.; Takita, T.; Umezawa, H. *J. Antibiot. (Tokyo)* **1978,** *31,* 1070–1072.
80. Williamson, D.; McLennan, I. J.; Bax, A.; Gamcsik, M. P.; Glickson, J. D. *J. Biomol. Struct. Dyn.* **1990,** *8,* 375–398.
81. Calafat, A. M.; Won, H.; Marzilli, L. G. *J. Am. Chem. Soc.* **1997,** *119,* 3656–3664.
82. Rajani, C.; Kincaid, J. R.; Petering, D. H. *J. Am. Chem. Soc.* **2004,** *126,* 3829–3836.
83. Kane, S. A.; Hecht, S. M. *Prog. Nucleic Acid Res. Mol. Biol.* **1994,** *49,* 313–352.
84. Hecht, S. M. *Acc. Chem. Res.* **1986,** *19,* 383–391.
85. Azambuja, E.; Fleck, J. F.; Batista, R. G.; Barreto, S. S. M. *Pulm. Pharmacol. Ther.* **2005,** *18,* 363–366.
86. Donohoe, T. J.; Jones, C. R.; Kornahrens, A. F.; Barbosa, L. C. A.; Walport, L. J.; Tatton, M. R.; O'Hagan, M.; Rathi, A. H.; Baker, D. B. *J. Org. Chem.* **2013,** *78,* 12338–12350.
87. Harding, M. M.; Long, G.V. *Curr. Med. Chem.* **1997,** *4,* 405–420.
88. Weiss, R. B.; Sarosy, G.; Clagettcarr, K.; Russo, M.; Leylandjones, B. *Cancer Chemother. Pharmacol.* **1986,** *18,* 185–197.
89. Marsh, W. S.; Garretson, A. L.; Wesel, E. M. *Antibiot. Chemother. N. Ill.)* **1961,** *11,* 151–157.
90. Niemi, J.; Metsa-Ketela, M.; Schneider, G.; Mantsala, P. Biosynthetic Anthracycline Variants. In Krohn, K., Ed.; *Anthracycline Chemistry and Biology I: Biological Occurence and Biosynthesis, Synthesis and Chemistry: Biological Occurence and Biosynthesis, Synthesis and Chemistry,* Vol. 282; Springer-Verlag Berlin: Berlin, 2008; pp 75–99.
91. Fujiwara, A.; Hoshino, T. *Crc Crit. Rev. Biotechnol.* **1986,** *3,* 133–157.
92. Lothstein, L.; Israel, M.; Sweatman, T. W. *Drug Resist. Updates* **2001,** *4,* 169–177.
93. Jensen, P. B.; Sorensen, B. S.; Sehested, M.; Demant, E. J. F.; Kjeldsen, E.; Friche, E.; Hansen, H. H. *Biochem. Pharmacol.* **1993,** *45,* 2025–2035.
94. White, J. R. *Biochem. Biophys. Res. Commun.* **1977,** *77,* 387–391.
95. Eizaguirre, A.; Yanez, M.; Eriksson, L. A. *PCCP* **2012,** *14,* 12505–12514.
96. Fiallo, M. M. L.; Drechsel, H.; Garnier-Suillerot, A.; Matzanke, B. F.; Kozlowski, H. *J. Med. Chem.* **1999,** *42,* 2844–2851.
97. Gregory, E. M.; Fridovic, I. *J. Bacteriol.* **1973,** *114,* 1193–1197.
98. White, H. L.; White, J. R. *Mol. Pharmacol..* **1968,** *4,* 549–565.
99. Bolzan, A. D.; Bianchi, M. S. *Mutat. Res. – Rev. Mut. Res.* **2001,** *488,* 25–37.
100. Sterba, M.; Popelova, O.; Vavrova, A.; Jirkovsky, E.; Kovarikova, P.; Gersl, V.; Simunek, T. *Antioxid. Redox Signaling* **2013,** *18,* 899–929.
101. Muindi, J. R. F.; Sinha, B. K.; Gianni, L.; Myers, C. E. *FEBS Lett.* **1984,** *172,* 226–230.
102. Asche, C. *Mini-Rev. Med. Chem.* **2005,** *5,* 449–467.
103. Cone, R.; Hasan, S. K.; Lown, J. W.; Morgan, A. R. *Can. J. Biochem.* **1976,** *54,* 219–223.
104. Gutteridge, J. M. C. *Biochem. Pharmacol.* **1984,** *33,* 3059–3062.
105. Yoshimoto, A.; Oki, T.; Inui, T. *J. Antibiot.* **1978,** *31,* 92–94.
106. Kim, B. S.; Moon, S. S.; Hwang, B. K. *J. Agric. Food Chem.* **2000,** *48,* 1875–1881.

107. Ming, L. J. *Med. Res. Rev.* **2003,** *23,* 697–762.
108. Rosenberg, B.; Vancamp, L.; Krigas, T. *Nature* **1965,** *205,* 698–699.
109. Budagumpi, S.; Endud, S. *Organometallics* **2013,** *32,* 1537–1562.
110. Özdemir, İ.; Denizci, A.; Öztürk, H. T.; Çetinkaya, B. *Appl. Organomet. Chem.* **2004,** *18,* 318–322.
111. Ray, S.; Mohan, R.; Singh, J. K.; Samantaray, M. K.; Shaikh, M. M.; Panda, D.; Ghosh, P. *J. Am. Chem. Soc.* **2007,** *129,* 15042–15053.
112. Fernández, G. A.; Gurovic, V. M. S.; Olivera, N. L.; Chopa, A. B.; Silbestri, G. F. *J. Inorg. Biochem.* **2014,** *135,* 54–57.
113. Samanta, T.; Roymahapatra, G.; Porto, W. F.; Seth, S.; Ghorai, S.; Saha, S.; Sengupta, J.; Franco, O. L.; Dinda, J.; Mandal, S. M. *PLoS One* **2013,** *8,* e58346.
114. Chang, E. L.; Simmers, C.; Knight, D. A. *Pharmaceuticals* **2010,** *3,* 1711.
115. Mann, B. E. Carbon Monoxide: An Essential Signalling Molecule. In Jaouen, G., MetzlerNolte, N., Eds.; *Medicinal Organometallic Chemistry*, Vol. 32; Springer-Verlag Berlin: Berlin, 2010; pp 247–285.
116. Nobre, L. S.; Seixas, J. D.; Romao, C. C.; Saraiva, L. M. *Antimicrob. Agents Chemother.* **2007,** *51,* 4303–4307.
117. Wareham, L. K.; Poole, R. K.; Tinajero-Trejo, M. *J. Biol. Chem.* **2015,** *290,* 18999–19007.
118. McLean, S.; Mann, B. E.; Poole, R. K. *Anal. Biochem.* **2012,** *427,* 36–40.
119. Davidge, K. S.; Sanguinetti, G.; Yee, C. H.; Cox, A. G.; McLeod, C. W.; Monk, C. E.; Mann, B. E.; Motterlini, R.; Poole, R. K. *J. Biol. Chem.* **2009,** *284,* 4516–4524.
120. Nagel, C.; McLean, S.; Poole, R. K.; Braunschweig, H.; Kramer, T.; Schatzschneider, U. *Dalton Trans.* **2014,** *43,* 9986–9997.
121. Smith, N. A.; Sadler, P. J. *Philos. Trans. R. Soc. A-Math. Phys. Eng. Sci.* **2013,** *371,* 13.
122. Donnelly, R. F.; Fletcher, N. C.; McCague, P. J.; Donnelly, J.; McCarron, P. A.; Tunney, M. M. *Lett. Drug Des. Discov.* **2007,** *4,* 175–179.
123. Celli, J. P.; Spring, B. Q.; Rizvi, I.; Evans, C. L.; Samkoe, K. S.; Verma, S.; Pogue, B. W.; Hasan, T. *Chem. Rev.* **2010,** *110,* 2795–2838.
124. Dwyer, F. P.; Gyarfas, E. C.; Rogers, W. P.; Koch, J. H. *Nature* **1952,** *170,* 190–191.
125. Dwyer, F. P.; Reid, I. K.; Shulman, A.; Laycock, G. M.; Dixson, S. *Aust. J. Exp. Biol. Med. Sci.* **1969,** *47,* 203–218.
126. Eriksson, M.; Leijon, M.; Hiort, C.; Norden, B.; Graslund, A. *Biochemistry* **1994,** *33,* 5031–5040.
127. Hiort, C.; Lincoln, P.; Norden, B. *J. Am. Chem. Soc.* **1993,** *115,* 3448–3454.
128. Song, H.; Kaiser, J. T.; Barton, J. K. *Nat. Chem.* **2012,** *4,* 615–620.
129. Chen, H. M.; Parkinson, J. A.; Parsons, S.; Coxall, R. A.; Gould, R. O.; Sadler, P. J. *J. Am. Chem. Soc.* **2002,** *124,* 3064–3082.
130. Krause-Heuer, A. M.; Grunert, R.; Kuhne, S.; Buczkowska, M.; Wheate, N. J.; Le Pevelen, D. D.; Boag, L. R.; Fisher, D. M.; Kasparkova, J.; Malina, J.; Bednarski, P. J.; Brabec, V.; Aldrich-Wright, J. R. *J. Med. Chem.* **2009,** *52,* 5474–5484.
131. Liu, H. K.; Sadler, P. J. *Acc. Chem. Res.* **2011,** *44,* 349–359.
132. Nhukeaw, T.; Temboot, P.; Hansongnern, K.; Ratanaphan, A. *BMC Cancer* **2014,** *14,* 19.
133. Collins, J. G.; Sleeman, A. D.; Aldrich-Wright, J. R.; Greguric, I.; Hambley, T. W. *Inorg. Chem.* **1998,** *37,* 3133–3141.
134. Greguric, I.; Aldrich-Wright, J. R.; Collins, J. G. *J. Am. Chem. Soc.* **1997,** *119,* 3621–3622.
135. Bolhuis, A.; Hand, L.; Marshall, J. E.; Richards, A. D.; Rodger, A.; Aldrich-Wright, J. *Eur. J. Pharm. Sci.* **2011,** *42,* 313–317.
136. O'Callaghan, D.; Vergunst, A. *Curr. Opin. Microbiol.* **2010,** *13,* 79–85.
137. Artal-Sanz, M.; de Jong, L.; Tavernarakis, N. *Biotechnol. J.* **2006,** *1,* 1405–1418.
138. Hiort, C.; Norden, B.; Rodger, A. *J. Am. Chem. Soc.* **1990,** *112,* 1971–1982.
139. Watson, R. T.; Jackson, J. L.; Harper, J. D.; KaneMaguire, K. A.; KaneMaguire, L. A. P.; KaneMaguire, N. A. P. *Inorg. Chim. Acta* **1996,** *249,* 5–7.

140. Kano, K.; Hasegawa, H. *Chem. Lett.* **2000,** 698–699.
141. Jiang, C. X.; Tong, M.Y.; Armstrong, D.W.; Perera, S.; Bao, Y.; MacDonnell, F. M. *Chirality* **2009,** *21,* 208–217.
142. Fletcher, N. C.; Brown, R. T.; Doherty, A. P. *Inorg. Chem.* **2006,** *45,* 6132–6134.
143. Lacour, J.; Torche-Haldimann, S.; Jodry, J. J.; Ginglinger, C.; Favarger, F. *Chem. Commun.* **1998,** 1733–1734.
144. O'Reilly, F. M.; Kelly, J. M. *J. Phys. Chem. B* **2000,** *104,* 7206–7213.
145. O'Reilly, F. M.; Kelly, J. M. *New J. Chem.* **1998,** *22,* 215–217.
146. O'Reilly, F.; Kelly, J.; Kirsch-De Mesmaeker, A. *Chem. Commun.* **1996,** 1013–1014.
147. Aldrich-Wright, J.; Brodie, C.; Glazer, E. C.; Luedtke, N. W.; Elson-Schwab, L.; Tor, Y. *Chem. Commun.* **2004,** 1018–1019.
148. Onfelt, B.; Gostring, L.; Lincoln, P.; Norden, B.; Onfelt, A. *Mutagenesis* **2002,** *17,* 317–320.
149. Nordell, P.; Lincoln, P. *J. Am. Chem. Soc.* **2005,** *127,* 9670–9671.
150. Rajput, C.; Rutkaite, R.; Swanson, L.; Haq, I.; Thomas, J. A. *Chemistry* **2006,** *12,* 4611–4619.
151. Lincoln, P.; Norden, B. *Chem. Commun.* **1996,** 2145–2146.
152. Li, F. F.; Collins, J. G.; Keene, F. R. *Chem. Soc. Rev.* **2015,** *44,* 2529–2542.
153. Morgan, J. L.; Spillane, C. B.; Smith, J. A.; Buck, D. P.; Collins, J. G.; Keene, R. *Dalton Trans.* **2007,** 4333–4342.
154. Li, F. F.; Mulyana, Y.; Feterl, M.; Warner, J. M.; Collins, J. G.; Keene, F. R. *Dalton Trans.* **2011,** *40,* 5032–5038.
155. Li, F. F.; Harry, E. J.; Bottomley, A. L.; Edstein, M. D.; Birrell, G. W.; Woodward, C. E.; Keene, F. R.; Collins, J. G. *Chem. Sci.* **2014,** *5,* 685–693.
156. Li, F. F.; Feterl, M.; Mulyana, Y.; Warner, J. M.; Collins, J. G.; Keene, F. R. *J. Antimicrob. Chemother.* **2012,** *67,* 2686–2695.
157. Li, F. F.; Feterl, M.; Warner, J. M.; Keene, F. R.; Collins, J. G. *J. Antimicrob. Chemother.* **2013,** *68,* 2825–2833.
158. Li, X.; Gorle, A. K.; Ainsworth, T. D.; Heimann, K.; Woodward, C. E.; Grant Collins, J.; Richard Keene, F. *Dalton Trans.* **2015,** *44,* 3594–3603.
159. Gorle, A. K.; Feterl, M.; Warner, J. M.; Wallace, L.; Keene, F. R.; Collins, J. G. *Dalton Trans.* **2014,** *43,* 16713–16725.
160. Andrews, S. C.; Robinson, A. K.; Rodriguez-Quinones, F. *FEMS Microbiol. Rev.* **2003,** *27,* 215–237.
161. Cassat, J. E.; Skaar, E. P. *Cell Host Microbe* **2013,** *13,* 509–519.
162. Hood, M. I.; Skaar, E. P. *Nat. Rev. Micro.* **2012,** *10,* 525–537.
163. Ecker, D. J.; Loomis, L. D.; Cass, M. E.; Raymond, K. N. *J. Am. Chem. Soc.* **1988,** *110,* 2457–2464.
164. Obrien, I. G.; Cox, G. B.; Gibson, F. *Biochim. Biophys. Acta* **1970,** *201,* 453.
165. Carrano, C. J.; Raymond, K. N. *J. Am. Chem. Soc.* **1979,** *101,* 5401–5404.
166. Stack, T. D. P.; Karpishin, T. B.; Raymond, K. N. *J. Am. Chem. Soc.* **1992,** *114,* 1512–1514.
167. Karpishin, T. B.; Raymond, K. N. *Angew. Chem. Int. Ed.* **1992,** *31,* 466–468.
168. Buchanan, S. K.; Smith, B. S.; Venkatramani, L.; Xia, D.; Esser, L.; Palnitkar, M.; Chakraborty, R.; van der Helm, D.; Deisenhofer, J. *Nat. Struct. Biol.* **1999,** *6,* 56–63.
169. Raymond, K. N.; Dertz, E. A.; Kim, S. S. *Proc. Natl. Acad. Sci. USA* **2003,** *100,* 3584–3588.
170. Rodgers, S. J.; Lee, C. W.; Ng, C. Y.; Raymond, K. N. *Inorg. Chem.* **1987,** *26,* 1622–1625.
171. Anderson, M. T.; Armstrong, S. K. *J. Bacteriol.* **2006,** *188,* 5731–5740.
172. Rabsch, W.; Voigt, W.; Reissbrodt, R.; Tsolis, R. M.; Baumler, A. J. *J. Bacteriol.* **1999,** *181,* 3610–3612.
173. Albrecht, M.; Franklin, S. J.; Raymond, K. N. *Inorg. Chem.* **1994,** *33,* 5785–5793.
174. Scherer, M.; Caulder, D. L.; Johnson, D.W.; Raymond, K. N. *Angew. Chem. Int. Ed.* **1999,** *38,* 1588–1592.

175. Albrecht, M. *Chemistry* **2000,** *6,* 3485–3489.
176. Albrecht, M.; Isaak, E.; Baumert, M.; Gossen,V.; Raabe, G.; Frohlich, R. *Angew. Chem. Int. Ed.* **2011,** *50,* 2850–2853.
177. Lehn, J. M.; Rigault, A.; Siegel, J.; Harrowfield, J.; Chevrier, B.; Moras, D. *Proc. Natl. Acad. Sci. USA* **1987,** *84,* 2565–2569.
178. Forgan, R. S.; Sauvage, J. P.; Stoddart, J. F. *Chem. Rev.* **2011,** *111,* 5434–5464.
179. Ruben, M.; Rojo, J.; Romero-Salguero, F. J.; Uppadine, L. H.; Lehn, J. M. *Angew. Chem. Int. Ed.* **2004,** *43,* 3644–3662.
180. Baxter, P. N. W.; Lehn, J. M.; Fischer, J.; Youinou, M. T. *Angew. Chem. Int. Ed.* **1994,** *33,* 2284–2287.
181. Ayme, J. F.; Lux, J.; Sauvage, J. P.; Sour, A. *Chemistry* **2012,** *18,* 5565–5573.
182. Chambron, J. C.; Heitz, V.; Sauvage, J. P. *J. Am. Chem. Soc.* **1993,** *115,* 12378–12384.
183. Stang, P. J.; Cao, D. H. *J. Am. Chem. Soc.* **1994,** *116,* 4981–4982.
184. Ousaka, N.; Grunder, S.; Castilla, A. M.; Whalley, A. C.; Stoddart, J. F.; Nitschke, J. R. *J. Am. Chem. Soc.* **2012,** *134,* 15528–15537.
185. Keene, F. R. *Dalton Trans.* **2011,** *40,* 2405–2418.
186. Maayan, G.; Albrecht, M. *Metallofoldamers: Supramolecular Architectures from Helicates to Biomimetics;* Wiley, 2013.
187. Kaner, R. A.; Scott, P. *Future Med. Chem.* **2015,** *7,* 1–4.
188. Garrett, T. M.; Koert, U.; Lehn, J. M.; Rigault, A.; Meyer, D.; Fischer, J. *J. Chem. Soc.-Chem. Commun.* **1990,** 557–558.
189. Hannon, M. J.; Painting, C. L.; Jackson, A.; Hamblin, J.; Errington, W. *Chem. Commun.* **1997,** 1807–1808.
190. Kerckhoffs, J.; Peberdy, J. C.; Meistermann, I.; Childs, L. J.; Isaac, C. J.; Pearmund, C. R.; Reudegger, V.; Khalid, S.; Alcock, N. W.; Hannon, M. J.; Rodger, A. *Dalton Trans.* **2007,** 734–742.
191. Peberdy, J. C.; Malina, J.; Khalid, S.; Hannon, M. J.; Rodger, A. *J. Inorg. Biochem.* **2007,** *101,* 1937–1945.
192. Hannon, M. J.; Moreno, V.; Prieto, M. J.; Moldrheim, E.; Sletten, E.; Meistermann, I.; Isaac, C. J.; Sanders, K. J.; Rodger, A. *Angew. Chem. Int. Ed.* **2001,** *40,* 880–884.
193. Hannon, M. J.; Meistermann, I.; Isaac, C. J.; Blomme, C.; Aldrich-Wright, J. R.; Rodger, A. *Chem. Commun.* **2001,** 1078–1079.
194. Lacour, J.; Ginglinger, C.; Favarger, F.; TorcheHaldimann, S. *Chem. Commun.* **1997,** 2285–2286.
195. Cardo, L.; Sadovnikova, V.; Phongtongpasuk, S.; Hodges, N. J.; Hannon, M. J. *Chem. Commun.* **2011,** *47,* 6575–6577.
196. Cardo, L.; Hannon, M. J. *Inorg. Chim. Acta* **2009,** *362,* 784–792.
197. Meistermann, I.; Moreno, V.; Prieto, M. J.; Moldrheim, E.; Sletten, E.; Khalid, S.; Rodger, P. M.; Peberdy, J. C.; Isaac, C. J.; Rodger, A.; Hannon, M. J. *Proc. Natl. Acad. Sci. USA* **2002,** *99,* 5069–5074.
198. Uerpmann, C.; Malina, J.; Pascu, M.; Clarkson, G. J.; Moreno, V.; Rodger, A.; Grandas, A.; Hannon, M. J. *Chemistry* **2005,** *11,* 1750–1756.
199. Oleksi, A.; Blanco, A. G.; Boer, R.; Uson, I.; Aymami, J.; Rodger, A.; Hannon, M. J.; Coll, M. *Angew. Chem. Int. Ed.* **2006,** *45,* 1227–1231.
200. Malina, J.; Hannon, M. J.; Brabec, V. *Chemistry* **2007,** *13,* 3871–3877.
201. Cerasino, L.; Hannon, M. J.; Sletten, E. *Inorg. Chem.* **2007,** *46,* 6245–6251.
202. Richards, A. D.; Rodger, A.; Hannon, M. J.; Bolhuis, A. *Int. J. Antimicrob. Agents* **2009,** *33,* 469–472.
203. Kenyon, S. H.; Bhattacharyya, J.; Benson, C. J.; Carmichael, P. L. *Toxicology* **2004,** *196,* 65–75.
204. Wellner, T.; Lueersen, L.; Schaller, K. H.; Angerer, J.; Drexler, H.; Korinth, G. *Food Chem. Toxicol.* **2008,** *46,* 1960–1968.

205. Munslow, I. J.; Clarke, A. J.; Deeth, R. J.; Westmoreland, I.; Scott, P. *Chem. Commun.* **2002,** 1868–1869.
206. Munslow, I. J.; Wade, A. R.; Deeth, R. J.; Scott, P. *Chem. Commun.* **2004,** 2596–2597.
207. Crust, E. J.; Clarke, A. J.; Deeth, R. J.; Morton, C.; Scott, P. *Dalton Trans.* **2004,** 4050–4058.
208. Westmoreland, I.; Munslow, I. J.; Clarke, A. J.; Clarkson, G.; Deeth, R. J.; Scott, P. *J. Organomet. Chem.* **2006,** *691,* 2228–2236.
209. Howson, S. E.; Allan, L. E. N.; Chmel, N. P.; Clarkson, G. J.; Deeth, R. J.; Faulkner, A. D.; Simpson, D. H.; Scott, P. *Dalton Trans.* **2011,** *40,* 10416–10433.
210. Sebli, C. P.; Howson, S. E.; Clarkson, G. J.; Scott, P. *Dalton Trans.* **2010,** *39,* 4447–4454.
211. Howson, S. E.; Allan, L. E. N.; Chmel, N. P.; Clarkson, G. J.; van Gorkum, R.; Scott, P. *Chem. Commun.* **2009,** 1727–1729.
212. Howson, S. E.; Bolhuis, A.; Brabec, V.; Clarkson, G. J.; Malina, J.; Rodger, A.; Scott, P. *Nat. Chem.* **2012,** *4,* 31–36.
213. Brabec, V.; Howson, S. E.; Kaner, R. A.; Lord, R. M.; Malina, J.; Phillips, R. M.; Abdallah, Q. M. A.; McGowan, P. C.; Rodger, A.; Scott, P. *Chem. Sci.* **2013,** *4,* 4407–4416.
214. Stahl, F. W. *Genetics* **1994,** *138,* 241–246.
215. Kaner, R. A.; Allison, S. J.; Faulkner, A. D.; Phillips, R. M.; Roper, D. I.; Shepherd, S. L.; Simpson, D. H.; Waterfield, N. R.; Scott, P. *Chem. Sci.* **2016**.
216. Zanetti, M.; Gennaro, R.; Romeo, D. *FEBS Lett.* **1995,** *374,* 1–5.
217. Durr, U. H. N.; Sudheendra, U. S.; Ramamoorthy, A. *Biochim. Biophys. Acta-Biomembr.* **2006,** *1758,* 1408–1425.
218. Kumar, S.V.; Lo, W. K. C.; Brooks, H. J. L.; Crowley, J. D. *Inorg. Chim. Acta* **2015,** *425,* 1–6.
219. Ivanov, A. I.; Christodoulou, J.; Parkinson, J. A.; Barnham, K. J.; Tucker, A.; Woodrow, J.; Sadler, P. J. *J. Biol. Chem.* **1998,** *273,* 14721–14730.
220. Mollmann, U.; Heinisch, L.; Bauernfeind, A.; Kohler, T.; Ankel-Fuchs, D. *BioMetals* **2009,** *22,* 615–624.
221. Hannon, M. J. *Chem. Soc. Rev.* **2007,** *36,* 280–295.
222. Kasiakou, S. K.; Michalopoulos, A.; Soteriades, E. S.; Samonis, G.; Sermaides, G. J.; Falagas, M. E. *Antimicrob. Agents Chemother.* **2005,** *49,* 3136–3146.
223. Cushnie, T. P. T.; Lamb, A. J. *Int. J. Antimicrob. Agents* **2011,** *38,* 99–107.
224. Ejim, L.; Farha, M. A.; Falconer, S. B.; Wildenhain, J.; Coombes, B. K.; Tyers, M.; Brown, E. D.; Wright, G. D. *Nat. Chem. Biol.* **2011,** *7,* 348–350.
225. Roemer, T.; Boone, C. *Nat. Chem. Biol.* **2013,** *9,* 222–231.
226. Abernethy, D. R.; DeStefano, A. J.; Cecil, T. L.; Zaidi, K.; Williams, R. L. *Pharm. Res.* **2010,** *27,* 750–755.

CHAPTER EIGHT

Carbon Nanotubes and Related Nanohybrids Incorporating Inorganic Transition Metal Compounds and Radioactive Species as Synthetic Scaffolds for Nanomedicine Design

F. Cortezon-Tamarit[1], H. Ge[1], V. Mirabello[1], M.B.M. Theobald[2], D.G. Calatayud[1,2], S.I. Pascu[1]
[1]University of Bath, Bath, United Kingdom; [2]University of Oxford, Oxford, United Kingdom

Contents

Inorganic and Organometallic Transition Metal Complexes with Biological Molecules and Living Cells
ISBN 978-0-12-803814-7
http://dx.doi.org/10.1016/B978-0-12-803814-7.00008-3

1. INTRODUCTION

1.1 The Problem

Cancer is one of the top 10 leading causes of death in the world. According to a cancer statistics report from the World Health Organization, 14.1 million adults in the world were diagnosed as having cancer and 8.2 million deaths (58.16%) occurred as a result of cancer in the world in 2012.[1] In the United Kingdom, cancer remains one of the top three common causes of death. In 2011, 331,487 people were diagnosed with cancer and there were 159,178 deaths from cancer in the United Kingdom (48.02%) (Cancer Research UK). Another statistic showed that 70% of patients with cancer had survived in the first year after diagnosis. Then this number reduced to 54.3% in the period of 5 years after diagnosis. Finally, among these, only 49.8% of patients with cancer had survived in 10 years' time.[2]

This survival rate was improved by an early diagnosis: 90% of patients had survived in 5 years when cancer was detected at the early stage (Table 8.1). Once tumor cells spread to adjacent tissue, i.e., middle stage, this rate dropped to 67%. Finally, only 10% of patients with cancer survived when diagnosed at the metastatic stages. This was particularly the case with the

Table 8.1 Five-Year Survival Rates in Colorectal Cancer by Stage at Diagnosis (1995–2005)[3]

Stage at Diagnosis	5-Year Survival Rate (%)
In early stage (local)	90
Spread to adjacent organ/lymph node	67
Spread to distant sites	10

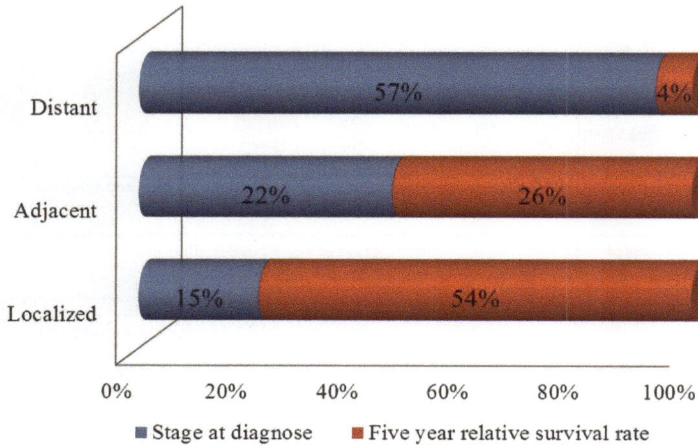

Figure 8.1 Lung cancer diagnosis and survival by stage, 2003–9.

cancer of lung, and another report from the Surveillance, Epidemiology, and End Results cancer statistics review showed that the earlier the stage at which the tumor was diagnosed, the higher the survival rate was[4] as shown in Fig. 8.1, only 15% of lung cancer could be diagnosed at localized stage, but 54% of patients survived. When the incidence of cancer was discovered at the spreading stage (22%), the survival rate dropped to 26%. However, most patients were diagnosed at late stages; hence, only 4% survived at the 5-year milestone. Thus, it can be concluded that early cancer diagnosis is equally important as rapid, effective, and targeted cancer treatment. It is generally accepted that the early and precise diagnosis of cancer can significantly improve patients' survival rate, and this has resulted in personalized medicine as a research field.

However, unlike other common diseases, early-stage cancer normally appears with an initial lack of symptoms, which makes early diagnosis of cancer extremely challenging. In fact, most patients with cancer are diagnosed at the middle or late stages, when a tumor mass is already grown as a result of tumor angiogenesis. Currently, four methods can identify cancer relatively early: medical imaging,[5] endoscopy,[6] biopsy,[7] and blood (other) sample tests[8,9] (Fig. 8.2).

Medical imaging techniques include X-ray imaging,[10] X-ray computed tomography (CT),[10,11] magnetic resonance imaging (MRI),[12] positron emission tomography (PET),[11] and single-photon emission CT (SPECT).[13] All techniques use one source of energy (X-ray, magnetism, or positron) to create a detailed view of the body to locate a tumor mass. Medical imaging techniques are matured techniques that can be generally

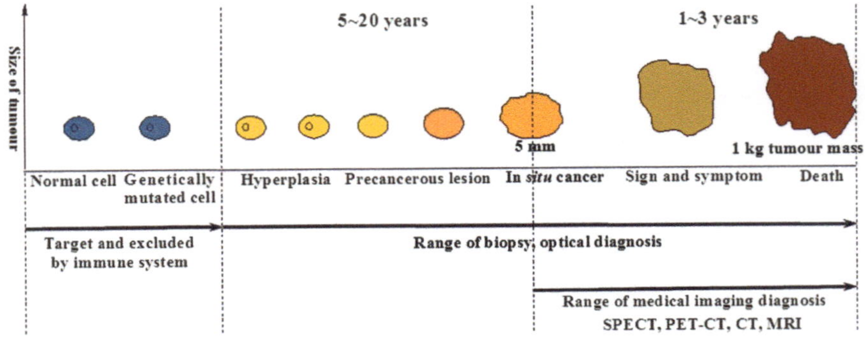

Figure 8.2 Stages of cancer development and range of imaging techniques that can be used. *CT*, computed tomography; *MRI*, magnetic resonance imaging; *PET-CT*, positron emission tomography–computed tomography; *SPECT*, single-photon emission computed tomography.

Table 8.2 Radioisotopes Used in Modern Positron Emission Tomography Chemistry and Imaging and Their Corresponding Half-Lives and Decay Types

Radioisotope	Decay Type	Half-Life	Radioisotope	Decay Type	Half-Life
^{14}O	β^+	1 min	^{62}Zn	β^+	9.2 h
^{15}O	β^+	2.1 min	^{65}Zn	β^+	243.7 days
^{13}N	β^+	10.0 min	^{38}K	β^+	7.6 min
^{11}C	β^+	20.3 min	^{82}Rb	β^+	1.3 min
^{18}F	β^+	109.7 min	^{59}Fe	β^-	44.5 days
^{32}P	β^-	14.3 days	^{64}Ga	β^+	2.6 min
^{61}Cu	β^+	3.35 h	^{67}Ga	EX	3.26 days
^{62}Cu	β^+	4.7 h	^{68}Ga	β^+	68.1 min
^{64}Cu	β^+	12.7 h	^{86}Y	β^+	14.7 h
^{67}Cu	β^-	61.9 h	^{90}Y	β^-	64 h

used to scan and diagnose all types of cancer. The quality of medical imaging methods rely heavily on radioisotopes applied as contrast agents. For example, PET is a modern nuclear imaging method that uses positron (β+)-emitting radioisotopes of elements, including ^{11}C, ^{15}O, ^{13}N, and ^{18}F, as well as some (unconventional) metallic radioisotopes of ^{64}Cu, ^{63}Zn, ^{38}K, ^{82}Rb, ^{32}P, ^{59}Fe, and ^{68}Ga, and the halogen series (Table 8.2). Some radiotracers including these radio-metals are at the clinical research stage. In PET, two gamma ray signals are emitted and detected, as they arrive in pairs to the detector (occurring at 180°). They are produced by a positron annihilation event by collision with an electron at the site of interest, with a well-defined image subsequently produced and interpreted computationally. A selection of some radioisotopes and their corresponding half-lives are listed in Table 8.2. Radioisotopes with a longer half-life are more appreciated for future diagnostic imaging because of their potential to emit longer during the circulation of the tracer in vivo, which can improve tumor–background ratio significantly.[14,15] Therefore, those radioisotopes with only minutes of half-life available will not be preferred for most targeted delivery in vivo imaging experiments. However, a long-lasting radioactivity can also damage living cells. Thus, those radioisotopes with greater than 10 days' half-life will be considered toxic to living organisms. Therefore, yttrium and gallium radioisotopes are popular in clinical research. When ^{68}Ga was anchored onto a suitable in vivo delivery system able to target tumors in precancerous stages, it selectively accumulated around infected sites and areas of inflammation to report the site of rapid cell division.[16] On the other hand, ^{90}Y and ^{86}Y were found in many multimodal early diagnosis and treatment approaches.[17]

The background and motivation for writing this chapter focused on inorganic and organometallic transition metal complexes supported onto carbon nanomaterial scaffolds and their functional role, were that molecular imaging and applications of nanotechnology in health care are interlinked research areas currently of high strategic importance to the United Kingdom and worldwide.[18] Furthermore, the interactions of transition metal complexes and materials with living systems, which are of relevance to tumor tracing by molecular imaging techniques, remains a "Holy Grail," particularly when complications emerge derived from targeting hypoxic tumors. The integrated probe and resulting features and functional roles will also be discussed. "Hot" (i.e., radioactive) metal ions have the possibility of being inserted for nuclear medicine testing in vivo, with the aim of generalizing the observed effects of the imaging probes on living systems to date, resulting in the new research fields of "radio-nanochemistry" and/or "nanoradioimaging," which may be viewed as subsets of nanomedicine research. This chapter also aims to close gaps, generalizing the design and assembly methods of multimodal imaging techniques with carbon nanomaterials, which have the ability to link imaging directly on the nano-, micro-, and millimeter scales.

As health care and technology advances lead to longer life expectancies, an ever-increasing and aging population faces growing risks from degenerative diseases such as cancer. As such, there is an increasing demand for molecular probes for imaging and early detection of cancers, such as, for example, prostate cancer. Research in chemistry and nanochemistry for imaging and sensing focuses on processes of relevance to the pharmaceutical industry and on biomolecule-based drugs under development or in clinical trials for personalized medicines including diagnostics and immunotherapeutics. Processes aiming toward integrated technologies based on the wider-spread approach using luminescent systems tagged to biomolecules such as antibodies and related targeted molecules are geared to attract attention regarding the need for early cancer detection, and tap into the approximately $3.3 billion medical diagnostics market estimated over the past few years.

Such probes are highlighted here and are also of relevance to multimodal imaging techniques in optical/nuclear medicine. These are important because it has been predicted that worldwide, this will reach a $7 billion market for radiochemicals alone by 2020.

Work focusing on multimodality imaging probe design and their delivery to hypoxic tumors will also be touched upon because these could lead to safe, extremely accurate, noninvasive, and affordable early diagnostics for tumors difficult to access noninvasively, and which are simultaneously resistant to therapy. This chapter highlights research in synthetic chemistry and nanochemistry carried out toward this objective, and focuses on subcellular studies that enhance our understanding of new imaging agents capable of bridging the gap between in vitro development and in vivo applications. Research that is fully embedded in the direction of biosensing and imaging/health care technology with aspects of nanomedicine and radionanochemistry will be highlighted here, with a particular emphasis on the detection and treatment of tumors including hypoxic tumors.

To date, little is understood about the most effective way to assemble and deliver multimodal contrast agents in a targeted manner (all-in-one imaging probes such as optical/radiopharmaceuticals and/or paramagnetic nanomaterials) necessary to achieve high-resolution images of processes taking place in cells and tissues, the precise mechanisms of probe uptake in cells and tissues, and the effect that these functional imaging tools may have on the targeted cells. Diagnostic imaging such as MRI and radiopharmaceutical imaging are crucial to health care and the health care industry. Many approaches have been developed for in vivo tracking of the fate of cells, but to date, only MRI has the resolution required to locate transplanted cells accurately in adult mammals.[19] PET is also emerging as a primary tool for cancer diagnosis and for monitoring the effects of therapy.[20] It is common knowledge that no single molecular imaging modality is sufficient to gain all necessary information both in vitro and in vivo. MRI has high resolution but low sensitivity and requires a large volume of sample to be administrated. PET has high sensitivity but poor resolution, although this continues to improve as image processing methods advance. Both PET and MRI require metal-based imaging agents, e.g., inorganic and organometallic transition metal complexes, to accumulate into tumors without loss of the active species en route to the target. Several drug delivery and targeting strategies have been developed based on supramolecular capsules, polymers, or lipid coat carriers.[21–28] Magnetic nanoparticles have been used for targeted delivery to the brain for MRI.[29] Optical fluorescence imaging does not have sufficient resolution in tissue more than a few millimeters thick, and is difficult to quantify. The combination of multiple molecular imaging techniques can offer synergistic advantages over single modality imaging techniques (optical, MRI, or PET alone). The use of carbon nanomaterials as scaffolds for imaging probes, equipped with long-lived, unconventional metallic radioisotopes incorporating ^{64}Cu for PET, or iron oxides, used as cytoplasmic labels for tracing cells in vivo using MRI is beginning to emerge. Detailed in vitro studies of such multimodal contrast agents are of interest to ensure that the different classes of imaging probes are merged together, generating a unique system in which the optical/MRI, optical/PET, and optical/MRI/PET probes have identical pharmacodynamics.[15] To date, it remains the case that limited systematic in vitro studies of the influence on cell toxicity mechanisms of the functionalities incorporated in imaging probes (built on carbon nanomaterial scaffolds) have been carried out, and a toolkit aiding the rational design of these new multimodal imaging agents has not yet been devised.[15] Modern approaches observe molecular imaging probes under conditions mimicking the physiologic environment in terms of pH and temperature, and, as the successful surface analysis tests narrow down the search for most stable imaging probe in vitro, which then allows their study in cells to rationalize their uptake, compartmentalization, mechanisms of interaction with cells and cells organelles. Single-walled carbon nanotubes (SWCNTs) are potentially promising delivery agents because their external surface, ends, and inner hollow space all can be functionalized (i.e., modified) by standard organic/inorganic chemistry methods. Carbon nanotubes (CNTs) are available from several highly reliable commercial sources (Elicarb from Thomas Swann Ltd.; Carbolex, from Rice University, etc.), with small batch-to-batch variations in structural parameters (diameter and size) and are well-characterized both experimentally and theoretically. A variety of methods allow their advanced purification and control of the length and diameter to levels suitable for biomedical imaging.[30,31] In addition, a number of attempts were employed to functionalize CNTs externally for various purposes.[32–35] Functionalized SWCNTs [as opposed to unfunctionalized but purified ones or to the long, rigid multiwalled CNTs (MWCNTs)[36,37]] were shown to be taken up preferentially into tumors with minimal toxic effects.[38–46] Their interactions with cells and the changes induced within cells treated with carbon nanomaterials are still a matter of much debate.[47] Radiolabeling (for PET or SPECT with either metallic and nonmetallic radioisotopes) was little demonstrated so far[48,49] and almost exclusively on the outer SWCNT surface, where the isotopes can be lost from the carriers, whereas their attachment procedure cannot be easily generalized and their mechanistic effects on cells virtually unexplored. All such studies employed only a subset of the nanotube's capabilities, and achieved only limited functions; the processes triggered by these in living systems are not fully understood. Limited information exists regarding the mechanisms of interactions between functional carbon nanomaterials and cells, although some studies on the cytotoxicity of the raw materials emerged, which is attributed to oxidative stress.[47] Compared with those, the approach in which current challenges are focused is entirely new and aimed at enhancing our understanding of the behavior of simultaneously filled and functionalized CNT carriers in standard (healthy as well as cancerous) cell lines and solid tumors. From here, a new class of entirely safe and reproducible class of nanocarriers for biomedical imaging applications may well emerge, with a major and significant advantage: potential high stability with respect to metal loss of metal-filled SWCNTs or transition metal complexes. CNT hybrids are highly attractive for the delivery of metal-based imaging reagents (both for MRI and PET) because their loss to nonspecific sites will be greatly minimized, thus allowing extremely low doses of the probe to be administrated. However, in the absence of a clear understanding of their interactions with cells, design considerations for new image probes are limited to guesswork. As such, several groups have adopted an interdisciplinary integral approach that ranges from assembling of imaging probes with complete functionality and understood physicochemical characteristics to cell delivery, uptake, and biodistribution testing and cytotoxicity evaluation to render the optimized probe (radiolabeled with longer-lived radioisotopes such as ^{64}Cu for PET) ready for in vivo evaluation. To date, only limited functions have been achieved, because only a subset of functionalities could be integrated. New approaches to develop a test-informed assessment protocol before designing and constructing a complete bioimaging probe and evaluating its functionality in cells as intact objects are needed. A major aspect of the references included here, yet virtually unachieved to date, is to rationalize the nature and strength of the connections among different components of the probe, and their impact on the overall functionality of the probe as a complementary approach to evaluating their interactions with cells. The use of such scaffolds has the potential to effect knowledge transfer rapidly as imaging reagents, in particular PET probes, are administered with minute amounts of compounds (nM) such that toxicologic testing may be rapid and the nanocarriers could progress quickly from the chemical bench to the clinic, and long half-life radioisotopes such as, for example, those available from cyclotrons such as ^{64}Cu [with a half-life of 12.7 h available via radioreaction $^{64}Ni(p,n)^{64}Cu$] or ^{89}Zr, with a half-life of 78.4 [available via the radioreaction $^{89}Y(p,n)^{89}Zr$], are particularly useful when it is necessary to visualize and image biological processes or radio-tagged biomolecules with long circulation time in the bloodstream.[50]

Cancer treatment currently uses intravenous immunotargeted therapeutics that have a low rate (1 in 100,000) of reaching their therapeutic targets in vivo. Nanosystems in clinical use are just a few, such as liposomes (for breast and ovarian cancer, and Kaposi sarcoma), or gadolinium or iron oxide nanoparticles as MRI contrast agents or lipid nanoparticles for ultrasound imaging. For high-resolution brain imaging, the state of the art is largely limited to ^{18} F-labeled glucose. The efficiency of all of these therapeutic or imaging agents in use is affected by their lack of the specific delivery of a large number of active drug molecules to the target.[51,52] Moreover, systems capable to deliver precisely, would administer reagents close to the target and avoid the cellular surveillance system.[53]

Current and future nanomedicine-focused research programs have the potential to affect the design and administration of nanotheranostic reagents greatly, because they aim to provide a general solution to drug delivery to the brain while keeping administered drug payloads extremely low. Prior studies based on various nanomaterials as nanocarriers achieved integration of only a subset of functionalities.[54–57] Moreover, targeting of brain diseases through this technology (using simultaneous imaging and therapy) is just emerging; only a handful of studies have attempted it.[58,59]

Multidisciplinary approaches to nanomedicine design go all the way from synthetic scaffolds design to cell uptake and biodistribution testing, rendering the final optimized probe ready for radiolabeling and in vivo testing for therapy and post-therapy imaging. Such approaches are typically modeled according to the life cycle of a medicine in general, as recommended by the Medicines and Health care Products Regulatory Agency.[60]

The following issues are typically being addressed in research programs with a focus on metals in medicine and in nanomedicine design across the community:

- A complete range of functionalities, as described earlier and in Fig. 8.3, are being integrated. A CNT-based approach to nanotheranostic design has a major advantage over other drug delivery strategies (based on polymers or lipid coat carriers or supramolecular capsules) owing to its high stability with respect to loss of the imaging reagent; this is prevented by its encapsulation inside the nanotube.
- Special attention is typically being paid to designing nanocarriers such that they can be recovered from the body. Water-soluble medicines tend to be excreted through the kidneys, whereas fat-soluble compounds tend to be excreted through liver and gall bladder into intestine.[61,62] The design therefore provides nanocarriers with groups modulating both their hydrophilicity and lipophilicity via a specially designed synthetic protocol to ensure their biocompatibility and integrity, as well as their facile recovery from the body after treatment.
- The probe design is generally being informed by multiple physical characterizations and cytotoxicity assays performed at each step, which assesses success in tailoring the probes with the required functionalities.
- Scale-up of both carbon and inorganic nanomaterial used is needed to track development changes and to produce small batches for lowest cost evaluation. Such protocols lead to functional SWCNT-based approaches and are likely to lead to a favorable balance between delivery-compatible size/payload capacity/minimizing toxicity/ease of recovery from the body for a nanoparticulate-based nanomedicine.

Blood sample tests are also common methods applied to reveal cancer, because characteristic markers such as overexpressed molecules, including sugars, fats, proteins, RNA, and DNA can be identified by in vitro assays.[63] However, selectivity or sensitivity problems remain in current testing reagents.[64,65] Biopsy continues to be the most certain way to diagnose cancer; in which a tissue sample is collected from the site of interest. Subsequently, the morphology and gene status are normally examined. However, it is still the case that it is usually difficult to locate the site of a tumor. As such, molecular imaging has the potential to speed the diagnosis in combination with blood sample testing to allow the location of a tumor or early cancer detection,[66,67] including near-infrared (NIR) imaging investigations using advanced endoscopy for colorectal cancer and breast imaging. Other methods, more commonly used in the diagnosis of tumors that are difficult to access, involve optical fluorescence imaging. These involve

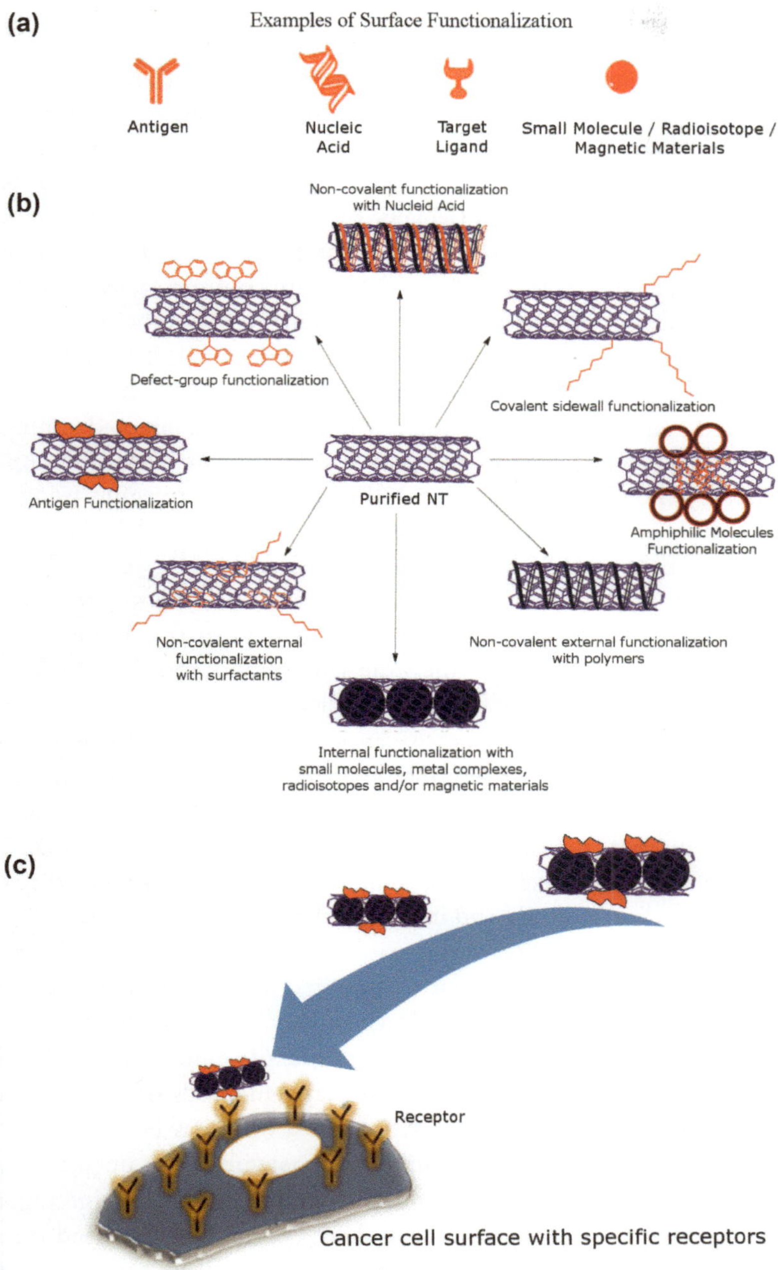

Figure 8.3 Aspects of "all-in-one" probe design: (a) Targeted functionalities, (b) schematic of noncovalent and covalent functionalization of nanotube (NT) scaffold, and (c) biocompatible/fluorescent coating single-walled carbon nanotube–encapsulated metallic radionuclides and/or magnetic materials targeting group receptor on the tumor cell surface.

the use of fluorophores and, for enhanced imaging,[68] quantum dots (QDs) which have been used as parts of endoscopes.[69,70] In general, endoscopy is applied at the middle or late stages of a cancer diagnosis and is combined with other imaging tools to confirm a cancer diagnosis, or with biopsy to collect a tissue sample for further investigations.[70] Interestingly, multi-modality imaging and targeted delivery of such systems were highlighted in frontier publications.[71,72] Of particular importance QDs were studied extensively because they are readily available from commercial sources: these have been modified with targeting RGD peptides and bombesin fragments as well as labeled with ^{18}F, resulting in dual probe for PET coupled with NIR fluorescence imaging. NIR–emitting commercial QDs have also been radiolabeled and $\alpha_v\beta_3$ integrin and GRP receptors were targeted aiming at the imaging of prostate and/or colon cancers for ultimate diagnosis coupled with therapy monitored by in vivo NIR fluorescence imaging using the far-red region of the emission spectra whereas the tumor delineation was followed by observed PET imaging.

Overall, it is commonly accepted that medical imaging techniques are not currently sufficiently developed to respond to the unmet clinical needs of early cancer diagnosis unless the sensitivity of tools can be further improved and radiotoxicity of tracers in medical use is considerably reduced. Biopsy is a method applicable in early cancer research, but owing to the lack of precision, it could perhaps be more effective if used together with more sensitive techniques as a confirmation of a diagnosis. Hence, blood sample tests would be ideal for early cancer diagnosis, and one long-term aim of research in this field remains the assembly of fluorescent probes that can target cancer cells selectively and induce minimal harm to normal cells.

The other interest of this overview is to focus on the precise delivery of a diagnostic or therapeutic reagent to a desired location in living cells and tissues. The design of a successful drug delivery system is determined by several factors, including the choice of the types of vehicle improved in the drug containment and delivery, size and shape, methods of functionalization, ease of incorporation of the contrast agent (this may be radioactive or fluorescent), issues of cellular and tumor microenvironment uptake, as well as general side effects induced if the body identifies it as a "foreign object." A biocompatible nanoparticulate drug delivery agent would need careful selection as the first step, and this could have a crucial role in establishing a drug delivery system. It would directly as well as indirectly affect subsequent steps in the design, synthesis, and imaging in vivo for a diagnostic nanomedicine. As vectors for drug delivery, nanoparticles offer a range of unique advantages, including their ability to modify drug pharmacokinetics in vivo and their capacity to be loaded with high concentrations of different drugs, for an advanced combination of diagnostic and therapeutic (denoted "nanotheranostics"). The properties of nanoparticles can also be modified to control drug release, alter blood circulation half-lives and improve biodistribution profiles, increase tissue permeability, and target specificity, as well as provide enhanced metabolic stability.[73]

Currently, at least nine different types of nanocarriers are widely used in biomedical research and have applications with relevance to tumor diagnosis and treatment[74] (Fig. 8.4).

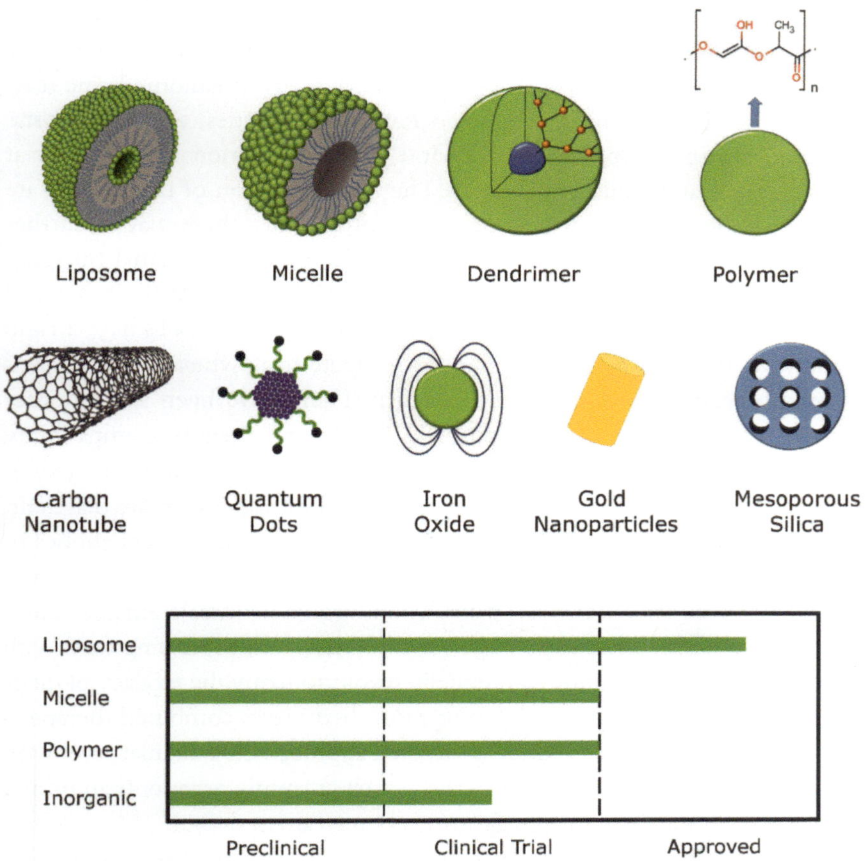

Figure 8.4 General categories of nanocarriers and current development stages for molecular imaging of cancer. *Reproduced from Mirabello, V.; Calatayud, D.G.; Arrowsmith, R. L.; Pascu, S.I. Metallic Nanoparticles as Synthetic Building Blocks for Cancer Nanodiagnostics: From Materials Design to Molecular Imaging Applications.* J. Mater. Chem. B. ***2015,*** http://dx.doi.org/10.1039/c5tb00841g, *with permission from The Royal Society of Chemistry.*

As shown in Fig. 8.4, most nanocarriers studied to date could be categorized into four groups: (a) lipid-based systems, (b) surfactant-based systems, (c) polymer-based systems, and (d) inorganic material–based systems. Liposomes were the first type of vesicles discovered in 1965.[76] Phospholipids were one of the few solubilizers that appeared to be well-tolerated in vivo, and an increase in the solubility of any drug trapped within was found to be 100- to 10,000-fold. Followed by nearly 50 years of intensive experiments and studies, the application of liposomes in the biomedical field was approved.[77] Both micelle and polymeric vesicles were less widely studied than the lipid-based systems; hence, the development and in vivo testing of these two types of carriers proceeded more slowly than that of liposomes.

Inorganic nanocarriers, particularly iron nanoparticles, have been intensely studied because of their sizes, shapes, and ability to act as nanomedicine scaffolds, as well as their unique physicochemical properties, e.g., magnetism, emission/absorption spectroscopies. Most inorganic carriers are currently at the preclinical and clinical trial stages. The most common of these, QDs, are intensely fluorescent, and it has been demonstrated that these may be further bound to anticancer agents to make the resulting dual functional (imaging and therapeutic) composition, which is traceable in tissues. At the other end of the spectrum in terms of proximity to clinical use, SWCNTs have a rigid framework and can absorb NIR light to generate heat, which could be used for photothermal therapy cancer treatment. There have been several other reviews with a focus on developing SWCNTs and their biocompatibility, which raises awareness of their inherent toxicity.[78,79] Furthermore, iron oxides are probably the most popular nanomaterials in clinical use; they are magnetic, can be rapidly derivatized, and can be combined with fluorescent probes to generate dual-modality imaging and/or a sensing complex or nanohybrid. For these, a core-shell design in nanotheranostics (a nanoscale amalgamation of diagnostics and therapeutics) is a relatively new but promising field with multidisciplinary applications, potentially revolutionizing the efficacy of diagnostic nanomedicines and, in the long-term their use as combined therapeutics, for patient benefit. This offers a targeted approach on a cellular level with future applications on a molecular scale, which currently represents an unmet clinical need for an improved diagnostic contrast agent design.[75,80]

Small molecule-based contrast agents can be incorporated in in vivo nanodiagnostics, thus greatly increasing the sensitivity and selectivity of otherwise undetectable pathologies.[81] Nanomedicine and radiolabeled nanomedicine advances further increase this diagnostic potential by opening up new possibilities of imaging technologies and applications, such as those using fluorescence imaging techniques. Many different types of nanomaterials are currently under development; some are even at the clinical phase, such as gadolinium chelates and salts (Magnevist), and iron oxide nanoparticles.[82] Nanomedicine offers the use of current imaging technologies, including new opportunities in fluorescent tagging, which to date has not been widely used owing to poor signal strength, wide emission bands in vivo, and poor tissue penetration.[81] It offers even more insight into disease progression and can give information about tumor uptake mechanisms.[83] This level of depth is required to probe diseased tissue on a molecular level and gain a full understanding of the mechanisms involved. Hence, longer-term, more precise patient diagnosis and treatment may become available through nanomedicine, particularly multimodality aspects of radio-nanochemistry and radio-nanoimaging.

Particularly for iron nanoparticles and related nanohybrids, there has been a vast amount of literature in relative terms, and radiolabeling options such as external surface labeling, internal encapsulation, core-shell trapping in polymer–nanoparticle matrices, and extensive use of chelating appendage strategies.[84–93] This latter method is mainly highlighted in the literature for using 1,4,7,10-tetraazacyclododecane-1,4,7,10-tetraacetic acid (DOTA) or 1,4,7-triazacyclononane-1,4,7-triacetic acid (NOTA), both of which are commercially available and suitable for copper-64 and gallium-68 coordination, and have relevant radiochemistry protocols that are well-developed, although acceptable radiochemical yields are mainly obtained under acidic pH (4.5–4.9) and harsh radio-nanochemistry conditions, such as 70–90°C heating for DOTA and 30–50°C for NOTA. Such tags may be coupled with the use of linkers [including polyethylene glycol (PEG) of varying lengths] and tailor-made chemistry involving standard amide couplings and peptide bond formation, thiols, alcohols, or "click chemistry" adapted for work on a nanoscale and targeting biomolecule appendages. Because this aspect of nanochemistry including radiochemistry approaches has been covered exhaustively, we will focus in this chapter on less usual nanomaterials and their tagging with imaging agents including their radio-nanochemistry.

2. POSITRON EMISSION TOMOGRAPHY–SINGLE-PHOTON EMISSION COMPUTED TOMOGRAPHY METAL COMPLEX BIOCONJUGATES

2.1 Positron Emission Tomography

As stated earlier, PET operates by detecting photons emitted from the annihilation of a positron emitted from a radionuclide and colliding with an

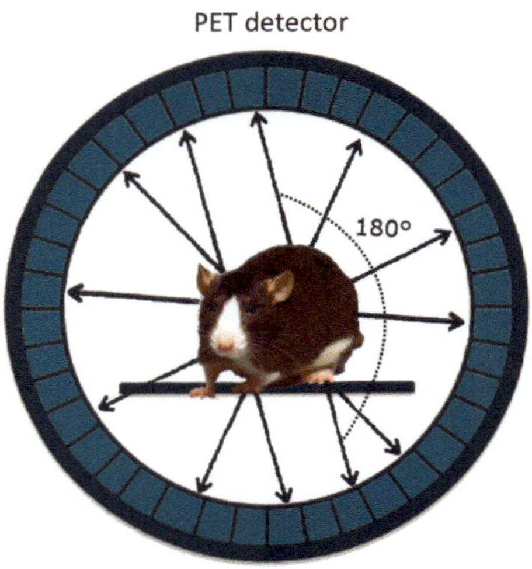

Figure 8.5 Representation of positron emission tomography (PET) detector.[94]

electron (Fig. 8.5). These photons are emitted and "travel" at 180 degrees to each other, and are detected by a circular detector array. Only coincident photons are counted, so background noise is low. PET radiotracer developments include a number of light, nonmetallic isotopes with short half-lives, as well as a number of heavier atoms (Table 8.2).

These short-lived radionuclides are most commonly used in PET because they can be introduced to biomolecules without changing the biological activity too much. In the case of ^{11}C, ^{13}N, and ^{15}O, these elements are naturally found in biomolecules, and so direct replacement of atoms can be done with minimal perturbation of biological activity. Fluorine, on the other hand, is rarely found in native biomolecules, but because it is roughly the same size and polarity as a hydroxyl group, it can be introduced; hence the extended interest in ^{18}F chemistry for PET.

2.2 Selected Metallic Radioisotopes and Their Metal Complexes of Relevance to Positron Emission Tomography

Copper has a number of characteristics that make it suitable for use in radiopharmaceuticals. It has a number of different radioisotopes that can be used as PET (^{60}Cu, ^{61}Cu, ^{62}Cu, and ^{64}Cu), SPECT (^{67}Cu), or therapeutic (^{64}Cu, ^{66}Cu, and ^{67}Cu) agents (Table 8.3).[95] The coordination chemistry of copper is also well-known and can be exploited when designing radiopharmaceuticals. Finally, because copper is the third most abundant transition metal within the body, its biochemistry is well-understood.[96]

Table 8.3 Copper Radioisotopes of Relevance to Positron Emission Tomography

Isotope	Half-Life	Imaging (Emission, Energy, Abundance)	Therapy (Energy, Range in Tissue)
^{60}Cu	20 min	PET (β^+, 873 keV; 93%)	
^{61}Cu	3.3 h	PET (β^+, 527 keV; 62%)	
^{62}Cu	9.74 min	PET (β^+, 1315 keV; 98%)	
^{64}Cu	12.7 h	PET (β^+, 278 keV; 19%)	B^-; 190 keV; 0.95 mm
^{66}Cu	5.4 min		B^-; 1109 keV; 5.6 mm
^{67}Cu	62 h	SPECT (γ, 91 keV; 7%; 93 keV, 16%; 185 keV, 48%)	B^-; 121 keV; 0.61 mm

^{64}Cu is the most versatile isotope because it combines electron capture (41%), β^- (40%), and β^+ (19%) decays, giving it both diagnostic and therapeutic properties.[97] ^{64}Cu can be generated via cyclotron reactions or by using a nuclear reactor. The half-life of ^{64}Cu (12.7 h), decay properties, and production in the absence of a biomedical cyclotron make this isotope a desirable radionuclide for chelation chemistry.[98] In nuclear reactors, Cu^{64} can be produced through the $Cu^{63}(n,y)^{64}Cu$ and $Zn^{64}(n,p)^{64}Cu$ pathways.[99] Currently the most commonly used route is the proton bombardment of an enriched ^{64}Ni target.[100,101] Anion exchange chromatography is then employed to process and purify ^{64}Cu from ^{64}Ni. ^{64}Ni can be recovered from this purification process and reused. However, the high cost of ^{64}Ni may limit the use of this method in the future.[102] Promising results of studies on ^{64}Cu production have been reported using targets such as enriched ^{66}Zn (27.9% natural abundance) and ^{68}Zn (18.8 natural abundance), which are much less expensive and therefore more accessible.[103–106]

Many chelating ligands can be used to coordinate copper ions; in this sense, DOTA and Cb-TE2A have shown high efficacy for generating chelators of copper-64.[107] A series of peptides capable of targeting somatostatin receptors, integrin-α and -β, gastrin-releasing peptide receptors, and formyl peptide receptors have also been developed.[108] The number of oxidation states available to copper in aqueous conditions is limited. Copper(I) and copper(III) complexes are accessible when using suitable donors, but most copper radiopharmaceuticals are in the copper(II) oxidation state.[109] The more common copper radiopharmaceutical chelators are the polyazamacrocycles. Copper(II) is of borderline hardness and has a high affinity for nitrogen donors.[109] Together with the stability of the complex owing to the macrocyclic effect,[110] this results in copper(II) adducts with high formation constants that are kinetically inert.[111] The use of macrocycles, such as cyclam and cyclen, for the ^{67}Cu-labeling of monoclonal antibodies, was first reported by Meares et al.[112] Since then, there have been numerous studies using DOTA and 1,4,8,11-tetraazacyclotetradecane-*N*′,*N*″,*N*‴-tetraacetic acid (TETA) chelators to label biomolecules with copper.[113–118] However, TETA has been widely used owing to the greater stability of the complexes.[119] Nonmacrocyclic ligands have also been studied. The use of monothiosemicarbazones,[120] *bis*(thiosemicarbazones),[121] dithiocarbamates,[122] and phosphines[123] was reported in several works. In particular, copper *bis*(thiosemicarbazonate) complexes were used to deliver the copper radionuclide to hypoxic tissue. *Bis*(thiosemicarbazone) ligands and their metal complexes were first reported almost 60 years ago. They were initially studied because of their antitumor properties.[124–128] Currently, over 50 mono- and *bis*(thiosemicarbazone) compounds are known. Among the most active, 3-ethoxy-2-oxobutyraldehyde *bis*(thiosemicarbazone), H_2KTS, and 3-ethoxy-2-oxobutyraldehyde *bis*(4-*N*-methyl-3-thiosemicarbazone), H_2KTSM deserve mention.[125] The copper(II) adduct of H_2KTS ligand has greater antitumor activity than does the ligand and is capable of inhibiting DNA synthesis and tumor growth in vivo. Studies on the mechanism of antitumor activity suggested that after uptake, the metal is reduced by intracellular thiol groups, and therefore the copper atom is dissociated from the ligand.[129] It was thus proposed that the free copper ions can potentially inhibit DNA synthesis or that thiol groups within enzymes might be damaged when reducing the complex and therefore become unable to conduct their original role of enzyme in cell respiration.[129,130] Interest in the use of copper(II) *bis*(thiosemicarbazonate) complexes as potential radiopharmaceuticals started in the 1990s. The ^{62}Cu(II)PTSM (Fig. 8.6) complex was studied for its potential as a blood perfusion PET tracer. Since then, other *bis*(thiosemicarbazonate) complexes began to be studied, and in 1997 the hypoxia selective uptake of ^{62}Cu(II)ATSM was first reported in an isolated rat heart model of ischemia.[131] Blower and collaborators investigated the correlation between the structural and chemical properties of 13 ^{64}Cu(II)*bis*(thiosemicarbazonate) complexes with hypoxia selectivity in vitro.[132] Such studies confirmed that hypoxia selectivity highly depends on the backbone substituents (R^1 and R^2) as well as the Cu(II)/Cu(I)

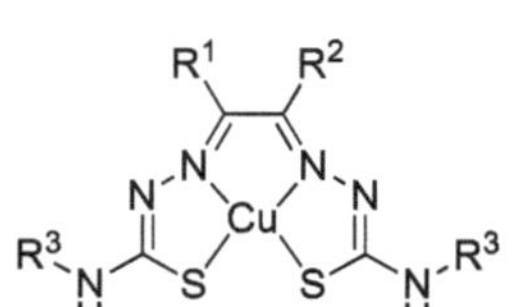

Cu(II)GTSM	$R^1 = R^2 = H$	$R^3 = Me$
Cu(II)PTSM	$R^1 = H$	$R^2 = R^3 = Me$
Cu(II)ATSM	$R^1 = R^2 = R^3 = Me$	

Figure 8.6 Structure of a selection of copper(II) *bis*(thiosemicarbazonate) complexes of relevance to hypoxia-related research.

reduction potential of the complex, while changing the groups on the pendent arms (R^3) has little effect on the reduction potential.

Cu(II)ATSM showed the best selectivity and its in vitro kinetics in EMT6 cells was compared against that of [^{18}F]fluoromisonidazole (^{18}F-FMISO).[133] It showed nearly threefold higher retention at lower oxygen concentrations [90% uptake at 0 parts per million (ppm) O_2] compared with higher concentrations (31% uptake at 2×10^5 ppm O_2), with the threshold of selectivity lying between 5×10^3 ppm oxygen (3.8 mmHg) and 1×10^3 ppm (0.8 mmHg). ^{18}F-FMISO also showed hypoxia selectivity, but uptake was much lower, with only 10% uptake at 0 ppm O_2. The washout kinetics of ^{64}Cu(II)ATSM were also superior compared with ^{18}F-FMISO.[134–136] The first preclinical in vivo studies of ^{64}Cu(II)ATSM in mice bearing EMT6 tumors showed optimal tumor uptake after 10 min postinjection with trapping within hypoxic regions of the tumor.[133] Further studies in 9L gliosarcoma tumors in rats suggested a strong correlation between tumor pO_2 levels and Cu(II)ATSM uptake.[134] In 2006, the uptake of ^{64}Cu(II) ATSM with the well-established hypoxia marker drug EF-5 in three tumor lines was compared by Yuan et al.[137] A number of clinical studies were carried out using Cu(II)ATSM, the first of which was published in 2000 for patients with lung cancer.[138] This study showed that ^{60}Cu(II)ATSM accumulated in hypoxic lung, rectal, and cervical tumors and that there was an inverse correlation between tumor uptake and therapy response.[138–143] As these initial studies drew the attention of the scientific community, the US Food and Drug Administration approved a multicenter clinical trial using ^{64}CuATSM.[144] The first clinical study proved a comparison between the image quality of ^{60}Cu(II)ATSM and ^{64}Cu(II)ATSM in patients with cervical carcinoma.[142] These studies demonstrated that the tumor uptake of ^{60}Cu(II)ATSM and ^{64}Cu(II)ATSM was comparable when agent deliveries were separated by 1–9 days; however, the imaging data showed ^{64}Cu(II) ATSM to have a better signal-to-noise ratio than for ^{60}Cu(II)ATSM.

Numerous research studies were carried out to determine the mechanism responsible for the retention of Cu(II)ATSM within hypoxic cells. The suggested mechanisms all pointed out that an initial Cu(II)/Cu(I) reduction followed by dissociation of the copper from the ligand occurs within hypoxic cells. The mechanism initially suggested by Fujibayashi et al. proposed that Cu(II)ATSM is only reduced in hypoxic cells and then irreversibly trapped.[131] Studies carried out on subcellular fractions showed that such a reduction reaction mainly occurs in the microsome and cytosol of a cell, is heat sensitive and enhanced by adding reducible nicotinamide adenine dinucleotide phosphate (NADPH), which thus indicates an enzymatic reduction process.[145] Dilworth and Holland refined the mechanism through spectroelectrochemical, computational, and in vitro kinetic studies and showed that Cu(I)ATSM is stable in aqueous buffer solutions and that the reversibility of metal reduction is highly pH dependent.[146–148] The mechanism shown in Fig. 8.7 was proposed based on these studies and included these protonation steps.

The copper derivative diacetyl *bis*(*N*-methyl-thiosemicarbazone), Cu(ATSM), has proved to be a highly effective imaging agent in PET studies to identify hypoxic tissues.[149] The bis(thiosemicarbazone) complexes are able to cross the blood–brain barrier to generate images of cerebral blood flow. As such, there is ongoing interest in synthesizing such complexes and studying their biological uptake by fluorescence imaging. More recently, the synthesis of Cu(II) complexes with bis(thiosemicarbazonate) ligands bearing a pendent Bodipy fluorophore has been proposed, together with their Ni(II) and Zn(II) analogs. Such

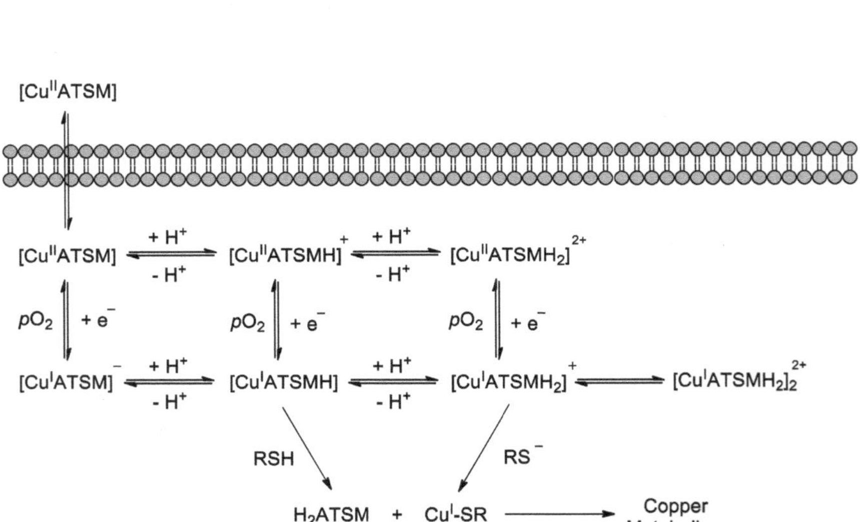

Figure 8.7 Currently accepted mechanism for the uptake of diacetyl *bis*(*N*-methyl-thiosemicarbazone) and subsequent copper delivery. *Adapted from Holland, J.P.; Giansiracusa, J.H.; Bell, S.G.; Wong, L.-L.; Dilworth, J.R. In Vitro Kinetic Studies on the Mechanism of Oxygen-Dependent Cellular Uptake of Copper Radiopharmaceuticals.* Phys. Med. Biol. ***2009,*** 54, *2103.* http://stacks.iop.org/0031-9155/54/i=7/a=017; *Holland, J.P.; Barnard, P.J.; Collison, D.; Dilworth, J.R.; Edge, R.; Green, J.C.; et al. Spectroelectrochemical and Computational Studies on the Mechanism of Hypoxia Selectivity of Copper Radiopharmaceuticals.* Chem. Eur. J. ***2008,*** 14, *5890–5907.* http://dx.doi.org/10.1002/chem.200800539; *Holland, J.P.; Green, J.C.; Dilworth, J.R. Probing the Mechanism of Hypoxia Selectivity of Copper bis(Thiosemicarbazonato) Complexes: DFT Calculation Of Redox Potentials and Absolute Acidities in Solution.* Dalton Trans. ***2006,*** *783–794.* http://dx.doi.org/10.1039/b512656h.

complexes were successfully used for in vitro confocal fluorescence cell imaging, whereas two-photon fluorescence lifetime imaging microscopy (FLIM) measurements were employed to determinate the stability of the complexes within cells (Fig. 8.8).[15b]

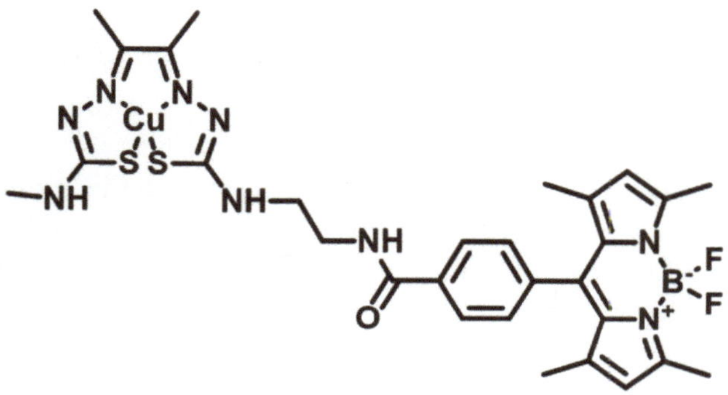

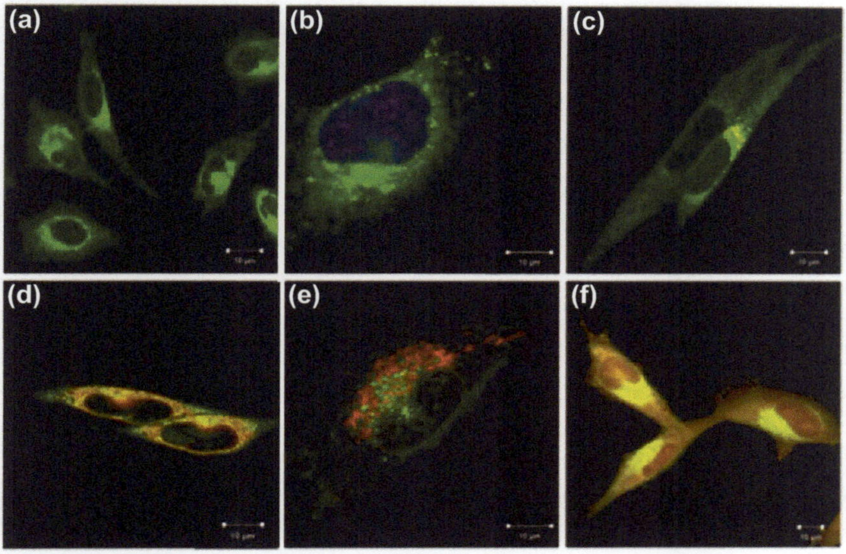

Figure 8.8 Overlays of confocal fluorescence uptake images of CuL1 in HeLa cell line, at 10 mM, 20 min incubation, $\lambda_{ex}=488$ nm with (a) CuL1 alone; (b) Hoechst nucleic acid stain, 1 μg/mL, 30 min, $\lambda_{ex}=405$ nm; (c) Lysotracker red DND-99, 200 nM, 60 min, $\lambda_{ex}=543$ nm; (d) ER-tracker red, 1 mM, 20 min, $\lambda_{ex}=543$ nm; (e) Mitotracker red FM, 100 nM, 30 min, $\lambda_{ex}=543$ nm; (f) Cell Tracker orange CMTR, 10 μM, 30 min, $\lambda_{ex}=543$ nm. All scale bars measure 10 μm. *From Waghorn, P.A.; Jones, M.W.; Theobald, M.B.M.; Arrowsmith, R.L.; Pascu, S.I.; Botchway, S.W.; Faulknera, S.; Dilworth, J.R. Shining Light on the Stability of Metal Thiosemicarbazonate Complexes in Living Cells by FLIM.* Chem. Sci. ***2013****, 4, 1430–1441.* http://dx.doi.org/10.1039/c2sc21489j.

For the first time, confocal fluorescence imaging coupled with FLIM showed that under aerobic conditions, Cu(ATSM)–Bodipy remained uncharged and was therefore intact after 20 min and 1 h. Instead, the Cu(GTSM) analog is completely dissociated after 20 min. A fluorescence tridentate ligand bearing Bodipy and capable of coordinated rhenium and technetium-99m (^{99m}Tc) metal centers proposed by Benny et al [ReCl(CO)3(triphos-Ph)(Bodipy)][OTf] was used as an effective cancer imaging agent in PC-3 cells using single-photon excitation between 460 and 500 nm and an emission filtered at 510 nm (Fig. 8.9).[15c]

The long half-life ^{89}Zr (78.41 h) makes it one of the most common metal tracers for antibody conjugation.[108,150] The most prominent chelator of Zr^{+4}, and therefore the most widely used, is deferoxamine (DFO), a hexadentate siderophore containing three hydroxamate groups for chelating metals (Fig. 8.10). There are two possible routes for its conjugation. The first is the peptide coupling of a tetrafluorophenyl ester of iron(III) DFO with any given antibody, followed by removal of the iron. The second is incubation of the antibody with the chelator thiocyanate deferoxamine (DFO-SCN). The most recent development of DFO antibody conjugation is the involvement of maleimide or halide-modified DFO.[108] Advanced purification was carried out by standard methods accepted in the field, using size exclusion chromatography, to remove the unreacted ^{89}Zr at incubation times of 30–120 min at room temperature.

^{89}Zr-Bevacizumab was compared with its ^{111}In analogous to evaluate the potential of ^{89}Zr complex as a radiopharmaceutical in molecular imaging.[151] Both radiopharmaceuticals were synthesized; labeling afforded yields of 98.0% ± 0.7% for the ^{89}Zr conjugate and 96.6% ± 0.5% for In^{111}. The specific activity for each was 58 and 50 MBq/mg, respectively. This result confirmed that labeling of ^{89}Zr and In^{111} indicated viable radiometals for binding to bevacizumab. The protein-bound radioactivity for each showed a small decrease (6% and <5%, respectively). This ensured optimal measurement and imaging during the 1-week period. ^{89}Zr is a desired isotope for conjugation with bevacizumab because of their similar half-lives of 3.27 and 17–21 days, respectively. However, the high radiation exposure values experienced by patients are a limitation. This is because of the high energy, high abundance of gamma rays (909 KeV), and low abundance of beta rays produced upon ^{89}Zr decay. Clinical doses have not yet been established, but the continuous development of increased insensitivity of PET cameras has a potential to reduce radiation exposure

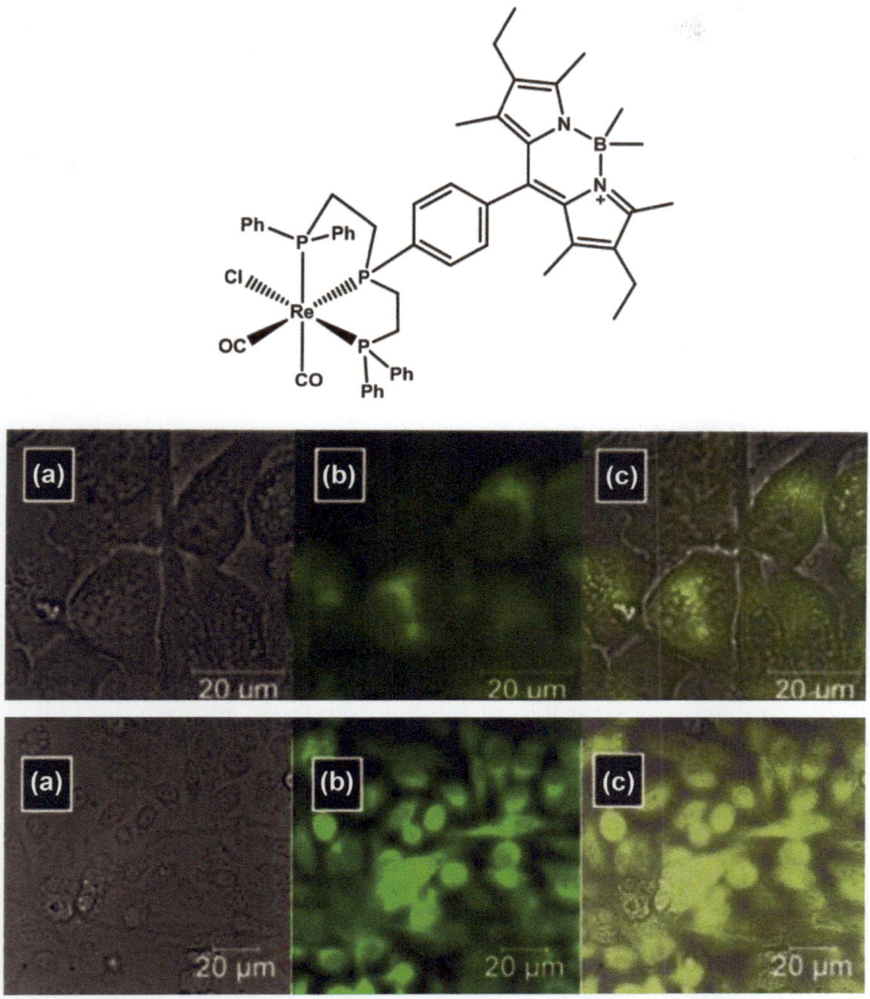

Figure 8.9 Epifluorescence imaging of PC-3 living cells with compounds $[ReCl(CO)_3(triphos\text{-}Ph)(Bodipy)][OTf]$ (*Top*) cis,mer-$[ReCl(CO)_2]$, 100 mM, 1% dimethylsulfoxide, 15 min. (*Bottom*) fac-$[Re(CO)_3]^+$, 5, 50 mM, 2% ethanol, 15 min. (a) Brightfield image. (b) Green channel λ_{ex} = 460–500 nm, long pass filtered at 510 nm. (c) Overlay of (a) and (b). *From Davies, L.H.; Kasten, B.B.; Benny, P.D.; Arrowsmith, R.L.; Ge, H.; Pascu, S.I.; Botchway, S.W.; Clegg, W.; Harringtona, R.W.; Higham, L.J. Re and* ^{99m}*Tc Complexes of BodP3 – Multi-Modality Imaging Probes.* Chem. Commun. ***2014,*** 50, *15503–15505.* http://dx.doi.org/10.1039/c4cc06367h.

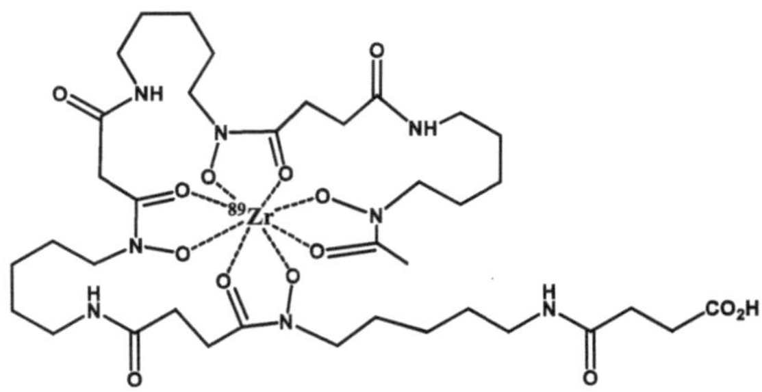

Figure 8.10 Structure of ^{89}Zr-DFO chelate.

limits in the near future. Experimental studies have established that the novel chelate-free, heat-induced metal ion binding reaction can be used to label ultrasmall superparamagnetic iron oxide nanoparticles with ^{89}Zr–Feraheme (Fig. 8.11).

Radiolabeling studies found that ^{89}Zr–Feraheme (^{89}Zr–FH) can be isolated with 93%±3% radiochemical yield and greater than 98% radiochemical purity using size-exclusion chromatography. Moreover, ^{89}Zr–FH was found to be thermodynamically and kinetically stable in vitro using a series of ligand challenge and plasma stability tests, and in vivo using PET/CT imaging and biodistribution studies in mice.[82] The development of a new PET radiotracer, ^{89}Zr-desferrioxamine–labeled transferrin (^{89}Zr-transferrin) was capable of binding transferrin receptor 1 (TFRC, CD71). These preclinical data established ^{89}Zr-transferrin as a sensitive tool for the noninvasive measurement of oncogene-driven TFRC expression in prostate cancer.[152]

2.3 Single-Photon Emission Computed Tomography

In a non-specialist approach, SPECT uses radionuclides that emit gamma rays during their decay, which are detected using a gamma camera (Fig. 8.12).

For a radionuclide to be suitable for SPECT (and PET), a number of factors must be considered. Its half-life must be long enough to allow synthesis, purification, and subsequent accumulation at the target, but not so long as to subject the patient to too much radiation. It must have a suitable mode of decay, ideally with only emission of gamma rays; the energy of the emitted gamma rays must be suitably high to escape the body and be readily available and at low cost.[153] A number of SPECT radionuclides are given in Table 8.4.[153,154]

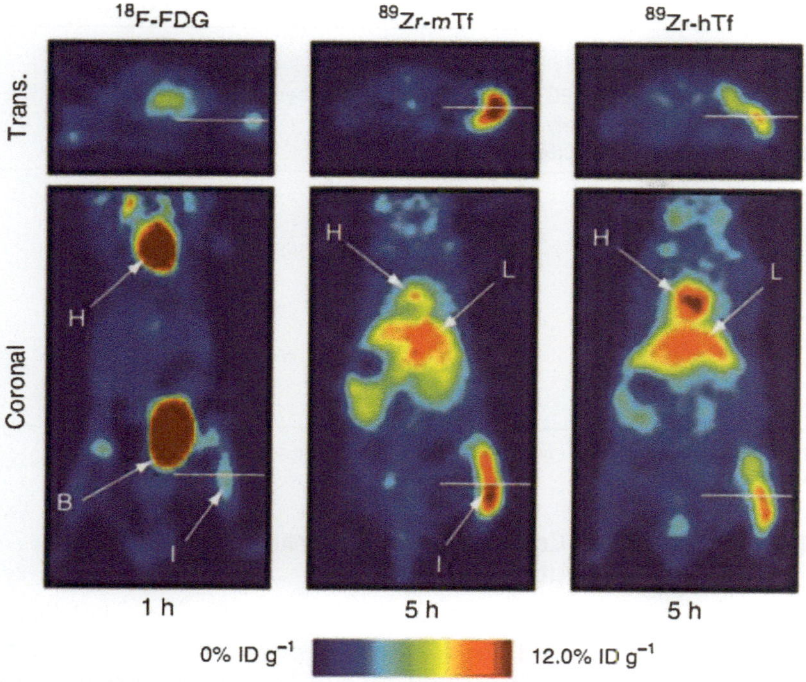

Figure 8.11 Positron emission tomography (PET) imaging of inflammation with ^{89}Zr-transferrin (^{89}Zr-mTf). Comparison of representative transverse (Trans.) and coronal PET images of ^{18}F-labeled glucose (^{18}F-FDG), mouse ^{89}Zr-mTf, and human ^{89}Zr-transferrin (^{89}Zr-hTf) in a subcutaneous, turpentine oil–induced model of inflammation in wild-type immunocompetent male mice. The color scale for all PET image data shows radiotracer uptake in units of %ID/g, with red corresponding to the highest activity and blue to the lowest activity. *B*, bladder; *H*, heart; *I*, inflamed tissue; *L*, liver.

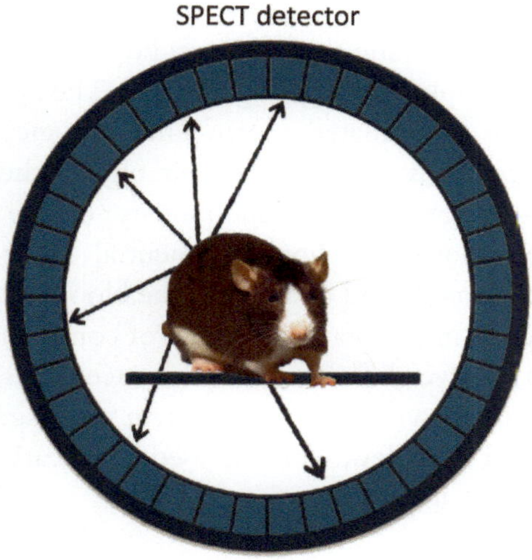

Figure 8.12 Representation of single-photon emission computed tomography detector.[94]

Table 8.4 A selection of SPECT radionuclides; out of these radioisotopes ^{99m}Tc, ^{188}Re and ^{186}Re are the most widely used

Radionuclide	Half-Life (h)	E_γ (keV)	Sources
^{99m}Tc	6.02	141	Generator
^{111}In	67.2	171, 245	Cyclotron
^{67}Ga	78.2	93, 185, 300	Cyclotron
^{67}Cu★	62.0	93, 185	Accelerator
^{186}Re★	89.3	137	Reactor
^{188}Re★	17.0	155	Generator
^{123}I	13.1	159	Cyclotron
^{131}I★	192	365	Reactor

Radioisotopes denoted ★ are also used, or under development, in compounds of simultaneous relevance for therapy.

2.4 Selected Metal Complexes of Relevant Single-Photon Emission Computed Tomography

In the case of ^{99m}Tc traces, their wide use results from its ideal lifetime and because it is available from a ^{99}Mo–^{99m}Tc generator at low cost, first developed in 1959.[155] ^{99m}Tc is widely used in nuclear medicine owing to its nearly ideal nuclear properties. It has a relatively low cost and is conveniently available from commercial $^{99}Mo/^{99m}Tc$ generators, which has led to the use of $^{99m}TcO_4^-$ for clinical applications. Studies showed that Tc in lower oxidation states could be labeled with different ligands to allow imaging. Thus, the use of ^{99m}Tc radiopharmaceuticals increased in medical imaging and a large number of compounds have been developed. There are three generations of ^{99m}Tc radiopharmaceuticals, depending on their physicochemical properties, site-specific targeting, and biological distribution. The first generation of ^{99m}Tc radiopharmaceuticals is used as perfusion agents that display no targeting function. The ligand is biologically inactive without the coordinated metal. The biological distribution of these radiopharmaceuticals depends on their physicochemical properties, such as size and charge, and the lipophilicity of the metal complex. Several studies have shown that lipophilic neutral ^{99m}Tc complexes with tetradentate amine oxime ligand could cross the blood–brain barrier (BBB) without being trapped in the brain. This kind of compounds has led to the commercialization of several ^{99m}Tc radiopharmaceuticals, such as Ceretec (Fig. 8.13) and Neurolite, which are used in brain imaging; Technescan, used in renal imaging, and Myoview, which is used as a heart imaging agent as well as in other clinical applications.

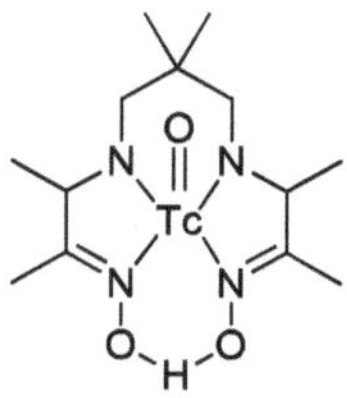

Figure 8.13 "First-generation" technetium radiopharmaceutical Ceretec.

Ceretec has been used to image the effects of stroke because its lipophilicity and lack of charge allow it to cross the BBB. It is trapped by conversion to a more hydrophilic species.[156] Heart imaging agents such as Cardiolite were initially designed as potassium cation mimics to exploit the Na/K ATPase mechanism of uptake within heart tissue, although subsequent studies suggested that a diffusion mechanism with electrostatic binding was the route of uptake. There have also been neutral technetium complexes developed for myocardial imaging, such as TcN-*N*-ethyl-*N*-ethoxydithiocarbamato, which contains a TcN^{2+} core, and which confirms that the monocationic charge is not essential. Other organs have been targeted using first-generation complexes, including the liver and kidney, whereas more recent second-generation complexes target receptor sites or biological functions.[157]

A "second-generation" approach has specific targeting properties in the biologically active molecule to deliver the radionuclide.[157] The development of this second generation requires the use of bifunctional chelating ligands to stabilize the metal center and couple receptor-targeting organic molecules, e.g., antibodies. One example of these new-generation compounds is [^{99m}Tc]-TRODAT-1, which is used as a dopamine transporter imaging agent for the early detection of Parkinson disease and Machado–Joseph disease.

Finally, the most recent (third) generation of ^{99m}Tc radiopharmaceuticals is radiolabeled biomolecules in which the metal core is conjugated to the essential parts of biomolecules as hormones or polypeptides, and in which the physicochemical properties of the biomolecule remain unaffected. Other common SPECT tracers are based on Group 13 elements such as ^{111}In and ^{67}Ga.

Unlike technetium, only one oxidation state is available to indium and gallium under physiologic conditions: the +3 oxidation state.[158] They tend to prefer nitrogen and oxygen donors, and a range of bifunctional

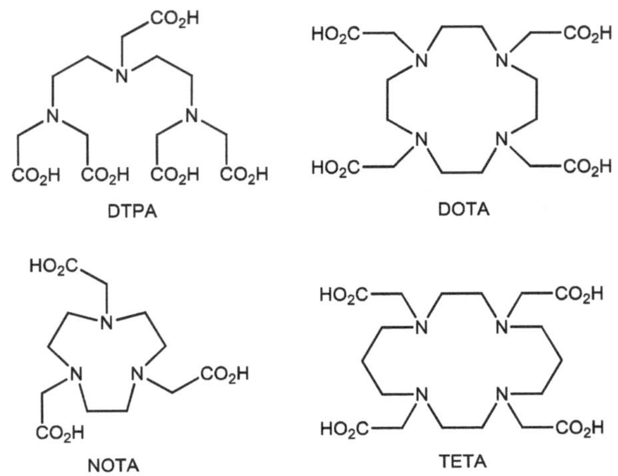

Figure 8.14 Bifunctional chelators used to conjugate gallium and indium. These coordinate in either an N_3O_3 or N_4O_2 manner.[109,111] *DOTA*, 1,4,7,10-tetraazacyclododec ane-1,4,7,10-tetraacetic acid; *DTPA*, diethylenetriaminepentaacetic acid; *NOTA*, 1,4,7-tri-azacyclononane-1,4,7-triacetic acid; *TETA*, 1,4,8,11-tetraazacyclotetradecane-*N′*,*N″*,*N‴*-tetraacetic acid.

chelators containing these have been developed (Fig. 8.14).[111] These overcome some of the problems presented by highly charged cations, such as hydrolysis at $pH > 4$ and ligand exchange with transferrin. This exchange is a particular problem for gallium because Ga(III) has physical properties similar to Fe(III), which is tightly bound by transferrin. This affinity to transferrin has been used to image tumors simply by injecting a gallium salt (such as gallium citrate) and allowing the gallium transferrin complex to form.[159]

2.5 Carbon Nanomaterials as Synthetic Scaffolds for Nanomedicine Design

Carbon nanomaterials are newly emerging allotropic forms of carbon in which at least one of their dimensions is measured on the nanoscale (1–100 nm).[160] The most well-known species in this category are fullerenes, discovered in the 1970s, CNTs, first reported in 1991, and graphene, reported in 2004.[161] Interest in this area has grown exponentially because of the superior properties of these materials and their importance demonstrated by the award of the Nobel Prize in Physics in 2010 to the team that discovered graphene.

Interestingly, there are few reports on in vivo magnetic nanoparticle delivery to the brain for MRI, but the integrated imaging (optical and

MRI, PET, or SPECT) and therapeutic targeting of brain diseases (including gliomas and hydrocephalus), in which multiple agents function synergistically and the efficiency of the chemotherapeutic payload is monitored noninvasively in real-time via a co-incorporated diagnostic probe, remains undocumented. A report is concerned with the BBB crossing (in vitro) by fluorescent high–surface area particles,[162] but there are only few reports on SWCNT uptake in brain cells for cancer therapy (Dai, 2007–08) (Fig. 8.15).[163] Several interesting reports of SWCNT-anchored imaging agents for optical, PET, or SPECT in vivo by Honjie Dai 2007–08 and Kostarelos 2007–08 have emerged.

This chapter also focuses on the use of metal-functionalized CNTs as synthetic scaffolds and their possible application as imaging probes in detecting cancer. Imaging uptake into cells of traceable nanoprobes will lead us to an understanding of processes that need to be considered for the successful targeted delivery of drugs to the sites of action (while minimizing systemic toxicity and enzymatic degradation of the probe). The imaging agents (suitable for MRI and/or PET) will be encapsulated into stable SWCNT carriers derivatized so that they can be guided to target cells. Using nanotube carriers prevents the loss of imaging agents to nonspecific sites, enabling metal ions to reach the target efficiently and simultaneously while keeping the doses necessary for use to obtain an image extremely low. For complete functionality, the nanotube probes are tailored here with groups modulating their lipophilicity, with fluorescent tags to monitor their delivery and uptake into tumors, and with biological "address" units to link to the target cells. A specially designed synthetic protocol aims to ensure a highly optimized biocompatible coating of the nanoprobes. These target functionalities are summarized in Fig. 8.16, which shows a schematic of a biocompatible "coated" nanotube probe scaffold. Success in tailoring the probes and the robustness of successive scaffolds can be evaluated at each step through physical characterization.

CNTs can be regarded as a wrapped sheet of graphene with diameters around 1 nm and lengths that can go above 1 μm.[164] The structure of a CNT is characterized by its chiral angle, θ, and two vectors (n, m); depending on the values of these parameters, a specific "helicity" is obtained when "wrapping" the graphene sheet.

According to this, three types of CNTs are observed when varying these values. Two are simpler: the first one with (n, 0) generates the "zigzag" structure, and the other when n=m (n, n) generates the "armchair" structure.

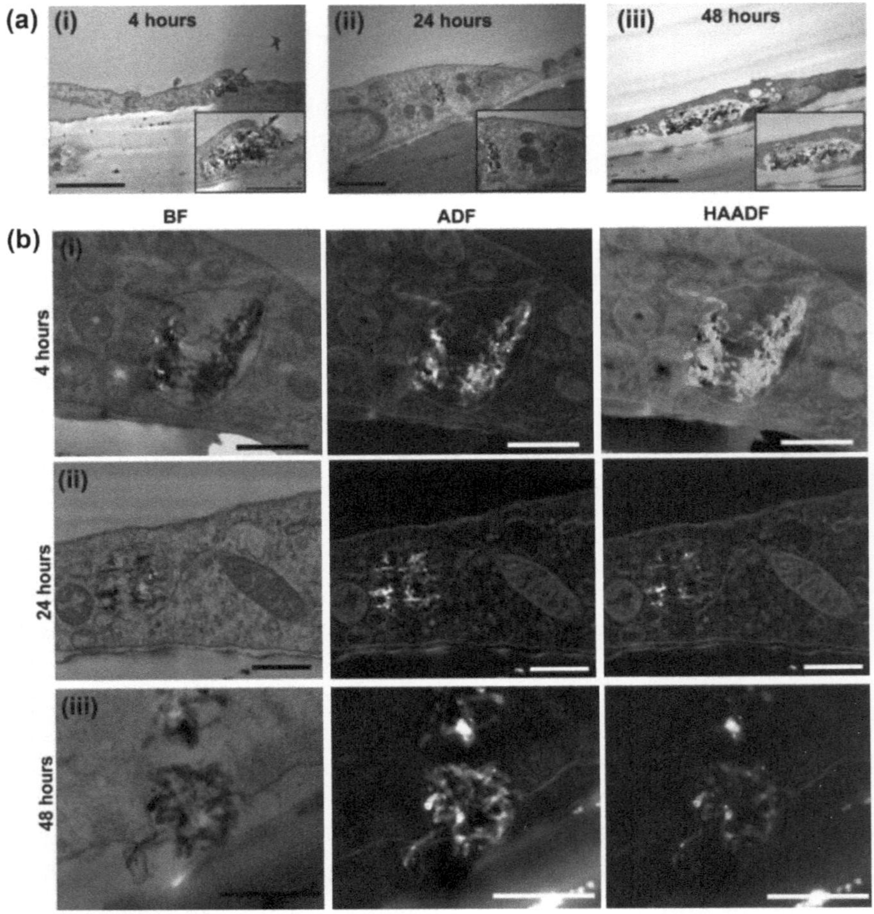

Figure 8.15 Transcytosis pattern of multiwalled carbon nanotubes (MWCNTs)–NH_3^+ across the primary bronchial epithelial cell (PBEC) monolayer after incubation of MWCNTs–NH_3^+ with an in vitro blood–brain barrier model. MWCNTs–NH_3^+ (20 μg/mL) were added to the apical chamber and incubated with the PBEC for (i) 4 h, (ii) 24 h, and (iii) 48 h. (a) Brightfield (BF) transmission electron microscopy images of polyester filters showing initial interaction of MWCNTs–NH_3^+ with the PBEC monolayer after 4 h (i). After 24 h (ii), the MWCNTs–NH_3^+ clusters appeared within endocytic vesicles or multivesicular bodies. The MWCNTs–NH_3^+-containing vesicles showed evidence of fusion with the abluminal plasma membrane and were partly open toward the basal chamber after 48 h (iii). (b) Low-voltage scanning transmission electron microscopy images of the polyester filters confirming the uptake and transcytosis of MWCNTs–NH_3^+ across the PBEC monolayer. MWCNTs–NH_3^+ appeared within endocytic vesicles after (i) 4 and (ii) 24 h. The MWCNTs–NH_3^+-containing vesicles were then imaged partly open toward the basal chamber, allowing the release of MWCNTs–NH_3^+ after 48 h (iii). The three detection modes: BF, annular darkfield (ADF), and high angular annular darkfield (HAADF) helped identify the electron-dense structures of MWCNTs–NH_3^+ within the cell body and the osmium-rich lipid membranes surrounding the nanostructures. Scale bars: (a) 1 μm and (*inset*) 500 nm. (b) First and second row: 500 nm; last row: 400 nm. *Reproduced with permission from Elsevier Kafa, H.; Wang, J.T.-W.; Rubio, N.; Venner, K.; Anderson, G.; Pach, E.; et al. The Interaction of Carbon Nanotubes with an In Vitro Blood–Brain Barrier Model and Mouse Brain In Vivo.* Biomaterials ***2015,*** 53 *437–452.* http://dx.doi.org/10.1016/j.biomaterials.2015.02.083.

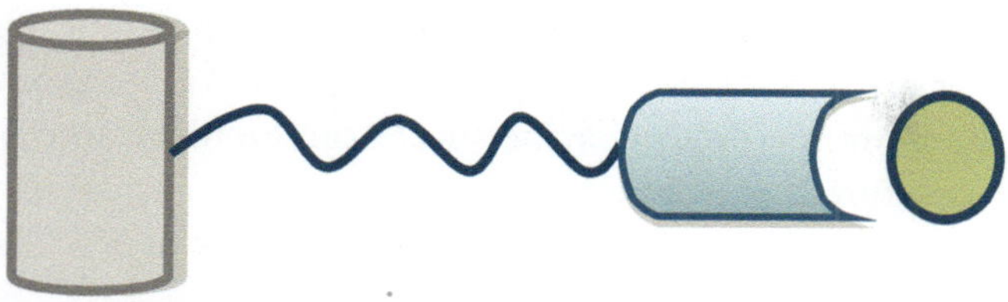

Figure 8.16 General representation of an imaging probe containing a nanomaterial and a targeting molecule to bind a specific analyte.

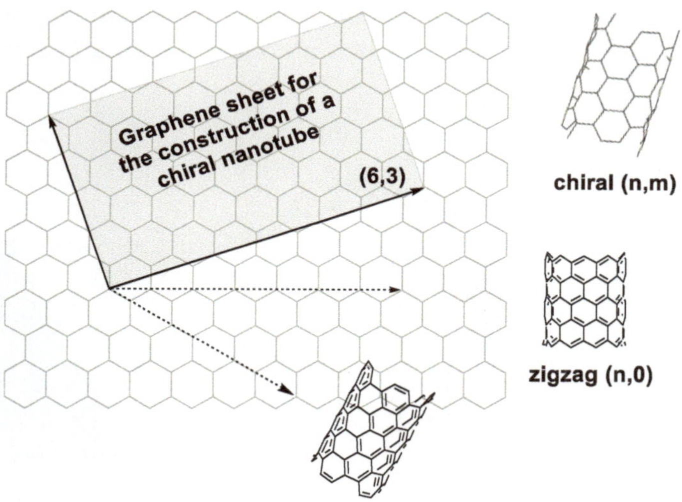

Figure 8.17 Different nanotube types emerging from a graphene sheet. Possible vectors arranged by pairs of integers (n, m) for a general structure of single-walled carbon nanotubes (the alternating double bonds in the sp^2 graphene layer have been omitted for clarity).[192] Finally, once the integer is any number between 0 and n, i.e. (n, m), the structure is known as chiral.[190,191]

The last nonstandard CNT is generated when $n > m > 0$, and the structure is called chiral[165] (Fig. 8.17). The chirality is a factor to be considered because it determines the final properties of the nanotubes. This is especially relevant for their conducting properties: Armchair nanotubes are metallic, zigzag tubes are metallic if n is a multiple of three, and chiral tubes can be either semiconducting or metallic.[164]

SWCNTs consist of a single graphene layer that is wrapped according to one of these methods, whereas MWCNTs are formed from more than one tube, placed one inside another. The simplest MWCNTs is a double-walled

CNT; when adding more tubes, the diameter can go up to tens of nanometers.[165] The shape of SWCNTs, seen as a single layer of hexagonal carbon atom ring frameworks (or a graphene sheet), can be rolled into a seamless cylindrical structure that normally has spherical ends (analogous to hemifullerenes). Many studies indicated that the properties of SWCNTs were correlated to their structure.[188,189]

MWCNTs were first reported by Ijima in 1991,[166] and SWCNTs were described 2 years later in 1993[167]; since their discovery, the range of reactions, applications, and potential applications of CNTs has grown exponentially. In Fig. 8.18, a representation of the number of publications including the terms "carbon nanotubes," "carbon nanotube and drug delivery," and

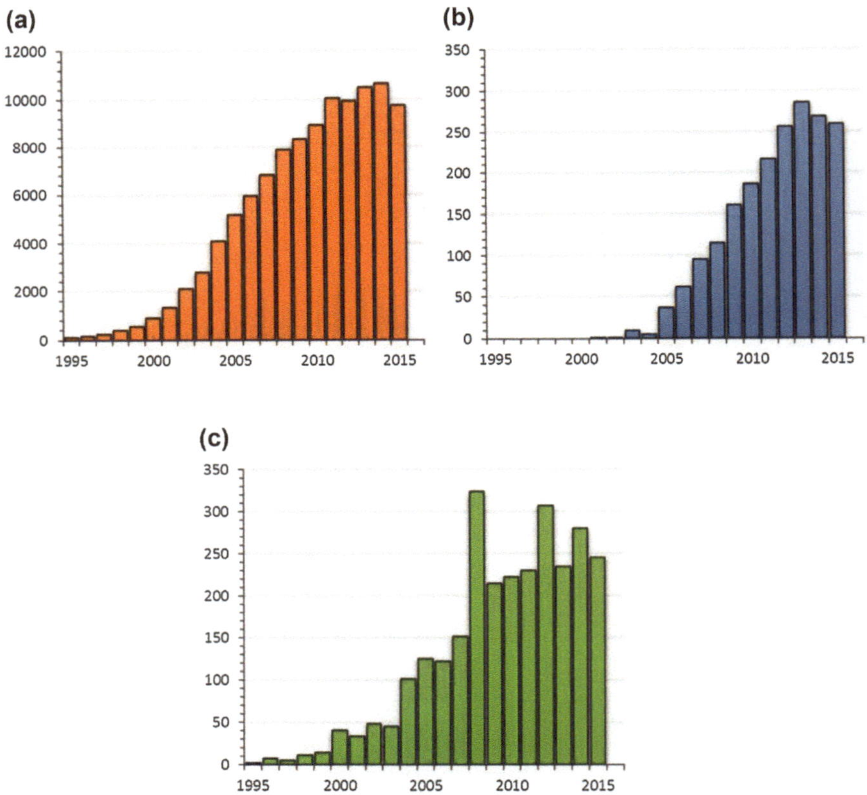

Figure 8.18 Publications containing the concepts: (a) "carbon nanotubes," (b) "carbon nanotubes and drug delivery," and (c) "carbon nanotubes and imaging" between 1995 and 2015. *Obtained using Scopus database, date of search May 4, 2016.*

"carbon nanotubes and imaging" can be observed. The number of publications reached a maximum in 2011 and has been maintained to date in approximately 10,000 research articles each year.

CNTs are synthesized primarily by three methods: two at high temperature and one at low temperature (<800°C). The first synthetic method is "arc discharge"; this method was employed by Ijima when he reported CNTs for the first time.[166] It consists on applying a voltage between two graphite electrodes in an inert atmosphere. The high temperature generated evaporates carbon from the electrodes, and when it is condensed again, the nanotubes are formed.[168] No catalyst is needed when synthesizing MWCNTs, but to obtain SWCNTs, a transition metal catalyst such as Co, Ni, or Fe is used.[169] The nanotubes obtained by applying this procedure have amorphous carbon and rest of the metal catalyst as impurities.[170]

The second synthetic method is a high-temperature method denoted laser ablation. It is based on the same principle as arc discharge, the condensation of carbon from the vaporization of graphite, but in this case it uses a laser source, normally a pulsed Nd:YAG laser, to vaporize the carbon present in graphite. The process is carried out inside the quartz reactor of a furnace that operates at high temperatures (900–1200°C) with a current of inert gas. Normally, the operation at higher temperatures (close to 1200°C), in the presence of a metal catalyst, favors the formation of SWCNTs.[171,172] Laser ablation is known to produce high-quality CNTs with a low impurity level; nevertheless, the yield is really low.[172]

The most widely applied procedure in the synthesis of CNTs is chemical vapor deposition (CVD) (method 3). The process involves heating a supported catalyst to 600–1000°C while passing a hydrocarbon gas. As in the previous technique, MWCNTs are formed when applying lower temperatures (300–800°C), and higher temperatures are needed to synthesize SWCNTs[168] using catalytic metallic nanoparticles.[173] Many precursors are normally employed in the CVD process, such as benzene, methane, ethylene, and xylene, but in the synthesis of SWCNTs carbon monoxide (high-pressure carbon monoxide) or ethanol is normally the source of carbon.[170] The precursor has to be chosen carefully because the nature of the precursor will have an effect on the final CNT morphology,[174] and thus the scaffolds' nanotoxicity in cells.

Polyhedral carbon and graphitic particles as impurities are more difficult to remove.[175] The most widespread purification methods involve refluxing with strong acids such as HCl, HNO_3, or even piranha mixture (a solution that contains a 4:1 concentration of concentrated sulfuric acid and hydrogen peroxide).[176] Although these methods help to purify CNTs from amorphous carbon and catalytic particles, they introduce oxygenated functions and cause oxidative damage to the sidewall (the nanotubes can be cut and open).[177]

There is increasing evidence that under certain conditions, appropriately functionalized SWCNTs can be rendered nontoxic while intracellular.[34] Increasingly, research communities became aware of the requirements of these nanomaterials when they were considered for drug delivery applications, and SWCNTs have emerged as the most favorably regarded in the family, unlike MWCNTs and double-walled CNTs. This stretch from the reproducibility of industrially purchased batches to the necessity of rigorous purification and consistency in diameter and length. Unfunctionalized SWCNTs with good batch-to-batch reproducibility need to be obtained from reliable commercial sources, and rigorous attention is typically being paid to purifying the nanotubes to eliminate toxic catalytic particles and other carbonaceous nanostructures emerging from the synthetic processes described earlier.[178] Established, reproducible procedures[178] exist and provide bulk physical inspections and local understanding of structures and purity by using Raman spectroscopy, high-resolution transmission electron microscopy (HR-TEM) and atomic force microscopy (AFM). The further requirement of homogeneous lengths for nanoprobes remains to be achieved, likely using standard cutting procedures and chromatographic separation and/or microfluidic chips[179] to fulfill the aim to employ CNTs of consistent length (maximum 100–200 nm and diameter of about 2 nm) in radio-nanochemistry research and/or multimodal imaging techniques that rely on SWCNTs as synthetic scaffolds.

3. POTENTIAL APPLICATION OF CARBON NANOMATERIAL NANOHYBRIDS INCORPORATING INORGANIC TRANSITION METAL COMPLEXES IN BIOMEDICAL APPLICATIONS

This application raises the concern of the possible toxic effects these materials can have upon contact with living systems. CNTs are highly hydrophobic and nonbiodegradable. There is no general agreement about CNTs' toxicity because their physical parameters strongly depend on the route of synthesis, purification, and functionalization applied.

Exposure routes to CNTs can be by the skin, oral, or aerial. The last is the most studied because it is considered the main exposure route leading to toxicity.[180] There is a great interest in determining whether CNTs act like asbestos fibers. Therefore, it has become essential to determine the structure in which CNTs are found in the moment of potential exposure. Normally, SWCNTs do not exist as individual tubes but tend to agglomerate in clumps of 20–50 nanotubes because of van der Waals attraction.[180] However, the effect of CNTs in the lungs has been extensively studied and there is general agreement that they can cause damage by oxidative stress or pulmonary inflammation.[45,180]

Related to this topic, the fiber pathogenicity paradigm is a structure to toxicity model that predicts the pathogenicity of fibers. This model mainly considers the thinness, length, and biopersistence of such fibers. It was applied to the toxicity of asbestos fibers and can be considered in nanotubes because they have a diameter small enough to enter the airways by inhalation and are biopersistent over the long term in the lungs. Furthermore, the effects in the lungs are length-dependent because longer-shaped nanotubes would not be completely enclosed by macrophages causing oxidative stress. Therefore, when CNTs are considered as particles, the adverse effects would be limited to the lungs, but when they act as fibers they can affect the pleura as well.[181]

As stated previously, the toxicity of CNTs is subjected to a list of physical properties such as composition, size and shape, diameter and length, agglomeration, structure, and surface modification. The presence of metals such as Fe, Co, and Ni, could be a source of toxicity if the metals are released in vivo, especially with SWCNTs, because they contain metals as an impurity generated during synthesis. The shape can also affect the toxic effects; the open ends of shorter SWCNTs have more possibilities of interacting with the membrane and cause cell damage.[180]

Despite intrinsic toxic effects, there is a large effort to make CNTs biocompatible, and many biomedical applications have appeared.

The application of CNTs in medicine passes by increasing solubility in aqueous solvents. This can be achieved by oxidation, covalent functionalization, or the use of some dispersant; some can enhance the solubility of CNTs up to the g/mL scale.[182] There are many examples in vitro and in vivo of functionalized CNTs that are not toxic. The nanomaterial is excreted in vivo and no cell death is observed in vitro. Even some examples of biodegradation of CNTs have appeared in the literature.[183]

CNTs have created great interest in nanomedicine because they have proven to cross cellular membranes at high efficiencies functionalized with specific molecules that conserve their activity once inside the cell.[184] For this reason, many applications have opened up for CNTs. They can be used for drug and gene delivery as carriers, cellular growth substrates, and tissue scaffolds, in vivo imaging, or disease treatment by hyperthermia.[184,185] The construction of biosensors based on CNTs is also being investigated.[160]

In treatment, they can be applied to destroy cancerous cells because they absorb in the NIR region where the tissues are almost transparent. When the CNTs are localized in malignant tissue and irradiated with NIR radiation, they can be heated and destroy the specific tissue.[186]

This same principle is applied for biomedical imaging where nanotubes can take advantage of their NIR fluorescent properties to act as imaging probes. Not just in NIR, CNTs can be applied to a range of imaging techniques such as Raman, MRI imaging, and even nuclear imaging when they are functionalized with a chelating agent. Several reviews have recently covered the use of CNTs in the bioimaging and biosensing fields.[187]

3.1 Impurities Typically Found in as-Made Single-Walled Carbon Nanotubes and Their Potential Health and Environment Risks

Regardless of the producing methods used, because synthesized SWCNT samples contain a certain amount of impurities, including amorphous carbon, fullerenes, metallic nanoparticles, multishell carbon nanoparticles, and nanocrystalline graphite, various types of contamination depend on the support, such as silicon coming from Si substrates. These SWCNTs cannot be used directly in most applications because the presence of those impurities after synthesis will cause many problems, especially in biomedical applications. Major problems come from metallic catalysts, which are highly reductive and lead to a significant rise in the reactive oxidative species (ROS) level in living systems.[194]

ROS refers to free radicals and other oxygen-related reactive compounds, including singlet oxygen ($^1O_2$), hydrogen peroxide (H_2O_2), and hydroperoxide (ROOH).[195,196] All of these compounds are highly reactive to nearby environments, which makes ROS able to initiate free radical reaction to convert a nonradical molecule to a radical. In the beginning of ROS generation, molecular oxygen is reduced by a reducing reagent, such as metallic particles, to form a superoxide radical ion O_2^-. Then a series of free radical chain reactions will be initialed by this superoxide.[197] Formation of hydrogen superoxide in cellular mitochondria may be described by the reaction in Scheme 8.1.

Scheme 8.1 shows the simple reaction at the basis of the formation of a hydrogen peroxide (H_2O_2) from superoxide (O_2^-). In mitochondria, superoxide is further reduced to hydrogen peroxide and molecular oxygen, which is catalyzed by superoxide dismutase. H_2O_2 is more kinetically and thermodynamically stable than superoxide ions and can diffuse to the cytoplasm to enhance radical reactions, unless it has been eliminated by catalase or glutathione scavenger biomolecules. At the end of the radical chain reaction, the most reactive and harmful radical, hydroxyl radical (OH·), is produced from H_2O_2 via the Fenton reaction and Haber–Weiss reaction catalyzed by transition metals, as shown in Scheme 8.2.[198] The hydroxyl radical has extremely short lifetimes (around 1 ns): this makes this radical species extraordinary reactive in living cells have the ability to damage all the biomolecules, including DNA, RNA,[199] lipids, proteins[200] and carbohydrates.[201] In fact, hydroxyl radicals are normally consumed at the site of formation, hence, it makes hydroxyl radical impossible to eliminate by enzymatic reactions.

As a result of ROS level upregulation, there are a series of consequences outlined in Fig. 8.19.[202] Both hydrogen peroxide and hydroxyl radical can damage DNA, and radicals can oxidize purine and pyrimidine bases to introduce strand breaks and mutations. Poly-ADP-ribose polymerase is subsequently activated to repair the damage via depleting malfunctioned DNA.[203] Radicals such as hydroxyl radicals can oxidize sulfhydryl groups and amino acids, leading to the formation of a defective protein. In the same way, hydroxyl radicals can also initiate lipid peroxidation; it converts unsaturated lipids to lipid peroxide, and as a consequence, it leads to the disruption of the lipid phosphate bilayer structure and a decrease in the membrane fluidity, while increasing the structure of membrane permeability and

$$O_2^{-\cdot} + O_2^{-\cdot} + 2H^+ \rightarrow H_2O_2 + O_2$$

Scheme 8.1 Formation of hydrogen superoxide in mitochondria.

$$Fe^{2+} + H_2O_2 \rightarrow Fe^{3+} + OH^{\cdot} + OH^-$$

or

$$O_2^{-\cdot} + H_2O_2 \xrightarrow{\text{metal catalyst}} O_2 + OH^{\cdot} + OH^-$$

Scheme 8.2 Formation of hydroxyl radical from hydrogen peroxide.

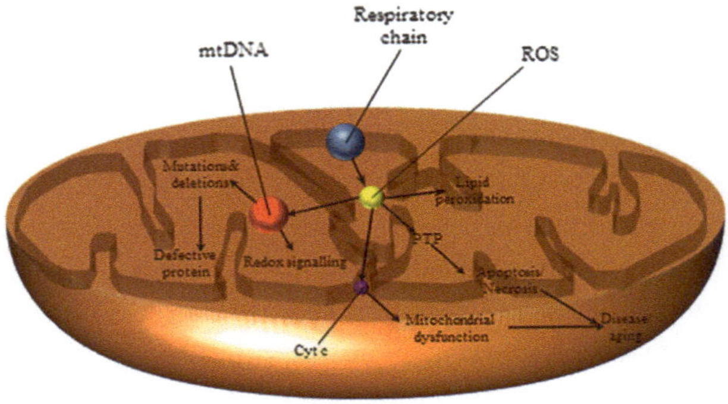

Figure 8.19 Schematic representation of effect of reactive oxidative species (ROS) in cellular mitochondria.[202] *Cyt c*, cytochrome complex.

modifying the membrane proteins.[204] All of these modifications will cause cell death by necrosis. On the other hand, ROS can also activate redox signaling, release protein cytochrome complex (e.g., cytochrome complex), and open permeability transition pores to initiate cell death via apoptosis. In general, both methods of cell death have been shown to lead to degenerative diseases or accelerated aging in living organisms.[205–207]

Carbonaceous nanomaterials such as amorphous carbon, fullerenes, multishell carbon nanoparticles, and nanocrystalline graphite are also known to cause problems, and their presence limits the applications of SWCNTs in nanomedicine. Although these carbonaceous nanomaterials will not cause direct damage to cells or tissues, they can attach to SWCNTs (or self-aggregates) to encapsulate metal catalysts. As aggregates, they reduce the solubility of the nanomaterials, which makes functionalization of SWCNTs impossible. Many articles have pointed out that pristine SWCNT strands in general will not cause drastic inflammatory or immune responses, until SWCNTs are aggregated and have a significant larger diameter.[208,209]

In that case, SWCNTs will be detected by the immune system with the aim of eliminating them from the living organisms. In addition, impure SWCNTs administrated in vivo can activate platelets in blood vessels to cause thrombosis.[210] Finally, carbonaceous impurities also affect the functionalization process of SWCNTs, leading to a decrease in the degree of functionalization and solubility of the functionalized SWCNTs. Therefore, purification of SWCNTs to remove metallic nanoparticles and carbonaceous impurities is an essential step before functionalization and further biomedical application of SWCNTs can be considered.

3.2 Design of Biocompatible Fluorescent Sensors

A fluorophore is itself normally nonselective, nonspecific, and nonsensitive to meet a given biomedical purpose; and as mentioned previously, many fluorophores can introduce ROS during the process of photodynamic photobleaching. Hence, further modification of the fluorophore is required to make fluorescent molecules more biocompatible.[211]

As it can be seen in Fig. 8.20, a typical fluorescent sensor designed by James and Pascu typically consists of three parts including a fluorescence group, a spacer, and a targeting group. The fluorescence group is the core part of the fluorescent sensor, which absorbs light energy excited to give a fluorescent signal; the wavelength of the light emission depends on the chosen fluorophore.[211–213]

There are six commonly used commercial fluorophores that absorb and emit light and that cross the whole visible spectrum (Fig. 8.21). These fluorophores can absorb light at individual excitation wavelengths and emit at a characteristic wavelength. For instance, in the case of fluorescein, an excited state is obtained by absorbing light at 488 nm. Subsequently, a fluorescent emission takes place at 524 nm in an aqueous environment. In general, most fluorophores used in research are excited between 300 and 450 nm, which gives a refined fluorescent signal. However, this is not perfect for biomedical imaging applications because the light source is high energy and will not only bleach fluorophores but also cause damage to the tissue of a living organism. As mentioned, high-power lasers increase oxidative stress levels in living cells and tissues, leading to mutations in cells and resulting in cell apoptosis or tumors. Hence, NIR fluorophores are preferred for labeling

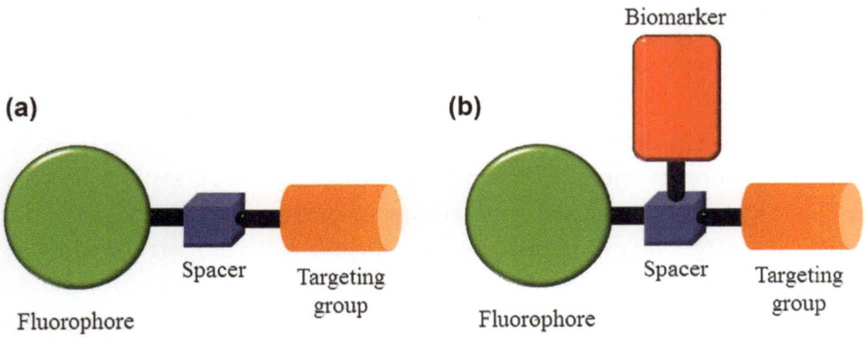

Figure 8.20 Schematic representation of a typical fluorescent sensor developed (a) and linked to a bio-targeting functionality (b).

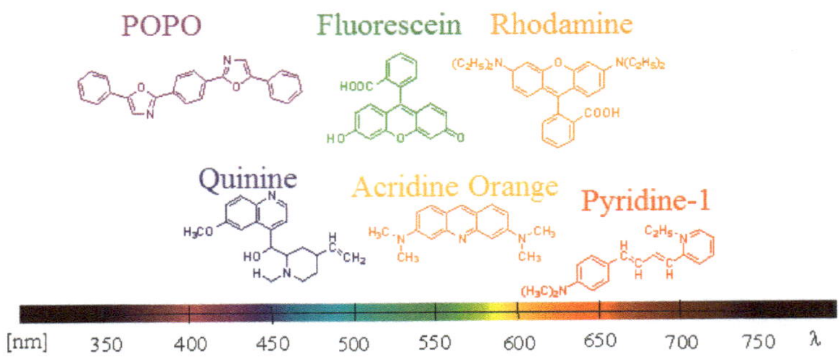

Types of fluorophore	Emission wavelength (nm)
POPOP	409
Quinine	451
Fluorescein	524
Acridine orange	526
Rhodamine B	615
Pyridine 1	650

Figure 8.21 Selection of common fluorophores used for biomedical applications and their different maximum emission wavelengths. POPOP, 1,4-*bis*(5-phenyloxazol-2-yl) benzene.

biological molecules whereas a low-energy light source raises another problem: most fluorescent molecules will not be excited; as a result, the intensity peak of the spectrum will be significantly lower than those of the UV side fluorophores. This also explains why a fluorescent sensor structure is developed, because anchoring a fluorophore to a spacer not only increases the distance between the fluorescent molecule and receptor molecule to avoid the interruption of the receptor molecule to the fluorophore; but also the signal of the fluorescence, which also varies depending on the length of the spacer. The targeting group on the sensor is applied to recognize a molecule of interest; for example, the molecule may be attached to a boronic acid group, which has a considerably high affinity for diols, and hence specificity for a variety of sugar-based biomolecules. Therefore, this type of molecular sensor will have the sensitivity to detect and target monosaccharides, polysaccharides, and glycoproteins in biological experiments.[214] Binding to these target molecules has been shown to lead to an electrochemical change of the receptor molecule. This alteration of the electronic state is conducted via the spacer to the fluorescent molecule and initiates either enhancement or quenching of the fluorescence, which in turn can be monitored by fluorescence microscopy.

In addition, there has been some work involving the use of an extra functional group, which was applied onto an earlier sensor design from the Pascu group. This is a biomarker and enables the fluorescent sensor to be applied for biological purposes, for example, to detect and label cancer cells. Biomarkers can vary depending on the cancer type and typically have high affinity to bind to those cancer cells.[215] Unlike common sensors based on photo-induced electron transfer, a sensor incorporated with a biomarker is a dual targeting sensor. Therefore, fluorescence can be switched on and off but also enhanced or quenched once the sensor binds to the selected biological target.

3.3 Applications of Single-Walled Carbon Nanotubes in Chemotherapeutic Drug Delivery

Nanomedicines, including those based on functionalized SWCNTs, have been considered excellent tools for cancer research because of their quasi–one-dimensional nanostructure; unique optical, mechanical, and electronic properties; ultrahigh surface area; remarkable cell membrane penetrability; and facile functionalization by different methods. Hence, they are highly biocompatible and have been used in biomedical detection and imaging and CNT-based therapeutics, including drug delivery, thermal therapy, and gene therapy.[216]

Fully functionalized SWCNTs are ideal vectors in many research applications, such as sensors, probes, actuators, composites, nanoelectronic devices, and drug delivery systems within biomedical applications.[217] In research SWCNTs have been applied as a sensor synthetic scaffold and drug carriers or nanocapsules for biomedical research. In terms of the molecular vector aspect of SWCNTs, their behavior depends on the method of functionalization. Functionalized SWCNTs have the potential to carry small molecules, peptides, proteins, genes, and DNA across cell membranes, often with little cytotoxicity.[218]

Epidermal growth factor (EGF) and EGF receptors, for example, are normally low on healthy cell surfaces; however, they are often overexpressed on cancer cells, and hence it is an ideal target site for cancer recognition. Table 8.5 lists a series of cancer receptors that may be

Table 8.5 Cancer Target Receptors and Their Corresponding Biomarkers and Anticancer Drugs

Target Receptors	Biomarker	Anticancer Drugs	Type of Carbon Nanotube	References
Early growth response receptor	Epidermal growth factor	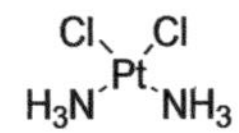 Cisplatin	SWCNTs	220
Tyrosine kinase-7	Sgc8c aptamer	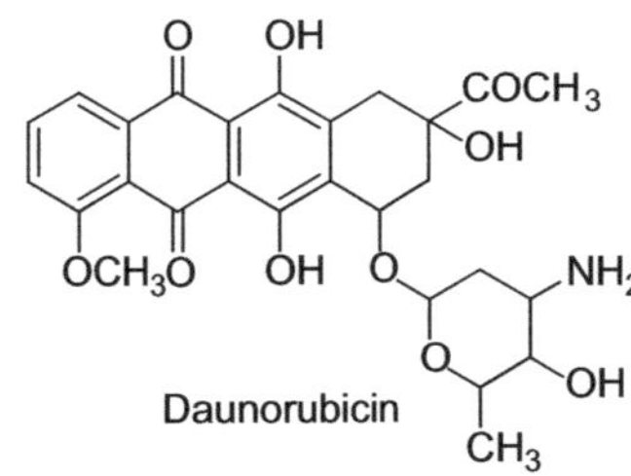 Daunorubicin	SWCNTs	221
Endothelial cells	Asparagine–glycine–arginine (NGR)	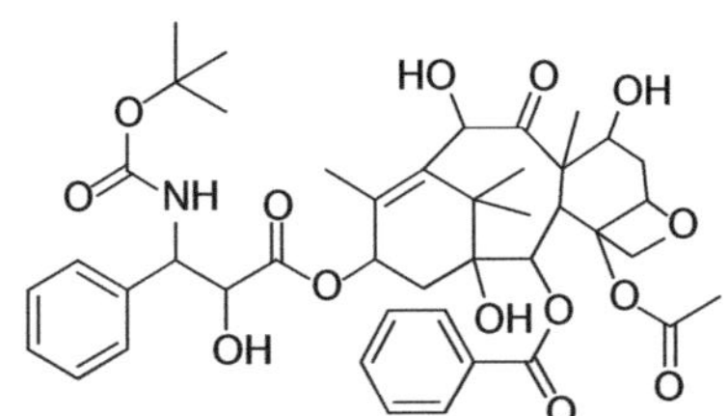 Docetaxel	SWCNTs	222
Folate receptor	Folate	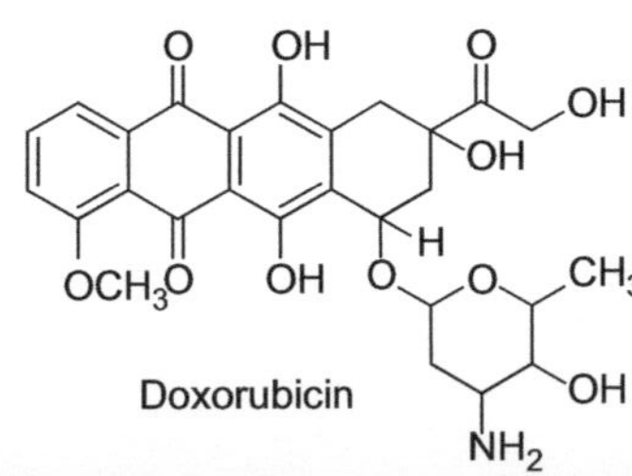 Doxorubicin	MWCNTs	223

Integrin $\alpha v\beta 3$	Arg-Gly-Asp (RGD)	Doxorubicin	SWCNTs	224
Folate receptor	Folate	Doxorubicin	SWCNTs	225
Folate receptor	Folate	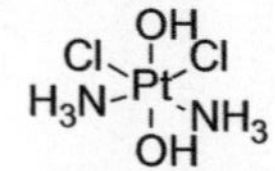 Platinum (IV) anticancer complex	SWCNTs	226

MWCNTs, multiwalled carbon nanotubes; *SWCNTs*, single-walled carbon nanotubes.

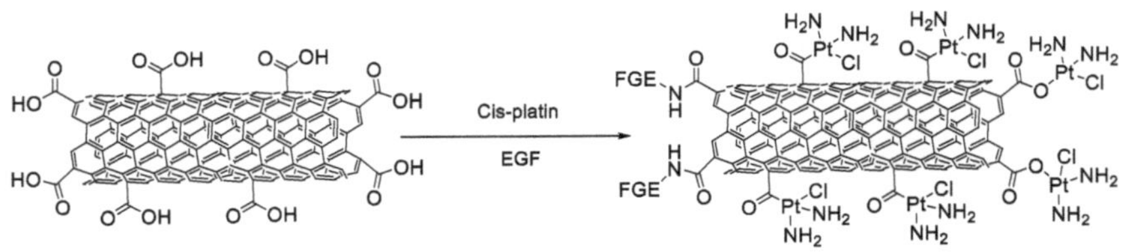

Figure 8.22 Schematic representation of the general functionalization of single-walled carbon nanotubes with *cis*-platin and epidermal growth factor (EGF) factors yielding a drug delivery system attached to a cancer cell.[219] *FGE,* formylglycine-generatingenzyme.

overexpressed in cancerous tissues and their corresponding biomarker and anticancer drugs. A research report was concerned with the modality in which a SWCNT-based device can be designed as a drug delivery system (DDS) toward cancer cell targeting. First, SWCNTs can be functionalized via a covalent method; carboxylic groups can subsequently be introduced onto the open ends and sidewalls of CNTs. Then, an EGF was added to SWCNTs via formation of carboxyl-amide bonds: this functionalized SWCNT, including its biomarker complex, is known as an immunosensor (Fig. 8.22). This can help in targeting of the recognition molecules on the surface of cancer cells.

Cis-platin, as well as the other anticancer drugs shown in Table 8.5, can be tagged onto SWCNTs and delivered to cancer cells. *Cis*-platin is attached to the defect site of CNTs via ester bonds. The whole DDS is modified to improve water solubility and to deliver selectively to cancer cells. Once EGF on the functionalized SWCNTs binds to the EGF receptors, the whole DDS will be endocytosed into the cancer cell. Anticancer drugs have been found to be released inside cells and subsequently terminate cancer cells.[219] Research work into the cellular biocompatibility of SWCNTs started almost immediately once SWCNTs were functionalized in the study of their potential application to the biomedical field. Lacerda et al. showed that SWCNT-based DDS can easily cross the BBB with no external transporter device.[227] The SWCNT DDS commonly enters cells rapidly in the first hour of internalization and distribute in the cytoplasm. There are generally two pathways for the cellular uptake of SWCNTs: (1) direct insertion or passive diffusion through the cell membrane, and (2) indirect incorporation via endocytosis. The choice of uptake route depends on the size of the SWCNTs. In general, individual SWCNTs can enter the cell membrane directly, whereas bundles of SWCNTs have been shown to enter cells by endocytosis.[228] For in vivo experiments, apart from conventional oral and injection administration routes, there are also two other ways: namely, transdermal and ophthalmic drug delivery. The former is designed based on the tiny needle shape of SWCNTs and super drug load capacity.

Im et al. assembled an electro-sensitive transdermal DDS that combines CNTS with polyethylene oxide and pentaerythritol triacrylate polymer and uses an electro-spinning method to control drug release.[229] As for the SWCNT-based DDS, these have been targeted for ophthalmologic applications designed for several eye diseases. Ophthalmic drugs commonly have a short residence time (<5 min) and only 1–5% of applied therapy may penetrate the cornea and be applied to intraocular tissue. However, CNT-enhanced ocular DDS has significantly improved the bioavailability of a few ophthalmic medications that have the potential to overcome the limitation of traditional ophthalmic DDS. CNTs could also be used to target different therapeutic agents to different ocular sites.[194,230]

3.4 Applications of Functional Single-Walled Carbon Nanotubes in Cellular Imaging

The unique physical properties of SWCNTs and their facile functionalization make them extremely popular in the biomedical imaging field. SWCNTs can be applied in bioimaging areas in two different ways: (1) SWCNTs can be used directly as a contrast reagent; (2) SWCNT materials are used as synthetic scaffolds to carry imaging probes to the site of interest. The former is applicable to imaging techniques, including fluorescence imaging, Raman and photoacoustic (PA) imaging. Semiconducting SWCNTs have a narrow band gap, which allows SWCNTs to emit fluorescence in the NIR regions of the spectrum, between 700 and 1400 nm.[231] Similarly, SWCNTs contain strong resonance Raman scattering with an extremely large scattering cross-section. This makes SWCNTs promising Raman imaging probes in biological sensing and imaging techniques.[231] PA imaging is an imaging technique in which non-ionizing laser pulses are delivered into biological tissues. Some laser energy will be absorbed and converted into heat, resulting in ultrasonic emission. This ultrasonic emission is collected by ultrasonic transducers and then analyzed to produce images.[232] The latter applications of SWCNTs in bioimaging areas include MRI,[233] PET,[48] and optical imaging. These techniques mainly depend on contrast agents as probes attached onto the surface of SWCNTs.

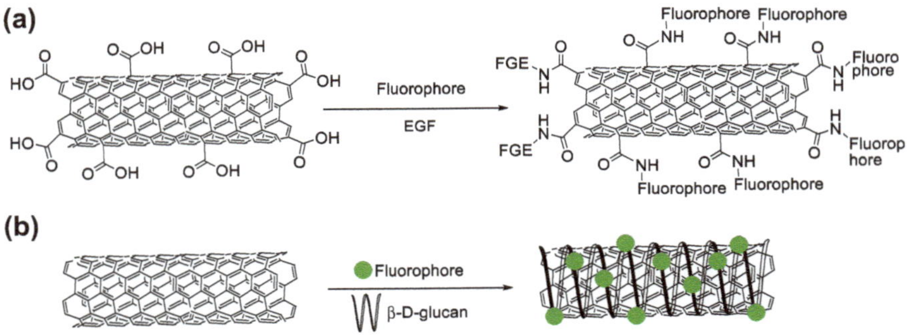

Figure 8.23 Schematic representations of fluorescent sensor–single-walled carbon nanotube complex used in cancer imaging. (a) Covalent method. (b) Noncovalent wrapping method with polysaccharides.[235] *EGF*, epidermal growth factor; *FGE*, formylglycine-generatingenzyme.

Fig. 8.23 represents a covalent method of assembling DDS probes, which is similar to the assembling method shown in Fig. 8.22. As such, SWCNTs are oxidized and carboxyl groups are introduced; subsequently, both EGF and fluorophores are loaded at the defect site and at the end of SWCNTs. Hence, the complex can selectively target cancer cells and is traceable in vitro. Instead of covalently inserting functional groups, a non-covalent wrapping method is normally adopted to coat SWCNTs with a water-soluble polymer, such as PEG[234] or β-D-glucan[235] (Fig. 8.23b). Compared with covalent functionalization, which allows the use of only a limited number of defect site and ends, glucans have many binding sites available on their surface. Hence, the loading capacity of fluorescent sensors onto glucan-functionalized SWCNTs is likely to be improved. In addition, SWCNTs have a strong quenching effect on many fluorophores; hence, the direct attachment of fluorescent molecules will cause the loss of fluorescence emission. Furthermore, pristine SWCNTs have been shown to be toxic to cells. Therefore, coating SWCNTs with biocompatible polymers has been of interest for biological experiments that use these materials.[236,237]

SWCNTs were covalently amine functionalized and the radio-metal-ion chelates DFO or DOTA were attached onto the surface (Fig. 8.24). This system was used to chelate ^{89}Zr produced in a synchrotron. Such a nanoprobe also included an antibody for targeted delivery to tumor tissue where new vasculature was being created. The work presented by Ruggero et al. demonstrated how incorporation of the SWCNT scaffold into the

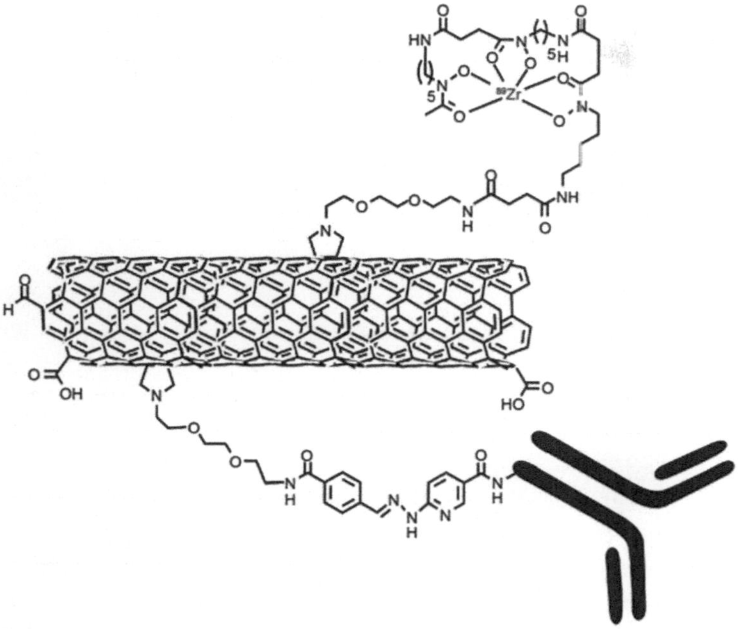

Figure 8.24 Graphical representations of key moieties appended to the water-soluble single-walled carbon nanotube–NH_2 by covalent functionalization with radionuclides and antibodies. *From Zhang, S.; Bao, A.; Sun, T.; Wang, E.; Wang, J. PEI/Zr^{4+}-Coated Nanopore for Selective and Sensitive Detection of ATP in Combination with Single-Walled Carbon Nanotubes.* Biosens. Bioelectron. ***2015,*** 63, *287–293.* http://dx.doi.org/10.1016/j.bios.2014.07.062.

radio-probe permitted specific activity to be amplified and improved the signal-to-noise ratio. The work included chelation of the alpha-emitter ^{225}Ac for therapy with DOTA.

Zr^{+4} has also been used in combination with SWCNT in cone-shaped nanopores of a polyethyleneimine system by Zhang et al. (Fig. 8.25). Such a label-free biosensor was designed to detect the concentration of adenosine triphosphate–bound aptamer, whereas other free aptamers are removed by SWCNTs.[238]

The application of bismuth and CNTs combined seems to be described in a number of patents, but the same applications are rarely described in peer-reviewed literature. The use of bismuth as a molecular probe for X-ray CT imaging offers advantages over conventional iodine-based agents. Bismuth possesses a higher atomic number and electron density (Z = 83)

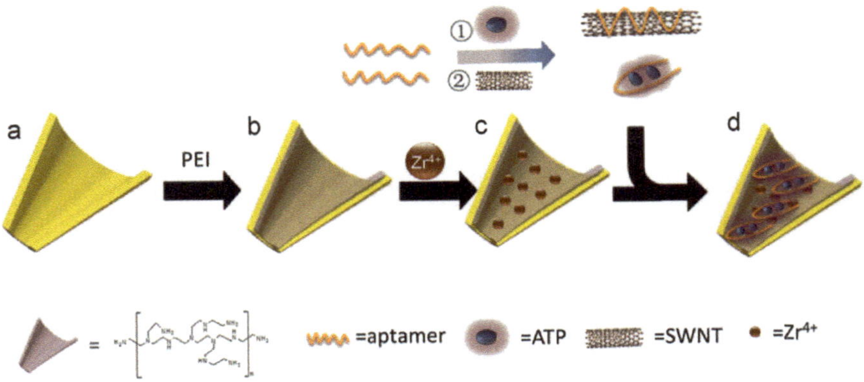

Figure 8.25 Schematic illustration of sensing strategy based on Zr^{4+}–polyethyleneimine (PEI)-modified single nanopore.[238]

and therefore greater X-ray attenuation intensity than iodine (Z = 53). Bismuth is also considered to be relatively nontoxic and has been used in cosmetics and medical applications.[239,240] Ultrashort CNTs (about 50 nm) were employed to encapsulate $BiOCl/Bi_2O_3$ and used as a contrast agent in CT.[239] The labeling and tracking of stem cells was mentioned but no in vivo imaging was presented.

To the best of our knowledge, in the literature no ^{188}Re-labeled nanotubes have been described as either nanodiagnostic or nanotherapeutic. However, the use of ^{99m}Tc and ^{188}Re-labeled lipid nanocapsules has been reported, as well as their kinetic biodistribution.[241] It is perhaps interesting that the functionalization of cold rhenium within SWCNT has been explored. The encapsulation of rhenium–fullerene complex such as $Re(\mu_5\text{-}C_{60}H_5)(CO)_3$ have been investigated to prove the reactivity of the inner wall of SWCNTs when filled with catalytically active atoms.[242] The reactivity of Re cluster toward SWCNTs was also explored and compared with those of osmium and tungsten. The authors reported that the order of reactivity $W < Re < Os$ is independent of the metal cluster size, shape, and orientation, and is related to the metal-nanotube bonding energy and amount of electronic density transferred between metal and SWCNTs.[243] The use of ^{99m}Tc in the functionalization of CNTs has been described for single imaging agents and targeted and multimodal imaging probes.[244–246] The preparation of a multimodal theranostics based on an MWCNT–hyaluronic acid system (HA-MWCNTs) was described by Jain et al. The nanotubes were noncovalently functionalized with doxorubicin (DOX) via π–π stacking interactions whereas the NIR dye Alexa-fluor 647 was attached covalently. The nanoprobe was radiolabeled with ^{99m}Tc in high efficiency but no details of the nature of the interaction between the nanotubes and the radioisotope were given. The cold probe was studied in vitro by confocal microscopy and in vivo, and showed the biodistribution profiles in mice.[244]

Remarkably, pharmacodynamic studies in breast cancer–induced rats showed that the tumor-growth inhibitory effect of DOX-loaded HA-MWCNT was five times more effective than free DOX in an equivalent concentration. DOX delivered through HA-MWCNTs lacks any detectable cardiotoxicity, hepatotoxicity, or nephrotoxicity. Jiang and collaborators also reported that MWCNT covalently functionalized with the antitumor agent 10-hydroxycamptothecin (HCPT) and successively attached to fluorescein isothiocyanate (FITC) fluorophore for in vitro confocal fluorescence microscopy (Fig. 8.26). The MWCNT-HCPT was further functionalized with diethylenetriaminepentaacetic dianhydride followed by labeling with radioactive nuclide ^{99m}Tc ($T_{1/2} = 6.02\,h$) for in vivo studies. The authors reported that the MWCNT-HCPT complex showed a superior antitumor effect of the conjugates compared with its native formulation (clinical HCPT injection) and low toxicity to living mice.[245]

MWCNT as dual MRI and SPECT agents were described.[246] The authors reported the synthesis and characterization of iron oxide–decorated MWCNTs for MRI imaging. The hybrid material was further functionalized and radiolabeled with ^{99m}Tc-dipicolylamine-alendronate for SPECT imaging. In vivo SPECT/CT and MRI imaging were conducted in mice in addition to ex vivo biodistribution studies. Transmission electron microscopy images of liver and spleen tissues show colocalization of iron oxide nanoparticles and MWCNTs within the same intracellular vesicles, which suggested the stability of the probe in vivo. ^{99m}Tc was also employed to radiolabel boron nitrate nanotubes (BNNTs). The functionalization of BNNTs with a glycol chitosan polymer followed by ^{99m}Tc labeling was reported by Ferriera Soares et al. The biodistribution of such BNNTs was also evaluated by ex vivo studies and scintigraphic imaging in healthy mice. Results showed that after 24 h, nanostructures were accumulated in the liver, spleen, and gut and eliminated via renal excretion.[247]

The biodistribution and tumor-targeting ability of SWCNT functionalized with phospholipid–PEG was investigated using copper-64 and DOTA, and was studied in vivo. The authors described the conjugation of SWCNTs to both ^{64}Cu and c(RGDyK), a potent integrin $\alpha_v\beta_3$ antagonist (Fig. 8.27).[48]

The covalent functionalization of SWCNTs certainly proved to be an efficient method for decorating the outer surface of CNTs with radionuclei.

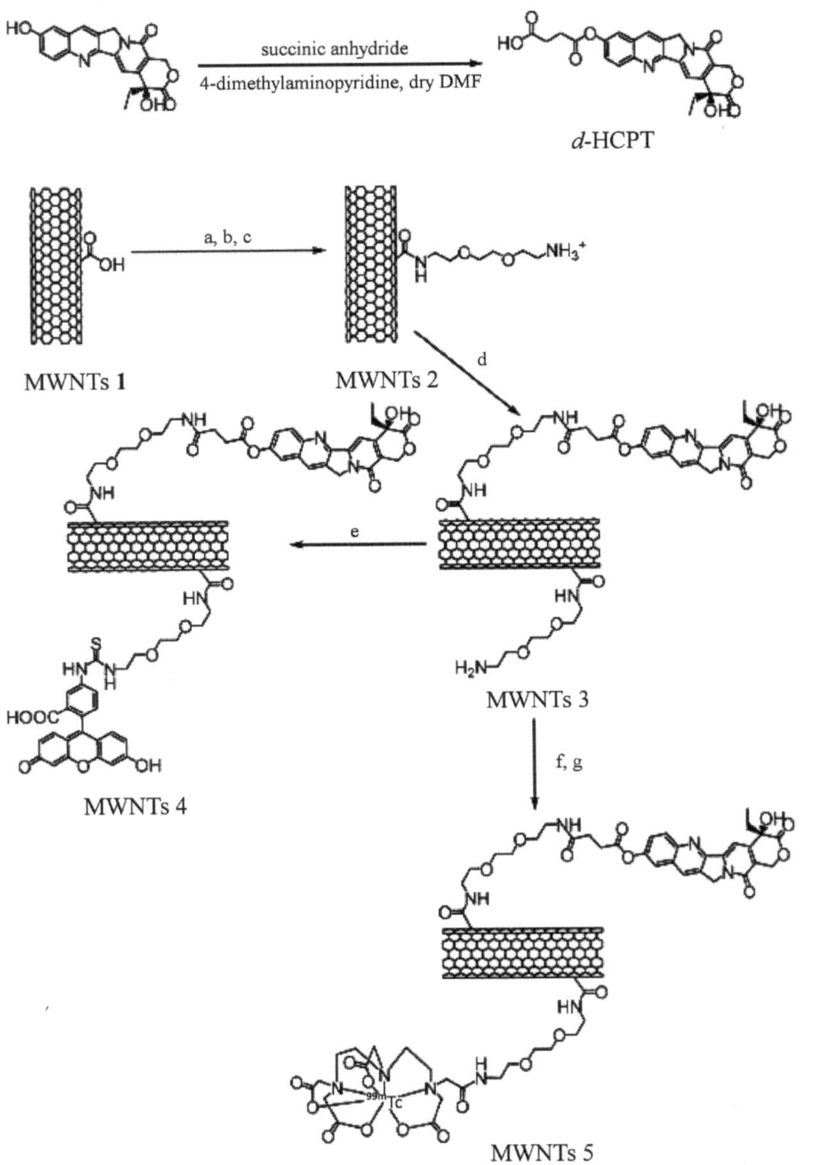

Figure 8.26 Synthesis route of the multiwalled carbon nanotube–10-hydroxycamptothecin (HCPT) conjugates. (a) Thionyl chloride, reflux. (b) Boc-$NH(CH_2CH_2O)_2$-$CH_2CH_2NH_2$, triethylamine, anhydrous tetrahydrofuran, reflux. (c) 4 M HCl in dioxane. (d) d-HCPT, EDC HCl, NHS, triethylamine, anhydrous dimethylformamide (DMF). (e) Fluorescein isothiocyanate, anhydrous DMF; (f) Diethylenetriaminepentaacetic dianhydride, triethylamine, anhydrous dimethyl sulfoxide; (g) stannous chloride, $^{99m}TcO^{4-}$.[245]

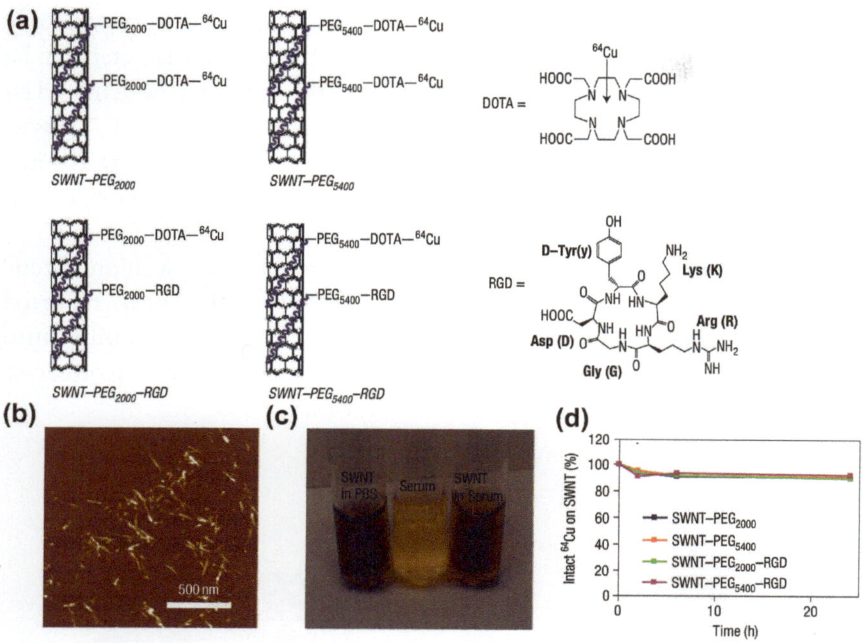

Figure 8.27 (a) Schematic drawings of noncovalently functionalized single-walled carbon nanotube (SWCNT)-phospholipid–polyethylene glycol (PEG) with 1,4,7,10-tetraazacyclododecane-1,4,7,10-tetraacetic acid (DOTA)–^{64}Cu. (b) Atomic force microscopy image of SWCNT–PEG5400 deposited onto a silicon substrate. (c) Photograph of stable SWCNT–PEG2000 suspensions in phosphate-buffered saline and full fetal bovine serum. (d) Serum stability test showing that ^{64}Cu remains intact on carbon nanotubes over 24-h incubation in full mouse serum.[48] *RGD*, Arg-Gly-Asp; *SWNT*, single-walled nanotube.

However, Wilson and coworkers reported an efficient strategy for generating a PET/MRI probe based on a CNT platform without the use of a chelating ligand. Gd^{3+} and $^{64}Cu^{2+}$ ions were loaded into ultrashort SWCNTs by sonication, and PET imaging confirmed the stability of the probe in vivo for up to 48 h. The authors reported that the probe was quickly cleared from circulation, with enhanced accumulation in the lungs.[248]

3.5 Overview of Common Functionalization Methods for Single-Walled Carbon Nanotubes

For biomedical applications, it has been reported that functionalization of SWCNTs has an equally important role as purification. After purification of the metallic, amorphous carbon and graphitic carbon impurities from the as-made SWCNTs, functionalization is used to modify the structure of SWCNTs to improve solubility and biocompatibility. SWCNT diameter control can easily be achieved through selective oxidation and purification; thus, diameters can be narrowed to a small range (e.g., 1.3–1.5 nm). Length control can be achieved by ultracentrifugation and size-exclusion chromatography. CVD-based CNT synthesis and purification leading to the Elicarb variety of SWCNTs by Thomas Swan, Ltd. is close to expectations from the synthetic scaffold to provide high dimensional reproducibility with low levels of defects.

The reliable purification methods cited previously allow high batch-to-batch reproducibility for commercial SWCNTs. It has been reported that the functionalization of SWCNTs enhances their dispensability into individual tubes in common solvents and further reduces the cytotoxicity of SWCNTs in biomedical applications (Fig. 8.28).[249]

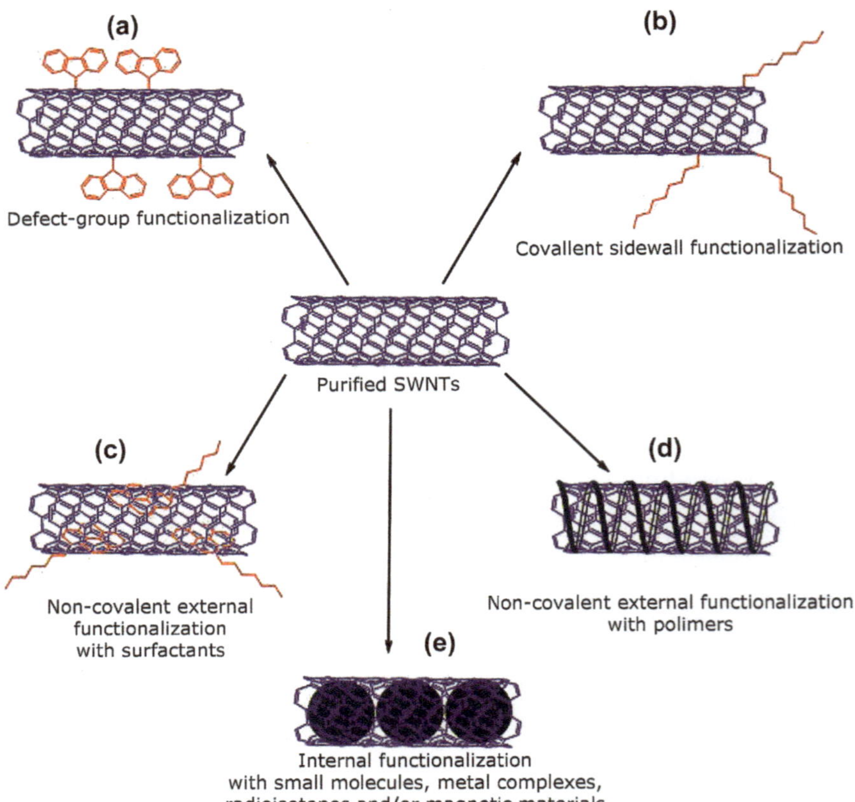

Figure 8.28 Schematic representations of different methods of functionalization of single-walled carbon nanotubes (SWCNTs).[193]

Fig. 8.28 shows the most common methods of functionalization used in complementary approaches to metal complexes incorporation techniques. In general, these may be broadly divided into two categories: covalent and noncovalent methods. Covalent functionalization involves breaking and forming new bonds on the defect and sidewall of SWCNTs, as can be seen in Fig. 8.28a and b. There are at least 10 common methods, described in Fig. 8.29, e.g., acid-based treatment, ozonolysis, halogenation, reductive coupling, oxidative coupling, polymer grafting, hydrogenation, electrophilic addition, cycloaddition, and nucleophilic addition to functionalize SWCNTs.[250]

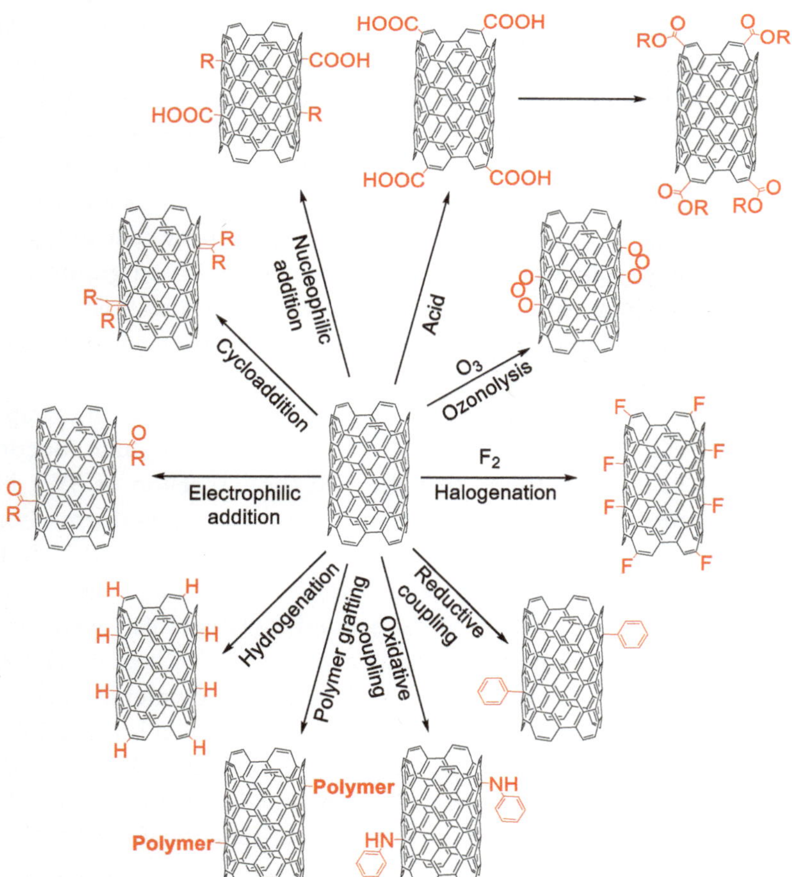

Figure 8.29 Covalent functionalization of single-walled carbon nanotubes allowing the first stages of scaffold preparation before metal complex incorporation.[250]

The most commonly used covalent method is acid (oxidative) treatment, which introduces carboxyl groups at the sidewall, defect sites, and open ends of SWCNTs.[251,252] Furthermore, carboxyl groups on SWCNTs can subsequently be treated with hydroxyl or amino groups to form an ester bond or carboxamide bond. This allows further modification by incorporating other organic groups molecules onto the surface of SWCNTs. An alternate way to add functional groups on the side of SWCNTs is, for example, through halogenation reactions. Once a halide is inserted onto the hexagonal framework surface of CNTs, in subsequent steps this can be converted through standard organic chemistry, e.g., by carrying out substitution reactions onto halide-functionalized SWCNTs.

As for noncovalent functionalization, this is based on supramolecular interactions between the surface of SWCNTs and other intact molecules, such as polymers, polysaccharides, porphyrin oligomers, etc. Instead of breaking and forming a new bond between the functional group and the sp^2 carbon of the SWCNTs (which causes local disruption of the synthetic structure), the functional groups or molecules attach onto or "wrap around" the SWCNTs via van der Waals interactions or aromatic stacking. PEG and pyrene have been commonly used for noncovalent functionalization to achieve dispersion of SWCNTs.[253] Pyrene-based groups attach to SWCNTs via aromatic stacking. Pyrene can act as a scaffold molecule whereas further modifications can be performed to the attached pyrene to functionalize future SWCNT constructs (Fig. 8.28c).

PEGs are commercially available polymer chains that can wrap or coat SWCNTs. PEGs have both hydrophobic and hydrophilic functionalities; thus it has been found that they aid the dispersion of SWCNTs in aqueous media.[254] Further functionalization can be performed onto PEG chains to introduce a variety of different molecules (Fig. 8.28d). The major advantage of the noncovalent method is that it does not damage the existing structure of SWCNTs. Second, polymers can wrap entire SWCNTs, and thus all of the surface area can be used to load much larger molecules than by the covalent method. Finally, the polymer layer acts not only as an anchor but also as a protecting layer. This shields the SWCNT surface from the environment and may also avoid further damage created by SWCNTs to the surrounding cells. Noncovalently wrapped SWCNTs have been shown to reduce the cytotoxicity of functionalized SWCNTs to cancer cells in the 3-(4,5-dimethylthiazol-2-yl)-2,5-diphenyltetrazolium bromide assays.[255]

Apart from covalent and noncovalent functionalization on the outer surface of SWCNTs, an alternative possibility for functionalizing them is the method known as internal functionalization. This refers to filling SWCNTs with ions or small molecules. The most important advantage is that the encapsulated molecules can be protected by the SWCNT walls; hence it can lead to the stabilization of reactive species inside the tubes as well as to the prevention of contamination or the loss of molecules. However, this method is still under development regarding the precise control of the filling yield.[256] There are still many problems: For example, the fill capacity is limited and the mechanism of controlled release of particles trapped within CNTs remains an intense area of investigation.[257,258]

3.6 General Developments and Applications of Functional Single-Walled Carbon Nanotubes as Synthetic Scaffolds for the Biomedical Field

As stated earlier, SWCNTs are a relatively new inorganic material. However, there has been intensive study and research based on it to explore the properties and applications of this novel material at the interface with biological research. There is a rich body of work (highlighted by us and others) on progress in CNT synthesis, purification, and functionalization to ensure biocompatibility and reproducibility.[228,230,259]

Fig. 8.30 highlights five basic properties with potential applications in the biomedical field: chemical, electrical, magnetic, optical, and field emission. Because raw, as-made, and even purified SWCNTs are difficult to dissolve into all of the ranges of the solvents, chemical properties were initially investigated with the aim of improving solubility. After chemical modification,

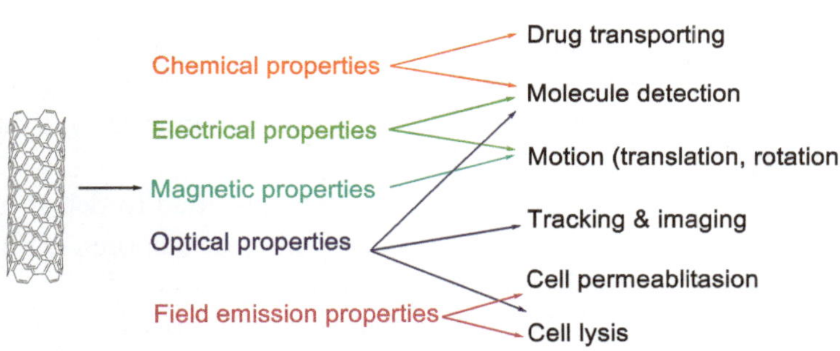

Figure 8.30 Properties and application of single-walled carbon nanotubes in biomedical field.

new bonds formed, for example, amide bonds, carboxyl, or hydroxyl groups are introduced on the surface or on the side of the SWCNTs via covalent functionalization. On the other hand, polymer or oligomers can be functionalized on SWCNTs by van der Waals intermolecular forces or π–π stacking via noncovalent methods. Both ways not only improve the solubility of the SWCNTs but also extend the potential application of SWCNTs toward biomedical areas.[260] For example, a carboxyl group formed from covalent functionalization can be reacted further with a hydroxyl group to form an ester bond, or with an amine group to form a peptide bond. As such, there are many different compounds or small molecules (such as radiolabeled probes, fluorescent molecules, DNA, and proteins) that can be introduced to process different biomedical tasks. For instance, SWCNTs can be functionalized and act as a vector to deliver drugs to a desired site, or modified to act as a sensor to detect overexpressed growth factors in cancer cells.[261]

There is a global interest in carbon nanomaterial nanotoxicity, and research interest has shifted toward the discovery of new methods to describe the role of CNTs toward gross cellular damage and cytocompatibility, which may be complemented by the specific evaluation of candidate complexes including proinflammatory cytokine production, ROS generation, stress protein activation, DNA damage, and eventually the determination of the full molecular/genetic signature using postgenomic approaches.[212]

Optical properties of the SWCNTs are another important factor that determines the application of SWCNTs in the biomedical field. There are two main optical properties of SWCNTs: (1) their ability to absorb all light in the visible region, hence resulting in the color of SWCNTs, which is completely black; and (2) the absorbed light at wavelength between 700 and 1100 nm, which is emitted as fluorescent signals, or leads to converting the energy into heat. The former situation is applied as an imaging tool to track and image cells, whereas in the latter case their application is killing the targeted cells.[262,263]

Electrical properties of SWCNTs have also been used to detect molecules, but rather than bind to target molecule and cause a fluorescent intensity change, these lead to changes in their electrical potential.[264,265]

Magnetic properties of SWCNTs can be normally introduced via internal cavity filling of metal; for instance, metal ions of Ti, Fe, Co, Ni, Cu, Ga, In, Zn, Gd, and Fe–Co alloys have been incorporated.[266] The external coating of SWCNTs with magnetic nanoparticles has also been reported.[267] These magnetic CNTs are generally employed as a scaffold to prepare more complex multifunctional vectors for biological applications. These vectors are able to carry fluorescent molecules, proteins, DNA, targeting ligands, and therapeutic drugs, and these multifunctional vectors can also be used in cellular imaging, cell tracking, lymphatic targeting, cancer lymph node metastasis treatment, and cancer gene therapy.[268]

The field emission property of SWCNTs is the last important property of SWCNTs. Electrons are readily emitted from the tips of the SWCNTs, either owing to oxidized tips or because of the curvature of the SWCNTs. When a potential is applied between a CNT surface and an anode, this is generated by continuous and pulsed X-rays using an SWCNT-based field emission cathode. Miniaturized SWCNT-based X-ray devices have been reported to be inserted into the body by endoscopy to achieve precise X-ray radiation therapy directly toward a desired area to be imaged and to minimize damage to surrounding healthy tissue.[269,270]

3.7 Photothermal Therapy Mediated by the Use of Functional Single-Walled Carbon Nanotubes

Photothermal therapy (PTT) is a physicochemical therapy for cancer treatment that uses optical radiation in the NIR wavelength range (700–2000 nm). When a laser is focused on a tissue, photons are absorbed by intercellular and intracellular areas and the energy of photons is converted into heat. As a result, the tissue temperature increases, leading to cell and tissue death. NIR light ranges between 700 and 1000 nm and tissue transparency is ideal for optical imaging and phototherapy. Both MWCNTs and SWCNTs can exhibit strong optical absorption in the NIR regions.[271] Once an SWCNT is exposed to NIR light, this generates heat by light absorption and induces the thermal destruction of cancer cells containing significant concentrations of SWCNTs (Fig. 8.31).[272]

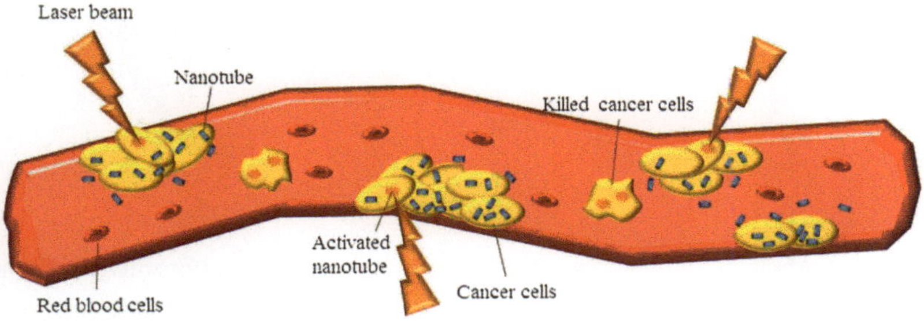

Figure 8.31 Application of single-walled carbon nanotubes in photothermal therapy to cancer site.

CNT-associated thermal therapy can be combined with time-resolved infrared imaging techniques, and thus enables PTT to be more selective and traceable for targeting individual cancer cells or their small clusters labeled with CNTs. This approach can be used for the early stages of cancer treatment and it is efficient for the treatment of small tumors, the edges of tumors, and micrometastases.[273] In research, PEG-functionalized SWCNTs as described are normally injected into the tumor mass site.[271] They are found to be accumulated in nearby muscle and skin tissue after tumors are destroyed and then slowly translocate into the liver and spleen, from which nanotubes are gradually excreted. This research shows that CNT-enhanced PTT can potentially become a noninvasive and harmless way to treat cancer.[274]

3.8 Photodynamic Therapy Using Functionalized Single-Walled Carbon Nanotubes

Photodynamic therapy (PDT) is a developing technique currently used for malignant tumors and microbial parasite treatment. It is similar to PTT and both techniques kill tumor cells from inside. The probes used are nontoxic in the absence of light and then become highly toxic under illumination (Fig. 8.32). However, they have fundamental differences from each other in that cancer cells are killed by heat radiation in PTT, whereas in PDT cells are eliminated via a rising level of ROS generated in situ.

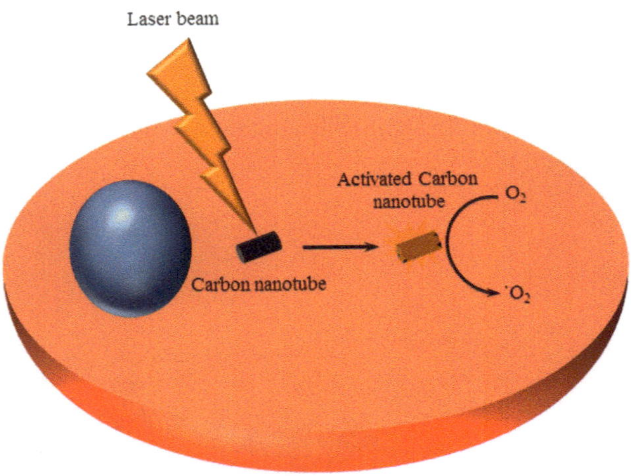

Figure 8.32 Application of single-walled carbon nanotubes in photothermal therapy at cancer site.

A typical PDT device consists of three components: a photosensitizer, a light source, and tissue oxygen. The photosensitizer is the core part of the system; it can generate ROS from tissue oxygen under illumination. Porphyrins are considered to be good photosensitizers owing to their low cytotoxicity in the dark, high absorption coefficients in the red spectral region, and high quantum yields for singlet oxygen production.[275] Furthermore, QDs have been functionalized onto porphyrin framework. These were used as a resonant compound[276] able to harvest light and transfer energy to the porphyrin to enhance ROS production further. Finally, a DDS consisting of SWCNTs, a porphyrin, and QDs has been assembled, and data have shown that this DDS had a higher capability of generating ROS than SWCNTs alone. It also made SWCNT-enhanced porphyrin-type DDSs good candidates as photosensitizers for PDT.[277]

3.9 Functional Single-Walled Carbon Nanotube–Enhanced Gene Therapy

Gene therapy is an alternative method to traditional chemotherapy in cancer treatment. The basic principle of gene therapy is to replace damaged or missing genes with healthy genes.[278] However, the major challenge for this technique is the gene delivery itself. In the past, researchers used nonpathogenic viruses as vectors, which occasionally led to a significant immune response.[279] Hence, an ideal vector design is important for this therapy. Many reports have shown that carboxyl-functionalized SWCNTs can act as a vector for gene-encoding nucleic acids and plasmid DNA that enhance the gene therapeutic capacity compared with DNA alone.[46] At the same time, some reports showed that short interfering RNA (siRNA) can combine with SWCNTs for efficient gene delivery. siRNA is only 20–25 nucleotides long and is known to be complementary to the mutated gene. In the cytoplasm, it can bind to the mutated gene and disable the cytoplasm's action. This process is called gene silencing.[280,281] From the presented works it is clear that direct injection of siRNA is not an effective approach, whereas the siRNA and SWCNT complex can easily be taken up by cancer cells and then revealed by immune-recognizing cells to induce an immune response for the particular gene, to kill cells with damaged genes.[281] Furthermore, SWCNTs can be functionalized by single-stranded DNA and subsequently coated with DNA probes. Thus, DNA probes are protected from enzymatic cleavage and interference from nucleic acid–binding proteins in the circulation. After being taken up by cells, most SWCNTs modified with DNA could target a specific mRNA inside living cells, with increased self-delivery potential and intracellular biostability compared with free DNA probes. Hence, this new conjugate provides great potential for applications in the field of genetic engineering.[282]

3.10 Metal Nanoparticle-Functionalized Single-Walled Carbon Nanotubes and Their Biomedical Applications

High yield filling of (pristine, highly purified) SWCNTs with inorganic compounds or organic molecules has been demonstrated, and several authors have reported on their encapsulation.[283–286] Opened SWCNTs[178,283,284] can be exposed to aqueous solutions of simple metal salts such as CuX_2 or GdX_3 (X = halides such as Br^- or Cl^-, or hydroxyl groups) or iron oxide nanoparticles. Although the filling of SWCNTs with metal species from saturated solutions is well-established (and can be easily extrapolated to MRI-active species), to date it is not known whether filling can also be achieved at the low concentrations at which metallic radionuclides suitable for PET are generated. (Generator or cyclotron-generated ions for PET or SPECT imaging are typically in concentrations of 10^{-6} to 10^{-8} M.) Any potential leaking of the encapsulated species from open SWCNTs can be prevented by the successive derivatization steps of polymer and supramolecular wrapping, which we envisage provides simultaneous "corking" of the nanotubes. As an alternative, some authors have also implemented a method in which C_{60} or C_{70} molecules are used for "corking."[287] Filling using molten-salt methods (i.e., yielding crystalline filling, as in Fig. 8.2a) have also been used,[288,289] with the metal halides mentioned earlier. Methods available for endohedral filling from molten salts may well be adapted in the future to ensure that conditions compatible with radiopharmaceutical work. This has led to the emergence of the field denoted radio-nanochemistry. In this case, the "cold" filling material acting as a support for the radionuclide may be a benign, water-soluble metal halide, e.g., NaCl, which can easily be removed from the outside of the tubes, or GdX_3 (with the dual advantage of generating probes for multimodal MRI/PET imaging techniques). For the endohedral filling methods, the ends of the nanotubes are expected to close under the reaction conditions used, which, so far, mainly involved thermal annealing of the CNTs.

3.11 Probing the Mechanisms of Interactions: Bioimaging Probes and Cells

Several tests on stress protein induction in certain raw SWCNTs were reported and used microarrays to look at global gene expression[290] and heme oxygenase (HO-1).[291] One approach to a comprehensive mechanistic study using the library of probes assembled as described would be a two-step process. A first step would be essentially to screen where you would use general markers of oxidative stress. For example, these would use fluorescent dyes that quantify ROS + HO-1 messenger RNA (mRNA) accumulation to select a small range of functionalized SWCNTs, which generate ROS to different degrees. At the second stage, a more detailed study of selected probes would be used to compare processes triggered by filled versus unfilled SWCNTs; functionalized (supramolecularly versus covalent), and tagged and targeted versus raw (including short, long, corked, open, pure, and as-made SWCNTs). Current research in our laboratories examines the relationship between the precise types of functionalization and oxidative stress response and the link between the level of oxidative stress and functional outcome to generate precise cell membrane damage and apoptosis altered gene expression profiles, in a line of healthy cells and in standard cancerous lines. To investigate mechanisms of toxicity, researchers typically select cell types and conditions under which significant toxicity is observed. For a comprehensive pharmacologic profile, a measure of ROS generation as a function of time and SWCNT concentration needs to be achieved; this is typically performed using the fluorescent probe carboxydichlorofluorescein, which is fairly nonspecific. The type of ROS involved uses chemical probes by previously established methodologies (e.g., deuterium oxide to enhance singlet oxygen lifetime; sodium azide and histidine, which does not quench superoxide anion, to quench singlet oxygen; hydroxyl radical scavengers such as mannitol, which do not affect singlet oxygen or superoxide; and specific ROS scavenging enzymes such as catalase and superoxide dismutase; and specific inhibition of enzymes such as inhibition of NADPH oxidase by diphenylene iodonium).[292] A mild iron chelator, which defines the role of iron while avoiding the interfering collateral effects of strong iron chelators can be used.[293] The mechanism can be further probed using the most sensitive diagnostic of oxidative stress, which is activation of gene expression (mRNA accumulation detected by reverse transcription polymerase chain reaction using specific probes). Activation of HO-1 can be examined. Such activation is influenced by various forms of oxidative stress but is not affected by iron and specific metalloproteinase induction, arising out of a different pathway dependent on free iron. To further understand the nature of oxidative stress upon treatment with the nanoprobes, glutathione depletion and mitochondrial membrane potential (by JC-1 fluorescence) can be measured. HO-1 activation can be used as the most sensitive method to detect low levels of oxidative stress. Overall, the quest aims is to understand, how filled and functionalized nanotubes can affect the mechanistic aspects of cytotoxicity.

4. CONCLUSIONS AND FUTURE PROSPECTS

A number of real and perceived difficulties and controversies remain in applying metal and metal complex–functionalized carbon nanomaterials in health care.[294] Most publications on the drug delivery aspects of functional CNTs convey confidence that the benefits of the new technology proposed outweigh the risks (real and perceived) involved, particularly where such developments are coupled with those of metals in medicine and nanomedicine.

In any such preclinical developments, relevant regulatory approvals are typically being sought at an early stage, and risks are minimized via stage-gated approaches. Given the expansion of SWCNT basic science and the focus on their applications, health care applications using nanotheranostics based on SWCNTs are highly likely to be at the forefront of the interest of researchers as well as regulatory agencies within the next decades.

Considerations of ethics, safety, and public involvement in nanotechnology for health care have become paramount. Current debate about the potential toxicity of CNTs is more relevant than ever because CNT production is growing by hundreds of metric tons a year as applications move from research into commercial products (e.g., Bayer had up to 3000 metric tons of production capacity in 2013). The global CNT industry turned over around $668.3 million in 2010, with an MWCNT production value of approximately $631.5 million and SWCNT production value of around $36.8 million, and was forecasted to grow to $1.1 billion by 2016 at a compound annual growth rate of 10.5%.[295] In view of this massive growth, CNTs need to be assessed seriously.[37] Bare-surface (unfunctionalized) MWCNTs of larger diameters than SWCNTs were found[296] to trigger innate and adaptive immune responses in mice, which led to the formation of chronic pulmonary inflammation and granulomas.

On the other hand, appropriately functionalized SWCNTs showed no acute toxicity[297] and no toxic effects in mice over 4 months' observation. These quasi-one-dimensional structures are ideal for modifications with biomolecules and have already been used successfully in tumor targeting.[298] Their relatively long circulation time and rapid renal clearance make SWCNTs suitable therapeutic and diagnostic scaffolds for nanomedicines.[299] It is clear that many toxic effects of nanoparticles come from their large surface area, which is more reactive and thus capable of triggering oxidative stresses in cells. However, CNT's sp^2-bonded surface is chemically inert. To assess CNTs' toxicity correctly, they have to be compared with other sp^2-bonded nanomaterials such as nanographite and C60 Buckyballs. Graphite is not acutely toxic but causes pneumoconiosis,[300] which is characterized by granulomas, emphysema, tissue death, and hardening of the blood vessels, among other symptoms, and has been long recognized in workers involved in mining and processing graphite. Furthermore, preclinical studies on C_{60}, which is available in pure form, found no toxic effects on rats and no apparent genetic damage in bacteria.[301] Thus, the reported toxicity of bare nanotubes might have nothing to do with their size or aspect ratio, but may simply be related to the general toxicity of unfunctionalized graphitic structures. Assessing CNTs' effects on living organisms is complicated by the fact that many nanotube samples still contain residues of catalytic metals (Co, Ni, Fe, Y, Mo, etc.) used for their production. Such metal nanoparticles and other impurities (polyaromatic hydrocarbons and graphitic nanoparticles) may have a greater effect on organisms than CNTs themselves. Preliminary indications have shown that it is difficult to test the toxicity of nanoparticles in general because they agglomerate so readily. Research programs of complex surface functionalization will continue to need to accomplish the total avoidance of agglomeration in solution, aiming for extremely pure scaffolds and their stable dispersions in biological media. Step-by step cytotoxicity tests will be crucial to help clarify under what conditions these CNTs are toxic. Currently there is no common understanding of (and no comprehensive approach in assessing) the risk profile of nanotechnology in general and carbon nanomaterials in particular (although there have been studies of progress made in addressing the toxicology of nanomaterials). The need for systematic risk assessment and categorization of nanomaterials has been acknowledged,[302,303] because there is no commonly accepted standard guidance on the physicochemical characterization of engineered nanoscale materials for the toxicologic assessment and safe handling and disposal of manufactured nanomaterials. Suggested guidelines for the safe handling of nanomaterials are available in the public domain.[304,305] Because the primary route for nanotube exposure is through the respiratory system, the preparation of all materials and all processing experiments typically need to be carried out in well-ventilated laboratories equipped with extracting fume hoods. Because the effects of CNTs on the environment are not entirely understood, all nanotube-containing products will be disposed as hazardous materials. Self-regulation, a long-standing tradition in the professions, is widespread when tackling safety issues relating to nanotechnology. The impact on the environment needs to be assessed during the life cycle of the nanomaterials synthesized, and when these incorporate transition metal species, the landscape is further complicated. In the longer term, this work may form the basis of creating a robust, standard system suitable for evaluating health and the environmental impact of therapeutic nanomaterials, especially when (radio)metal species are also involved. Significant research funding needs to be channeled toward research aiming to enhance our understanding of CNT toxicity in general terms, and on estimating their impact on the environment, and steps must be determined that need to control or regulate nanotechnology.[306]

Animal tests are controversial, and these need to be kept to a minimum. The main aim would need to be first to develop high-throughput in vitro tests on a large variety of cell types. The advantage of this methodology is that it will require extremely small quantities of administered drug, in sub-pharmacologic quantities (dispersion less than 10^{-6} to 10^{-8} g/L in biocompatible media would give a suitable signal-to-noise ratio via PET imaging).

Addressing the public perception of nanotechnology means that it remains a constant effort to tackle ongoing controversies affecting the public perception of nanotechnology.[307] Much of the fear over new technologies concerns the risk of the effects that new, less well-known, and less-understood products may have on human tissue and on the environment. To appreciate these risks requires an understanding of the properties of nanostructures and their interaction with the living cells. As a result of rigorous cytotoxicity tests after each derivatization step, the behavior in cells will be better understood for the nanomedicines designed, and an evaluation of the real risks can be made during fabrication or synthesis in the laboratory and during the lifetime of the product development and accidental human or environmental exposure. Issues of ethics, safety, environmental impact, and exposure will be considered using multicriteria decision analysis versus constructive technology assessment tools.[308,309] Clinical (safety, efficacy, and effectiveness), economic (cost-effectiveness), patient-related (quality of life, ethical/juridical, and psychosocial), and organizational aspects (diffusion and adoption) and scenario drafting need to be considered holistically such that such projects situated at the interface between metals in medicine and nanomedicine coupled with radio-nanochemistry and radiobiology can succeed.

ACKNOWLEDGMENTS

The authors thank the Science and Technologies Facilities Council, Centre for Doctoral Training in Sustainable Chemical Technologies, EC and ERC Consolidator grant "O2SENSE" and University of Bath. This work was funded by the European Commission FP7 Programme through the Marie Curie Initial Training Network PROSENSE (grant no. 317420, 2012–2016). The authors would also like to thank Prof. Malcolm Green, FRS, Prof. Jon Dilworth, and Prof. Stanley Botchway for invaluable discussions and training in single-walled carbon nanotube chemistry and nanomedicine, as well as their molecular imaging applications.

REFERENCES

1. Ren, J. S.; Bray, F.; Masuyer, E.; Ferlay, J. Global Estimates of Cancer Prevalence for 27 Sites in the Adult Population in 2008. *Int. J. Cancer* **2013,** *5,* 1133–1145.
2. *All Cancers Combined Key Stats;* Cancer Research: UK, 2015.
3. *U.S.N.I. Of Health, SEER Cancer Statistics Review 1975–2005;* National Cancer Institute, 2010.
4. Howlader, N.; Noone, A. M.; Krapcho, M.; Garshell, J.; Neyman, N.; Altekruse, S. F.; et al. *SEER Cancer Statistics Review 1975–2010;* National Cancer Institute: Bethesda, MD, 2013; p 2013. Available at: http//seer.Cancer.Gov/csr.
5. Bach, P. B.; Mirkin, J. N.; Oliver, T. K.; et al. Benefits and Harms of CT Screening for Lung Cancer: A Systematic Review. *JAMA* **2012,** *307,* 2418–2429. http://dx.doi.org/10.1001/jama.2012.5521.
6. Xin, L.; Liao, Z.; Jiang, Y.-P.; Li, Z.-S. Indications, Detectability, Positive Findings, Total Enteroscopy, and Complications of Diagnostic Double-Balloon Endoscopy: A Systematic Review of Data over the First Decade of Use. *Gastrointest. Endosc.* **2011,** *74,* 563–570.
7. Sausville, E. A.; Longo, D. L. Principles of Cancer Treatment: Surgery, Chemotherapy, and Biologic Therapy. *Harrisons Princ. Intern. Med.* **2005,** *16,* 464.
8. Smith, R. A.; Manassaram-Baptiste, D.; Brooks, D.; Cokkinides, V.; Doroshenk, M.; Saslow, D.; et al. Cancer screening in the United States, 2014: A Review of Current American Cancer Society Guidelines and Current Issues in Cancer Screening. *CA Cancer J. Clin.* **2014,** *64,* 30–51.
9. Dennis, R.; Tou, S.; Miller, R. Colorectal Cancer: Prevention and Early Diagnosis. *Med. Baltim.* **2011,** *39,* 243–249.
10. Zhu, H.; Wang, M.; Feng, D.; Feng, Y.; Ren, Y.; Chen, J.; et al. Ultrasonography, X-Ray and CT Imaging Findings of a Giant Pericardial Lipoma: Imaging Diagnosis and Review of the Literature. *Oncol. Lett.* **2014,** 7, 195–198.

11. Townsend, D. W.; Carney, J. P. J.; Yap, J. T.; Hall, N. C. PET/CT Today and Tomorrow. *J. Nucl. Med.* **2004,** *45,* 4S–14S.
12. Pilewskie, M.; King, T. A. Magnetic Resonance Imaging in Patients with Newly Diagnosed Breast Cancer: A Review of the Literature. *Cancer* **2014,** *120,* 2080–2089.
13. Jambor, I.; Kuisma, A.; Ramadan, S.; Huovinen, R.; Sandell, M.; Kajander, S.; et al. Prospective Evaluation of Planar Bone Scintigraphy, SPECT, SPECT/CT, 18F-naf PET/CT and Whole Body 1.5 T MRI, Including DWI, for the Detection of Bone Metastases in High Risk Breast and Prostate cancer Patients: SKELETA Clinical Trial. *Acta Oncol. Madr.* **2015,** 1–9.
14. Sun, X. L.; Cai, W. B.; Chen, X. Y. Positron Emission Tomography Imaging Using Radio Labeled Inorganic Nanomaterials. *Acc. Chem. Res.* **2015,** *48,* 286–294. http://dx.doi.org/10.1021/ar500362y.
15. a. Pascu, S. I.; Waghorn, P. A.; Conry, T.; Lin, B.; James, C.; Zayed, J. M. Design Considerations towards Simultaneously Radiolabeled and Fluorescent Imaging Probes Incorporating Metallic Species. *Adv. Inorg. Chem.* **2009,** *61,* 131–178. http://dx.doi.org/10.1016/S0898-8838(09)00203-7.
 b. Waghorn, P. A.; Jones, M. W.; Theobald, M. B. M.; Arrowsmith, R. L.; Pascu, S. I.; Botchway, S. W.; Faulknera, S.; Dilworth, J. R. Shining Light on the Stability of Metal Thiosemicarbazonate Complexes in Living Cells by FLIM. *Chem. Sci.* **2013,** *4,* 1430–1441. http://dx.doi.org/10.1039/c2sc21489j.
 c. Davies, L. H.; Kasten, B. B.; Benny, P. D.; Arrowsmith, R. L.; Ge, H.; Pascu, S. I.; Botchway, S. W.; Clegg, W.; Harringtona, R. W.; Higham, L. J. Re and ^{99m}Tc Complexes of BodP$_3$ – Multi-Modality Imaging Probes. *Chem. Commun.* **2014,** *50,* 15503–15505. http://dx.doi.org/10.1039/c4cc06367h.
16. Mokaleng, B. B.; Ebenhan, T.; Ramesh, S.; Govender, T.; Kruger, H. G.; Parboosing, R.; et al. Synthesis, Ga-68-Radiolabeling, and Preliminary In Vivo Assessment of a Depsipeptide-Derived Compound as a Potential PET/CT Infection Imaging Agent. *Biomed. Res. Int.* **2015,** 12. http://dx.doi.org/10.1155/2015/284354.
17. Kang, C. S.; Chen, Y. W.; Lee, H.; Liu, D. J.; Sun, X.; Kweon, J.; et al. Synthesis and Evaluation of a New Bifunctional NETA Chelate for Molecular Targeted Radiotherapy Using Y-90 or Lu-177. *Nucl. Med. Biol.* **2015,** *42,* 242–249. <Go to ISI>://CCC:000350782200006.
18. RSC. *Nanotechnology – The Issues.* http://www.rsc.org/pdf/policy/nanotechnology.pdf, (n.d.).
19. Dunning, M. D.; Lakatos, A.; Loizou, L.; Kettunen, M.; Ffrench-Constant, C.; Brindle, K. M.; et al. Superparamagnetic Iron Oxide-Labeled Schwann Cells and Olfactory Ensheathing Cells Can Be Traced In Vivo by Magnetic Resonance Imaging and Retain Functional Properties after Transplantation into the CNS. *J. Neurosci.* **2004,** *24,* 9799–9810.
20. NCRI PET. *Strategic Planning Group, a Framework for PET Research in the UK;* , 2007; pp 1–71.
21. Conn, M. M.; Rebek, J. Self-Assembling Capsules. *Chem. Rev.* **1997,** *97,* 1647–1668. http://dx.doi.org/10.1021/cr9603800.
22. Hof, F.; Craig, S. L.; Nuckolls, C.; Rebek, J. Molecular Encapsulation. *Angew. Chem. Int. Ed. Engl.* **2002,** *41,* 1488–1508. http://www.ncbi.nlm.nih.gov/pubmed/19750648.
23. Caulder, D. L.; Raymond, K. N. Supermolecules by Design. *Acc. Chem. Res.* **1999,** *32,* 975–982. http://dx.doi.org/10.1021/ar970224v.
24. Davis, A.V.; Yeh, R. M.; Raymond, K. N. Supramolecular Assembly Dynamics. *Proc. Natl. Acad. Sci. USA* **2002,** *99,* 4793–4796. http://dx.doi.org/10.1073/pnas.052018299.
25. Hof, F.; Rebek, J. Molecules within Molecules: Recognition through Self-Assembly. *Proc. Natl. Acad. Sci. USA* **2002,** *99,* 4775–4777. http://dx.doi.org/10.1073/pnas.042663699.
26. Swiegers, G. F.; Malefetse, T. J. Classification of Coordination Polygons and Polyhedra According to Their Mode of Self-Assembly. *Chem. Eur. J.* **2001,** 7, 3636–3643. http://dx.doi.org/10.1002/1521-3765(20010903)7. 17<3636::AID-CHEM3636>3.0.CO;2-#.

27. Zielhuis, S.W.; Seppenwoolde, J.-H.; Mateus, V.A. P.; Bakker, C. J. G.; Krijger, G. C.; Storm, G.; et al. Lanthanide-Loaded Liposomes for Multimodality Imaging and Therapy. *Cancer Biother. Radiopharm.* **2006,** *21,* 520–527. http://dx.doi.org/10.1089/cbr.2006.21.520.
28. Koning, G. A.; Krijger, G. C. Targeted Multifunctional Lipid-Based Nanocarriers for Image-Guided Drug Delivery. *Anticancer. Agents Med. Chem.* **2007,** *7,* 425–440. http://www.ncbi.nlm.nih.gov/pubmed/17630918.
29. van Kasteren, S. I.; Campbell, S. J.; Serres, S.; Anthony, D. C.; Sibson, N. R.; Davis, B. G. Glyconanoparticles Allow Pre-Symptomatic In Vivo Imaging of Brain Disease. *Proc. Natl. Acad. Sci. USA* **2009,** *106,* 18–23. http://dx.doi.org/10.1073/pnas.0806787106.
30. Salzmann, C. G.; Lee, G. K. -C.; Ward, M. A. H.; Chu, B. T. T.; Green, M. L. H. Highly Hydrophilic and Stable Polypeptide/Single-Wall Carbon Nanotube Conjugates. *J. Mater. Chem.* **2008,** *18,* 1977. http://dx.doi.org/10.1039/b800494c.
31. Salzmann, C. G.; Ward, M. A. H.; Jacobs, R. M. J.; Tobias, G.; Green, M. L. H. Interaction of Tyrosine-, Tryptophan-, and Lysine-containing Polypeptides with Single-Wall Carbon Nanotubes and its Relevance for the Rational Design of Dispersing Agents. *J. Phys. Chem. C* **2007,** *111,* 18520–18524. http://dx.doi.org/10.1021/jp076013h.
32. Coleman, K. S. Nanotubes. *Annu. Rep. Sect. A Inorg. Chem.* **2007,** *103,* 379. http://dx.doi.org/10.1039/b705543a.
33. Balasubramanian, K.; Burghard, M. Chemically Functionalized Carbon Nanotubes. *Small* **2005,** *1,* 180–192. http://dx.doi.org/10.1002/smll.200400118.
34. Yang, W.; Thordarson, P.; Gooding, J. J.; Ringer, S. P.; Braet, F. Carbon Nanotubes for Biological and Biomedical Applications. *Nanotechnology* **2007,** *18,* 412001. http://dx.doi.org/10.1088/0957-4484/18/41/412001.
35. Coleman, K. S.; Chakraborty, A. K.; Bailey, S. R.; Sloan, J.; Alexander, M. Iodination of Single-Walled Carbon Nanotubes. *Chem. Mater.* **2007,** *19,* 1076–1081. http://dx.doi.org/10.1021/cm062730x.
36. Porter, A. E.; Gass, M.; Muller, K.; Skepper, J. N.; Midgley, P. A.; Welland, M. Direct Imaging of Single-Walled Carbon Nanotubes in Cells. *Nat. Nanotechnol.* **2007,** *2,* 713–717. http://dx.doi.org/10.1038/nnano.2007.347.
37. Poland, C. A.; Duffin, R.; Kinloch, I.; Maynard, A.; Wallace, W. A. H.; Seaton, A.; et al. Carbon Nanotubes Introduced into the Abdominal Cavity of Mice Show Asbestos-like Pathogenicity in a Pilot Study. *Nat. Nanotechnol.* **2008,** *3,* 423–428. http://dx.doi.org/10.1038/nnano.2008.111.
38. Zavaleta, C.; de la Zerda, A.; Liu, Z.; Keren, S.; Cheng, Z.; Schipper, M.; et al. Noninvasive Raman Spectroscopy in Living Mice for Evaluation of Tumor Targeting with Carbon Nanotubes. *Nano Lett.* **2008,** *8,* 2800–2805. http://dx.doi.org/10.1021/nl801362a.
39. Cherukuri, P.; Gannon, C. J.; Leeuw, T. K.; Schmidt, H. K.; Smalley, R. E.; Curley, S. A.; et al. Mammalian Pharmacokinetics of Carbon Nanotubes Using Intrinsic Near-infrared Fluorescence. *Proc. Natl. Acad. Sci. USA* **2006,** *103,* 18882–18886. http://dx.doi.org/10.1073/pnas.0609265103.
40. Lacerda, L.; Bianco, A.; Prato, M.; Kostarelos, K. Carbon Nanotubes as Nanomedicines: From Toxicology to Pharmacology. *Adv. Drug Deliv. Rev.* **2006,** *58,* 1460–1470. http://dx.doi.org/10.1016/j.addr.2006.09.015.
41. Salvador-Morales, C.; Basiuk, E. V.; Basiuk, V. A.; Green, M. L. H.; Sim, R. B. Effects of Covalent Functionalization on the Biocompatibility Characteristics of Multi-Walled Carbon Nanotubes. *J. Nanosci. Nanotechnol.* **2008,** *8,* 2347–2356. http://dx.doi.org/10.1166/jnn.2008.090.
42. Shi Kam, N. W.; Jessop, T. C.; Wender, P. A.; Dai, H. Nanotube Molecular Transporters: Internalization of Carbon Nanotube–Protein Conjugates into Mammalian Cells. *J. Am. Chem. Soc.* **2004,** *126,* 6850–6851. http://dx.doi.org/10.1021/ja0486059.

43. Lacerda, L.; Bianco, A.; Prato, M.; Kostarelos, K. Carbon Nanotube Cell Translocation and Delivery of Nucleic Acids In Vitro and In Vivo. *J. Mater. Chem.* **2008,** *18,* 17–22. http://dx.doi.org/10.1039/B711554G.
44. Ali-Boucetta, H.; Al-Jamal, K. T.; McCarthy, D.; Prato, M.; Bianco, A.; Kostarelos, K. Multiwalled Carbon Nanotube–Doxorubicin Supramolecular Complexes for Cancer Therapeutics. *Chem. Commun.* **2008,** 459–461. http://dx.doi.org/10.1039/B712350G.
45. Kostarelos, K. The Long and Short of Carbon Nanotube Toxicity. *Nat. Biotechnol.* **2008,** *26,* 774–776. http://dx.doi.org/10.1038/nbt0708-774.
46. Prato, M.; Kostarelos, K.; Bianco, A. Functionalized Carbon Nanotubes in Drug Design and Discovery. *Acc. Chem. Res.* **2008,** *41,* 60–68. http://dx.doi.org/10.1021/ar700089b.
47. Yang, H.; Liu, C.; Yang, D.; Zhang, H.; Xi, Z. Comparative Study of Cytotoxicity, Oxidative Stress and Genotoxicity Induced by Four Typical Nanomaterials: the Role of Particle Size, Shape and Composition. *J. Appl. Toxicol.* **2009,** *29,* 69–78. http://dx.doi.org/10.1002/jat.1385.
48. Liu, Z.; Cai, W.; He, L.; Nakayama, N.; Chen, K.; Sun, X.; et al. In Vivo Biodistribution and Highly Efficient Tumour Targeting of Carbon Nanotubes in Mice. *Nat. Nanotechnol.* **2007,** *2,* 47–52. http://dx.doi.org/10.1038/nnano.2006.170.
49. Lacerda, L.; Soundararajan, A.; Singh, R.; Pastorin, G.; Al-Jamal, K. T.; Turton, J.; et al. Dynamic Imaging of Functionalized Multi-Walled Carbon Nanotube Systemic Circulation and Urinary Excretion. *Adv. Mater.* **2008,** *20,* 225–230. http://dx.doi.org/10.1002/adma.200702334.
50. Bergström, M.; Grahnén, A.; Långström, B. Positron Emission Tomography Microdosing: a New Concept with Application in Tracer and Early Clinical Drug Development. *Eur. J. Clin. Pharmacol.* **2003,** *59,* 357–366. http://dx.doi.org/10.1007/s00228-003-0643-x.
51. Peer, D.; Karp, J. M.; Hong, S.; Farokhzad, O. C.; Margalit, R.; Langer, R. Nanocarriers as an Emerging Platform for Cancer Therapy. *Nat. Nanotechnol.* **2007,** *2,* 751–760. http://dx.doi.org/10.1038/nnano.2007.387.
52. Bianco, A.; Kostarelos, K.; Prato, M. Opportunities and Challenges of Carbon-Based Nanomaterials for Cancer Therapy. *Expert Opin. Drug Deliv.* **2008,** *5,* 331–342. http://dx.doi.org/10.1517/17425247.5.3.331.
53. Aulton, M. E. *Aulton's Pharmaceutics: The Design and Manufacture of Medicines;* Churchill Livingstone: London, United Kingdom, 2007.
54. Liu, Z.; Winters, M.; Holodniy, M.; Dai, H. siRNA Delivery into Human T Cells and Primary Cells with Carbon-Nanotube Transporters. *Angew. Chem. Int. Ed. Engl.* **2007,** *46,* 2023–2027. http://dx.doi.org/10.1002/anie.200604295.
55. Yang, R.; Yang, X.; Zhang, Z.; Zhang, Y.; Wang, S.; Cai, Z.; et al. Single-Walled Carbon Nanotubes-Mediated In Vivo and In Vitro Delivery of siRNA into Antigen-Presenting Cells. *Gene Ther.* **2006,** *13,* 1714–1723. http://dx.doi.org/10.1038/sj.gt.3302808.
56. Jaggi, J. S.; V Seshan, S.; McDevitt, M. R.; Sgouros, G.; Hyjek, E.; Scheinberg, D. A. Mitigation of Radiation Nephropathy after Internal Alpha-Particle Irradiation of Kidneys. *Int. J. Radiat. Oncol. Biol. Phys.* **2006,** *64,* 1503–1512. http://dx.doi.org/10.1016/j.ijrobp.2005.11.036.
57. McDevitt, M. R.; Chattopadhyay, D.; Kappel, B. J.; Jaggi, J. S.; Schiffman, S. R.; Antczak, C.; et al. Tumor Targeting with Antibody-Functionalized, Radiolabeled Carbon Nanotubes. *J. Nucl. Med.* **2007,** *48,* 1180–1189. http://dx.doi.org/10.2967/jnumed.106.039131.
58. Liu, Z.; Chen, K.; Davis, C.; Sherlock, S.; Cao, Q.; Chen, X.; et al. Drug Delivery with Carbon Nanotubes for In Vivo Cancer Treatment. *Cancer Res.* **2008,** *68,* 6652–6660. http://dx.doi.org/10.1158/0008-5472.CAN-08-1468.

59. Régina, A.; Demeule, M.; Ché, C.; Lavallée, I.; Poirier, J.; Gabathuler, R.; et al. Antitumour Activity of ANG1005, a Conjugate between Paclitaxel and the New Brain Delivery Vector Angiopep-2. *Br. J. Pharmacol.* **2009,** *155,* 185–197. http://dx.doi.org/10.1038/bjp.2008.260.
60. *No Title.* http://www.mhra.gov.uk/index.htm, (n.d.).
61. Gratton, S. E. A.; Ropp, P. A.; Pohlhaus, P. D.; Luft, J. C.; Madden, V. J.; Napier, M. E.; et al. The Effect of Particle Design on Cellular Internalization Pathways. *Proc. Natl. Acad. Sci. USA* **2008,** *105,* 11613–11618. http://dx.doi.org/10.1073/pnas.0801763105.
62. Ferrari, M. Nanogeometry: beyond Drug Delivery. *Nat. Nanotechnol.* **2008,** *3,* 131–132. http://dx.doi.org/10.1038/nnano.2008.46.
63. Gorelik, E.; Lokshin, A. *Multifactorial Assay for Cancer Detection;* , 2004.
64. Diamandis, E. P. Analysis of Serum Proteomic Patterns for Early Cancer Diagnosis: Drawing Attention to Potential Problems. *J. Natl. Cancer Inst.* **2004,** *96,* 353–356.
65. Sawyers, C. L. The Cancer Biomarker Problem. *Nature* **2008,** *452,* 548–552.
66. Eloubeidi, M. A.; Chen, V. K.; Eltoum, I. A.; Jhala, D.; Chhieng, D. C.; Jhala, N.; et al. Endoscopic Ultrasound–Guided Fine Needle Aspiration Biopsy of Patients with Suspected Pancreatic Cancer: Diagnostic Accuracy and Acute and 30-day Complications. *Am. J. Gastroenterol.* **2003,** *98,* 2663–2668.
67. Ono, H. Early Gastric Cancer: Diagnosis, Pathology, Treatment Techniques and Treatment Outcomes. *Eur. J. Gastroenterol. Hepatol.* **2006,** *18,* 863–866.
68. Huang, S.; Liu, S. Y.; Wang, K.; Yang, C. J.; Luo, Y. M.; Zhang, Y. D.; et al. Highly Fluorescent and Bioresorbable Polymeric Nanoparticles with Enhanced Photostability for Cell Imaging. *Nanoscale* **2015,** *7,* 889–895. http://dx.doi.org/10.1039/c4nr05576d.
69. Oh, G.; Yoo, S. W.; Jung, Y.; Ryu, Y. M.; Park, Y.; Kim, S. Y.; et al. Intravital Imaging of Mouse Colonic Adenoma Using MMP-Based Molecular Probes with Multi-Channel Fluorescence Endoscopy. *Biomed. Opt. Express* **2014,** *5,* 1677–1689. http://dx.doi.org/10.1364/boe.5.001677.
70. Kells, K. R.; Kong, K. Y.; White, W. B.; Kaddi, C.; Wang, M. D. IEEE, LED Light Source for Fluorescence Endoscopy Using Quantum Dots. *2013 IEEE Point-of-Care Healthc. Technol.* **2013,** 9–12. <Go to ISI>://WOS:000318858100003.
71. Glover, S. C.; Tretiakova, M. S.; Carroll, R. E.; Benya, R. V. Increased Frequency of Gastrin-releasing Peptide Receptor Gene Mutations during Colon-Adenocarcinoma Progression. *Mol. Carcinog.* **2003,** *37,* 5–15. http://dx.doi.org/10.1002/mc.10117.
72. Hu, K.; Wang, H.; Tang, G.; Huang, T.; Tang, X.; Liang, X.; et al. In Vivo Cancer Dual-Targeting and Dual-Modality Imaging with Functionalized Quantum Dots. *J. Nucl. Med.* **2015,** *56,* 1278–1284. http://dx.doi.org/10.2967/jnumed.115.158873.
73. Li, S.-D.; Huang, L. Pharmacokinetics and Biodistribution of Nanoparticles. *Mol. Pharm.* **2008,** *5,* 496–504.
74. Mai, W. X.; Meng, H. Mesoporous Silica Nanoparticles: a Multifunctional Nano Therapeutic System. *Integr. Biol.* **2013,** *5,* 19–28.
75. Mirabello, V.; Calatayud, D. G.; Arrowsmith, R. L.; Pascu, S. I. Metallic Nanoparticles as Synthetic Building Blocks for Cancer Nanodiagnostics: From Materials Design to Molecular Imaging Applications. *J. Mater. Chem. B* **2015**. http://dx.doi.org/10.1039/c5tb00841g.
76. Bangham, A. D.; Standish, M. M.; Watkins, J. C. Diffusion of Univalent Ions across the Lamellae of Swollen Phospholipids. *J. Mol. Biol.* **1965,** *13,* 238–252.
77. Kroon, J.; Metselaar, J. M.; Storm, G.; van der Pluijm, G. Liposomal Nanomedicines in the Treatment of Prostate Cancer. *Cancer Treat. Rev.* **2014,** *40,* 578–584. http://dx.doi.org/10.1016/j.ctrv.2013.10.005.
78. Liang, F.; Chen, B. A Review on Biomedical Applications of Single-Walled Carbon Nanotubes. *Curr. Med. Chem.* **2010,** *17,* 10–24.

79. Li, J.; Chang, X.; Chen, X.; Gu, Z.; Zhao, F.; Chai, Z.; et al. Toxicity of Inorganic Nanomaterials in Biomedical Imaging. *Biotechnol. Adv.* **2014,** *32,* 727–743.
80. Ling, D.; Lee, N.; Hyeon, T. Chemical Synthesis and Assembly of Uniformly Sized Iron Oxide Nanoparticles for Medical Applications. *Acc. Chem. Res.* **2015,** *48,* 1276–1285. http://dx.doi.org/10.1021/acs.accounts.5b00038.
81. Gu, F. X. Nanotechnology and Diagnostic Imaging: New Advances in Contrast Agent Technology. *J. Nanomed. Nanotechnol.* **2011.**
82. Boros, E.; Bowen, A. M.; Josephson, L.; Vasdev, N.; Holland, J. P. Chelate-Free Metal Ion Binding and Heat-Induced Radiolabeling of Iron Oxide Nanoparticles. *Chem. Sci.* **2015,** *6,* 225–236. http://dx.doi.org/10.1039/C4SC02778G.
83. Tréhin, R.; Figueiredo, J.-L.; Pittet, M. J.; Weissleder, R.; Josephson, L.; Mahmood, U. Fluorescent Nanoparticle Uptake for Brain Tumor Visualization. *Neoplasia* **2006,** *8,* 302–311.
84. Chen, F.; Ellison, P. A.; Lewis, C. M.; Hong, H.; Zhang, Y.; Shi, S.; et al. Chelator-Free Synthesis of a Dual-Modality PET/MRI Agent. *Angew. Chem. Int. Ed.* **2013,** *52,* 13319–13323. http://dx.doi.org/10.1002/anie.201306306.
85. Chakravarty, R.; Shukla, R.; Ram, R.; Tyagi, A. K.; Dash, A.; Venkatesh, M. Development of a Nano-Zirconia Based $^{68}Ge/^{68}Ga$ Generator for Biomedical Applications. *Nucl. Med. Biol.* **2011,** *38,* 575–583. http://dx.doi.org/10.1016/j.nucmedbio.2010.10.007.
86. de Blois, E.; Sze Chan, H.; Naidoo, C.; Prince, D.; Krenning, E. P.; Breeman, W. A. P. Characteristics of SnO_2-Based $^{68}Ge/^{68}Ga$ Generator and Aspects of Radiolabelling DOTA-Peptides. *Appl. Radiat. Isot.* **2011,** *69,* 308–315. http://dx.doi.org/10.1016/j.apradiso.2010.11.015.
87. Chakravarty, R.; Valdovinos, H. F.; Chen, F.; Lewis, C. M.; Ellison, P. A.; Luo, H.; et al. Intrinsically Germanium-69-Labeled Iron Oxide Nanoparticles: Synthesis and In-Vivo Dual-Modality PET/MR Imaging. *Adv. Mater.* **2014,** *26,* 5119–5123. http://dx.doi.org/10.1002/adma.201401372.
88. Kim, D. K.; Zhang, Y.; Voit, W.; Rao, K. V.; Muhammed, M. Synthesis and Characterization of Surfactant-Coated Superparamagnetic Monodispersed Iron Oxide Nanoparticles. *J. Magn. Magn. Mater.* **2001,** *225,* 30–36. http://dx.doi.org/10.1016/S0304-8853(00)01224-5.
89. Sun, S.; Zeng, H. Size-controlled Synthesis of Magnetite Nanoparticles. *J. Am. Chem. Soc.* **2002,** *124,* 8204–8205. http://dx.doi.org/10.1021/ja026501x.
90. Liao, X.; Zhu, J.; Zhong, W.; Chen, H.-Y. Synthesis of Amorphous Fe_2O_3 Nanoparticles by Microwave Irradiation. *Mater. Lett.* **2001,** *50,* 341–346. http://dx.doi.org/10.1016/S0167-577X(01)00251-8.
91. Wang, W.-W.; Zhu, Y.-J.; Ruan, M.-L. Microwave-Assisted Synthesis and Magnetic Property of Magnetite and Hematite Nanoparticles. *J. Nanoparticle Res.* **2007,** *9,* 419–426. http://dx.doi.org/10.1007/s11051-005-9051-8.
92. Pellico, J.; Lechuga-Vieco, A. V.; Benito, M.; García-Segura, J. M.; Fuster, V.; Ruiz-Cabello, J.; et al. Microwave-Driven Synthesis of Bisphosphonate Nanoparticles Allows In Vivo Visualisation of Atherosclerotic Plaque. *RSC Adv.* **2015,** *5,* 1661–1665. http://dx.doi.org/10.1039/C4RA13824D.
93. Wong, R. M.; Gilbert, D. A.; Liu, K.; Louie, A. Y. Rapid Size-Controlled Synthesis of Dextran-Coated, 64 Cu-Doped Iron Oxide Nanoparticles. *ACS Nano* **2012,** *6,* 3461–3467. http://dx.doi.org/10.1021/nn300494k.
94. Sarpaki, S. *Synthesis of Multimodal PET and SPECT Imaging Agents – Towards Hypoxia Selective Inhibitors*, Unpublished doctoral thesis; University of Bath: Bath, United Kingdom, 2015.
95. Smith, S. V. Molecular Imaging with Copper-64. *J. Inorg. Biochem.* **2004,** *98,* 1874–1901. http://www.sciencedirect.com/science/article/B6TGG-4DBKGV1-1/2/aa90d30469f8d893af4a61ed3b9e76b0.

96. Linder, M. C.; Hazegh-Azam, M. Copper Biochemistry and Molecular Biology. *Am. J. Clin. Nutr.* **1996,** *63,* 797S–811S. http://www.ajcn.org/cgi/content/abstract/63/5/797S.
97. Blower, P. J.; Lewis, J. S.; Zweit, J. Copper Radionuclides and Radiopharmaceuticals in Nuclear Medicine. *Nucl. Med. Biol.* **1996,** *23,* 957–980. http://www.sciencedirect.com/science/article/B6T9Y-3W31862-5/2/78517f4df2886fc95dc507e094186657.
98. Hoffman, T. J.; Smith, C. J. True Radiotracers: Cu-64 Targeting Vectors Based upon Bombesin Peptide. *Nucl. Med. Biol.* **2009,** *36,* 579–585. http://dx.doi.org/10.1016/j.nucmedbio.2009.03.007.
99. Kagadis, G. C.; Loudos, G.; Katsanos, K.; Langer, S. G.; Nikiforidis, G. C. In Vivo Small Animal Imaging: Current Status and Future Prospects. *Med. Phys.* **2010,** *37,* 6421–6442. http://dx.doi.org/10.1118/1.3515456.
100. Obata, A.; Kasamatsu, S.; McCarthy, D. W.; Welch, M. J.; Saji, H.; Yonekura, Y.; et al. Production of Therapeutic Quantities of 64Cu Using a 12MeV Cyclotron. *Nucl. Med. Biol.* **2003,** *30,* 535–539. http://www.sciencedirect.com/science/article/B6T9Y-48WNCV7-C/2/177ee53d48a4c33d48813c3b1a96ade5.
101. Szelecsényi, F.; Blessing, G.; Qaim, S. M. Excitation Functions of Proton Induced Nuclear Reactions on Enriched 61Ni and 64Ni: Possibility of Production of No-Carrier-Added 61Cu and 64Cu at a Small Cyclotron. *Appl. Radiat. Isot.* **1993,** *44,* 575–580. http://www.sciencedirect.com/science/article/B6TJ0-46X3K2D-M/2/a4e76f856a31e3b41b3e2cfa4cad6ae6.
102. Smith, S. V. Molecular Imaging with Copper-64 in the Drug Discovery and Development Arena. *Expert Opin. Drug Discov.* **2007,** *2,* 659–672. http://dx.doi.org/10.1517/17460441.2.5.659.
103. Bonardi, M. L.; Groppi, F.; Mainardi, H. S.; Kokhanyuk, V. M.; V Lapshina, E.; V Mebel, M.; et al. Cross Section Studies on 64Cu with Zinc Target in the Proton Energy Range from 141 Down to 31MeV. *J. Radioanal. Nucl. Chem.* **2005,** *264,* 101–105. http://dx.doi.org/10.1007/s10967-005-0681-1.
104. Szelecsényi, F.; Steyn, G. F.; Kovács, Z.; Vermeulen, C.; van der Meulen, N. P.; Dolley, S. G.; et al. Investigation of the 66Zn(p,2pn)64Cu and 68Zn(p,x)64Cu Nuclear Processes up to 100MeV: Production of 64Cu. *Nucl. Instrum. Methods Phys. Res. Sect. B Beam Interact. Mater. Atoms* **2005,** *240,* 625–637. http://www.sciencedirect.com/science/article/B6TJN-4GNCFVH-5/2/674f7a9937b0820f3c333a3b29cf1425.
105. Tárkányi, F.; Ditrói, F.; Csikai, J.; Takács, S.; Uddin, M. S.; Hagiwara, M.; et al. Activation Cross-Sections of Long-Lived Products of Proton-Induced Nuclear Reactions on Zinc. *Appl. Radiat. Isot.* **2005,** *62,* 73–81. http://www.sciencedirect.com/science/article/B6TJ0-4D3WFMY-2/2/5e5b3a00c68f81f15698c7bc0405d5dd.
106. Jalilian, A. R.; Sabet, M.; Rowshanfarzad, P.; Kamali-dehghan, M.; Akhlaghi, M.; Mirzaii, M. Production of Copper-64 Diacetyl-bis (N4-methylthiosemicarbazone) for Therapeutic Purposes. *Radiopharm. Radiopharmacol.* **2008,** *7,* 166–171.
107. Jensen, A. I.; Binderup, T.; Kumar Ek, P.; Kjær, A.; Rasmussen, P. H.; Andresen, T. L. Positron Emission Tomography Based Analysis of Long-Circulating Cross-Linked Triblock Polymeric Micelles in a U87MG Mouse Xenograft Model and Comparison of DOTA and CB-TE2A as Chelators of Copper-64. *Biomacromolecules* **2014,** *15,* 1625–1633. http://dx.doi.org/10.1021/bm401871w.
108. Zeglis, B. M.; Lewis, J. S. A Practical Guide to the Construction of Radiometallated Bioconjugates for Positron Emission Tomography. *Dalton Trans.* **2011,** *40,* 6168–6195. http://dx.doi.org/10.1039/C0DT01595D.
109. Wadas, T. J.; Wong, E. H.; Weisman, G. R.; Anderson, C. J. Coordinating Radiometals of Copper, Gallium, Indium, Yttrium, and Zirconium for PET and SPECT Imaging of Disease. *Chem. Rev.* **2010,** *110,* 2858–2902. http://dx.doi.org/10.1021/cr900325h.

110. Cabbiness, D. K.; Margerum, D. W. Macrocyclic Effect on the Stability of Copper(II) Tetramine Complexes. *J. Am. Chem. Soc.* **1969,** *91,* 6540–6541. http://dx.doi.org/10.1021/ja01051a091.
111. Liu, S. Bifunctional Coupling Agents for Radiolabeling of Biomolecules and Target-Specific Delivery of Metallic Radionuclides. *Adv. Drug Deliv. Rev.* **2008,** *60,* 1347–1370. http://www.sciencedirect.com/science/article/B6T3R-4SBSDCS-3/2/0fb16df56a2a1fe8fe9cbc1a03d3442c.
112. Meares, C. F. Chelating Agents for the Binding of Metal Ions to Antibodies. *Int. J. Radiat. Appl. Instrum. Part B Nucl. Med. Biol.* **1986,** *13,* 311–318. http://www.sciencedirect.com/science/article/B7GH9-4D4KGSS-3/2/e1ab2b34fa25bb5ae84c6277c9e01214.
113. Chen, X.; Liu, S.; Hou, Y.; Tohme, M.; Park, R.; Bading, J. R.; et al. MicroPET Imaging of Breast Cancer [Alpha]v-Integrin Expression with 64Cu-Labeled Dimeric RGD Peptides. *Mol. Imaging Biol.* **2004,** *6,* 350–359. http://www.sciencedirect.com/science/article/B6X15-4DCDW84-9/2/72c4c59cd25f46328a6b7fd4966f951c.
114. Chen, X.; Park, R.; Tohme, M.; Shahinian, A. H.; Bading, J. R.; Conti, P. S. MicroPET and Autoradiographic Imaging of Breast Cancer Î±v-Integrin Expression Using 18F- and 64Cu-Labeled RGD Peptide. *Bioconjug. Chem.* **2003,** *15,* 41–49. http://dx.doi.org/10.1021/bc0300403.
115. Lewis, J. S.; Srinivasan, A.; Schmidt, M. A.; Anderson, C. J. In Vitro and In Vivo Evaluation of 64Cu-TETA-Tyr3-Octreotate. A New Somatostatin Analog with Improved Target Tissue Uptake. *Nucl. Med. Biol.* **1999,** *26,* 267–273. http://www.sciencedirect.com/science/article/B6T9Y-3W4XHXR-3/2/f2fcf08f65992d492c337a939153b8cd.
116. Li, W. P.; Lewis, J. S.; Kim, J.; Bugaj, J. E.; Johnson, M. A.; Erion, J. L.; et al. DOTA-D-Tyr1-Octreotate: A Somatostatin Analogue for Labeling with Metal and Halogen Radionuclides for Cancer Imaging and Therapy. *Bioconjug. Chem.* **2002,** *13,* 721–728. http://dx.doi.org/10.1021/bc015590k.
117. Parry, J.; Andrews, R.; Rogers, B. MicroPET Imaging of Breast Cancer Using Radiolabeled Bombesin Analogs Targeting the Gastrin-Releasing Peptide Receptor. *Breast Cancer Res. Treat.* **2007,** *101,* 175–183. http://dx.doi.org/10.1007/s10549-006-9287-8.
118. Parry, J. J.; Kelly, T. S.; Andrews, R.; Rogers, B. E. In Vitro and in Vivo Evaluation of ^{64}Cu-Labeled DOTA−Linker−Bombesin(7−14) Analogues Containing Different Amino Acid Linker Moieties. *Bioconjug. Chem.* **2007,** *18,* 1110–1117. http://dx.doi.org/10.1021/bc0603788.
119. Wadas, T. J.; Anderson, C. J. Radiolabeling of TETA- and CB-TE2A-Conjugated Peptides with Copper-64. *Nat. Protoc.* **2007,** *1,* 3062–3068. http://dx.doi.org/10.1038/nprot.2006.431.
120. Wei, L.; Easmon, J.; Nagi, R. K.; Muegge, B. D.; Meyer, L. A.; Lewis, J. S. 64Cu-Azabicyclo[3.2.2]Nonane Thiosemicarbazone Complexes: Radiopharmaceuticals for PET of Topoisomerase II Expression in Tumors. *J. Nucl. Med.* **2006,** *47,* 2034–2041. http://jnm.snmjournals.org/content/47/12/2034.
121. Laforest, R.; Dehdashti, F.; Lewis, J. S.; Schwarz, S. W. Dosimetry of 60/61/62/64Cu-ATSM: A Hypoxia Imaging Agent for PET. *Eur. J. Nucl. Med. Mol. Imaging* **2005,** *32,* 764–770.
122. Matsumoto, K.; Fujibayashi, Y.; Konishi, J.; Yokoyama, A. Radiolabelling and Biodistribution of 62Cu-Dithiocarbamate – An Application for the New 62Zn/62Cu Generator. *Radioisotopes* **1990,** *39,* 482–486.
123. Alidori, S.; Gioia Lobbia, G.; Papini, G.; Pellei, M.; Porchia, M.; Refosco, F.; et al. Synthesis, In Vitro and In Vivo Characterization of 64Cu(I) Complexes Derived from Hydrophilic Tris(hydroxymethyl)phosphane and 1,3,5-triaza-7-phosphaadamantane Ligands. *J. Biol. Inorg. Chem.* **2008,** *13,* 307–315. http://dx.doi.org/10.1007/s00775-007-0322-y.
124. Bähr, G.; Edith, H. Über emprotide Schwermetall-Innerkomplexe der alpha-Diketon-dithiosemicarbazone (Thiazone). I, Zeitschrift Für Anorg. *Und Allg. Chem.* **1952,** *268,* 351–363. http://dx.doi.org/10.1002/zaac.19522680414.

125. French, F. A.; Freedlander, B. L.; Hoskino, A.; French, J. Carcinostatic Action of Polycarbonyl Compounds and Their Derivatives. *Cancer Res.* **1958,** *18,* 1290–1300. http://cancerres.aacrjournals.org/content/18/11/1290.abstract.
126. Gingras, B.A.; Hornal, R.W.; Bayley, C. H.The Preparation of Some Thiosemicarbazones and Their Copper Complexes. Part I. *Can. J. Chem.* **1960,** *38,* 712–719.
127. Gingras, B.A.; Somorjai, R. L.; Bayley, C. H.The Preparation of Some Thiosemicarbazones and Their Copper Complexes. *Can. J. Chem.* **1961,** *39,* 973–985.
128. Gingras, B.A.; Suprunchuk, T.; Bayley, C. H.The Preparation of Some Thiosemicarbazones and Their Copper Complexes: Part III. *Can. J. Chem.* **1962,** *40,* 1053–1059.
129. Petering, D. H. The Reaction Of 3-Ethoxy-2-Oxobutyraldehyde bis (Thiosemicarbazonato) Copper(Ii) with Thiols. *Bioinorg. Chem.* **1972,** *1,* 273–288. http://www.sciencedirect.com/science/article/B6W7Y-4344GXW-2/2/717ac8d0a40906e1b90c6d7b07890870.
130. Kessel, D.; McElhinney, R. S. The Role of Metals in the Antitumor Action of 1,5-Bisthiosemicarbazones. *Mol. Pharmacol.* **1975,** *11,* 298–309. http://molpharm.aspetjournals.org/content/11/3/298.abstract.
131. Fujibayashi, Y.; Taniuchi, H.; Yonekura, Y.; Ohtani, H.; Konishi, J.; Yokoyama, A. Copper-62-ATSM: A New Hypoxia Imaging Agent with High Membrane Permeability and Low Redox Potential. *J. Nucl. Med.* **1997,** *38,* 1155–1160. http://jnm.snmjournals.org/cgi/content/abstract/38/7/1155.
132. Dearling, J. L. J.; Lewis, J. S.; Mullen, G. E. D.; Rae, M. T.; Zweit, J.; Blower, P. J. Design of Hypoxia-Targeting Radiopharmaceuticals: Selective Uptake of Copper-64 Complexes in Hypoxic Cells In Vitro. *Eur. J. Nucl. Med. Mol. Imaging* **1998,** *25,* 788–792. http://dx.doi.org/10.1007/s002590050283.
133. Lewis, J. S.; McCarthy, D. W.; McCarthy, T. J.; Fujibayashi, Y.; Welch, M. J. Evaluation of 64Cu-ATSM In Vitro and In Vivo in a Hypoxic Tumor Model. *J. Nucl. Med.* **1999,** *40,* 177–183.
134. Lewis, J. S.; Sharp, T. L.; Laforest, R.; Fujibayashi, Y.; Welch, M. J. Tumor Uptake of Copper-diacetyl-bis(n4-methylthiosemicarbazone): Effect of Changes in Tissue Oxygenation. *J. Nucl. Med.* **2001,** *42,* 655–661. http://jnm.snmjournals.org/cgi/content/abstract/42/4/655.
135. O'Donoghue, J. A.; Zanzonico, P.; Pugachev, A.; Wen, B.; Smith-Jones, P.; Cai, S.; et al. Assessment of Regional Tumor Hypoxia Using ^{18}F-Fluoromisonidazole and ^{64}Cu(II)-Diacetyl-bis(N4-Methylthiosemicarbazone) Positron Emission Tomography: Comparative Study Featuring Micropet Imaging, Po_2 Probe Measurement, Autoradiography, and Fluorescent Microscopy in the R3327-AT and FaDu Rat Tumor Models. *Int. J. Radiat. Oncol.* **2005,** *61,* 1493–1502. http://www.sciencedirect.com/science/article/B6T7X-4FW7F5J-9/2/a5fca7c69ecfc2321e9622723ffa168a.
136. Burgman, P.; O'Donoghue, J. A.; Lewis, J. S.; Welch, M. J.; Humm, J. L.; Ling, C. C. Cell Line-Dependent Differences in Uptake and Retention of the Hypoxia-Selective Nuclear Imaging Agent Cu-ATSM. *Nucl. Med. Biol.* **2005,** *32,* 623–630. http://www.sciencedirect.com/science/article/B6T9Y-4GMY0BW-8/2/936822304f68fdddb0e4b999c99a5a0a.
137. Yuan, H.; Schroeder, T.; Bowsher, J. E.; Hedlund, L. W.; Wong, T.; Dewhirst, M. W. Intertumoral Differences in Hypoxia Selectivity of the PET Imaging Agent 64Cu(II)-Diacetyl-bis(N4-Methylthiosemicarbazone). *J. Nucl. Med.* **2006,** *47,* 989–998. http://jnm.snmjournals.org/cgi/content/abstract/47/6/989.
138. Takahashi, N.; Fujibayashi, Y.; Yonekura, Y.; Welch, M. J.; Waki, A.; Tsuchda, T.; et al. Evaluation of 62Cu Labeled Diacetyl-bis(N4-Methylthiosemicarbazone) as a Hypoxic Tissue Tracer in Patients with Lung Cancer. *Ann. Nucl. Med.* **2000,** *14,* 323–328.
139. Dehdashti, F.; Grigsby, P. W.; Lewis, J. S.; Laforest, R.; Siegel, B. A.; Welch, M. J. Assessing Tumor Hypoxia in Cervical Cancer by Positron Emission Tomography with 60Cu-ATSM. *J. Nucl. Med.* **2008,** *49,* 201–205.

140. Dehdashti, F.; Mintun, M. A.; Lewis, J. S.; Bradley, J.; Govindan, R.; Laforest, R.; et al. In Vivo Assessment of Tumor Hypoxia in Lung Cancer with 60Cu-ATSM. *Eur. J. Nucl. Med. Mol. Imaging* **2003,** *30,* 844–850. http://dx.doi.org/10.1007/s00259-003-1130-4.
141. Dietz, D.W.; Dehdashti, F.; Grigsby, P.W.; Malyapa, R. S.; Myerson, R. J.; Picus, J.; et al. Tumor Hypoxia Detected by Positron Emission Tomography with 60Cu-ATSM as a Predictor of Response and Survival in Patients Undergoing Neoadjuvant Chemoradiotherapy for Rectal Carcinoma: A Pilot Study. *Dis. Colon Rectum* **2008,** *51,* 1641–1643. http://journals.lww.com/dcrjournal/Fulltext/2008/51110/Tumor_Hypoxia_Detected_by_Positron_Emission.7.aspx.
142. Lewis, J. S.; Laforest, R.; Dehdashti, F.; Grigsby, P. W.; Welch, M. J.; Siegel, B. A. An Imaging Comparison of 64Cu-ATSM and 60Cu-ATSM in Cancer of the Uterine Cervix. *J. Nucl. Med.* **2008,** *49,* 1177–1182.
143. Dehdashti, F.; Grigsby, P. W.; Mintun, M. A.; Lewis, J. S.; Siegel, B. A.; Welch, M. J. Assessing Tumor Hypoxia in Cervical Cancer by Positron Emission Tomography with 60Cu-ATSM: Relationship to Therapeutic Response—A Preliminary Report. *Int. J. Radiat. Oncol.* **2003,** *55,* 1233–1238.
144. Dunphy, M. P. S.; Lewis, J. S. Radiopharmaceuticals in Preclinical and Clinical Development for Monitoring of Therapy with PET. *J. Nucl. Med.* **2009,** *50,* 106S–121S. http://jnm.snmjournals.org/cgi/content/abstract/50/Suppl_1/106S.
145. Obata, A.; Yoshimi, E.; Waki, A.; Lewis, J. S.; Oyama, N.; Welch, M. J.; et al. Retention Mechanism of Hypoxia Selective Nuclear Imaging/Radiotherapeutic Agent Cu-Diacetyl-bis(N4-Methylthiosemicarbazone) (Cu-ATSM) in Tumor Cells. *Ann. Nucl. Med.* **2001,** *15,* 499–504.
146. Holland, J. P.; Giansiracusa, J. H.; Bell, S. G.; Wong, L.-L.; Dilworth, J. R. In Vitro Kinetic Studies on the Mechanism of Oxygen-Dependent Cellular Uptake of Copper Radiopharmaceuticals. *Phys. Med. Biol.* **2009,** *54,* 2103. http://stacks.iop.org/0031-9155/54/i=7/a=017.
147. Holland, J. P.; Barnard, P. J.; Collison, D.; Dilworth, J. R.; Edge, R.; Green, J. C.; et al. Spectroelectrochemical and Computational Studies on the Mechanism of Hypoxia Selectivity of Copper Radiopharmaceuticals. *Chem. Eur. J.* **2008,** *14,* 5890–5907. http://dx.doi.org/10.1002/chem.200800539.
148. Holland, J. P.; Green, J. C.; Dilworth, J. R. Probing the Mechanism of Hypoxia Selectivity of Copper bis(Thiosemicarbazonato) Complexes: DFT Calculation Of Redox Potentials and Absolute Acidities in Solution. *Dalton Trans.* **2006,** 783–794. http://dx.doi.org/10.1039/b512656h.
149. Bonnitcha, P. D.; Vavere, A. L.; Lewis, J. S.; Dilworth, J. R. In Vitro and In Vivo Evaluation of Bifunctional Bisthiosemicarbazone 64Cu-Complexes for the Positron Emission Tomography Imaging of Hypoxia. *J. Med. Chem.* **2008,** *51,* 2985–2991. http://dx.doi.org/10.1021/jm800031x.
150. Holland, J. P.; Divilov, V.; Bander, N. H.; Smith-Jones, P. M.; Larson, S. M.; Lewis, J. S. 89Zr-DFO-J591 for ImmunoPET of Prostate-Specific Membrane Antigen Expression In Vivo. *J. Nucl. Med.* **2010,** *51,* 1293–1300. http://dx.doi.org/10.2967/jnumed.110.076174.
151. Nagengast, W. B.; de Vries, E. G.; Hospers, G. A.; Mulder, N. H.; de Jong, J. R.; Hollema, H.; et al. In Vivo VEGF Imaging with Radiolabeled Bevacizumab in a Human Ovarian Tumor Xenograft. *J. Nucl. Med.* **2007,** *48,* 1313–1319. http://dx.doi.org/10.2967/jnumed.107.041301.
152. Holland, J. P.; Evans, M. J.; Rice, S. L.; Wongvipat, J.; Sawyers, C. L.; Lewis, J. S. Annotating MYC Status with 89Zr-Transferrin Imaging. *Nat. Med.* **2012,** *18,* 1586–1591. http://www.nature.com/nm/journal/v18/n10/abs/nm.2935.html#supplementary-information.
153. Reichert, D. E.; Lewis, J. S.; Anderson, C. J. Metal Complexes as Diagnostic Tools. *Coord. Chem. Rev.* **1999,** *184,* 3–66. http://www.sciencedirect.com/science/article/B6TFW-3XG177K-2/2/5185411cf47d61e0d56a2268b3f9740f.

154. Alford, R.; Ogawa, M.; Choyke, P. L.; Kobayashi, H. Molecular Probes for the In Vivo Imaging of Cancer. *Mol. Biosyst.* **2009,** *5,* 1279–1291. http://dx.doi.org/10.1039/b911307j.
155. Liu, S. The Role of Coordination Chemistry in the Development of Target-Specific Radiopharmaceuticals. *Chem. Soc. Rev.* **2004,** *33,* 445–461. http://dx.doi.org/10.1039/b309961j.
156. Guo, Z.; Sadler, P. J. Metals in Medicine. *Angew. Chem. Int. Ed.* **1999,** *38,* 1512–1531. http://dx.doi.org/10.1002/(SICI)1521-3773(19990601)38:11<1512::AID-ANIE1512>3.0.CO;2-Y.
157. Dilworth, J. R.; Parrott, S. J. The Biomedical Chemistry of Technetium and Rhenium. *Chem. Soc. Rev.* **1998,** *27,* 43–55.
158. Anderson, C. J.; Welch, M. J. Radiometal-Labeled Agents (Non-Technetium) for Diagnostic Imaging. *Chem. Rev.* **1999,** *99,* 2219–2234. http://dx.doi.org/10.1021/cr980451q.
159. Green, M. A. *Gallium and Copper Radiopharmaceutical Chemistry, New York;* , 1991.
160. Yang, W.; Ratinac, K. R.; Ringer, S. P.; Thordarson, P.; Gooding, J. J.; Braet, F. Carbon Nanomaterials in Biosensors: Should You Use Nanotubes or Graphene? *Angew. Chem. Int. Ed.* **2010,** *49,* 2114–2138. http://dx.doi.org/10.1002/anie.200903463.
161. Novoselov, K. S.; Geim, A. K.; V Morozov, S.; Jiang, D.; Zhang, Y.; Dubonos, S. V.; et al. Electric Field Effect in Atomically Thin Carbon Films. *Science* **2004,** *306,* 666–669. http://dx.doi.org/10.1126/science.1102896.
162. Xu, G.; Yong, K.-T.; Roy, I.; Mahajan, S. D.; Ding, H.; Schwartz, S. A.; et al. Bioconjugated Quantum Rods as Targeted Probes for Efficient Transmigration across an In Vitro Blood−Brain Barrier. *Bioconjug. Chem.* **2008,** *19,* 1179–1185. http://dx.doi.org/10.1021/bc700477u.
163. Kafa, H.; Wang, J. T. -W.; Rubio, N.; Venner, K.; Anderson, G.; Pach, E.; et al. The Interaction of Carbon Nanotubes with an In Vitro Blood–Brain Barrier Model and Mouse Brain In Vivo. *Biomaterials* **2015,** *53,* 437–452. http://dx.doi.org/10.1016/j.biomaterials.2015.02.083.
164. Hodge, S. A.; Bayazit, M. K.; Coleman, K. S.; Shaffer, M. S. P. Unweaving the Rainbow: A Review of the Relationship between Single-Walled Carbon Nanotube Molecular Structures and Their Chemical Reactivity. *Chem. Soc. Rev.* **2012,** *41,* 4409–4429. http://dx.doi.org/10.1039/c2cs15334c.
165. Prasek, J.; Drbohlavova, J.; Chomoucka, J.; Hubalek, J.; Jasek, O.; Adam, V.; et al. Methods for Carbon Nanotubes Synthesis-Review. *J. Mater. Chem.* **2011,** *21,* 15872–15884. http://dx.doi.org/10.1039/c1jm12254a.
166. Iijima, S. Helical Microtubules of Graphitic Carbon. *Nature* **1991,** *354,* 56–58. http://dx.doi.org/10.1038/354056a0.
167. Iijima, S.; Ichihashi, T. Single-Shell Carbon Nanotubes of 1-nm Diameter. *Nature* **1993,** *363,* 603–605. http://dx.doi.org/10.1038/363603a0.
168. Liu, W.-W.; Chai, S.-P.; Mohamed, A. R.; Hashim, U. Synthesis and Characterization of Graphene and Carbon Nanotubes: A Review on the Past and Recent Developments. *J. Ind. Eng. Chem.* **2013**. http://dx.doi.org/10.1016/j.jiec.2013.08.028. http://dx.doi.org/10.1016/j.jiec.2013.08.028.
169. Saito, R. Dresselhaus, M.S.; Dresselhaus, G. *Physical Properties of Carbon Nanotubes;* Imperial College Press: London, 1998. http://app.knovel.com/web/toc.v/cid:kpPPCN0006/viewerType:toc/root_slug:physical_properties_of_carbon_nanotubes.
170. Kumar, M. *Carbon Nanotube Synthesis and Growth Mechanism;* InTech, 2011. 16802.
171. Maser, W. K.; Benito, A. M.; Martínez, M. T. Production of Carbon Nanotubes: The Light Approach. *Carbon N. Y.* **2002,** *40,* 1685–1695. http://dx.doi.org/10.1016/S0008-6223(02)00009-X.
172. Paradise, M.; Goswami, T. Carbon Nanotubes – Production and Industrial Applications. *Mater. Des.* **2007,** *28,* 1477–1489. http://dx.doi.org/10.1016/j.matdes.2006.03.008.
173. Dai, H. Carbon Nanotubes: Opportunities and Challenges. *Surf. Sci.* **2002,** *500,* 218–241. http://dx.doi.org/10.1016/S0039-6028(01)01558-8.

174. Ghosh, K.; Kumar, M.; Maruyama, T.; Ando, Y. Micro-Structural, Electron-Spectroscopic and Field-Emission Studies of Carbon Nitride Nanotubes Grown from Cage-like and Linear Carbon Sources. *Carbon N. Y.* **2009,** *47,* 1565–1575. http://dx.doi.org/10.1016/j.carbon.2009.02.007.
175. Hou, P.-X.; Liu, C.; Cheng, H.-M. Purification of Carbon Nanotubes. *Carbon N. Y.* **2008,** *46,* 2003–2025. http://dx.doi.org/10.1016/j.carbon.2008.09.009.
176. Kruusenberg, I.; Alexeyeva, N.; Tammeveski, K.; Kozlova, J.; Matisen, L.; Sammelselg, V.; et al. Effect of Purification of Carbon Nanotubes on Their Electrocatalytic Properties for Oxygen Reduction in Acid Solution. *Carbon N. Y.* **2011,** *49,* 4031–4039. http://dx.doi.org/10.1016/j.carbon.2011.05.048.
177. Datsyuk, V.; Kalyva, M.; Papagelis, K.; Parthenios, J.; Tasis, D.; Siokou, A.; et al. Chemical Oxidation of Multiwalled Carbon Nanotubes. *Carbon N.Y.* **2008,** *46,* 833–840. http://dx.doi.org/10.1016/j.carbon.2008.02.012.
178. Tobias, G.; Shao, L.; Salzmann, C. G.; Huh, Y.; Green, M. L. H. Purification and Opening of Carbon Nanotubes Using Steam. *J. Phys. Chem. B* **2006,** *110,* 22318–22322. http://dx.doi.org/10.1021/jp0631883.
179. Shin, D. H.; Kim, J.-E.; Shim, H. C.; Song, J.-W.; Yoon, J.-H.; Kim, J.; et al. Continuous Extraction of Highly Pure Metallic Single-Walled Carbon Nanotubes in a Microfluidic Channel. *Nano Lett.* **2008,** *8,* 4380–4385. http://dx.doi.org/10.1021/nl802237m.
180. Rodriguez-Yañez, Y.; Muñoz, B.; Albores, A. Mechanisms of Toxicity by Carbon Nanotubes. *Toxicol. Mech. Methods* **2013,** *23,* 178–195. http://dx.doi.org/10.3109/15376516.2012.754534.
181. Donaldson, K.; Poland, C. A.; Murphy, F. A.; MacFarlane, M.; Chernova, T.; Schinwald, A. Pulmonary Toxicity of Carbon Nanotubes and Asbestos—Similarities and Differences. *Adv. Drug Deliv. Rev.* **2013,** *65,* 2078–2086. http://dx.doi.org/10.1016/j.addr.2013.07.014.
182. Firme, C. P., III; Bandaru, P. R. Toxicity Issues in the Application of Carbon Nanotubes to Biological Systems. *Nanomed. Nanotechnol. Biol. Med.* **2010,** *6,* 245–256. http://dx.doi.org/10.1016/j.nano.2009.07.003.
183. Kotchey, G. P.; Zhao, Y.; Kagan, V. E.; Star, A. Peroxidase-mediated Biodegradation of Carbon Nanotubes In Vitro and In Vivo. *Adv. Drug Deliv. Rev.* **2013,** *65,* 1921–1932. http://dx.doi.org/10.1016/j.addr.2013.07.007.
184. Heister, E.; Brunner, E. W.; Dieckmann, G. R.; Jurewicz, I.; Dalton, A. B. Are Carbon Nanotubes a Natural Solution? Applications in Biology and Medicine. *ACS Appl. Mater. Interfaces* **2013,** *5,* 1870–1891. http://dx.doi.org/10.1021/am302902d.
185. Battigelli, A.; Ménard-Moyon, C.; Da Ros, T.; Prato, M.; Bianco, A. Endowing Carbon Nanotubes with Biological and Biomedical Properties by Chemical Modifications. *Adv. Drug Deliv. Rev.* **2013,** *65,* 1899–1920. http://dx.doi.org/10.1016/j.addr.2013.07.006.
186. Kam, N. W. S.; O'Connell, M.; Wisdom, J. A.; Dai, H. Carbon Nanotubes as Multifunctional Biological Transporters and Near-Infrared Agents for Selective Cancer Cell Destruction. *Proc. Natl. Acad. Sci. USA* **2005,** *102,* 11600–11605. http://dx.doi.org/10.1073/pnas.0502680102.
187. (a) Gong, H.; Peng, R.; Liu, Z. Carbon Nanotubes for Biomedical Imaging: The Recent Advances. *Adv. Drug Deliv. Rev.* **2013,** *65,* 1951–1963. http://dx.doi.org/10.1016/j.addr.2013.10.002.
(b) Wen, J.; Xu, Y.; Li, H.; Lu, A.; Sun, S. Recent Applications of Carbon Nanomaterials in Fluorescence Biosensing and Bioimaging. *Chem. Commun.* **2015,** *51,* 11346–11358.
(c) Bartelmess, J.; Quinn, S. J.; Giordani, S. Carbon Nanomaterials: Multi-Functional Agents for Biomedical Fluorescence and Raman Imaging. *Chem. Soc. Rev.* **2015,** *44,* 4672–4698.
(d) Hong, G.; Diao, S.; Antaris, A. L.; Dai, H. Carbon Nanomaterials for Biological Imaging and Nanomedicinal Therapy. *Chem. Rev.* **2015,** *115,* 10816–10906.

188. Zhu, H.W.; Xu, C. L.; Wu, D. H.; Wei, B. Q.; Vajtai, R.; Ajayan, P. M. Direct Synthesis of Long Single-Walled Carbon Nanotube Strands. *Science* **2002,** *296,* 884–886.
189. Madani, S.Y.; Naderi, N.; Dissanayake, O.; Tan, A.; Seifalian, A. M. A New Era of Cancer Treatment: Carbon Nanotubes as Drug Delivery Tools. *Int. J. Nanomed.* **2011,** *6,* 2963.
190. Imtani, A. N.; Jindal, V. K. Structure of Chiral Single-Walled Carbon Nanotubes under Hydrostatic Pressure. *Comput. Mater. Sci.* **2009,** *46,* 297–302.
191. Hamada, N.; Sawada, S.; Oshiyama, A. New One-dimensional Conductors: Graphitic Microtubules. *Phys. Rev. Lett.* **1992,** *68,* 1579.
192. Aqel, A.; El-Nour, K. M. M.A.; Ammar, R. A. A.; Al-Warthan, A. Carbon Nanotubes, Science and Technology Part (I) Structure, Synthesis and Characterisation. *Arab. J. Chem.* **2012,** *5,* 1–23. http://dx.doi.org/10.1016/j.arabjc.2010.08.022.
193. Jeon, I.-Y.; Chang, D.W.; Kumar, N. A.; Baek, J.-B. *Functionalization of Carbon Nanotubes*, 2011.
194. Beg, S.; Rizwan, M.; Sheikh, A. M.; Hasnain, M. S.; Anwer, K.; Kohli, K. Advancement in Carbon Nanotubes: Basics, Biomedical Applications and Toxicity. *J. Pharm. Pharmacol.* **2011,** *63,* 141–163.
195. Zhao, E. H.; Ergul, B.; Zhao, W. Caffeine's Antioxidant Potency Optically Sensed with Double-Stranded DNA-Encased Single-Walled Carbon Nanotubes. *J. Phys. Chem. B* **2015,** *119,* 4068–4075. http://dx.doi.org/10.1021/acs.jpcb.5b00708.
196. Zhao, W. Lighting up H_2O_2: the Molecule that Is a "Necessary Evil" in the Cell. *Angew. Chem. Int. Ed.* **2009,** *48,* 3022–3024.
197. Dickinson, B. C.; Chang, C. J. Chemistry and Biology of Reactive Oxygen Species in Signaling or Stress Responses. *Nat. Chem. Biol.* **2011,** 7, 504–511.
198. Liochev, S. I.; Fridovich, I. Superoxide and Iron: Partners in Crime. *IUBMB Life* **1999,** *48,* 157–161.
199. Rubbo, H.; Radi, R.; Trujillo, M.; Telleri, R.; Kalyanaraman, B.; Barnes, S.; et al. Nitric Oxide Regulation of Superoxide and Peroxynitrite-Dependent Lipid Peroxidation. Formation of Novel Nitrogen-Containing Oxidized Lipid Derivatives. *J. Biol. Chem.* **1994,** *269,* 26066–26075.
200. Stadtman, E. R.; Levine, R. L. Protein Oxidation. *Ann. N. Y. Acad. Sci.* **2000,** *899,* 191–208.
201. Kaur, H.; Halliwell, B. Evidence for Nitric Oxide-mediated Oxidative Damage in Chronic Inflammation Nitrotyrosine in Serum and Synovial Fluid from Rheumatoid Patients. *FEBS Lett.* **1994,** *350,* 9–12.
202. Turrens, J. F. Mitochondrial Formation of Reactive Oxygen Species. *J. Physiol.* **2003,** *552,* 335–344.
203. Rodriguez-Rocha, H.; Garcia-Garcia, A.; Panayiotidis, M. I.; Franco, R. DNA Damage and Autophagy. *Mutat. Res. Mol. Mech. Mutagen* **2011,** *711,* 158–166. http://dx.doi.org/10.1016/j.mrfmmm.2011.03.007.
204. Nicolson, G. L.; Ash, M. E. Lipid Replacement Therapy: A Natural Medicine Approach to Replacing Damaged Lipids in Cellular Membranes and Organelles and Restoring Function. *Biochim. Biophys. Acta (BBA) Biomembr.* **2014,** *1838,* 1657–1679.
205. Wang, J.; Sun, P.; Bao, Y.; Liu, J.; An, L. Cytotoxicity of Single-Walled Carbon Nanotubes on PC12 Cells. *Toxicol. Vitr.* **2011,** *25,* 242–250.
206. Xue, Y.; Chen, H. Synthesis and Cytotoxicity of POSS Modified Single Walled Carbon Nanotubes. *J. Nanomater.* **2015**. http://dx.doi.org/10.1155/2015/407437.
207. Yu, L.; Wan, F.; Dutta, S.; Welsh, S.; Liu, Z.; Freundt, E.; et al. Autophagic Programmed Cell Death by Selective Catalase Degradation. *Proc. Natl. Acad. Sci. USA* **2006,** *103,* 4952–4957.
208. Qu, C.; Wang, L.; He, J.; Tan, J.; Liu, W.; Zhang, S.; et al. Carbon Nanotubes Provoke Inflammation by Inducing the Pro-Inflammatory Genes IL-1β and IL-6. *Gene* **2012,** *493,* 9–12. http://dx.doi.org/10.1016/j.gene.2011.11.046.

209. Yu, Y.; Zhang, Q.; Mu, Q.; Zhang, B.; Yan, B. Exploring the Immunotoxicity of Carbon Nanotubes. *Nanoscale Res. Lett.* **2008,** *3,* 271–277.
210. Bihari, P.; Holzer, M.; Praetner, M.; Fent, J.; Lerchenberger, M.; Reichel, C. A.; et al. Single-Walled Carbon Nanotubes Activate Platelets and Accelerate Thrombus Formation in the Microcirculation. *Toxicology* **2010,** *269,* 148–154. http://dx.doi.org/10.1016/j.tox.2009.08.011.
211. Larkin, J. D.; Frimat, K. A.; Fyles, T. M.; Flower, S. E.; James, T. D. Boronic Acid Based Photoinduced Electron Transfer (PET) Fluorescence Sensors for Saccharides. *New J. Chem.* **2010,** *34,* 2922–2931.
212. Pascu, S. I.; Arrowsmith, R. L.; Bayly, S. R.; Brayshaw, S.; Hu, Z. Towards Nanomedicines: Design Protocols to Assemble, Visualize and Test Carbon Nanotube Probes for Multi-Modality Biomedical Imaging. *Philos. Trans. R. Soc. A Math. Phys. Eng. Sci.* **2010,** *368,* 3683–3712.
213. Bull, S. D.; Davidson, M. G.; Van den Elsen, J. M. H.; Fossey, J. S.; Jenkins, A. T. A.; Jiang, Y.-B.; et al. Exploiting the Reversible Covalent Bonding of Boronic Acids: Recognition, Sensing, and Assembly. *Acc. Chem. Res.* **2012,** *46,* 312–326.
214. Nishiyabu, R.; Kubo, Y.; James, T. D.; Fossey, J. S. Boronic Acid Building Blocks: Tools for Self Assembly. *Chem. Commun.* **2011,** *47,* 1124–1150.
215. Fossey, J. S.; D'Hooge, F.; van den Elsen, J. M. H.; Pereira Morais, M. P.; Pascu, S. I.; Bull, S. D.; et al. The Development of Boronic Acids as Sensors and Separation Tools. *Chem. Rec.* **2012,** *12,* 464–478.
216. Shao, W.; Arghya, P.; Yiyong, M.; Rodes, L.; Prakash, S. *Carbon Nanotubes for Use in Medicine: Potentials and Limitations;*, 2013.
217. Sang, S.; Wang, Y.; Feng, Q.; Wei, Y.; Ji, J.; Zhang, W. Progress of New Label-Free Techniques for Biosensors: A Review. *Crit. Rev. Biotechnol.* **2015,** 1–17.
218. Kotagiri, N.; Kim, J.-W. Stealth Nanotubes: Strategies of Shielding Carbon Nanotubes to Evade Opsonization and Improve Biodistribution. *Int. J. Nanomed.* **2014,** *9,* 85.
219. Bhirde, A. A.; Patel, S.; Sousa, A. A.; Patel, V.; Molinolo, A. A.; Ji, Y.; et al. Distribution and Clearance of PEG-Single-Walled Carbon Nanotube cancer Drug Delivery Vehicles in Mice. *Nanomedicine* **2010,** *5,* 1535–1546. http://dx.doi.org/10.2217/nnm.10.90.
220. Bhirde, A. A.; Patel, V.; Gavard, J.; Zhang, G.; Sousa, A. A.; Masedunskas, A.; et al. Targeted Killing of Cancer Cells In Vivo and In Vitro with EGF-Directed Carbon Nanotube-Based Drug Delivery. *ACS Nano* **2009,** *3,* 307–316.
221. Taghdisi, S. M.; Lavaee, P.; Ramezani, M.; Abnous, K. Reversible Targeting and Controlled Release Delivery of Daunorubicin to Cancer Cells by Aptamer-Wrapped Carbon Nanotubes. *Eur. J. Pharm. Biopharm.* **2011,** *77,* 200–206.
222. Wang, L.; Zhang, M.; Zhang, N.; Shi, J.; Zhang, H.; Li, M.; et al. Synergistic Enhancement of cancer Therapy Using a Combination of Docetaxel and Photothermal Ablation Induced by Single-Walled Carbon Nanotubes. *Int. J. Nanomed.* **2011,** *6,* 2641.
223. Lu, Y.-J.; Wei, K.-C.; Ma, C.-C. M.; Yang, S.-Y.; Chen, J.-P. Dual Targeted Delivery of Doxorubicin to Cancer Cells Using Folate-Conjugated Magnetic Multi-Walled Carbon Nanotubes. *Colloids Surf. B Biointerfaces* **2012,** *89,* 1–9.
224. Liu, Z.; Sun, X.; Nakayama-Ratchford, N.; Dai, H. Supramolecular Chemistry on Water-Soluble Carbon Nanotubes for Drug Loading and Delivery. *ACS Nano* **2007,** *1,* 50–56.
225. Ji, Z.; Lin, G.; Lu, Q.; Meng, L.; Shen, X.; Dong, L.; et al. Targeted Therapy of SMMC-7721 Liver cancer Invitro and Invivo with Carbon Nanotubes Based Drug Delivery System. *J. Colloid Interface Sci.* **2012,** *365,* 143–149.
226. Dhar, S.; Liu, Z.; Thomale, J.; Dai, H.; Lippard, S. J. Targeted Single-Wall Carbon Nanotube-mediated Pt (IV) Prodrug Delivery Using Folate as a Homing Device. *J. Am. Chem. Soc.* **2008,** *130,* 11467–11476.

227. Lacerda, L.; Ali-Boucetta, H.; Kraszewski, S.; Tarek, M.; Prato, M.; Ramseyer, C.; et al. How Do Functionalized Carbon Nanotubes Land on, Bind to and Pierce through Model and Plasma Membranes. *Nanoscale* **2013,** *5,* 10242–10250.
228. Lacerda, L.; Russier, J.; Pastorin, G.; Herrero, M. A.; Venturelli, E.; Dumortier, H.; et al. Translocation Mechanisms of Chemically Functionalised Carbon Nanotubes across Plasma Membranes. *Biomaterials* **2012,** *33,* 3334–3343.
229. Im, J. S.; Bai, B. C.; Lee, Y.-S. The Effect of Carbon Nanotubes on Drug Delivery in an Electro-Sensitive Transdermal Drug Delivery System. *Biomaterials* **2010,** *31,* 1414–1419.
230. Mehra, N. K.; Jain, K.; Jain, N. K. Pharmaceutical and Biomedical Applications of Surface Engineered Carbon Nanotubes. *Drug Discov. Today 20,* **2015,** 750–759.
231. Welsher, K.; Sherlock, S. P.; Dai, H. Deep-tissue Anatomical Imaging of Mice Using Carbon Nanotube Fluorophores in the Second Near-Infrared Window. *Proc. Natl. Acad. Sci. USA* **2011,** *108,* 8943–8948.
232. Pramanik, M.; Song, K. H.; Swierczewska, M.; Green, D.; Sitharaman, B.; V Wang, L. H. In Vivo Carbon Nanotube-Enhanced Non-invasive Photoacoustic Mapping of the Sentinel Lymph Node. *Phys. Med. Biol.* **2009,** *54,* 3291–3301. http://dx.doi.org/10.1088/0031-9155/54/11/001.
233. Wang, C.; Ma, X.; Ye, S.; Cheng, L.; Yang, K.; Guo, L.; et al. Protamine Functionalized Single-walled Carbon Nanotubes for Stem Cell Labeling and In Vivo Raman/Magnetic Resonance/Photoacoustic Triple-Modal Imaging. *Adv. Funct. Mater.* **2012,** *22,* 2363–2375.
234. Liu, J. J.; Wang, C.; Wang, X. J.; Wang, X.; Cheng, L.; Li, Y. G.; et al. Mesoporous Silica Coated Single-Walled Carbon Nanotubes as a Multifunctional Light-Responsive Platform for Cancer Combination Therapy. *Adv. Funct. Mater.* **2015,** *25,* 384–392. http://dx.doi.org/10.1002/adfm.201403079.
235. Tamesue, S.; Numata, M.; Kaneko, K.; James, T. D.; Shinkai, S. Hierarchical Carbon Nanotube Assemblies Created by Sugar–Boric or Boronic Acid Interactions. *Chem. Commun.* **2008,** 4478–4480.
236. Numata, M.; Shinkai, S. Self-Assembled Polysaccharide Nanotubes Generated from B-1, 3-Glucan Polysaccharides. In *Self-Assembled Nanomater. II;* Springer, 2008; pp 65–121.
237. Jia, X.; Chen, P.; Xu, X.; Zhang, L. Lentinan Greatly Enhances the Dispersibility of Single-Walled Carbon Nanotubes in Water and Decreases the Cytotoxicity. *Bioact. Carbohydr. Diet. Fibre* **2013,** *1,* 111–119.
238. Zhang, S.; Bao, A.; Sun, T.; Wang, E.; Wang, J. PEI/Zr^{4+}-Coated Nanopore for Selective and Sensitive Detection of ATP in Combination with Single-Walled Carbon Nanotubes. *Biosens. Bioelectron.* **2015,** *63,* 287–293. http://dx.doi.org/10.1016/j.bios.2014.07.062.
239. Rivera, E. J.; Tran, L. A.; Hernandez-Rivera, M.; Yoon, D.; Mikos, A. G.; Rusakova, I. A.; et al. as a Potential Contrast Agent for X-Ray Imaging Applications. *J. Mater. Chem. B* **2013,** *1,* 4792–4800. http://dx.doi.org/10.1039/C3TB20742K. Bismuth@US-Tubes.
240. Briand, G. G.; Burford, N. Bismuth Compounds and Preparations with Biological or Medicinal Relevance. *Chem. Rev.* **1999,** *99,* 2601–2658. http://dx.doi.org/10.1021/cr980425s.
241. Ballot, S.; Noiret, N.; Hindré, F.; Denizot, B.; Garin, E.; Rajerison, H.; et al. ^{99m}Tc/^{188}Re-labelled Lipid Nanocapsules as Promising Radiotracers for Imaging and Therapy: Formulation and Biodistribution. *Eur. J. Nucl. Med. Mol. Imaging* **2006,** *33,* 602–607. http://dx.doi.org/10.1007/s00259-005-0007-0.
242. Chamberlain, T. W.; Meyer, J. C.; Biskupek, J.; Leschner, J.; Santana, A.; Besley, N. A.; et al. Reactions of the Inner Surface of Carbon Nanotubes and Nanoprotrusion Processes Imaged at the Atomic Scale. *Nat. Chem.* **2011,** *3,* 732–737. http://www.nature.com/nchem/journal/v3/n9/abs/nchem.1115.html#supplementary-information.

243. Zoberbier, T.; Chamberlain, T. W.; Biskupek, J.; Kuganathan, N.; Eyhusen, S.; Bichoutskaia, E.; et al. Interactions and Reactions of Transition Metal Clusters with the Interior of Single-walled Carbon Nanotubes Imaged at the Atomic Scale. *J. Am. Chem. Soc.* **2012,** *134,* 3073–3079. http://dx.doi.org/10.1021/ja208746z.
244. Datir, S. R.; Das, M.; Singh, R. P.; Jain, S. Hyaluronate Tethered, "Smart" Multiwalled Carbon Nanotubes for Tumor-Targeted Delivery of Doxorubicin. *Bioconjug. Chem.* **2012,** *23,* 2201–2213. http://dx.doi.org/10.1021/bc300248t.
245. Wu, W.; Li, R.; Bian, X.; Zhu, Z.; Ding, D.; Li, X.; et al. Covalently Combining Carbon Nanotubes with Anticancer Agent: Preparation and Antitumor Activity. *ACS Nano* **2009,** *3,* 2740–2750. http://dx.doi.org/10.1021/nn9005686.
246. Wang, J. T. -W.; Cabana, L.; Bourgognon, M.; Kafa, H.; Protti, A.; Venner, K.; et al. Magnetically Decorated Multiwalled Carbon Nanotubes as Dual MRI and SPECT Contrast Agents. *Adv. Funct. Mater.* **2014,** *24,* 1880–1894. http://dx.doi.org/10.1002/adfm.201302892.
247. Soares, D. C. F.; Ferreira, T. H.; de Aguiar Ferreira, C.; Cardoso, V. N.; de Sousa, E. M. B. Boron nitride Nanotubes Radiolabeled with ^{99m}Tc: Preparation, Physicochemical Characterization, Biodistribution Study, and Scintigraphic Imaging in Swiss Mice. *Int. J. Pharm.* **2012,** *423,* 489–495. http://dx.doi.org/10.1016/j.ijpharm.2011.12.002.
248. Cisneros, B. T.; Law, J. J.; Matson, M. L.; Azhdarinia, A.; Sevick-Muraca, E. M.; Wilson, L. J. Stable Confinement of Positron Emission Tomography and Magnetic Resonance Agents within Carbon Nanotubes for Bimodal Imaging. *Nanomedicine* **2014,** *9,* 2499–2509. http://dx.doi.org/10.2217/nnm.14.26.
249. Han, B.; Zhang, M.; Tang, T.; Zheng, Q.; Wang, K.; Li, L.; et al. The Long-term Fate and Toxicity of PEG-modified Single-Walled Carbon Nanotube Isoliquiritigenin Delivery Vehicles in Rats. *J. Nanomater.* **2014**. http://dx.doi.org/10.1155/2014/257391.
250. Wu, H.-C.; Chang, X.; Liu, L.; Zhao, F.; Zhao, Y. Chemistry of Carbon Nanotubes in Biomedical Applications. *J. Mater. Chem.* **2010,** *20,* 1036–1052.
251. Hudson, J. L.; Casavant, M. J.; Tour, J. M. Water-soluble, Exfoliated, Nonroping Single-Wall Carbon Nanotubes. *J. Am. Chem. Soc.* **2004,** *126,* 11158–11159.
252. Usrey, M. L.; Lippmann, E. S.; Strano, M. S. Evidence for a Two-Step Mechanism in Electronically Selective Single-Walled Carbon Nanotube Reactions. *J. Am. Chem. Soc.* **2005,** *127,* 16129–16135.
253. Behnam, B.; Shier, W. T.; Nia, A. H.; Abnous, K.; Ramezani, M. Non-Covalent Functionalization of Single-Walled Carbon Nanotubes with Modified Polyethyleneimines for Efficient Gene Delivery. *Int. J. Pharm.* **2013,** *454,* 204–215. http://dx.doi.org/10.1016/j.ijpharm.2013.06.057.
254. Lopez-Moreno, A.; Perez, E. M. Pyrene-Based Mechanically Interlocked SWNTs. *Chem. Commun. (Camb)* **2015,** *51,* 5421–5424. http://dx.doi.org/10.1039/c4cc08970g.
255. Tabet, L.; Bussy, C.; Setyan, A.; Simon-Deckers, A.; Rossi, M. J.; Boczkowski, J.; et al. Coating Carbon Nanotubes with a Polystyrene-Based Polymer Protects against Pulmonary Toxicity. *Part. Fibre Toxicol.* **2011,** *8.*
256. Hong, S.Y.; Tobias, G.; Al-Jamal, K.T.; Ballesteros, B.; Ali-Boucetta, H.; Lozano-Perez, S.; et al. Filled and Glycosylated Carbon Nanotubes for In Vivo Radioemitter Localization and Imaging. *Nat. Mater.* **2010,** *9,* 485–490. http://www.nature.com/nmat/journal/v9/n6/suppinfo/nmat2766_S1.html.
257. (a) Ge, H. *New Functionalised Carbon Based Nanomaterials for Biomedical Imaging Applications* (PhD thesis); University of Bath, 2015.
(b) Aigbirhio, F. I.; Allwein, S.; Anwar, A.; Atzrodt, J.; Audisio, D.; Badman, G.; Bakale, R.; Berthon, F.; Bragg, R.; Brindle, K. M.; et al. Abstracts of the 24th International Isotope Society (UK Group) Symposium: Synthesis and Applications of Labelled Compounds 2015. *Journal of Labelled Compounds and Radiopharmaceuticals* **2016,** *59,* 175–186.

258. Monthioux, M. Filling Single-Wall Carbon Nanotubes. *Carbon N. Y.* **2002,** *40,* 1809–1823.
259. Vittorio, O.; Duce, S. L.; Raffa, V.; Cuschieri, A. *Imaging and Biomedical Application of Magnetic Carbon Nanotubes;* , 2011. 17070.
260. Kumar, R.; Dhanawat, M.; Kumar, S.; N Singh, B.; K Pandit, J.; Sinha, V. R. Carbon Nanotubes: A Potential Concept for Drug Delivery Applications. *Recent Pat. Drug Deliv. Formul.* **2014,** *8,* 12–26.
261. Brahmachari, S.; Ghosh, M.; Dutta, S.; Das, P. K. Biotinylated Amphiphile-Single Walled Carbon Nanotube Conjugate for Target-Specific Delivery to Cancer Cells. *J. Mater. Chem. B* **2014,** *2,* 1160–1173. http://dx.doi.org/10.1039/c3tb21334j.
262. Jain, A.; Homayoun, A.; Bannister, C. W.; Yum, K. Single-Walled Carbon Nanotubes as Near-Infrared Optical Biosensors for Life Sciences and Biomedicine. *Biotechnol. J.* **2015,** *10,* 447–459.
263. Jung, H. S.; Kong, W. H.; Sung, D. K.; Lee, M.-Y.; Beack, S. E.; Keum, D. H.; et al. Nanographene Oxide–Hyaluronic Acid Conjugate for Photothermal Ablation Therapy of Skin Cancer. *ACS Nano* **2014,** *8,* 260–268.
264. Deria, P.; Olivier, J.-H.; Park, J.; Therien, M. J. Potentiometric, Electronic, and Transient Absorptive Spectroscopic Properties of Oxidized Single-Walled Carbon Nanotubes Helically Wrapped by Ionic, Semiconducting Polymers in Aqueous and Organic Media. *J. Am. Chem. Soc.* **2014,** *136,* 14193–14199.
265. Hu, Z.; Pantoş, G. D.; Kuganathan, N.; Arrowsmith, R. L.; Jacobs, R. M. J.; Kociok-Köhn, G.; et al. Interactions between Amino Acid-Tagged Naphthalenediimide and Single Walled Carbon Nanotubes for the Design and Construction of New Bioimaging Probes. *Adv. Funct. Mater.* **2012,** *22,* 503–518. http://dx.doi.org/10.1002/adfm.201101932.
266. Tao, X.; Dong, L.; Zhang, W.; Zhang, X.; Cheng, J.; Huang, H.; et al. Controllable Melting and Flow of β-Sn in Flexible Amorphous Carbon Nanotubes. *Carbon N. Y.* **2009,** *47,* 3122–3127.
267. Yu, F.; Chen, J.; Chen, L.; Huai, J.; Gong, W.; Yuan, Z.; et al. Magnetic Carbon Nanotubes Synthesis by Fenton's Reagent Method and Their Potential Application for Removal of Azo Dye from Aqueous Solution. *J. Colloid Interface Sci.* **2012,** *378,* 175–183.
268. Masotti, A. Multifunctional Nanoparticles, Nanocages and Degradable Polymers as a Potential Novel Generation of Non-Invasive Molecular and Cellular Imaging Systems. *Recent Pat. Nanotechnol.* **2011,** *5,* 163–177.
269. Bonard, J.; Salvetat, J.-P.; Stockli, T.; de Heer, W. A.; Forró, L.; Châtelain, A. Field Emission from Single-wall Carbon Nanotube Films. *Appl. Phys. Lett.* **1998,** *73,* 918–920.
270. Vaseashta, A.; Dimova-Malinovska, D. Nanostructured and Nanoscale Devices, Sensors and Detectors. *Sci. Technol. Adv. Mater.* **2005,** *6,* 312–318.
271. Wang, L.; Shi, J.; Zhang, H.; Li, H.; Gao, Y.; Wang, Z.; et al. Synergistic Anticancer Effect of RNAi and Photothermal Therapy Mediated by Functionalized Single-Walled Carbon Nanotubes. *Biomaterials* **2013,** *34,* 262–274.
272. Hashida, Y.; Tanaka, H.; Zhou, S.; Kawakami, S.; Yamashita, F.; Murakami, T.; et al. Photothermal Ablation of Tumor Cells Using a Single-Walled Carbon Nanotube–Peptide Composite. *J. Control. Release* **2014,** *173,* 59–66.
273. Ji, S.; Liu, C.; Zhang, B.; Yang, F.; Xu, J.; Long, J.; et al. Carbon Nanotubes in Cancer Diagnosis and Therapy. *Biochim. Biophys. Acta Rev. Cancer* **2010,** *1806,* 29–35. http://dx.doi.org/10.1016/j.bbcan.2010.02.004.
274. Liu, Z.; Robinson, J. T.; Tabakman, S. M.; Yang, K.; Dai, H. Carbon Materials for Drug Delivery & Cancer Therapy. *Mater. Today* **2011,** *14,* 316–323.
275. Khurana, M.; Ulrich, S.; Kim, A.; Moriyama, Y.; Netchev, G.; Akens, M. K.; et al. Biodistribution and Pharmacokinetic Studies of a Porphyrin Dimer Photosensitizer (Oxdime) by Fluorescence Imaging and Spectroscopy in Mice Bearing Xenograft Tumors. *Photochem. Photobiol.* **2012,** *88,* 1531–1538.

276. Kilin, D. S.; Tsemekhman, K.; V Prezhdo, O.; Zenkevich, E. I.; von Borczyskowski, C. *Ab Initio* Study of Exciton Transfer Dynamics from a Core–Shell Semiconductor Quantum Dot to a Porphyrin-Sensitizer. *J. Photochem. Photobiol. A Chem.* **2007,** *190,* 342–351.
277. Safar, G. A. M.; Gontijo, R. N.; Fantini, C.; Martins, D. C. S.; Idemori, Y. M.; Pinheiro, M. V. B.; et al. Enhanced Oxygen Singlet Production by Hybrid System of Porphyrin and Enriched (6,5) Single-Walled Carbon Nanotubes for Photodynamic Therapy. *J. Phys. Chem. C* **2015,** *119,* 4344–4350. http://dx.doi.org/10.1021/jp5111289.
278. Sundaramoorthy, R.; Vuyyuru, M.; Dhanaraju, M. D. Carbon Nanotube: A Flexible Approach for Nanomedicine and Drug Delivery. *Asian J. Pharm. Clin. Res.* **2015,** *8.*
279. Mingozzi, F.; Büning, H. Adeno-Associated Viral Vectors at the Frontier between Tolerance and Immunity. *Front. Immunol.* **2015,** *6,* 120. http://dx.doi.org/10.3389/fimmu.2015.00120.
280. Mohammadi, M.; Salmasi, Z.; Hashemi, M.; Mosaffa, F.; Abnous, K.; Ramezani, M. Single-Walled Carbon Nanotubes Functionalized with Aptamer and Piperazine–Polyethylenimine Derivative for Targeted siRNA Delivery into Breast Cancer Cells. *Int. J. Pharm.* **2015,** *485,* 50–60.
281. Siu, K. S.; Chen, D.; Zheng, X.; Zhang, X.; Johnston, N.; Liu, Y.; et al. Non-Covalently Functionalized Single-Walled Carbon Nanotube for Topical siRNA Delivery into Melanoma. *Biomaterials* **2014,** *35,* 3435–3442.
282. S.A.P. and I.C. Acta Chemica Scandinavica; Choi, Y. S.; Lee, M. Y.; David, A. E.; Park, Y. S. Nanoparticles for Gene Delivery: Therapeutic and Toxic Effects. *Mol. Cell. Toxicol.* **2014,** *10,* 1–8.
283. Thamavaranukup, N.; Höppe, H. A.; Ruiz-Gonzalez, L.; Costa, P. M. F.J.; Sloan, J.; Kirkland, A.; et al. Single-walled Carbon Nanotubes Filled with M OH (M = K, Cs) and Then Washed and Refilled with Clusters and Molecules. *Chem. Commun.* **2004,** 1686–1687. http://dx.doi.org/10.1039/B405956E.
284. Huh, Y.; Shao, L.; Tobias, G.; Green, M. L. H. Opening and Reversible Filling of Single-Walled Carbon Nanotubes with Various Materials. *J. Nanosci. Nanotechnol.* **2006,** *6,* 3360–3363. http://www.ncbi.nlm.nih.gov/pubmed/17252765.
285. Ilie, A.; Bendall, J. S.; Kubo, O.; Sloan, J.; Green, M. L. H. Force and Energy Dissipation Variations in Noncontact Atomic Force Spectroscopy of Composite Carbon Nanotube Systems. *Phys. Rev. B* **2006,** *74,* 235418. http://dx.doi.org/10.1103/PhysRevB.74.235418.
286. Bendall, J. S.; Ilie, A.; Welland, M. E.; Sloan, J.; Green, M. L. H. Thermal Stability and Reactivity of Metal Halide Filled Single-Walled Carbon Nanotubes. *J. Phys. Chem. B* **2006,** *110,* 6569–6573. http://dx.doi.org/10.1021/jp056405t.
287. Shao, L.; Lin, T.-W.; Tobias, G.; Green, M. L. H. A Simple Method for the Containment and Purification of Filled Open-Ended Single Wall Carbon Nanotubes Using C60 Molecules. *Chem. Commun.* **2008,** 2164–2166. http://dx.doi.org/10.1039/b800881g.
288. Sloan, J.; Kirkland, A. I.; Hutchison, J. L.; Green, M. L. H. Structural Characterization of Atomically Regulated Nanocrystals Formed within Single-Walled Carbon Nanotubes Using Electron Microscopy. *Acc. Chem. Res.* **2002,** *35,* 1054–1062. http://dx.doi.org/10.1021/ar010169x.
289. Shao, L.; Tobias, G.; Huh, Y.; Green, M. L. H. Reversible Filling of Single Walled Carbon Nanotubes Opened by Alkali Hydroxides. *Carbon N.Y.* **2006,** *44,* 2855–2858. http://dx.doi.org/10.1016/j.carbon.2006.06.010.
290. Sarkar, S.; Sharma, C.; Yog, R.; Periakaruppan, A.; Jejelowo, O.; Thomas, R.; et al. Analysis of Stress Responsive Genes Induced by Single-Walled Carbon Nanotubes in BJ Foreskin Cells. *J. Nanosci. Nanotechnol.* **2007,** *7,* 584–592. http://dx.doi.org/10.1016/j.bbi.2008.05.010.
291. Li, Z.; Hulderman, T.; Salmen, R.; Chapman, R.; Leonard, S. S.; Young, S.-H.; et al. Cardiovascular Effects of Pulmonary Exposure to Single-Wall Carbon Nanotubes. *Environ. Health Perspect.* **2007,** *115,* 377–382. http://dx.doi.org/10.1289/ehp.9688.

292. Tyrrell, R. M. [27] Role for Singlet Oxygen in Biological Effects of Ultraviolet a Radiation. *Methods Enzymol.* **2000,** 290–296. http://dx.doi.org/10.1016/S0076-6879(00)19029-9.
293. Creighton-Gutteridge, M.; Tyrrell, R. M. A Novel Iron Chelator that Does Not Induce HIF-1 Activity. *Free Radic. Biol. Med.* **2002,** *33,* 356–363. http://www.ncbi.nlm.nih.gov/pubmed/12126757.
294. Riehemann, K.; Schneider, S. W.; Luger, T. A.; Godin, B.; Ferrari, M.; Fuchs, H. Nanomedicine – Challenge and Perspectives. *Angew. Chem. Int. Ed.* **2009,** *48,* 872–897.
295. *No Title.* www.nanowerk.com, (n.d.).
296. Chou, C.-C.; Hsiao, H.-Y.; Hong, Q.-S.; Chen, C.-H.; Peng, Y.-W.; Chen, H.-W.; et al. Single-Walled Carbon Nanotubes Can Induce Pulmonary Injury in Mouse Model. *Nano Lett.* **2008,** *8,* 437–445. http://dx.doi.org/10.1021/nl0723634.
297. Schipper, M. L.; Nakayama-Ratchford, N.; Davis, C. R.; Kam, N. W. S.; Chu, P.; Liu, Z.; et al. A Pilot Toxicology Study of Single-Walled Carbon Nanotubes in a Small Sample of Mice. *Nat. Nanotechnol.* **2008,** *3,* 216–221. http://dx.doi.org/10.1038/nnano.2008.68.
298. Kostarelos, K.; Bianco, A.; Prato, M. Hype Around Nanotubes Creates Unrealistic Hopes. *Nature* **2008,** *453,* 280. http://dx.doi.org/10.1038/453280c.
299. Singh, R.; Pantarotto, D.; Lacerda, L.; Pastorin, G.; Klumpp, C.; Prato, M.; et al. Tissue Biodistribution and Blood Clearance Rates of Intravenously Administered Carbon Nanotube Radiotracers. *Proc. Natl. Acad. Sci. USA* **2006,** *103,* 3357–3362. http://dx.doi.org/10.1073/pnas.0509009103.
300. Ranasinha, K. W.; Uragoda, C. G. Graphite Pneumoconiosis. *Br. J. Ind. Med.* **1972,** *7,* 909–923.
301. Mori, T.; Takada, H.; Ito, S.; Matsubayashi, K.; Miwa, N.; Sawaguchi, T. Preclinical Studies on Safety of Fullerene upon Acute Oral Administration and Evaluation for No Mutagenesis. *Toxicology* **2006,** *225,* 48–54. http://dx.doi.org/10.1016/j.tox.2006.05.001.
302. *Lux Research;* Lux Research Report, 2007. n.d.
303. Maynard, A. D.; Aitken, R. J.; Butz, T.; Colvin, V.; Donaldson, K.; Oberdörster, G.; et al. Safe Handling of Nanotechnology. *Nature* **2006,** *444,* 267–269. http://dx.doi.org/10.1038/444267a.
304. www.bsigroup.com/nano, n.d. 6699-2.
305. www.bsigroup.com/nano, n.d. PD 6699-1.
306. *No Title.* http://www.nanotec.org.uk/PressMediaMar042.htm, (n.d.).
307. Burri, R. V.; Bellucci, S. Public Perception of Nanotechnology. *J. Nanoparticle Res.* **2008,** *10,* 387–391. http://dx.doi.org/10.1007/s11051-007-9286-7.
308. Douma, K. F. L.; Karsenberg, K.; Hummel, M. J. M.; Bueno-de-Mesquita, J. M.; van Harten, W. H. Methodology of Constructive Technology Assessment in Health Care. *Int. J. Technol. Assess. Health Care* **2007,** *23,* 162–168. http://dx.doi.org/10.1017/S0266462307070262.
309. Linkov, I.; Satterstrom, F. K.; Steevens, J.; Ferguson, E.; Pleus, R. C. Multi-Criteria Decision Analysis and Environmental Risk Assessment for Nanomaterials. *J. Nanopart. Res.* **2007,** *9,* 543–554. http://dx.doi.org/10.1007/s11051-007-9211-0.

CHAPTER NINE

Immobilized Metal Affinity Chromatography (IMAC) for Metalloproteomics and Phosphoproteomics

Y.-Y. Chang, H. Li, H. Sun
The University of Hong Kong, Hong Kong, P.R. China

Contents

1. INTRODUCTION

Immobilized metal affinity chromatography (IMAC) was initially developed to purify native proteins with an intrinsic affinity to metal ions.[1] This is achieved by manipulating the coordination chemistry of metal ions, which are immobilized on a solid resin, and the ligand groups on biomolecules with metal-binding ability.[2] IMAC has been used for decades, in

Inorganic and Organometallic Transition Metal Complexes with Biological Molecules and Living Cells
ISBN 978-0-12-803814-7
http://dx.doi.org/10.1016/B978-0-12-803814-7.00009-5

particular to purify recombinant proteins with a polyhistidine tag (His-tag) encoded at either the N- or C- terminus of the target proteins. This technology represents a breakthrough in the purification of proteins and other biomolecules.[2] Bacterial metabolites or metallo-antibiotics, which are compounds that demand metal ions for full functionality as antibacterials and antivirals and exhibit antitumor activities,[3] can be extracted from bacteria and plants because IMAC has the ability to enrich metal-binding molecules. This is exemplified by Cu(II)- or Fe(III)-IMAC, which show success in retaining antibiotics including quinolones, β-lactams, tetracyclines, and aminoglycosides.[4,5] Besides, multidimensional IMAC, which is composed of serial IMAC columns functionalized with different metal ions, may allow the extraction of multiple components in a single sample entity to facilitate drug discovery processes. For example, using two in-series columns Yb(III)- and Cu(II)-IMAC, bleomycin and desferrioxamine B, two clinically used anticancer agents, have been isolated from *Streptomyces verticillus* fermentation culture with an improved yield, which has reduced environmental and financial costs in processing pharmaceutics and drug discovery.[6]

IMAC has been applied extensively to detect proteins that possess intrinsic metal-binding abilities, facilitating further understanding of the roles of metals in life processes as well as elucidating the mechanisms of action of metallo-agents in biological inorganic chemistry. Metalloproteins or metal-binding proteins account for one-quarter to one-third of the proteins in proteomes.[7,8] Although the identification of metalloproteins and their specific roles in life processes is challenging and warrants further exploration, metalloproteins have been found to be crucial for regulating metal homeostasis in living organisms. Moreover, the use of metallodrugs for therapeutic purposes [e.g., Pt(II)-based anticancer drugs] has received considerable attention. To accelerate the rational design of more metal-based therapeutics and diagnostic agents [e.g., Gd(III)-based magnetic resonance imaging contrasting agents], it is crucial to understand the mechanisms of action of metallodrugs, whereas the identification of potential drug targets at a proteome-wide scale by various approaches including IMAC will provide invaluable information.[9–11]

In addition, IMAC has been further extended to isolate phosphoproteins[12] and enrich substoichiometric phosphopeptides from proteolytically digested proteins[13] before analysis by mass spectrometry (MS). IMAC combined with MS allows the investigation of phosphoproteomes, which is critical in understanding how cells function as a result of the involvement of phosphorylated proteins in multiple cellular processes: for example, cell signaling, cell–cell communications, and regulating cell division.[14–16]

This chapter summarizes advances of IMAC in identifying metalloproteomes for a deeper understanding of the mechanisms of action of metallodrugs and the toxicity of certain metal ions. Practical guidance and improvements for the study of protein phosphorylation will also be discussed.

2. PRINCIPLES AND DESIGN OF IMMOBILIZED METAL AFFINITY CHROMATOGRAPHY COLUMNS

The separation principle of IMAC is to use the differential affinities of proteins (or other biomolecules) for immobilized metal ions that influence the separation efficiency. In general, a polydentate metal-chelating group is covalently linked onto the solid resin of IMAC, which binds to a metal ion [e.g., Ni(II), Co(II), Cu(II), Mn(II), Zn(II), Fe(III) and Ga(III)]. Upon interaction with the analyte molecules in the sample, the water molecules in the metal coordination site would undergo an exchange with the surface-exposed amino acid residues of the protein targets or other metal-coordinating molecules.[2] Because the interaction between the metal ions and the residues of proteins is reversible, it can be manipulated to allow the proteins to be adsorbed and desorbed under different conditions. The availability of coordination sites in the metal center for the retention of analyte molecules from cell culture and cellular extract has to be considered when choosing the chelator moieties. Nitrilotriacetate (NTA) and iminodiacetate (IDA) are the most commonly used metal chelators in IMAC, because there are two or three vacant sites available when the octahedral complexes of metal–NTA and metal–IDA, respectively, are formed.

Metal ions, especially transition metal ions, have been covalently linked to the resin of IMAC to identify different metal-binding proteins. The stability of the metal complexes, in particular first-row transition metal complexes, follows a trend known as the Irving–Williams series[17]: Mn(II) < Fe(II) < Co(II) < Ni(II) < Cu(II) > Zn(II), which suggests that the most stable complex should be formed by Cu(II) ions and a chelator ligand. Sometimes the target molecules compete with the chelator moiety for the metal ions, possibly removing the immobilized metal ions from the solid resin. Thus, the stability constants of metal–NTA complexes should be noted. The stability constants of the commonly used metal–NTA complexes for IMAC are shown in Fig. 9.1, in which the metal ions are also grouped according to their roles in IMAC analysis (for metalloproteomic or phosphoproteomic studies) discussed in this chapter.[18]

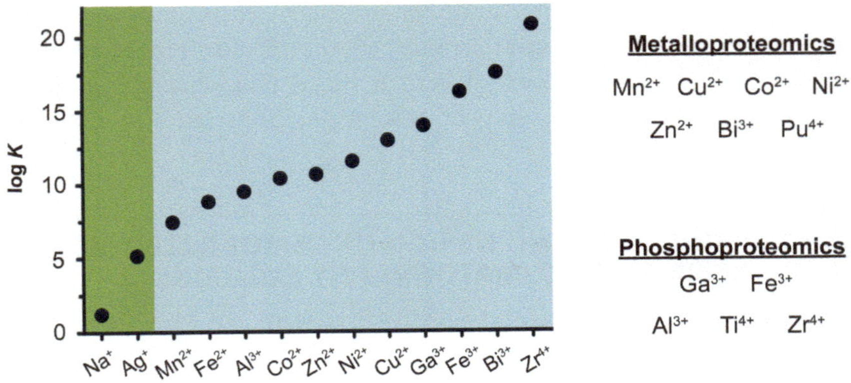

Figure 9.1 Plot of the stability constants (log *K*) of metal–nitrilotriacetate (NTA) complexes of different metal ions (left). Metal ions that are used in NTA-based immobilized metal affinity chromatography (IMAC) are highlighted in *blue* whereas those not used in IMAC are in *green*. Metal ions that are applied in metalloproteomics and phosphoproteomics are also listed in the figure (right).

On the other hand, affinities between the interacting amino acids and the immobilized metal ions can be predicted based on the hard–soft-acid–base theory.[19] For instance, the hard metal ions Fe(III) and Ti(IV) show a higher preference toward hard amino acid residues such as the oxygen from aspartate and glutamate residues, whereas borderline metal ions Ni(II) and Cu(II) show considerable affinities toward borderline amino acids such as nitrogen from histidine and soft ligand sulfur from cysteine.[20] Moreover, the coordination number of the immobilized metal ions and the geometry preference should also be taken into account when interpreting the results of IMAC analyses.

Fig. 9.2 illustrates a typical experimental procedure involved in IMAC analyses. Protein mixtures are loaded onto a metal-immobilized column and unbound molecules are washed out. Metal-binding proteins are captured on the IMAC column and then subjected to successive washing steps. Protein targets can be eluted by supplementing a buffer containing a chelate ligand, e.g., imidazole and ethylenediamine tetraacetic acid (EDTA), which displaces the protein targets or elute the bound molecules in concert with removal of the immobilized metal ions from the resin matrix. The metal-binding proteins are then identified through separation by two-dimensional gel electrophoresis (2-DE) according to the isoelectric points (in the first dimension) and the molecular weights (in the second dimension) before peptide mass fingerprinting by matrix-assisted laser desorption/ionization time-of-flight/time-of-flight mass spectrometry.[21] Alternatively, on-column trypsin digestion can be performed to analyze the metal-binding peptides

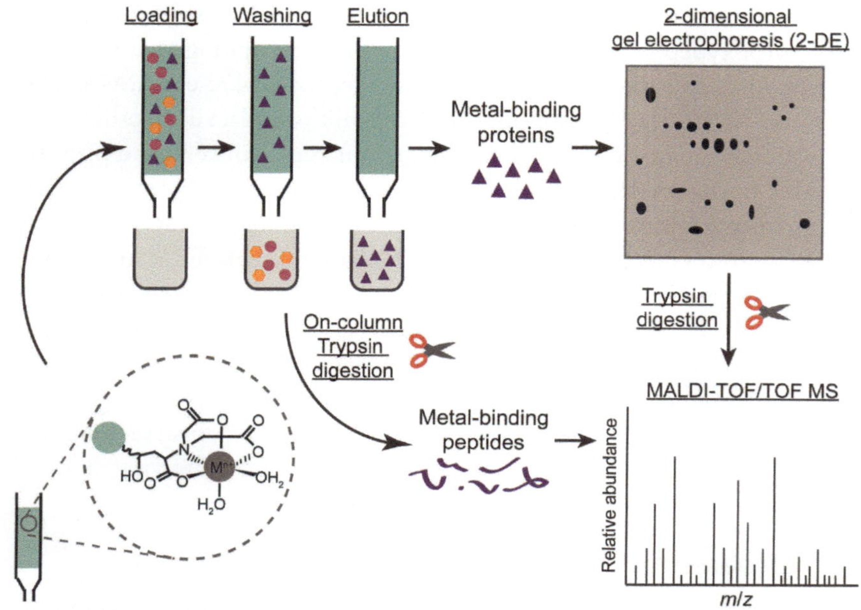

Figure 9.2 Overview of an immobilized metal affinity chromatography (IMAC) experiment for metalloproteomic analysis. Polydentate metal chelator moiety exemplified by nitrilotriacetate (NTA) is covalently conjugated to the column resin, which is coordinated to a metal ion. The sample is first loaded onto the IMAC column. Proteins bound to the immobilized metal ions are retained on the IMAC column whereas unbound molecules are removed after a few steps of washing. The bound molecules are digested by trypsin on the IMAC column and are recovered based on competitive exchange with a chelator, which are then subjected to mass spectrometry (MS)-based analysis. Alternatively, metal-binding proteins are subsequently eluted and separated by two-dimensional gel electrophoresis (2-DE) and then subjected to trypsin digestion before analysis by peptide mass fingerprinting [matrix-assisted laser desorption/ionization time-of-flight/time-of-flight mass spectrometry (MALDI-TOF/TOF MS)]. *m/z*, mass to charge ratio.

further after unbound molecules are washed out, which provides information about the amino acid composition (or binding motifs) in the metal coordination sphere.

3. IMMOBILIZED METAL AFFINITY CHROMATOGRAPHY FOR METALLOPROTEOMIC STUDY

IMAC has been regarded as one of the most convenient methods for purifying recombinant proteins, e.g., using an Ni-bound NTA chelating group for the His-tagged proteins, and has been extensively used to purify proteins from complex biological samples, such as superoxide dismutase,[22] serine carboxypeptidases,[23] and interferon.[24] This technology has facilitated the functional characterization of proteins in biomedical research. Moreover, when the Ni(II) on the resin of IMAC column is replaced by other metal ions, different metalloproteomes in bacteria, mammalian cells, and plant tissues have been identified.[25–29] To track the metal-binding proteins in samples, any bound metal ions from proteins can be removed by loading the samples onto the chelate resin of the IMAC column before IMAC analysis. A drawback of IMAC analysis is that the metal-binding events cannot be examined in situ. Besides, proteins that possess strong binding affinities (i.e., stronger than NTA and/or IDA) to the metal ions are usually present in their metal-bound form in cell lysate, which will not be captured by IMAC. Nonetheless, IMAC has been widely used as a prefractionation technique in the field of proteomics. Some IMAC-based metalloproteomic research will be summarized in the following sections (Table 9.1).

3.1 Mechanism of Toxicity of Plutonium(IV) to Humans

Heavy metal (e.g., arsenic, cadmium, plutonium, lead and mercury) pollution to the environment often occurs as a result of waste from power plants, pesticides, fertilizers, industrialization, and other human activities.[30–32] In general, these heavy metals possess deleterious effects on human health, such as renal dysfunction, severe damage to the nervous system, and even the development of cancer and its progression.[31,33] To understand the toxic mechanism of heavy metals, it is crucial to identify proteins that interact with heavy metal ions, further analyze those proteins by protein–protein interactions, and validate the disruption of protein functions before the mechanism of toxicity.

Plutonium has been known to be carcinogenic; it can enter the human body via different routes and is accumulated inside the body. Exposure to plutonium is largely attributed to nuclear power plants. Although Fe(III)-binding proteins human transferrin and ferritin have been found to be associated with Pu(IV), the specific biochemical interactions of Pu(IV) are not fully understood.[34,35] The causes of diseases relating to plutonium exposure and the potential therapy for treating these diseases remain vague. Using Pu(IV)-NTA IMAC in combination with 2-DE and MS, seven Pu(IV)-binding proteins were identified in mammalian cells.[36] Bioinformatic analysis of the subcellular localization and functions of the putative Pu(IV)-binding proteins revealed that Pu(IV) ions were predominantly accumulated in the cytoplasm of cells whereas the protein targets (e.g., 78 kDa

Table 9.1 Selected Metalloproteomic Studies Using Immobilized Metal Affinity Chromatography (IMAC)

IMAC	Sample Type	Results	Comments	References
Pu(IV)-NTA	PC12 cells (from rat adrenal gland)	7 potential Pu(IV)-binding cytosolic proteins were identified that were involved in antiapoptosis	Pu(IV) ions were retained on an IMAC column, which facilitated the subsequent analysis without concerning the radioactivity of samples	36
Cu(II)-IDA	Rice (*Oryza sativa* L.), B1139 (Cu-tolerant) and B1195 (Cu-sensitive)	27 differentially expressed Cu(II)-binding proteins were found, which elucidates the mechanism of heavy metal stress response and tolerance in plants	Samples were pretreated with a metal-unloaded IMAC column to remove any bound metals on proteins	27
Bi(III)-NTA	*Helicobacter pylori*	Proteins involved in various cellular processes were indicated to be potential drug targets, which corroborated the multitargeting mechanism of action of bismuth-based antimicrobials	IMAC served as an excellent method to characterize the biocoordination chemistry of metal in biological samples	49,50
Co(II)-IDA and Ni(II)-IDA	*Streptococcus pneumoniae*	IMAC identified Co(II)-binding proteins and Ni(II)-binding proteins that offered insight into metal homeostasis and metal-linked bacterial virulence	IMAC column without metal ions was used as a control to exclude nonspecific binding	25

glucose-regulated protein precursor and α-crystalline B chain) were correlated with antiapoptotic processes, which suggests that the carcinogenicity of Pu(IV) is attributed to its involvement in cancer development. Moreover, most identified Pu(IV)-binding proteins exhibit metal-binding ability including calcium, magnesium, or divalent transition metal ions; this implies that interaction with Pu(IV) may perturb the coordination of these metals to metalloproteins or may disturb the homeostasis of these metals. Although characterization of Pu(IV)–protein interactions warrants further investigation, this study postulates the mechanism of carcinogenesis of Pu(IV).

3.2 Copper Tolerance in Plants

Minerals, including nonmetal elements such as carbon, hydrogen, oxygen, nitrogen, sulfur, and phosphorus and metal ions such as potassium, magnesium, and calcium, are essential for the growth and development of plants. These elements are considered to be macroelements, which indicates that they are needed in relatively large quantities.[37] On the other hand, trace elements refer to the mineral nutrients that are required in small amounts, such as chlorine, copper, zinc, and molybdenum.[37] The availability of nutrients varies significantly, depending on abiotic factors such as climate and soil properties as well as all living organisms in the same habitats. Excess metal ions may induce the production of reactive oxidative species, resulting in perturbation of the metal homeostasis. Hence, mineral elements, in particular metal nutrients, have to be tightly regulated because of the potential toxicity exerted by the accumulation of excess metal ions.

IMAC is renowned for examining the changes in metalloproteomes in response to different stress conditions by analyzing the gel images of 2-DE after IMAC enrichment (e.g., appearance or disappearance of certain protein spots, increase or decrease in intensities of protein spots) (Fig. 9.3). Comparative proteomic analysis of copper proteomes in the Cu(II)-sensitive and Cu(II)-tolerant varieties of rice (*Oryza sativa* L.) by Cu(II)-IDA IMAC and 2-DE enabled the identification of molecular targets.[27,38] By comparing the intensities of the protein spots on 2-DE gel, more than 20 proteins were found to be expressed differently (either upregulated or downregulated) under copper stress. Among them, 16 proteins have not been previously reported to be involved in copper binding in animals or plants.[27] Functional analysis of the novel copper proteomes indicated that oxidative defense, detoxification, and pathogenesis may be disturbed upon exposure to high copper concentrations, whereas putative cold-shock domain proteins (also known to be RNA chaperones) and proteins involved in translation

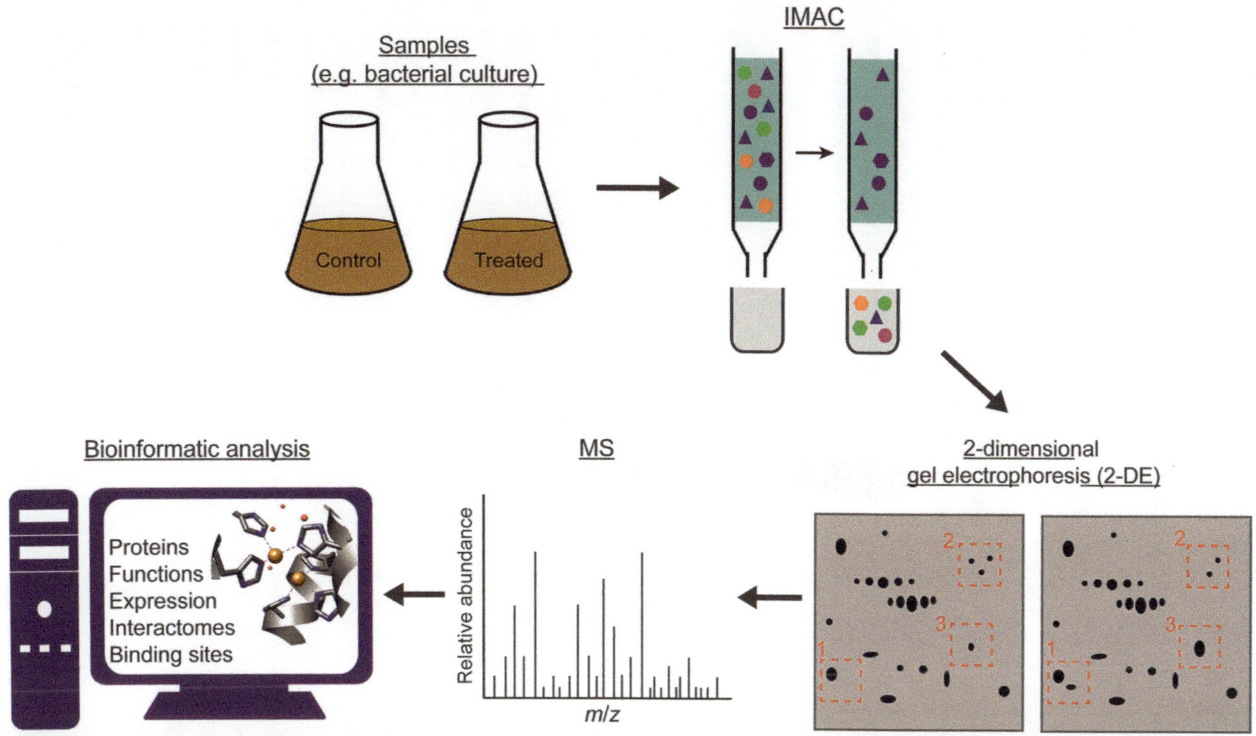

Figure 9.3 Comparative proteomic analysis of metal-binding proteins using immobilized metal affinity chromatography (IMAC). Metal-binding proteins of the control and the treated samples are enriched by IMAC, followed by separation in two-dimensional gel electrophoresis (2-DE) and protein identification by mass spectrometry (MS)-based analysis. The difference in expression of metalloproteomes is quantified by comparing the intensities of the protein spots. Further analysis by bioinformatics provides more comprehensive information on the metalloproteomes, such as the interacting partners and the geometry and preferences of amino acid residues at the metal-binding sites. *m/z*, mass to charge ratio.

initiation were also implicated to be linked to excess copper stress response and tolerance in plants, shedding light on the molecular mechanism of actions of the metal.

3.3 Molecular Mechanism of Bismuth(III)–Based Antiulcer Drugs

Bismuth-based antimicrobials have been used to treat bacterial infections for over 100 years, in particular for the eradication of *Helicobacter pylori* when used with antibiotics as triple or quadruple therapy.[39] Incorporation of bismuth-based drugs into the therapies slowed the relapse rate of *H. pylori* infection in addition to the synergy observed for clinically used bismuth drugs with clarithromycin and tetracycline against antibiotic-resistant *H. pylori* strains.[39–41] Until now, resistance to bismuth drugs has not been reported for *H. pylori* and other pathogenic microbes. Sustained susceptibility of pathogens to bismuth drugs and the low toxicity of bismuth compounds in humans attributable to the glutathione and multidrug resistance protein transporter-mediated disposition in human hosts[42] aroused further exploration of bismuth-based compounds for antimicrobial and other purposes.[43–45]

It is believed that proteins and enzymes are the targets of bismuth drugs.[46–48] To understand the molecular mechanisms of action of commonly prescribed bismuth drugs for accelerating novel drug design, it is important to identify the proteins that bind to the drugs or are differentially expressed. Comparative proteome analysis of *H. pylori*, with and without the clinically used bismuth-based antiulcer drug colloidal bismuth subcitrate, based on 2-DE and peptide mass fingerprinting, revealed several bismuth-binding proteins. Using immobilized-bismuth affinity chromatography (Bi-IMAC), proteins that were significantly upregulated or downregulated were identified from *H. pylori*, including chaperonin HspA and HspB, neutrophil-activating protein NapA, as well as three enzymes (alkyl hydroperoxide reductase TsaA, fumarase, and the urease subunit proteins UreB) and a translational factor (Ef-Tu).[49] This study suggests that bismuth may act against *H. pylori* via multiple functions such as the inhibition of proteases, modulation of cellular oxidative stress, and interference with nickel homeostasis.

A study revealed more than 300 bismuth-binding peptides in *H. pylori* based on Bi-IMAC. Bismuth exhibited much higher selectivity toward peptides enriched with cysteine and histidine residues with CX_nC, CX_nH, and HX_nH as the dominated motifs (Fig. 9.4).[50] Bioinformatic analysis of the

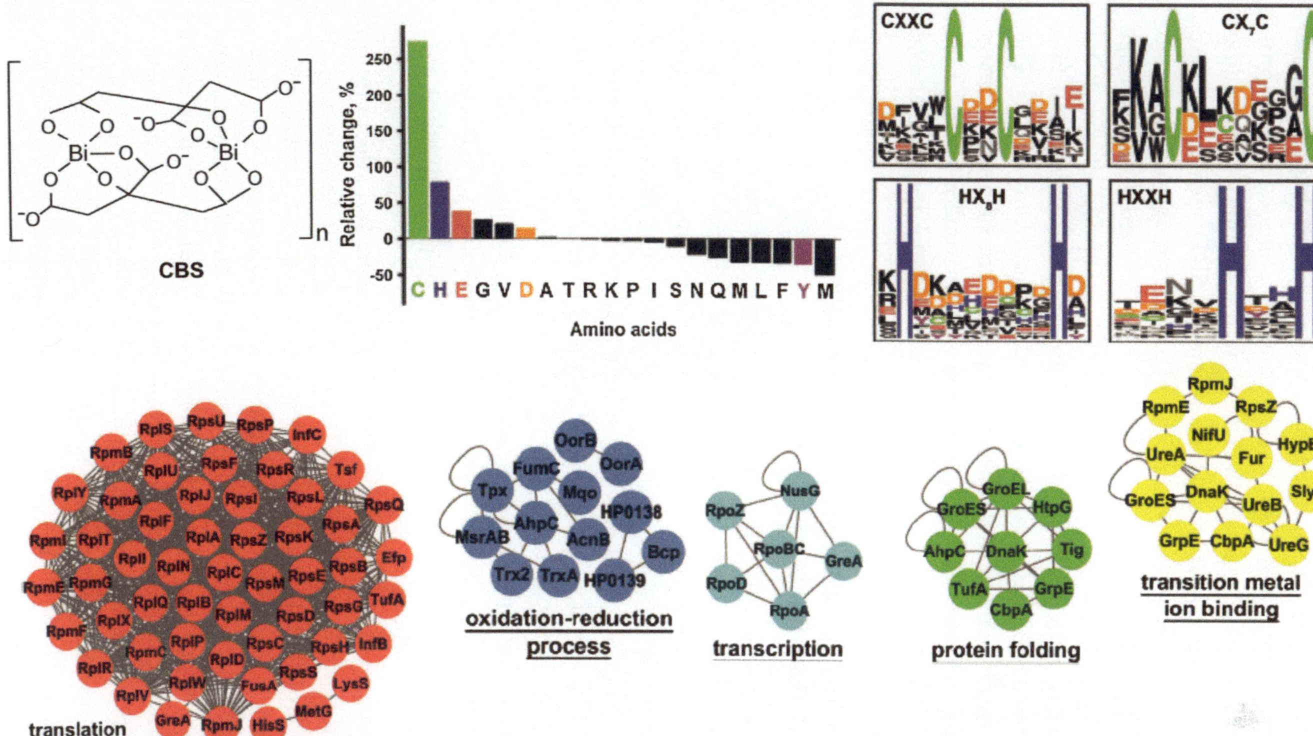

Figure 9.4 Analysis of bismuth-binding proteins/motifs in *Helicobacter pylori* based on immobilized-bismuth affinity chromatography and mass spectrometry as well as the protein–protein interaction network from bioinformatics. Binding sites consisting of cysteine and histidine, with the motifs sequences of CX_nC, CX_nH, and HX_nH, are predominantly identified. The protein–protein interaction network reveals that the metallodrug disturbs multiple cellular processes, elucidating the possible mechanisms of action of the antimicrobial drug. *CBS*, colloidal bismuth subcitrate. *Adapted from Wang, Y.; Tsang, C.-N.; Xu, F.; Kong, P.-W.; Hu, L.; Wang, J.; et al. Bio-coordination of Bismuth in Helicobacter pylori Revealed by Immobilized Metal Affinity Chromatography. Chem. Commun.* ***2015****, 51, 16479–16482 © 2016 Royal Society of Chemistry.*

Bi-binding proteins disclosed that bismuth perturbs numerous metabolic processes that are critical to the pathogen, including the defense mechanism against oxidative stress, transcription and translation processes, and protein folding. Importantly, proteins involved in transition metal ion binding (e.g., FKBP-type peptidyl-prolyl cis-trans isomerase SlyD, metallo-GTPase HypB, ribosomal proteins RpmE and RpmJ, urease subunit α UreA and urease subunit β UreB, urease accessory subunit UreG) were found to bind bismuth, which implied the potential functional disruption of the metal homeostasis in the pathogen. Indeed, an in vitro study has shown that bismuth inhibited the GTPase activity of *Hp*TufA and *Hp*HypB.[51,52] The study further confirmed the multitargeting mechanism of bismuth drugs through binding to important metalloenzymes and metalloproteins, thus disrupting multiple biological processes that are essential for the survival and pathogenesis of the pathogen.

3.4 Potential Roles of Cobalt(II) and Nickel(II) in Gram-Positive Bacteria

Pathogenic microbes require metal ion as nutrients to stabilize the structures and for the full functions of metalloproteins.[53] Transition metal ions in particular serve as cofactors to activate metalloenzymes that are critical for the survival and pathogenesis of the microbes. On the other hand, toxicity can be induced at high levels of metals in the niche. Pathogenic microbes therefore employ a highly sophisticated system of proteins that regulates the availability of metal nutrients and circumvents the toxicity of metals.[54] By capturing the metal-binding proteins through IMAC and MS-based analysis, 208 Ni(II)-binding proteins and 223 Co(II)-binding proteins from *Streptococcus pneumonia* were identified.[25] Although a certain extent of overlapping in the Co(II)- and Ni(II)-binding proteins was observed, distinctive functions of the identified proteins and selectivity on the amino acid residues were revealed. Gene ontology enrichment analysis suggested that Co(II)-binding proteins were involved in structural stabilization of ribosome and interaction with RNA, whereas proteins with ligase or oxidoreductase activity were found to bind Ni(II) ions. Moreover, lysine was exclusively recognized in peptides for Co(II) binding whereas valine and aspartic acid were more frequently associated with Ni(II) ions, despite similar preference in histidine and methionine toward Co(II) and Ni(II) coordination. This metalloproteomic study by IMAC provides a molecular basis for metal homeostasis, and the mechanisms of toxicity of metals to the bacterium are likely caused by the inhibition of essential enzymes.

4. IMMOBILIZED METAL AFFINITY CHROMATOGRAPHY FOR THE ANALYSIS OF PROTEIN PHOSPHORYLATION

Posttranslational modifications (PTMs) of proteins, consisting of phosphorylation, acetylation, acylation, alkylation, glycosylation, sumoylation, oxidation, and ubiquitination, can be considered covalent modifications of the side chains of specific amino acids that modulate the functions, cellular localization, and activities of proteins by altering the tertiary and quaternary structures of biomolecules to regulate diverse cellular processes.[55] Approximately 5% of the genomes of eukaryotes encode enzymes that participate in PTMs in proteomes, whereas about 75% of common amino acids may undergo these diversified modifications.[14]

One extensively studied PTM is protein phosphorylation, which involves histidine and aspartic acid in bacterial and fungal proteomes in addition to serine, threonine, and tyrosine in mammalian proteins.[14] Enzymes such as kinases and phosphatases are responsible for the phosphorylation and dephosphorylation of proteins, respectively. Phosphorylation introduces a negatively charged phosphate group to proteins, leading to conformational changes in the local microenvironment of proteins, subsequently resulting in reorganization of a domain that initiates cellular signaling pathways.[56]

Owing to the heterogeneity and scarcity of the modified proteins (about 1%) as well as dynamic modification processes in vivo, enormous effort has been made to detect and enrich specific modifications in cells and cell lysates. To capture phosphoproteomes, different approaches such as strong cation exchange chromatography, strong anion exchange chromatography, and antibody-based immunoprecipitation have been established to separate phosphorylated peptides from nonphosphorylated ones.[57] In addition, considering the nature of phosphate anions, hard metal ions including Fe(III), Ga(III), Zr(IV), and Ti(IV) are loaded to IMAC resin to study phosphoproteomes.[58–62] The antibody-based method (immunoprecipitation) preferentially isolates targets based on the recognition of the specific amino acid sequences by antibodies, whereas the IMAC technique may be more applicable to phosphopeptides with higher abundance. IMAC is one of the widely used techniques for proteomic analysis because the column resin shows high chemical stability and because of the relatively low cost, which is compatible with commonly used detergents such as Brij35, Tween-20, Tween-80, and Triton-X-100 compared with immunoaffinity-based methods. Moreover, the column resin exhibits high tolerance toward a wide range of reducing agents, e.g., dithiothreitol, dithioerythritol, and β-mercaptoethanol.

Ti(IV) is highly acidic, with a pK_a of about 0.5; therefore, it signifies a higher affinity toward oxygen, i.e., phosphoproteomes compared with Fe(III) and Ga(III) ions. Besides, differences in redox potential result in the deviation in identified phosphoproteomes when applying different metal ions in IMAC analysis. In contrast, nonspecific capture of peptides with amino acids such as glutamic acid and aspartic acid on the solid resin of IMAC has been reported as a result of the coordination of the carboxylate ($R{-}COO^-$) with immobilized metal ions such as Ti(IV) ions. Furthermore, hydrophobic interaction with the resin has been noted.[63] To improve specificity, the acidic residues can be modified by methyl esterification before IMAC enrichment; incorporation of organic solvents during sample loading will also minimize nonspecific interaction with the column matrix.[64] Alternatively, metal-oxide affinity chromatography, such as $Al(OH)_3$, ZrO_2, and TiO_2, can be used.[57,65]

An overview of IMAC-based phosphoproteomic analysis is illustrated in Fig. 9.5. Because of the small sample volume, and to reduce the sample loss, IMAC columns are often packed in commercial pipette tips (e.g., GELoader tip). With reference to the preparation of a StageTip, enrichment is performed by centrifugation.[58,66,67] In general, samples containing phosphoproteins or phosphopeptides are first alkylated via iodoacetamide or esterified via methanol, followed by a desalting step using a C18 reverse-phase column. To minimize nonspecific binding to the IMAC and increase the specificity of the method toward phosphopeptides, the samples are loaded onto an IMAC column in buffers at pH < 3.0 (e.g., 0.1% trifluoroacetic acid and 50% acetonitrile).[63] Under such a condition, acidic peptides in the sample will be neutralized whereas phosphopeptides will retain their negative charge and binding affinities toward the IMAC resins. The phosphopeptides are eluted from IMAC resin using alkaline buffers (e.g., 10% v/v ammonium solution at pH 10–11) or with chelate ligands (e.g., EDTA or ascorbic acid) or buffers containing a high concentration of phosphate.[66,68]

To overcome the drawback of an IMAC-based phosphoproteomic study, several improvements to IMAC enrichment techniques have been made through the rational design of the experimental setup. First, to improve the matrix material to reduce nonspecific adsorption, monodisperse microspheres made of polystyrene with a uniform size (about 13 μm) and column packing have been made to provide higher chemical stability and column

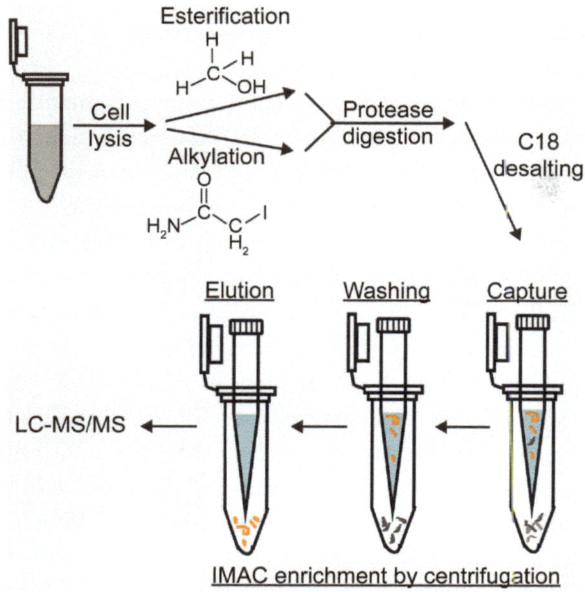

Figure 9.5 Overview of phosphopeptide enrichment using immobilized metal affinity chromatography (IMAC). Cells are first lysed and esterified by methanol or alkylated by iodoacetamide. The samples are then digested by proteases and desalted by C18 column. Enrichment by IMAC is achieved through centrifugation, after which mass spectrometry (MS) technique is used for qualitative and quantitative analysis. *LC*, liquid chromatography.

efficiency as well as a larger surface area of hydrophilic matrix to interact with phosphopeptides.[69,70] Besides, graphene has been suggested as a matrix material for IMAC, considering the possibility of synthesizing functionalized graphene material with large surface areas.[71] Second, the orientation of the metal–chelator complexes in the column resin can be optimized by incorporating a flexible linker between the microsphere matrix and the chelator moiety, generating poly(glycidyl methacrylate-*co*-trimethylopropane trimethacrylate) microspheres that offer spatial separation between the immobilized Ti(IV) ions and the microsphere matrix.[70] Third, conventional NTA and IDA chelator moieties can be replaced to enhance metal immobilization. Oxidative self-polymerization of dopamine to form hydrophilic polydopamine has been proposed, owing to its excellent biocompatibility, serving as a potential ligand to coordinate Fe(III) or Ti(IV).[71,72] Moreover, the use of phosphonate group as the ligand group to bind Ti(IV) ions, with two oxygen atoms from two phosphonate groups coordinating each Ti(IV) ion, has been recommended.[66] Fourth, the chemical stability of the matrix material is crucial for phosphopeptide enrichment because of the necessity in sample loading and elution under acidic pH ($pH < 3.0$) to preserve the negative charge of phosphorylated peptides while protonating the peptides with glutamate or aspartate residues.[73] Trifluoroacetic acid (6%v/v) may be used to provide an acidic condition to restrain nonspecific interaction between acidic amino acid residues and the column resin.[65,66]

Furthermore, it has been documented that the detection of multiphosphorylated peptides may be interfered by monophosphorylated peptides because of charge differences that contribute to differentiated electrostatic interactions between positively charged metal ions and negatively charged phosphate groups.[74] Moreover, monophosphorylated proteins and multiphosphorylated proteins possess discriminated ionization efficiency, resulting in suppression of signals in MS for multiphosphorylated peptides and acidophilic phosphopeptides by monophosphoproteomes.[75] Multidimensional IMAC, consisting of two or more IMAC columns in sequence, offers a feasible solution for the separation of monophosphorylated and multiphosphorylated peptides in samples to enhance capture efficiency. Sequential elution from IMAC, in which multiphosphorylated peptides are first separated from the monophosphorylated and nonphosphorylated peptides in general, may greatly improve efficiency in identifying phosphorylation sites, especially when analyzing samples with low amounts of sample input.[75] It has been reported that loading samples to Ga(III)-IMAC and subsequently to Fe(III)-IMAC demonstrated a higher efficiency in enrichment.[74] Almost all of the tested multiphosphorylated peptides (>90%) were isolated by Ga(III)-IMAC, predominantly consisting of peptides with acidophilic phosphorylation sites (phosphorylation sites consisting of aspartic acid or glutamic acid). On the other hand, Fe(III)-IMAC exhibited a higher preference toward proline-directed or basophilic phosphorylation sites (sites with arginine or lysine).

An integrated strategy to analyze different PTMs sequentially in a single sample flow has been proposed, enabling the analysis of protein phosphorylation, ubiquitination, and acetylation in succession.[76] The technique used Fe(III)-NTA-based IMAC to enrich phosphoproteomes and antibodies to isolate ubiquitinated and acetylated proteins. Taking advantage of volatile solvents used in phosphopeptide enrichment, it is convenient first to sequester phosphorylated proteins before examining other PTMs by immunoaffinity-based techniques. The serial analysis of several PTMs offers a methodology for understanding signal transduction pathways systemically.

4.1 Development of Immunotherapies

The immune system is sophisticated, constituting of a diverse network of pathways and processes that may affect immune responses. Antigen molecules (whether self or foreign) are recognized by receptors on the cell surface, after which immune responses are elicited. Balance in the costimulatory and inhibitory signals thus regulates the amplitudes of immune responses. This is exemplified by the actions of cytotoxic T-lymphocyte–associated protein 4 (CTLA4) and cluster of differentiation 28 (CD28). CTLA4 is a receptor protein that suppresses immune response in T-lymphocytes (also called T-cells) whereas CD28 counteracts its action by generating costimulatory signals to activate T-cells.[77] The exact mechanism of T-cell activation–deactivation by CD28 and CTLA4 is obscure; however, it is believed that protein phosphatases SHP2 and PP2A are involved against the actions of kinases initiated by CD28.[78]

Cancer immunotherapies rely on the presentation of cancer-specific antigens to specific receptors to stimulate cytotoxic T-cells to eliminate tumor cells selectively in cancer treatment. Development of immunotherapies has become increasingly popular owing to the potential discrimination between cancerous cells and normal healthy cells. Dysregulation of phosphorylation and dephosphorylation of certain proteins have been revealed to be correlated with some disease conditions, especially in malignant transformation, and have a crucial role in signal transduction in oncogenes and abnormal cell differentiation and growth.[79] In addition, alterations in phosphorylation of proteins are implicated as serving as characteristic biomarkers for cancer progression. Phosphopeptides from various types of cancers, such as ovarian cancer, colorectal cancer, skin cancer, and even leukemia, have been determined.[79] Tumor-specific phosphopeptides have therefore been considered excellent targets for immunotherapies. Methodologies that can effectively segregate and identify phosphopeptides from mammalian cell lines and from clinical samples are urgently demanded.[78,79]

A study showed that Fe(III)-IDA- and Fe(III)-NTA-based IMAC workflow using a fused silica microcapillary column can be used to enrich a subfemtomolar level of human leukocyte antigen (HLA) class I–associated phosphopeptides from ovarian cancer cells and colorectal cancer tissue.[68] Over 150 and 80 HLA-associated phosphopeptides were uniquely identified in cancerous cell lines and tissue from a cancer patient, respectively. A combination of collision-activated dissociation and electron transfer dissociation for peptide fragmentation has been suggested to preserve the labile phosphorylation sites. The authors also affirmed the use of anhydrous methanol and anhydrous acetyl chloride to esterify acid residues; phosphopeptide standards were included to monitor the capture efficiency and sample recoveries of the proposed protocol. The efficiency of the two chelator moieties (IDA and NTA) was compared, with a higher selectivity when

Fe(III)-NTA-based setup was used. This implied that NTA might be superior to IDA for capturing phosphopeptides. This protocol will facilitate the enrichment of low-input samples (10^{-6} to 10^{-9} g of peptides), enabling the investigation of phosphoproteomes in clinical samples.

4.2 Proteomic Study of Drug Responses

Kinase inhibitors such as SU11274 and staurosporine are clinically effective against various kinds of cancers. The former is an inhibitor of c-Met tyrosine kinase, which competes with the substrate adenosine triphosphate (ATP) to interact with c-Met kinase, whereas the latter originated naturally from the bacterium *Streptomyces staurosporeus* and inhibits binding of ATP to protein kinase C (PKC). c-Met and PKC signaling are highly linked with gastric carcinomas.[80,81] For instance, c-Met inhibitors may interfere with the signaling in β-catenin, apoptosis, and the mitogen-activated protein kinase/Erk pathways, disturbing the growth and survival of cancer cells.

The response of SU11274 and staurosporine in human gastric cancer cells investigated by IMAC and immunoaffinity methods has allowed the profiling of PTMs of proteins as well as the analysis of changes in total proteomes.[82] Total proteome analysis indicated trivial changes in total protein levels; however, considerable deviations in phosphoproteomes were discovered after treatment by the two drugs. Western blotting analysis of proteins Erk1 and Erk2 confirmed the relatively unchanged protein level but a significant reduction in the phosphorylation of Thr185/202 and Tyr187/204 in response to treatment with SU11274, which confirmed the importance of analyzing alterations in PTMs, together with the total protein level, for a better understanding of the biological responses against a drug treatment.[83] Analysis of the phosphorylated proteomes by Fe(III)-IMAC uncovered as much as 7000 phosphopeptides, whereas about 1500–3500 peptides were yielded by different antibodies. Surprisingly, little overlap in identified phosphopeptides between IMAC and antibody-based methods suggested that both methods were complementary, and it might be beneficial to employ different experimental techniques to obtain complete phosphoproteome coverage. It was also discovered that compared with SU11274, staurosporine led to a larger change in the phosphoproteomes (either an increase or decrease in abundance) upon treatment of gastric cancer cells, which signifies that SU11274 and staurosporine may interrupt different cellular processes despite the similar actions of the two drugs in inhibiting ATP binding to kinases. This study offers a robust label-free quantitative strategy for the analysis of potential biomarkers and drug targets, and improves our understanding of the signaling responses of drugs in biological systems.

5. NOVEL POLYMER-BASED METAL ION AFFINITY CAPTURE: ALTERNATIVE TO IMMOBILIZED METAL AFFINITY CHROMATOGRAPHY

Polymer-based metal ion affinity capture (PolyMAC) uses soluble nanopolymers (e.g., polyamidoamine dendrimers) functionalized with metal ions to allow the chelation and subsequent isolation of phosphopeptides in a homogeneous environment.[84] The concept was first proposed in 2005 to capture and characterize phosphorylated proteins, in which soluble dendrimers were used, and was further developed using polyamidoamine dendrimers functionalized with titanium or iron to enrich and analyze phosphopeptides.[84–86]

Most phosphoproteomic approaches including IMAC have several drawbacks such as low reproducibility and selectivity, largely owing to the heterogeneity of solid-phase extraction-based isolation during phosphopeptide enrichment steps, which makes it difficult to recover low-abundance

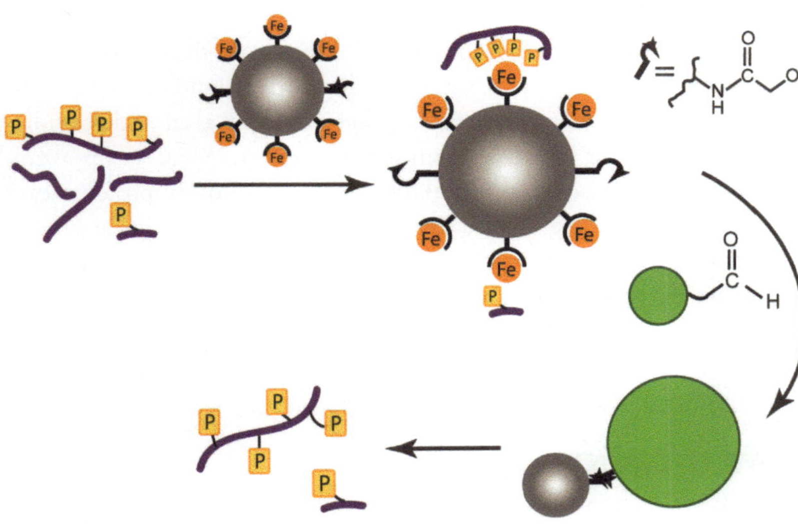

Figure 9.6 Schematic illustration of polymer-based metal ion affinity capture for phosphoproteomics. Phosphopeptides are incubated with Fe(III)-functionalized nanopolymers (*gray sphere*), which are then captured by aldehyde-functionalized agarose beads (*green*) via amide linkage. Phosphopeptides are finally recovered in alkaline buffer.

phosphopeptides from complex samples. However, the PolyMAC approach overcomes these constraints. PolyMAC-based phosphopeptide enrichment is performed in a homogeneous environment instead of solid-phase extraction, reducing the operation time and improving the selectivity, robustness, and reproducibility of capture processes.[86] Until now, Ti(IV) and Fe(III) have been chelated on soluble polymers to identify phosphoproteomes. As illustrated in Fig. 9.6, in the latest generation of PolyMAC, samples were first incubated with PolyMAC reagents, after which the PolyMAC-bound phosphopeptides were covalently linked to aldehyde-functionalized agarose beads via the linker molecule hydroxylamine to isolate the PolyMAC-coordinated phosphopeptides from the bulk solution. Those captured phosphorylated peptides were eventually retrieved in buffers of alkaline pH (e.g., 400 mM ammonium hydroxide). The selectivity (i.e., ratio of phosphopeptides to the total number of peptides) was doubled using PolyMAC-Fe over Fe-IMAC in the same sample set, which indicated that PolyMAC may offer a powerful and widely applicable technique for phosphoproteomics and molecular signaling.[86]

6. CONCLUSION

IMAC serves as an important proteomic approach and has been used extensively on metalloproteomics and phosphoproteomics in past decades in addition to its wide use in the pharmaceutical industry for drug development. Combined with other techniques, IMAC facilitates the elucidation of mechanisms of action of metallodrugs and metallo-agents as well as an understanding of the role of metal ions in living organisms. Moreover, IMAC is a powerful tool for enriching and isolating phosphopeptides for phosphoproteomics. It provides an insight into the diverse signaling pathways in biological systems. Improvements to IMAC through advancements in the device designs and sample preparation procedures have been under vigorous investigation. Given its high compatibility and simplicity, IMAC coupled with various biochemical techniques as integrative methodologies will have a continuously important role in proteomic and chemical biology research.

ACKNOWLEDGMENTS

We thank the Research Grants Council of Hong Kong, the Livzon Pharmaceutical Group, and the University of Hong Kong [an Emerging Strategic Research Theme (e-SRT) on Integrative Biology] for their support of our research on medicinal inorganic chemistry and metalloproteomics of bismuth. We thank Haibo Wang for helpful comments. YYC acknowledges a scholarship from the university.

REFERENCES

1. Porath, J.; Carlsson, J. A.N. ; Olsson, I.; Belfrage, G. Metal Chelate Affinity Chromatography, a New Approach to Protein Fractionation. *Nature* **1975,** *258,* 598–599.
2. Cheung, R. C.; Wong, J. H.; Ng, T. B. Immobilized Metal Ion Affinity Chromatography: a Review on Its Applications. *Appl. Microbiol. Biotechnol.* **2012,** *96,* 1411–1420.
3. Ming, L. J. Structure and Function of "Metalloantibiotics". *Med. Res. Rev.* **2003,** *23,* 697–762.
4. Takeda, N.; Matsuoka, T.; Gotoh, M. Potentiality of IMAC as Sample Retreatment Tool in Food Analysis for Veterinary Drugs. *Chromatographia* **2010,** *72,* 127–131.
5. Gokhale, N.; Patwardhan, A.; Cowan, J. A. Metalloaminoglycosides: Chemistry and Biological Relevance. In *Aminoglycoside Antibiotics;* John Wiley & Sons, Inc., 2007; pp 235–254.
6. Gu, J.; Codd, R. The Resolution of Two Clinical Agents, Bleomycin and Desferrioxamine B, from a *Streptomyces verticillus* Fermentation Mixture Using Multi-dimensional Immobilised Metal Ion Affinity Chromatography. *RSC Adv.* **2015,** *5,* 3443–3453.
7. Finney, L. A.; O'Halloran, T. V. Transition Metal Speciation in the Cell: Insights From the Chemistry of Metal Ion Receptors. *Science* **2003,** *300,* 931–936.
8. Martínez-Fábregas, J.; Rubio, S.; Díaz-Quintana, A.; Díaz-Moreno, I.; De la Rosa, M. Proteomic Tools for the Analysis of Transient Interactions between Metalloproteins. *FEBS J.* **2011,** *278,* 1401–1410.
9. Sun, X.; Tsang, C.-N.; Sun, H. Identification and Characterization of Metallodrug Binding Proteins by (Metallo)proteomics. *Metallomics* **2009,** *1,* 25–31.
10. Mjos, K. D.; Orvig, C. Metallodrugs in Medicinal Inorganic Chemistry. *Chem. Rev.* **2014,** *114,* 4540–4563.
11. Barry, N. P. E.; Sadler, P. J. Exploration of the Medical Periodic Table: towards New Targets. *Chem. Commun.* **2013,** *49,* 5106–5131.
12. Andersson, L.; Porath, J. Isolation of Phosphoproteins by Immobilized Metal (Fe^{3+}) Affinity Chromatography. *Anal. Biochem.* **1986,** *154,* 250–254.
13. Neville, D. C.; Rozanas, C. R.; Price, E. M.; Gruis, D. B.; Verkman, A. S.; Townsend, R. R. Evidence for Phosphorylation of Serine 753 in CFTR Using a Novel Metal-Ion Affinity Resin and Matrix-assisted Laser Desorption Mass Spectrometry. *Protein Sci.* **1997,** *6,* 2436–2445.
14. Walsh, C. T.; Garneau-Tsodikova, S.; Gatto, G. J. Jr. Protein Posttranslational Modifications: the Chemistry of Proteome Diversifications. *Angew. Chem. Int. Ed.* **2005,** *44,* 7342–7372.
15. Hunter, T. Signaling – 2000 and beyond. *Cell* **2000,** *100,* 113–127.
16. Pawson, T.; Scott, J. D. Protein Phosphorylation in Signaling – 50 Years and Counting. *Trends Biochem. Sci.* **2005,** *30,* 286–290.
17. Irving, H.; Williams, R. J. P. 637. The Stability of Transition-Metal Complexes. *J. Chem. Soc.* **1953,** 3192–3210.
18. Anderegg, G. Critical Survey of Stability Constants of NTA Complexes. *Pure Appl. Chem.* **1982,** *54,* 2693–2758.
19. Pearson, R. G. Hard and Soft Acids and Bases. *J. Am. Chem. Soc.* **1963,** *85,* 3533–3539.
20. Lemire, J. A.; Harrison, J. J.; Turner, R. J. Antimicrobial Activity of Metals: Mechanisms, Molecular Targets and Applications. *Nat. Rev. Microbiol.* **2013,** *11,* 371–384.
21. Wang, Y.; Wang, H.; Li, H.; Sun, H. Metallomic and Metalloproteomic Strategies in Elucidating the Molecular Mechanisms of Metallodrugs. *Dalton Trans.* **2015,** *44,* 437–447.
22. Kågedal, L., Ed. *Immobilized Metal Ion Affinity Chromatography.* In *Protein Purification: Principles, High-resolution Methods, and Applications*; Janson, J. C., Rydén, L., Eds.; John Wiley & Sons: New York, 1998.

23. Krishnan, S.; Vijayalakshmi, M. A. Purification and Some Properties of Three Serine Carboxypeptidases from *Aspergillus niger*. *J. Chromatogr. A* **1986,** *370,* 315–326.
24. Edy, V. G.; Billiau, A.; de Somer, P. Purification of Human Fibroblast Interferon by Zinc Chelate Affinity Chromatography. *J. Biol. Chem.* **1977,** *252,* 5934–5935.
25. Sun, X.; Yu, G.; Xu, Q.; Li, N.; Xiao, C.; Yin, X.; et al. Putative Cobalt- and Nickel-Binding Proteins and Motifs in *Streptococcus pneumoniae*. *Metallomics* **2013,** *5,* 928–935.
26. Smith, S. D.; She, Y. M.; Roberts, E. A.; Sarkar, B. Using Immobilized Metal Affinity Chromatography, Two-Dimensional Electrophoresis and Mass Spectrometry to Identify Hepatocellular Proteins with Copper-Binding Ability. *J. Proteome Res.* **2004,** *3,* 834–840.
27. Chen, C.; Song, Y.; Zhuang, K.; Li, L.; Xia, Y.; Shen, Z. Proteomic Analysis of Copper-Binding Proteins in Excess Copper-Stressed Roots of Two Rice (*Oryza Sativa* L.) Varieties with Different Cu Tolerances. *PLoS One* **2015,** *10,* e0125367.
28. Barnett, J. P.; Scanlan, D. J.; Blindauer, C. A. Identification of Major Zinc-Binding Proteins from a Marine Cyanobacterium: Insight into Metal Uptake in Oligotrophic Environments. *Metallomics* **2014,** *6,* 1254–1268.
29. Thierse, H. J.; Helm, S.; Pankert, P. Metalloproteomics in the Molecular Study of Cell Physiology and Disease. *Methods Mol. Biol.* **2008,** *425,* 139–147.
30. Förstner, U.; Wittmann, G. T. W. *Metal Pollution in the Aquatic Environment;* Springer-Verlag: Heidelberg, 2012.
31. Tchounwou, P. B.; Yedjou, C. G.; Patlolla, A. K.; Sutton, D. J. Heavy Metal Toxicity and the Environment. In *Molecular, Clinical and Environmental Toxicology;* Luch, A., Ed.; Springer Basel: Basel, 2012; pp 133–164.
32. Järup, L. Hazards of Heavy Metal Contamination. *Br. Med. Bull.* **2003,** *68,* 167–182.
33. Johri, N.; Jacquillet, G.; Unwin, R. Heavy Metal Poisoning: the Effects of Cadmium on the Kidney. *Biometals* **2010,** *23,* 783–792.
34. Jensen, M. P.; Gorman-Lewis, D.; Aryal, B.; Paunesku, T.; Vogt, S.; Rickert, P. G.; et al. An Iron-dependent and Transferrin-Mediated Cellular Uptake Pathway for Plutonium. *Nat. Chem. Biol.* **2011,** *7,* 560–565.
35. Taylor, D. M. The Bioinorganic Chemistry of Actinides in Blood. *J. Alloys Compd.* **1998,** *271–273,* 6–10.
36. Aryal, B. P.; Paunesku, T.; Woloschak, G. E.; He, C.; Jensen, M. P. A Proteomic Approach to Identification of Plutonium-Binding Proteins in Mammalian Cells. *J. Proteomics* **2012,** *75,* 1505–1514.
37. DalCorso, G.; Manara, A.; Piasentin, S.; Furini, A. Nutrient Metal Elements in Plants. *Metallomics* **2014,** *6,* 1770–1788.
38. Song, Y.; Zhang, H.; Chen, C.; Wang, G.; Zhuang, K.; Cui, J.; et al. Proteomic Analysis of Copper-Binding Proteins in Excess Copper-Stressed Rice Roots by Immobilized Metal Affinity Chromatography and Two-Dimensional Electrophoresis. *Biometals* **2014,** *27,* 265–276.
39. Fock, K. M.; Graham, D.Y.; Malfertheiner, P. *Helicobacter pylori* research: Historical Insights and Future Directions. *Nat. Rev. Gastroenterol. Hepatol.* **2013,** *10,* 495–500.
40. Midolo, P. D.; Lambert, J. R.; Kerr, T. G.; Tee, W. In Vitro Synergy between Ranitidine Bismuth Citrate and Tetracycline or Clarithromycin against Resistant Strains of *Helicobacter pylori*. *Eur. J. Clin. Microbiol. Infect. Dis.* **1999,** *18,* 832–834.
41. Mendis, A. H. W.; Marshall, B. J. *Helicobacter pylori* and Bismuth. In *Biological Chemistry of Arsenic, Antimony and Bismuth;* John Wiley & Sons, Ltd, 2010; pp 241–262.
42. Hong, Y.; Lai, Y.-T.; Chan, G. C.-F.; Sun, H. Glutathione and Multidrug Resistance Protein Transporter Mediate a Self-propelled Disposal of Bismuth in Human Cells. *Proc. Natl. Acad. Sci. U.S.A.* **2015,** *112,* 3211–3216.
43. Andrews, P. C.; Busse, M.; Deacon, G. B.; Ferrero, R. L.; Junk, P. C.; MacLellan, J. G.; et al. Remarkable in Vitro Bactericidal Activity of Bismuth(III) Sulfonates against *Helicobacter pylori*. *Dalton Trans.* **2012,** *41,* 11798–11806.

44. Busse, M.; Trinh, I.; Junk, P. C.; Ferrero, R. L.; Andrews, P. C. Synthesis and Characterisation of Bismuth(III) Aminoarenesulfonate Complexes and Their Powerful Bactericidal Activity Against *Helicobacter pylori*. *Chem. Eur. J.* **2013,** *19,* 5264–5275.
45. Andrews, P. C.; Blair, V. L.; Ferrero, R. L.; Junk, P. C.; Kumar, I. Making Bispirin: Synthesis, Structure and Activity Against *Helicobacter pylori* of Bismuth(III) Acetylsalicylate. *Chem. Commun.* **2013,** *49,* 2870–2872.
46. Li, H.; Sun, H. Recent Advances in Bioinorganic Chemistry of Bismuth. *Curr. Opin. Chem. Biol.* **2012,** *16,* 74–83.
47. Ge, R.; Sun, H. Bioinorganic Chemistry of Bismuth and Antimony: Target Sites of Metallodrugs. *Acc. Chem. Res.* **2007,** *40,* 267–274.
48. Yang, N.; Sun, H. Biocoordination Chemistry of Bismuth: Recent Advances. *Coord. Chem. Rev.* **2007,** *251,* 2354–2366.
49. Ge, R.; Sun, X.; Gu, Q.; Watt, R. M.; Tanner, J. A.; Wong, B. C.; et al. A Proteomic Approach for the Identification of Bismuth-Binding Proteins in *Helicobacter pylori*. *J. Biol. Inorg. Chem.* **2007,** *12,* 831–842.
50. Wang, Y.; Tsang, C.-N.; Xu, F.; Kong, P.-W.; Hu, L.; Wang, J.; et al. Bio-coordination of Bismuth in *Helicobacter pylori* Revealed by Immobilized Metal Affinity Chromatography. *Chem. Commun.* **2015,** *51,* 16479–16482.
51. Wang, D.; Luo, B.; Shan, W.; Hao, M.; Sun, X.; Ge, R. The Effects of EF-Ts and Bismuth on EF-Tu in *Helicobacter pylori*: Implications for an Elegant Timing for the Introduction of EF-Ts in the Elongation and EF-Tu as a Potential Drug Target. *Metallomics* **2013,** *5,* 888–895.
52. Xia, W.; Li, H.; Sun, H. Functional Disruption of HypB, a GTPase of *Helicobacter pylori*, by Bismuth. *Chem. Commun.* **2014,** *50,* 1611–1614.
53. Skaar, E. P.; Raffatellu, M. Metals in Infectious Diseases and Nutritional Immunity. *Metallomics* **2015,** *7,* 926–928.
54. Hood, M. I.; Skaar, E. P. Nutritional Immunity: Transition Metals at the Pathogen-Host Interface. *Nat. Rev. Microbiol.* **2012,** *10,* 525–537.
55. Seo, J.; Lee, K. J. Post-translational Modifications and Their Biological Functions: Proteomic Analysis and Systematic Approaches. *J. Biochem. Mol. Biol.* **2004,** *37,* 35–44.
56. Johnson, L. N.; Lewis, R. J. Structural Basis for Control by Phosphorylation. *Chem. Rev.* **2001,** *101,* 2209–2242.
57. Villen, J.; Gygi, S. P. The SCX/IMAC Enrichment Approach for Global Phosphorylation Analysis by Mass Spectrometry. *Nat. Protoc.* **2008,** *3,* 1630–1638.
58. Thingholm, T. E.; Larsen, M. R. Phosphopeptide Enrichment by Immobilized Metal Affinity Chromatography. *Methods Mol. Biol.* **2016,** *1355,* 123–133.
59. Yue, X.; Schunter, A.; Hummon, A. B. Comparing Multistep Immobilized Metal Affinity Chromatography and Multistep TiO_2 Methods for Phosphopeptide Enrichment. *Anal. Chem.* **2015,** *87,* 8837–8844.
60. Kennedy, J. J.; Yan, P.; Zhao, L.; Ivey, R. G.; Voytovich, U. J.; Moore, H. D.; et al. Immobilized Metal Affinity Chromatography Coupled to Multiple Reaction Monitoring Enables Reproducible Quantification of Phospho-Signaling. *Mol. Cell. Proteomics* **2016,** *15,* 726–739.
61. Feng, S.; Ye, M.; Zhou, H.; Jiang, X.; Jiang, X.; Zou, H.; et al. Immobilized Zirconium Ion Affinity Chromatography for Specific Enrichment of Phosphopeptides in Phosphoproteome Analysis. *Mol. Cell. Proteomics* **2007,** *6,* 1656–1665.
62. Swaney, D. L.; Villén, J. Enrichment of Phosphopeptides via Immobilized Metal Affinity Chromatography. *Cold Spring Harb. Protoc.* **2016,** *2016.* http://dx.doi.org/10.1101/pdb.prot088005.
63. Kokubu, M.; Ishihama, Y.; Sato, T.; Nagasu, T.; Oda, Y. Specificity of Immobilized Metal Affinity-Based IMAC/C18 Tip Enrichment of Phosphopeptides for Protein Phosphorylation Analysis. *Anal. Chem.* **2005,** *77,* 5144–5154.
64. Ficarro, S. B.; McCleland, M. L.; Stukenberg, P. T.; Burke, D. J.; Ross, M. M.; Shabanowitz, J.; et al. Phosphoproteome Analysis by Mass Spectrometry and Its Application to *Saccharomyces cerevisiae*. *Nat. Biotechnol.* **2002,** *20,* 301–305.

65. Zhou, H.; Low, T. Y.; Hennrich, M. L.; van der Toorn, H.; Schwend, T.; Zou, H.; et al. Enhancing the Identification of Phosphopeptides From Putative Basophilic Kinase Substrates Using Ti(IV) Based IMAC Enrichment. *Mol. Cell. Proteomics* **2011,** *10.* http://dx.doi.org/10.1074/mcp.M110.006452.
66. Zhou, H.; Ye, M.; Dong, J.; Corradini, E.; Cristobal, A.; Heck, A. J.; et al. Robust Phosphoproteome Enrichment Using Monodisperse Microsphere-Based Immobilized Titanium (IV) Ion Affinity Chromatography. *Nat. Protoc.* **2013,** *8,* 461–480.
67. Rappsilber, J.; Mann, M.; Ishihama, Y. Protocol for Micro-purification, Enrichment, Pre-fractionation and Storage of Peptides for Proteomics Using StageTips. *Nat. Protoc.* **2007,** *2,* 1896–1906.
68. Abelin, J. G.; Trantham, P. D.; Penny, S. A.; Patterson, A. M.; Ward, S. T.; Hildebrand, W. H.; et al. Complementary IMAC Enrichment Methods for HLA-Associated Phosphopeptide Identification by Mass Spectrometry. *Nat. Protoc.* **2015,** *10,* 1308–1318.
69. Ugelstad, J.; Soderberg, L.; Berge, A.; Bergstrom, J. Monodisperse Polymer Particles – a Step Forward for Chromatography. *Nature* **1983,** *303,* 95–96.
70. Yu, Z.; Han, G.; Sun, S.; Jiang, X.; Chen, R.; Wang, F.; et al. Preparation of Monodisperse Immobilized Ti^{4+} Affinity Chromatography Microspheres for Specific Enrichment of Phosphopeptides. *Anal. Chim. Acta* **2009,** *636,* 34–41.
71. Yan, Y.; Zheng, Z.; Deng, C.; Li, Y.; Zhang, X.; Yang, P. Hydrophilic Polydopamine-coated Graphene for Metal Ion Immobilization as a Novel Immobilized Metal Ion Affinity Chromatography Platform for Phosphoproteome Analysis. *Anal. Chem.* **2013,** *85,* 8483–8487.
72. Piovesana, S.; Capriotti, A. L.; Cavaliere, C.; Ferraris, F.; Samperi, R.; Ventura, S.; et al. Phosphopeptide Enrichment: Development of Magnetic Solid Phase Extraction Method Based on Polydopamine Coating and Ti^{4+}-IMAC. *Anal. Chim. Acta* **2016,** *909,* 67–74.
73. Beausoleil, S. A.; Jedrychowski, M.; Schwartz, D.; Elias, J. E.; Villen, J.; Li, J.; et al. Large-scale Characterization of HeLa Cell Nuclear Phosphoproteins. *Proc. Natl. Acad. Sci. U.S.A.* **2004,** *101,* 12130–12135.
74. Tsai, C.-F.; Hsu, C.-C.; Hung, J.-N.; Wang, Y.-T.; Choong, W.-K.; Zeng, M.-Y.; et al. Sequential Phosphoproteomic Enrichment through Complementary Metal-Directed Immobilized Metal Ion Affinity Chromatography. *Anal. Chem.* **2014,** *86,* 685–693.
75. Thingholm, T. E.; Larsen, M. R. Sequential Elution from IMAC (SIMAC): an Efficient Method for Enrichment and Separation of Mono- and Multi-phosphorylated Peptides. *Methods Mol. Biol.* **2016,** *1355,* 147–160.
76. Mertins, P.; Qiao, J. W.; Patel, J.; Udeshi, N. D.; Clauser, K. R.; Mani, D. R.; et al. Integrated Proteomic Analysis of Post-translational Modifications by Serial Enrichment. *Nat. Methods* **2013,** *10,* 634–637.
77. Alegre, M.-L.; Frauwirth, K. A.; Thompson, C. B. T-cell Regulation by CD28 and CTLA-4. *Nat. Rev. Immunol.* **2001,** *1,* 220–228.
78. Pardoll, D. M. The Blockade of Immune Checkpoints in Cancer Immunotherapy. *Nat. Rev. Cancer* **2012,** *12,* 252–264.
79. Mohammed, F.; Cobbold, M.; Zarling, A. L.; Salim, M.; Barrett-Wilt, G. A.; Shabanowitz, J.; et al. Phosphorylation-Dependent Interaction between Antigenic Peptides and MHC Class I: A Molecular Basis for the Presentation of Transformed Self. *Nat. Immunol.* **2008,** *9,* 1236–1243.
80. Fahrmann, M. Targeting Protein Kinase C (PKC) in Physiology and Cancer of the Gastric Cell System. *Curr. Med. Chem.* **2008,** *15,* 1175–1191.
81. Guo, A.; Villen, J.; Kornhauser, J.; Lee, K. A.; Stokes, M. P.; Rikova, K.; et al. Signaling Networks Assembled by Oncogenic EGFR and c-Met. *Proc. Natl. Acad. Sci. U.S.A.* **2008,** *105,* 692–697.
82. Stokes, M.; Farnsworth, C.; Gu, H.; Jia, X.; Worsfold, C.; Yang, V.; et al. Complementary PTM Profiling of Drug Response in Human Gastric Carcinoma by Immunoaffinity and IMAC Methods with Total Proteome Analysis. *Proteomes* **2015,** *3,* 160–183.

83. Bodenmiller, B.; Mueller, L. N.; Mueller, M.; Domon, B.; Aebersold, R. Reproducible Isolation of Distinct, Overlapping Segments of the Phosphoproteome. *Nat. Methods* **2007,** *4,* 231–237.
84. Iliuk, A. B.; Martin, V. A.; Alicie, B. M.; Geahlen, R. L.; Tao, W. A. In-depth Analyses of Kinase-Dependent Tyrosine Phosphoproteomes Based on Metal Ion-Functionalized Soluble Nanopolymers. *Mol. Cell. Proteomics* **2010,** *9,* 2162–2172.
85. Tao, W. A.; Wollscheid, B.; O'Brien, R.; Eng, J. K.; Li, X.; Bodenmiller, B.; et al. Quantitative Phosphoproteome Analysis Using a Dendrimer Conjugation Chemistry and Tandem Mass Spectrometry. *Nat. Methods* **2005,** *2,* 591–598.
86. Iliuk, A.; Jayasundera, K.; Wang, W.-H.; Schluttenhofer, R.; Geahlen, R. L.; Tao, W. A. In-depth Analyses of B Cell Signaling through Tandem Mass Spectrometry of Phosphopeptides Enriched by PolyMAC. *Int. J. Mass Spectrom.* **2015,** *377,* 744–753.

CHAPTER TEN

The Analysis of Therapeutic Metal Complexes and Their Biomolecular Interactions

M.P. Sullivan[1,a], H.U. Holtkamp[1,a], S.M. Meier[2], C.G. Hartinger[1]
[1]University of Auckland, Auckland, New Zealand; [2]University of Vienna, Vienna, Austria

Contents

1. INTRODUCTION

The use of transition metals as medicines has a long history and extends back to ancient times. The success of cisplatin (Fig. 10.1), which has been used as an anticancer drug since 1978, stimulated a vigorous resurgence in their development.[1] The chemical and biological properties of transition

[a] These authors contributed equally to this work.

Inorganic and Organometallic Transition Metal Complexes with Biological Molecules and Living Cells
ISBN 978-0-12-803814-7
http://dx.doi.org/10.1016/B978-0-12-803814-7.00010-1

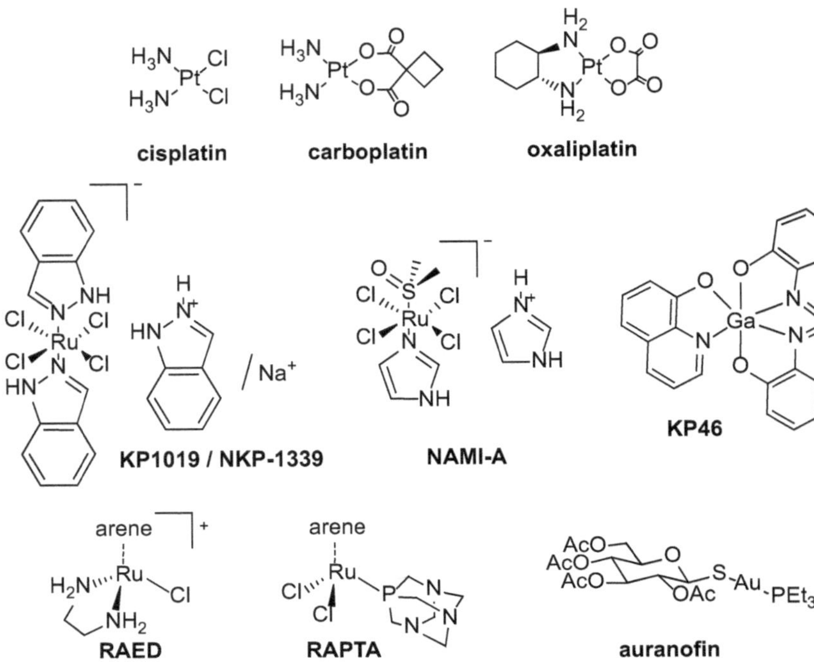

Figure 10.1 Chemical structures of representative Pt, Ru, Ga, and Au anticancer drugs (arene = *p*-cymene, toluene, biphenyl).

metal-based drugs can be modulated by varying the metal, and therewith its redox properties and coordination geometry, and/or the ligands. Now there are innumerable Pt as well as non-Pt transition metal anticancer agents based on metal centers such as Ru, Rh, Os, Ga, and Ti.[2] The ruthenium(III) complexes KP1019/NKP-1339 (the latter also known as KP1339 or IT139) and imidazolium [*trans*-$RuCl_4$(1*H*-imidazole)(DMSO)] (NAMI-A; Fig. 10.1) successfully completed phase I clinical trials, although the latter has failed phase I/II studies.[3] The range of medical applications of transition metal-based drugs has greatly expanded and now includes contrast agents, and antibacterial and antiparasitic medicines, to name a few.

It is crucial for the advancement of metallodrugs into clinical studies to understand the mechanism by which they interact with biomolecular targets, particularly in the matrices present in biological environments. An enormous array of bioanalytical methods and techniques have been developed with the aim of characterizing metallodrug pharmacological properties from the point of administration through to subsequent intracellular localization and the identification of biomolecular targets within the cancer cell.[4]

Metallodrugs can be metabolized to a wide variety of species in biological environments and speciation can be extremely difficult because of the high number of potential interaction partners. In this context, the term "metabolized" refers to enzymatic metabolization and ligand exchange reactions. The latter is a crucial event in the activation mechanism of metal-based drugs, since most of them are prodrugs. While element-specific methods for quantification can be employed, methods to characterize the molecular structures of the formed species are required to give a more complete picture.

The study of transition metal drugs in biological matrices is a pertinent challenge and this chapter will outline important analytical methods used to study coordination and organometallic compounds within tissue, serum, and cell matrices, whether qualitatively or quantitatively.[5] Quintessential bioanalytical techniques will be introduced and important advantages and disadvantages with regard to metallodrug research are discussed.

2. METALLODRUG BINDING TO BLOOD SERUM PROTEINS

Serum proteins in the blood plasma have many functions including transport of lipids, hormones, vitamins, and minerals in the circulatory system. The major serum proteins are human serum albumin (HSA) and transferrin (Tf) both of which are involved in endogenous ligand transport, delivery, and storage and are therefore of high importance also for the modes of action of metallodrugs.[6]

2.1 Serum Protein Metalation and Separation

Transition metal ions are known to have high affinity for electron donors or nucleophiles present in many biomolecules, in particular in the side chains of the amino acids L-histidine (imine), L-cysteine (thiol), L-methionine (thioether), and L-glutamate (carboxylic acid). Binding of metallodrugs to serum proteins is considered an important part in metallodrug mode of action. While some metallodrugs have been suggested to be deactivated by interaction with proteins, others have been specifically designed to exploit them as biological vectors to the tumor site or to target proteins overexpressed in tumor cells.

After intravenous administration, blood serum proteins, including HSA and Tf, are among the first biological binding partners of metallodrugs and are of greatest interest as potential metallodrug carriers.[6] HSA is taken up into the tumor at increased levels relative to normal tissue because of the enhanced permeability and retention (EPR) effect.[7] The principal binding regions for biomolecules on HSA are located in subdomains IIA and IIIA.

Tf is capable of binding two Fe^{III} ions and acts as an iron transporter via receptor-mediated endocytosis.[8] Notably, the Fe^{III} binding sites are not fully occupied and other metal ions may bind to Tf in the same fashion as Fe^{III} does. Considering that tumor cells have high iron demand, this has led to the suggestion that Tf could act as a transporter for anticancer metallodrugs. This is in contrast to the historical view that serum protein binding of drugs is linked to side effects and deactivation. It is therefore important to differentiate between noncovalent interactions and coordinative binding modes of metallodrugs to proteins and to further assess whether the latter allow transport or cause deactivation.

Several methods have been used to study the binding of metallodrugs to serum proteins. Size exclusion chromatography (SEC) is an ideal separation technique for proteins based on their size in physiologically relevant conditions and has been used in studies on many metallodrugs.[9] An alternative separation technique that has increased in popularity for the study of metal drugs, particularly in biological matrices, is capillary electrophoresis (CE).[10,11] The most common form of CE separation is known as capillary zone electrophoresis (CZE), and allows the separation of neutral molecules from ionic species because of their differing mobilities in the electric field depending on their charge and size.[12]

SEC and CZE have frequently been hyphenated to inductively coupled plasma mass spectrometry (ICP-MS) and used to study the binding interactions of metallodrugs such as cisplatin, the Ru compounds KP1019/NKP-1339 and $[Ru(\eta^6\text{-}C_6H_5Me)(PTA)Cl_2]$ (RAPTA-T), or tris(8-oxyquinolinato)gallium(III) (KP46; Fig. 10.1) to HSA or Tf.[13–15] SEC-ICP-MS studies demonstrated that cisplatin would bind to HSA and Tf to the same extent.[13] The side effects in cisplatin treatment are associated with high serum protein binding,[16] despite clinical trials of cisplatin conjugates with HSA and Tf showing anticancer activity in humans.[7,17] SEC in combination with ICP atomic emission spectroscopy (AES) was used to analyze the reactivity of cisplatin and carboplatin toward serum samples obtained from patients. ICP-AES is about two orders of magnitude less sensitive than ICP-MS. However, the fact that Pt is not endogenous in biological specimens still makes it a sensitive and cheaper alternative to ICP-MS. SEC-ICP-AES was used to determine the total serum protein-bound cisplatin after 3 and 24 h as 76% and 96%, respectively, while for carboplatin values of 7% and 40% were observed.[18] Carboplatin has been developed as an alternative to cisplatin with fewer side effects and reduced protein binding may be the reason for these improved pharmacological properties.

One of the limitations of SEC is the insufficient separation of albumin and transferrin in serum protein studies.[14] By coupling SEC to an anion-exchange column [ion chromatography (IC)] in a 2D approach, HSA- and Tf-Ru adducts were successfully separated from each other based on their different isoelectric points ($pI_{HSA}=4.9$, $pI_{Tf}=5.4–5.9$). The orthogonal array of SEC-IC still allowed for online fraction transfer and online analysis by ICP-MS (Fig. 10.2). In contrast to cisplatin, Ru^{III} compounds appear to kinetically prefer Tf binding while the majority of Ru complexes are finally detected on HSA.[15] The protein interaction for such Ru compounds is considered an important factor to achieve selective delivery to tumor cells and reduce side effects.[19] In line with these studies, only a KP1019–HSA adduct was detected in the analysis of KP1019/serum incubations and in in vivo samples from patients receiving KP1019 in clinical phase I trials (Fig. 10.2).

Electron paramagnetic resonance spectroscopy requires the presence of unpaired electrons as found in Ru^{III} ($3d^5$) complexes. This property facilitates the characterization of their chemical environments and revealed that KP1019 initially interacts noncovalently with HSA and serum but over time forms coordinative bonds.[20] In this setup, a reaction with HSA was observed in serum rather than binding to Tf. It was shown that the axial heterocyclic ligand is a large determinant for the strength of the noncovalent

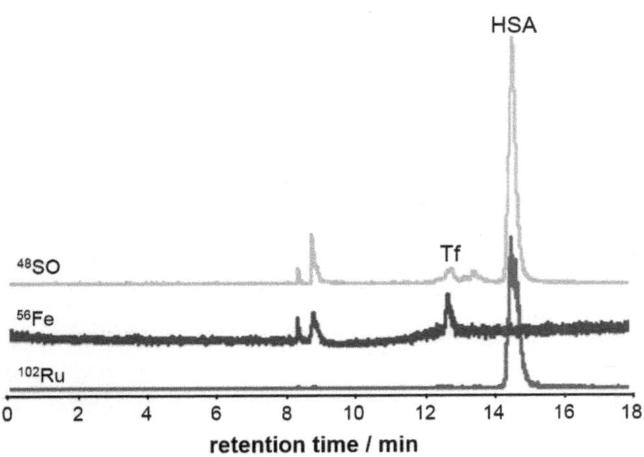

Figure 10.2 A plasma sample from a patient treated with KP1019 analyzed with size exclusion chromatography ion chromatography inductively coupled plasma mass spectrometry (SEC-IC-ICP-MS). *HSA*, human serum albumin; *Tf*, transferrin. *Reproduced from Sulyok, M.; Hann, S.; Hartinger, C. G.; Keppler, B. K.; Stingeder, G.; Koellensperger, G. Two Dimensional Separation Schemes for Investigation of the Interaction of an Anticancer Ruthenium(III) Compound with Plasma Proteins.* J. Anal. At. Spectrom. ***2005,*** 20, *856–863 with permission of the RSC © 2005.*

attachment to the hydrophobic binding sites of HSA.[21] A series of pyridine-based NAMI-A derivatives was synthesized to select and optimize properties that promote binding to HSA.[22] It was shown that increased derivatization prevented formation of covalent bonds, but instead stabilized noncovalent interactions.

As an alternative to SEC-based separations, CE was used for analysis of similar sets of samples and comparison of CE and SEC performance showed similar levels of protein ruthenation in either method.[15] CZE demonstrated notable advantages including low sample and solvent consumption, short analysis times, extremely high resolving power, and it is perfectly compatible with physiological conditions. However, SEC has higher robustness in regard to hyphenation to the ICP-MS and better detection limits.[23] Some of the limitations of CE, such as potential of the proteins to interact with the silanol groups on the capillary surface, can be overcome, e.g., by coating the surface of the capillary. Groessl et al. coated the capillary with polybrene which allowed elution of distinct protein peaks and the separation of proteins from free KP1019.[15]

Bytzek et al. compared the hyphenation of SEC-IC and CZE with ICP-MS for the analysis of in vivo samples of NKP-1339 in mouse plasma.[24,25] Both analysis methods found that in mice treated with high concentrations of NKP-1339 both the albumin and total protein content of plasma would drop significantly. While the high serum protein binding has been suggested to be responsible for the low adverse effects of NKP-1339, hypoalbuminemic patients may be at risk of experiencing adverse effects. Complementary competitive binding studies with HSA and established, specific binding site markers bilirubin (site I), warfarin (site I), and dansylglycine (site II) were used to assess the binding site specificity of KP1019/NKP-1339.[26] Through a combination of spectrofluorimetry, ultrafiltration UV–Vis spectrophotometry, and CZE, it was shown that KP1019 and NKP-1339 will bind to either binding site with no preference. This may cause problems for patients suffering from hyperbilirubinemia as bilirubin may interfere with drug binding.

Combination therapy is commonly practiced and employs more than one drug for the treatment of cancer patients. A potential combination treatment consisting of cisplatin and NAMI-A was investigated on serum protein interactions.[27] In an in vivo study, additive effects for this potential combination of drugs was found, independent of the treatment order. Native-polyacrylamide gel electrophoresis (PAGE) and sodium dodecylsulfate (SDS)-PAGE were used to separate the serum proteins and protein adducts. Native-PAGE maintains proteins in their native state while SDS-PAGE uses denaturants. Metallation may result in denaturation and therefore changes the electrophoretic mobility of the proteins. Imaging gels from native-PAGE and SDS-PAGE experiments for metal distribution with laser ablation (LA)-ICP-MS showed that both compounds were attached to virtually identical proteins. LA involves ablating a solid surface using pulses from a laser and the resulting released material is carried by a gas flow to the ICP-MS.[28] When applied in combination with ICP-MS to native-PAGE,[28] this method allowed the detection of HSA and Tf dimers both carrying Ru and Pt, while such dimers were absent under the denaturing conditions in SDS-PAGE. However, satisfactory separation of serum proteins cannot be achieved using native 1D gel methods. Therefore 2D PAGE was used to effectively separate the proteins in the first dimension by isoelectric point and then by molecular weight in the second dimension.[29,30] 2D PAGE allowed for conclusive separation and identification of albumin and transferrin as the main binding partners for cisplatin, among others.[27] In addition, the separation of their isoforms can be achieved by 2D, which are also implicated in cancer.

The functionalization of drugs with maleimide has emerged as a drug development trend since HSA–drug adducts may take advantage of the EPR effect. Maleimide reacts with the thiol group of Cys34 to form a covalent bond. The functionalizing of a Ru(arene) complex with a maleimide moiety resulted in the specific modification of HSA as shown by SEC with different detection methods, NMR spectroscopy, and electrospray ionization (ESI)-MS. In vitro anticancer activity data show that some of the maleimide-functionalized ruthenium complexes were significantly more cytotoxic than the established parent compound RAPTA-C (Fig. 10.1).[31]

2.2 Characterization of the Metal Coordination on Serum Proteins

Element-specific methods allow for the quantification of the distribution of metal-containing species among serum components. However, to identify the type and site of modification of a protein, bioanalytical methods are required that deliver structural information. As mentioned, metal ions have high affinity for biological nucleophiles upon activation, which are found in many amino acid side chains. In addition to the coordinative binding event on single amino acids, the binding may occur in an actual protein binding site and may also be of noncovalent nature.

The modern MS-based fragmentation method electron capture dissociation (ECD) in a high-resolution Fourier transform ion cyclotron resonance-MS can be used to characterize the interaction of cisplatin with Met, His, Cys, Asp, and Glu.[32] ECD fragmentation resulted in radical-mediated side-chain losses for platinum binding sites and to the detection of small molecular ions containing platinum and the characteristic amino acid fragment. These species proved to be characteristic markers for cisplatin binding sites within a peptide or protein. However, the method is only useful to proteins of up to around 30 kDa, which limits its use for relevant serum proteins such as HSA and Tf (66 and 76 kDa, respectively). Therefore other methods are required to allow binding site characterization on such proteins. In case of HSA, the involvement of the sulfhydryl group of Cys34 was studied by Gronias and Pizzo.[33] They attempted an indirect approach to determine the involvement of Cys34 in Pt binding and carried out a carboxyamidomethylation of the SH group, which resulted in a four- to five-fold decrease in Pt coordination detected photometrically.

More sophisticated experiments were based on 2D ^{1}H-^{15}N heteronuclear single quantum coherence NMR spectroscopy. Using ^{15}N-labeled platinum amine complexes, Ivanov et al.[34] demonstrated that there are sites other than Cys34 that play an important role in cisplatin binding. They report cisplatin coordinates to Met298 in the form of an *S*,*N*-macrochelate, which shows that more than one amino acid in close proximity may be involved in the binding event.

MS is routinely employed to determine the amino acid sequence of proteins. This can be done in top-down or bottom-up approaches, the former involving the fragmentation capability of multistage MS instruments while the latter employs enzymatic digestion of the intact protein adduct, e.g., with trypsin, and the subsequent ESI-MS/MS analysis of smaller peptides. The latter is a common approach in classical proteomics experiments.

By applying bottom-up MS on apo-Tf and apo-Tf incubated with cisplatin it was found that it coordinates to the hydroxyl functional group of Thr457.[35] Similarly, the reaction of cisplatin with recombinant HSA (rHSA) allowed the identification of bifunctional cisplatin–rHSA adducts. After activation, cisplatin was shown to form intramolecular interdomain crosslinks between His67 of domain I and His247 of domain II.[36] This is also the major binding site of the essential element zinc and cisplatin may interfere with the zinc homeostasis, resulting in hyperzincuria and hypozincemia observed for patients treated with cisplatin.

Multidimensional liquid chromatography coupled to an MS operated in tandem MS (MS/MS) mode has emerged as a powerful means of determining the binding partner of metallodrugs in complex matrices and even whole cells. Such an MS/MS-based method is also known as multidimensional protein identification technology (MudPIT) and comprises strong cation exchange and reversed-phase (RP) chromatography. MudPIT has a large dynamic range while maintaining high sensitivity with the detection of tryptic peptides at concentrations as low as ~10 fmol.[37] MudPIT studies allowed comprehensive analysis of the preferred binding partners of cisplatin in serum, from which HSA, Tf, and other abundant serum proteins were found platinated.[37] Furthermore, information on the binding sites was extracted from the MS/MS data and Cys34, Met329, Met548, Tyr150 (or Tyr148), and Asp375 (or Glu376) were established as the cisplatin HSA coordination sites while Glu265, Tyr314, Glu385, and Thr457 were found for Tf, confirming in part the finding of others.[35] This shows that Pt can be found coordinated to both sulfur (thiolate, thioether) and oxygen donor (carboxylate) atoms but nitrogen (imine) has also been suggested as a binding partner. Low sequence coverage of 54% for HSA and 51% for Tf did not prevent identification of cisplatin-binding partners. However, the bottom-up approach risks ligand exchange reactions during digestion and LC-MS analysis.

X-ray crystallography gives an insight into biomolecule–metal binding sites at atomic resolution; however, it requires a single crystal of a protein to be grown, which is often not easily obtainable and not necessarily representative of the structure in solution. A crystal structure of the cisplatin–HSA adduct revealed multiple binding sites (Fig. 10.3) with both

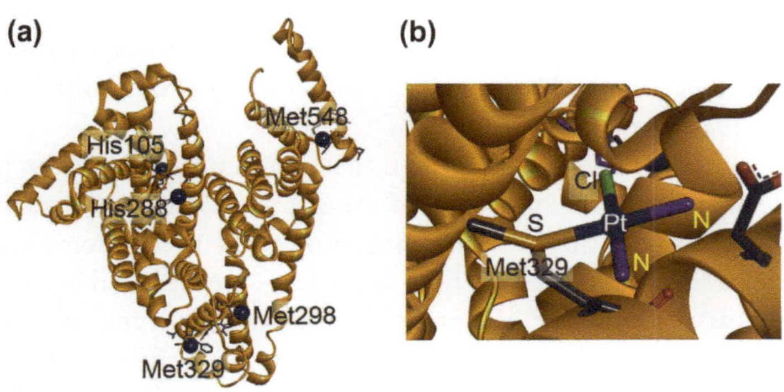

Figure 10.3 (a) Crystal structure of human serum albumin (HSA) with five bound platinum species; (b) zoom into the $Pt(NH_3)_2Cl$ adduct formed with the sulfur atom of Met329 (PDB ID 4S1Y).[38]

mono- and bifunctional Pt adducts formed (i.e., the Pt center being coordinated to one or two protein donor atoms).[38] Similar to the MudPIT study, Met329 and Met548 were identified as binding partners but no further overlaps between the two studies were found. The ancillary ligands at the Pt center could not always be unambiguously identified. The difference in binding partners detected may be related to the different incubation and crystal growth conditions. Notably, binding of cisplatin caused minimal change in the overall protein conformation, possibly through use of the soaking method.[38]

Kratz et al. monitored the very rapid reaction between KP1019 and Tf using SEC with UV–Vis detection. In the presence of bicarbonate, KP1019 binds to Tf through an intermediate complex. Further X-ray crystallography studies with KP1019 using the Tf sequence homolog lactoferrin showed that a ligand exchange reaction between a chlorido ligand and His253 occurs and the side chain of Lys301 may help stabilize the interaction.[39] Furthermore, a crystal structure of an adduct between HSA and the KP1019-analogous complex anion $[RuCl_5(1H\text{-indazole})]^{2-}$ was obtained and monofunctional adduct formation with His146 in subdomain IB as well as in a bidentate modality at His242 and Lys199 in subdomain IIB was detected.[40] Interestingly such interaction and loading on HSA resulted in significantly enhanced cytotoxic activity of the Ru compound.

Similar to Ru^{III}, Ga^{III} was suggested to be transported by Tf through binding to the iron binding sites. CZE-ICP-MS was used to study the binding behavior of gallium nitrate and the anticancer drug candidate KP46 (Fig. 10.1) to Tf and HSA.[41] Both species displayed a high affinity for Tf in the presence of carbonate. The study demonstrated that Ga^{III} was bound reversibly and the addition of Fe^{III} or a pH change displaced gallium. In contrast to KP1019, the binding affinity of KP46 to HSA was much lower, even in a 20-fold excess.

These analyses have clearly shown that serum proteins play an important role in the modes of action of transition metal complexes within the body. However, the exact binding mode and the biological role of such interactions for many of these complexes are still under discussion, especially as the number of potential binding sites, even on a protein like HSA, complicates the accurate localization of the metal center(s). The localization is further complicated by the kinetics of the reaction and the lability of some of the adducts formed as well as the possibility of noncovalent interactions.

3. TISSUE DISTRIBUTION AND CELLULAR ACCUMULATION

3.1 Metal Distribution Studies in Biological Tissue

Severe side effects of metallodrugs are associated with unspecific accumulation in healthy tissue, such as liver, kidney, or nerve ends. Bioimaging provides a means for visual assessment of both healthy and tumorigenic tissue. Imaging techniques, such as LA-ICP-MS, matrix-assisted laser desorption/ionization (MALDI)-MS, X-ray fluorescence microscopy, and nanoscale secondary ion mass spectrometry (nanoSIMS), allow for the study of metal distribution in tissues or even more complex higher organisms. Modern nanoSIMS profits from lateral resolution in the sub-μm range and permits imaging subcellular compartments. In combination with element-specific methods, this allows speciation and tracking of metallodrug distribution in tissues and cells.

LA-ICP-MS analysis of 3-μm sagittal sections from rats treated with pharmacologically relevant doses of cisplatin showed accumulation in the cortex and corticomedullary regions (Fig. 10.4), both being areas that suffer from cisplatin-induced nephrotoxicity.[42] This is of biological relevance as cisplatin treatment is correlated with severe nephrotoxicity.[43] By also measuring Zn and Cu distributions, it was revealed that Pt displaces these elements relative to untreated tissue. Furthermore, administration in combination with cilastatin/imipenem resulted in a decrease in Pt accumulation within the cortex, demonstrating that protective measures can be successful in a combination therapy.

LA-ICP-MS was also applied for assessing the distribution within viscera (liver, kidney, spleen, and muscle) from mice treated with relevant pharmacological doses of NKP-1339 and cisplatin.[44] The results were independently validated through microwave digestion and direct infusion ICP-MS. Although LA-ICP-MS requires significantly less sample preparation, it is more challenging to obtain absolute quantitative results and often relative quantitation is performed. This is mainly because of a lack of certified reference materials in tissue-based matrices.[44] To overcome this issue, homogenized pig liver was spiked with the internal standards In and Re, as well as Ru and/or Pt standards, which, however, did not result in significantly improved data quality. Both NKP-1339 and cisplatin had the same distribution pattern with the metals found to be largely concentrated in the kidney cortex, with a homogeneous distribution in the liver and the muscle with areas of enrichment shown in the spleen.

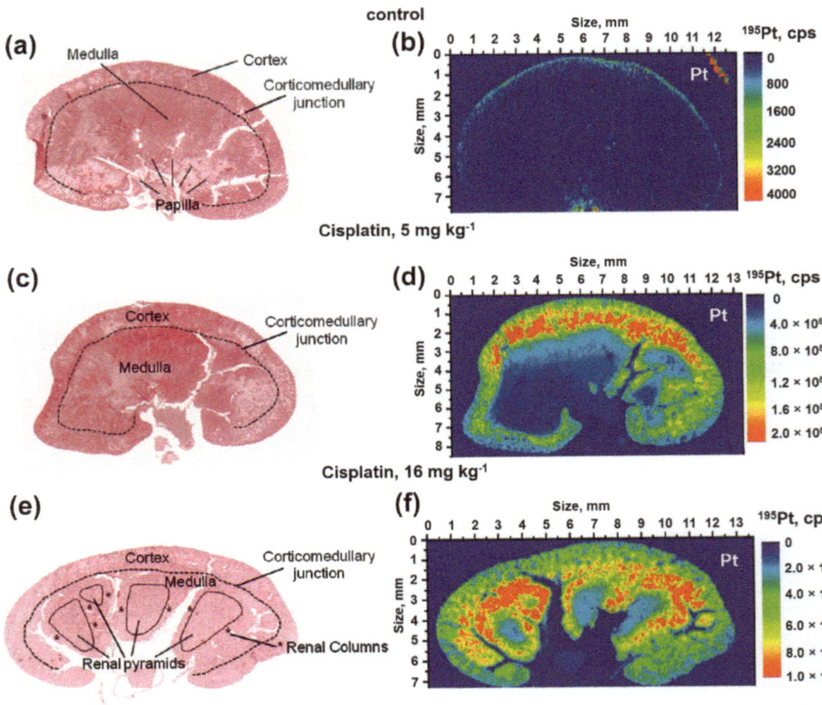

Figure 10.4 Histological micrographs of rat kidneys stained with hematoxylin and eosin (a, c, and e) compared with the equivalent laser ablation inductively coupled plasma mass spectrometry (LA-ICP-MS) images monitoring for ^{195}Pt (b, d, and f). Samples are from a control rat (a and b) and rats treated with 5 mg/kg (c and d) and 16 mg/kg cisplatin (e and f). *Reprinted from Moreno-Gordaliza, E.; Giesen, C.; Lázaro, A.; Esteban-Fernández, D.; Humanes, B.; Cañas, B.; Panne, U.; Tejedor, A.; Jakubowski, N.; Gómez-Gómez, M. M. Elemental Bioimaging in Kidney by LA–ICP–MS as a Tool to Study Nephrotoxicity and Renal Protective Strategies in Cisplatin Therapies.* Anal. Chem. ***2011,*** 83, *7933–7940 with permission from ACS © 2011.*

LA-ICP-MS was also applied for studying damaged tissue that needed to be removed after accidental extravasation of Pt-based drugs into the tissue surrounding the infusion site, allowing the Pt distribution to be analyzed in muscle, fat, nerve, and connective tissue types in patients.[45] The damaged tissue had Pt concentrations 10 to 50 times higher than the surrounding tissue. Nerve and connective tissue from the extravasation site showed the highest platinum concentrations and this was pervasive as it remained five times higher up to 4 weeks after the extravasation. Fat tissue was found difficult to ablate and formed confluent drops, which resulted in an underrepresentation of Pt concentration.

In an orthogonal approach, LA-ICP-MS and MALDI imaging were used to study cisplatin and oxaliplatin distribution in tumor samples of colorectal or ovarian peritoneal carcinomatosis.[46] MALDI imaging was found to suffer from significant signal suppression resulting from interference from the matrix while LA-ICP-MS was not influenced by the matrix and had higher sensitivity. Similarly, a colon cancer CT-26 tumor model was used to study the distribution of satraplatin and oxaliplatin when administered over a period of two weeks.[47] The LA-ICP-MS distribution revealed that the majority of the intratumoral Pt was located in the surrounding tissues of the tumor rather than the malignant part. The kidney is the major clearance point prior to excretion and similar to cisplatin a large proportion of the Pt was found in the cortex with only a small amount detected in the medulla.

3.2 Metallodrug Accumulation and Active Transport Into the Cell

Serum proteins are considered important in the transport of anticancer metallodrugs and have been suggested to be involved in the cellular uptake. In particular, Tf has often been compared to a Trojan horse able to hide cytotoxic metal species and to deliver them into tumor cells, which overexpress transferrin receptors. Atomic absorption spectroscopy (AAS), another sensitive element-specific detection method, was used to compare the cellular accumulation of KP1019, KP1019 conjugated with Tf (2:1), and KP1019 loaded onto Tf in the presence of Fe^{III} (1:1) in SW480 cancer cells.[48] The addition of Fe^{III} facilitated the cellular uptake of Ru, which may be related to Tf maintaining a more natural conformation as compared to a full loading with KP1019.

3.3 Intracellular Compartment Distribution

X-ray fluorescence imaging (XRF) is an element-specific analysis method and allows for the determination of the intracellular distribution of metals.[5] Microprobe XRF imaging of cells treated in vitro with KP1019 or NAMI-A was used to identify their intracellular distribution and colocalization with other biologically relevant elements.[49] The requirement of synchrotron radiation limits access to such experiments for many research groups and the analysis of Ru is challenging. The study demonstrated that the cellular fates of KP1019 and NAMI-A are different. For KP1019, Ru was found to be colocalized with Fe and together with the observed change in iron distribution after KP1019 administration supports the hypothesis that ruthenium may enter the cell via transferrin binding. Furthermore, colocalization of Ru with P is indicative of uptake into the nucleus. In contrast, it was suggested that NAMI-A interacts with the cell membranes and is not internalized.[49]

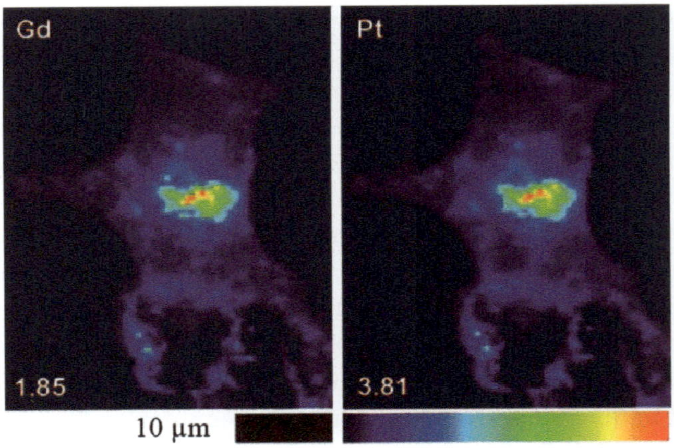

Figure 10.5 X-ray fluorescence imaging (XRF) maps showing the elemental distribution of Gd and Pt for an A549 human lung carcinoma cell treated with 5 μM of Pt–Gd complex. The maximal area density is reported on the bottom left in μg cm^{-2}. *Reproduced from Crossley, E. L.; Aitken, J. B.; Vogt, S.; Harris, H. H.; Rendina, L. M. Selective Aggregation of a Platinum–Gadolinium Complex within a Tumor-cell Nucleus.* Angew. Chem. Int. Ed. Engl. ***2010****, 49, 1231–1233 with permission from Wiley © 2010.*

XRF imaging was also used to follow the fate of a molecule that comprises one Gd^{III} and two Pt^{II} centers in A549 human lung carcinoma cells.[50] Gd^{III} complexes are well known as contrast agents in MRI while the Pt^{II} center was equipped with a planar ligand system to act as a selective intercalator into DNA. XRF imaging was used to determine the elemental distribution in a single cell, and it showed that Pt and Gd are colocalized at a 2:1 ratio (Fig. 10.5). These findings support that the complex stays intact through the uptake and accumulation process and the overlap with high levels of Zn and P demonstrated that there was an accumulation in the nucleus.

NanoSIMS combined with confocal laser-scanning microscopy is an imaging technique with the ability to detect multiple elements with isotope selectivity at high spatial resolution. This technique was used to analyze human colon cancer cells treated with ^{15}N-labeled cisplatin at subcellular levels. Cisplatin was located within sulfur-rich structures such as those of the nucleus and cytoplasmic organelles and the phosphate-rich chromatin (Fig. 10.6). At high concentrations dissociation of $^{194}Pt–^{15}N$ was observed in the nucleoli, and may have an effect on the mode of action of cisplatin.[51]

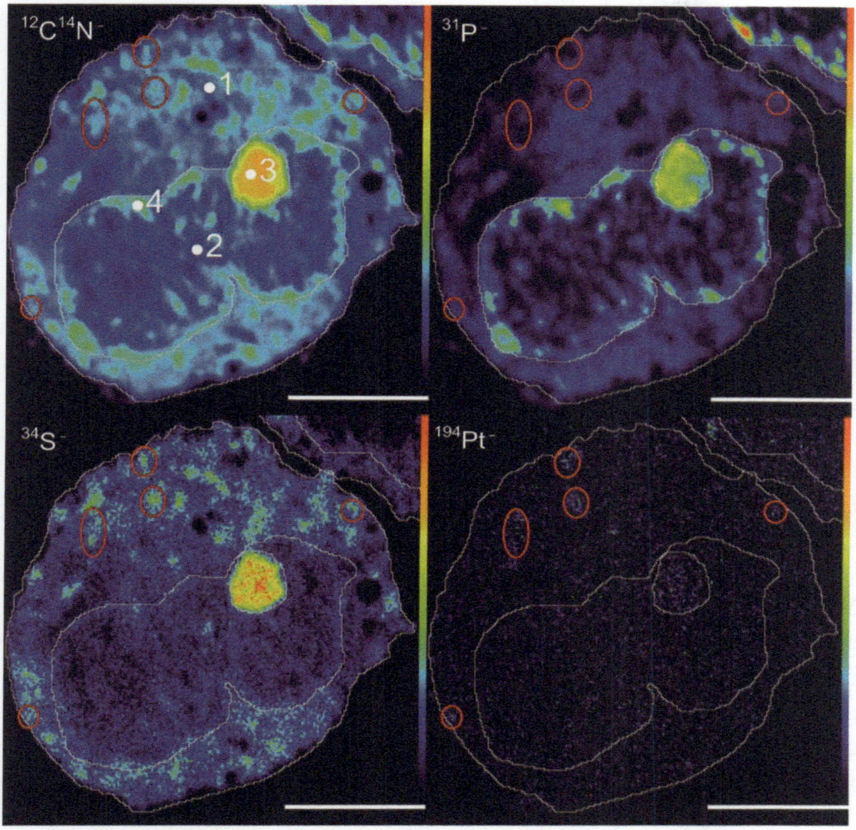

Figure 10.6 $^{12}C^{14}N$, ^{34}S, ^{31}P, and ^{194}Pt secondary ion signal intensity distribution maps of semithin sections of SW480 cells treated with 25 µM cisplatin. Identified cellular locations are (1) the cytoplasm; (2) the nucleus; (3) the nucleolus; and (4) the chromatin. ^{194}Pt and ^{34}S signals indicate Pt–S aggregates form in the cytoplasm (*circled in red*). Scale bar = 5 µm. *Reproduced from Legin, A. A.; Schintlmeister, A.; Jakupec, M. A.; Galanski, M.; Lichtscheidl, I.; Wagner, M.; Keppler, B.K. NanoSIMS Combined with Fluorescence Microscopy as a Tool for Subcellular Imaging of Isotopically Labeled Platinum-Based Anticancer Drugs.* Chem. Sci. ***2014***, *5, 3135–3143 with permission from RSC © 2014.*

4. CELLULAR TARGETS

4.1 Cancer Cells

Cancer cells feature an enormous number of potential targets for metal-based anticancer agents. Inspired by the success of cisplatin, many anticancer agents were initially designed to mimic the mode of action of cisplatin by targeting DNA. The identification of protein targets for organic anticancer drugs has inspired new research directions and in particular the design of protein-targeted drugs.

4.1.1 DNA as a Target in Cancer Chemotherapy—Cisplatin and Its Successors

DNA has been identified as the target of Pt-based anticancer agents and many metal-based anticancer agents may share this target. The high affinity of Pt anticancer agents to the *N7* of guanine leads to the formation of intra- and interstrand crosslinks mainly through coordination of a Pt moiety to adjacent guanine residues. This binding mode has inspired the development of new generations of Pt and non-Pt DNA targeting anticancer agents, along with the development of methods to study their interactions.

The nucleobases adenine, cytosine, guanine, and thymine and their respective deoxyribonucleotides (dNMPs) are the building blocks of DNA and have often been used as model systems to compare the affinity of metal-based compounds to DNA. To study the interaction of cisplatin with DNA and DNA models, a reversed-phase high-performance liquid chromatography (RP-HPLC) method hyphenated to ICP-MS was developed. Cisplatin was incubated with the four individual deoxyribonucleotides dAMP, dCMP, dGMP, and dTMP and the presence of phosphorous in the DNA deoxyribonucleotides was exploited and ^{31}P was simultaneously scanned for along with ^{195}Pt from cisplatin. An $[(NH_3)_2Pt(dGMP)_2]$ adduct was separable from the nucleotides using a C_8 narrow bore column. An oligonucleotide featuring adjacent guanine residues, i.e., TCCGGTCC, and calf thymus DNA were incubated with cisplatin, digested, and analyzed, which showed that the platinum was bound to the GG motif (Fig. 10.7).[52] In vitro studies using HPLC-AAS and ^{32}P-postlabeling on calf thymus cells treated with cisplatin and oxaliplatin showed a similar trend for both studies with cisplatin generating a higher number of GG and AG adducts compared to oxaliplatin.[53]

Murine fibroblast cells were exposed to cisplatin, and after incubation the DNA was extracted and hydrolyzed. Analysis with LC-ESI-MS/MS revealed that crosslinked species are present with a large majority found as GG adducts, as in previous studies.[54] A subsequent study showed that oxaliplatin was less selective in forming GG over AG adducts.[55] NMR solution structures confirmed the formation of GG adducts in cisplatin[56,57] and oxaliplatin.[58] Other studies on the same interactions using LC-ESI-MS/MS allowed discrimination between GG and AG adducts, which provided information on the specifics of metallodrug binding.

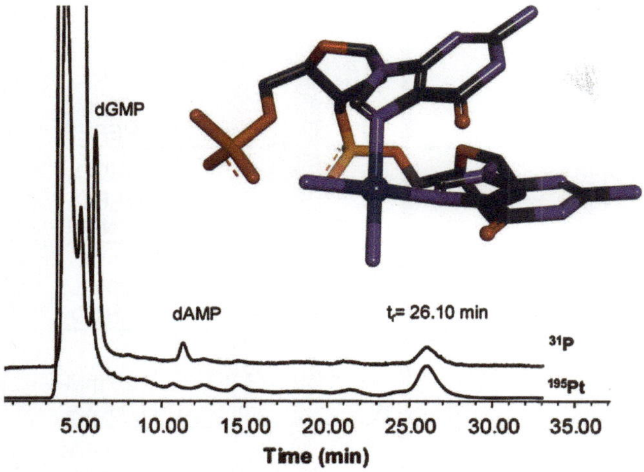

Figure 10.7 High-performance liquid chromatography inductively coupled plasma mass spectrometry (HPLC-ICP-MS) chromatogram showing the traces of ^{31}P and ^{195}Pt from the incubation with cisplatin and calf thymus DNA, recorded after enzymatic digestion. The peak at a retention time t_r of 26.10 min indicates the formation of the GG cisplatin adduct and the proposed structure is shown in the inset of the figure. *Modified from Garcia Sar, D.; Montes-Bayon, M.; Blanco Gonzalez, E.; Sanz-Medel, A. Speciation Studies of Cis-Platin Adducts with DNA Nucleotides Via Elemental Specific Detection (P and Pt) Using Liquid Chromatography-Inductively Coupled Plasma-Mass Spectrometry and Structural Characterization by Electrospray Mass Spectrometry.* J. Anal. At. Spectrom. ***2006,*** *21, 861–868 with permission from the RSC © 2006.*

CZE-UV has been presented as a convenient way of monitoring the kinetics of the reaction between cisplatin and 5′-GMP under simulated physiological conditions because of its low sample requirements, ease of use, rapid adjustment of parameters and buffer conditions and no need for organic solvents compared with HPLC.[59,60] Used as a screening method it allows the binding affinities of newly developed metal complexes to be determined and compared to cisplatin.[61] Hyphenation with ICP-MS detectors allows detection at even higher sensitivity.[62] CZE studies have shown that while the actual formation of the cisplatin–GMP adduct is unimpeded by matrix components, the half-life, and therefore peak distribution between reactants and products is much more susceptible (Fig. 10.8). The presence of salts including phosphate, chloride, and bicarbonate can result in the half-life nearly doubling relative to a water matrix.[63]

The combination of CZE with a light induced fluorescence detector gives a very sensitive detection method for fluorescent molecules. To detect cisplatin, a fluorescent analog needed to be prepared. However, this

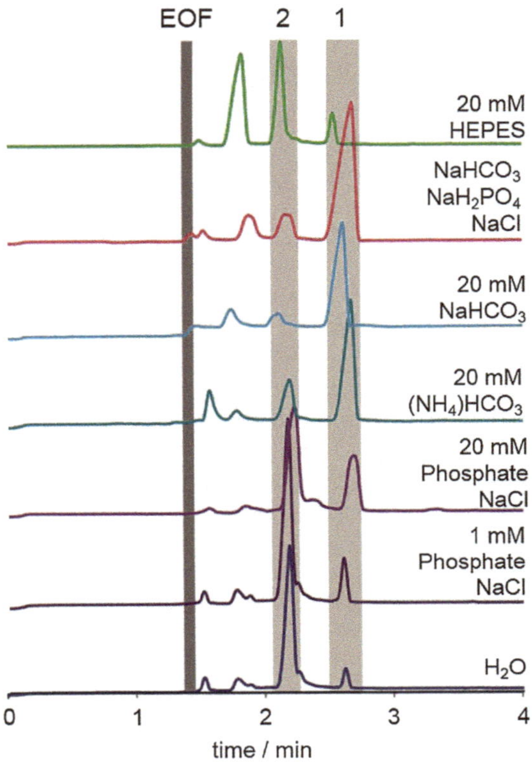

Figure 10.8 Capillary zone electrophoresis ultraviolet (CZE-UV) electropherograms of the reaction of cisplatin with deoxyguanosine monophosphate (dGMP) in different incubation solutions recorded after 24 h. The electroosmotic flow (EOF), the main adduct (2, $[Pt(NH_3)_2(dGMP)_2]^{2-}$), and free dGMP (1) are indicated. *HEPES*, 4-(2-hydroxyethyl)-1-piperazineethanesulfonic acid. *Reproduced from Grabmann, G.; Keppler, B. K.; Hartinger, C. G. A Systematic Capillary Electrophoresis Study on the Effect of the Buffer Composition on the Reactivity of the Anticancer Drug Cisplatin to the DNA Model 2′-Deoxyguanosine 5′-Monophosphate (dGMP).* Anal. Bioanal. Chem. ***2013,*** 405, *6417–6424 with permission from Springer © 2013.*

modification changes the intrinsic properties and reactivity of the Pt complex. This study reported reactions of the Pt species with noninnocent buffer components, such as Tris, and together with MS enabled the characterization of Pt adducts with low molecular weight components in cell culture experiments.[64]

The disadvantage of using nucleotides as model systems is that the impact of the 3D conformation of the DNA double helix on the binding event is neglected.[65] This was demonstrated for a set of pH-sensitive cisplatin analogs and their binding to nucleotides as compared to plasmid DNA using capillary

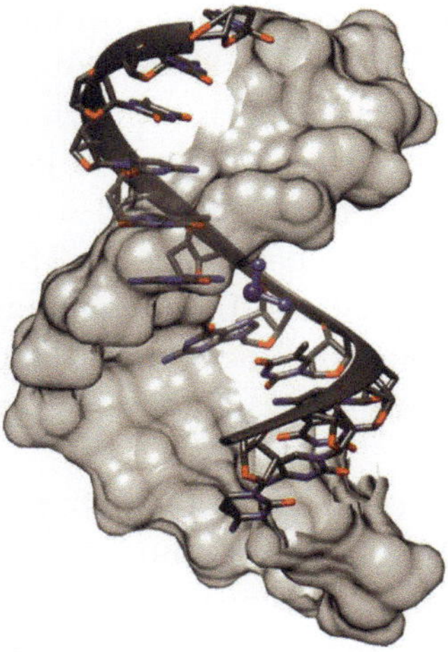

Figure 10.9 DNA oligomer double helix with cisplatin bound through two guanine residues (PDB ID 3LPV).[66]

and agarose gel electrophoresis, respectively. This issue was also addressed when a dodecamer of DNA was incubated with cisplatin and a crystal structure was obtained (Fig. 10.9) with the $Pt(NH_3)_2$ fragment coordinated to adjacent guanine residues forming an intrastrand crosslink (GG).[66] While the type of adduct may be the same, the 3D structure of the DNA may impact the kinetics of the reaction and the binding site. As shown by X-ray crystallography and NMR spectroscopy, cisplatin binds to the GG or GA motif in the major groove, while the formation of interstrand crosslinks was found in the minor groove.[67]

Similar to Pt anticancer agents, Ru complexes also tend to form DNA adducts upon ligand exchange reactions and most often guanine-*N7* is the target.[68] While Ru compounds have often been hypothesized as not having DNA but proteins as the target, DNA has been suggested as the main target for RAED-type compounds of the general formula $[Ru(\eta^6\text{-arene})(\text{ethylenediamine})X]^+$ (X = halide, Fig. 10.1).[69] RAED-C (arene = *p*-cymene) reacts most rapidly with GMP as compared to the other nucleotides. Further, time-resolved measurements show a reaction scheme that proceeds through an intermediate, coordinated to the oxygen of the phosphate, and then proceeds to the most stable species, which is bonded again to the *N7* of the guanine moiety.[70] However, other Ru compounds such as RAPTA-C and KP1019 also react with nucleotides in a similar manner, but only if solely incubated with the DNA or DNA models.[68,71,72]

4.1.2 Nucleosome Core Particle Binding—At the Interface Between Protein and DNA World

The nucleosome core particle (NCP) is the basic repeating element of histone-packaged DNA comprising chromatin. It holds a unique position in biomolecule-binding studies for metal-based drugs as chromatin metalation may have a significant biological impact. This is an excellent model system comprising both protein and DNA components, where the DNA is wrapped around the proteinaceous octamer of the nucleosome core. Using this model allows for the identification of protein versus DNA affinity in a single experiment.

Crystallographic studies on cisplatin binding to the NCP showed that when platinated, the platinum-bound nucleotides dictate the rotational placement on the octamer. The platinated nucleotides also affect the sliding of DNA around the octamer core and further studies show that RNA polymerases may traverse the cisplatin damaged template strand but stall when the platinated nucleotides are reached.[73] Interestingly, platinum binds not only to the DNA double helix, but also to histone proteins to a lower extent.

RAPTA-C shows specific binding to the protein octamer of the NCP at exposed His, Glu, and Lys residues.[74] A comparison between recent Ni^{II}- and Co^{II}-containing structures shows Ru^{II} binds in similar sites.[74a] Further studies with RAPTA-C and the analogous organoruthenium compound RAED-C on their NCP interaction with ICP-MS revealed that 77% of the RAED-C is located on the DNA portion of the chromatin (Fig. 10.10), while 61% of the RAPTA-C is associated with the protein component.[75]

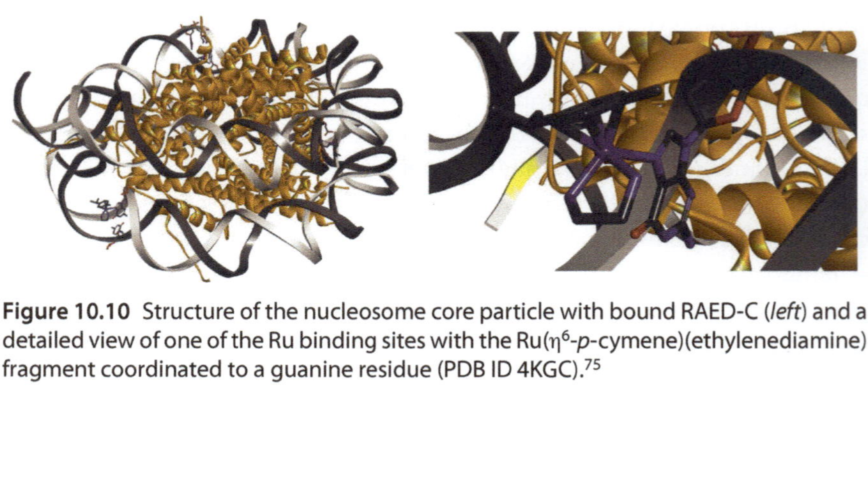

Figure 10.10 Structure of the nucleosome core particle with bound RAED-C (*left*) and a detailed view of one of the Ru binding sites with the Ru(η^6-*p*-cymene)(ethylenediamine) fragment coordinated to a guanine residue (PDB ID 4KGC).[75]

4.1.3 Protein Interactions—From Model Proteins to Protein-Targeting Compounds

Metals tend to form coordination bonds with nucleophiles of amino acid side chains upon hydrolysis of a labile metal–ligand bond, e.g., halides. However, often these amino acid side chains are buried deep in the core and are inaccessible. To estimate the reactivity of metallodrugs, model proteins are used to identify the types of adducts formed and the stoichiometry. Prominent examples of model proteins are cytochrome *c* (cyt *c*), ubiquitin (Ub), myoglobin (Mb), and superoxide dismutase (SOD). All of these proteins can be analyzed with modern mass spectrometers, by employing soft ESI and MALDI methods, which allows characterization of their covalent metal adducts.[76]

Gibson and Costello[77] carried out pioneering work using Ub as a model protein to study the protein binding behavior of cisplatin and other Pt complexes. It was possible to postulate a mechanism of binding for cisplatin that involved the consecutive formation of mono-, di-, and trifunctional adducts because of ligand exchange reactions, i.e., in the latter case the protein occupies three of four possible ligand binding sites on the square-planar platinum(II). In contrast, the major peak observed for the anticancer inactive isomer transplatin was assigned to a monodentately coordinated monoadduct. This supports different modes of action for both cisplatin and transplatin, also demonstrated by biological experiments. Cisplatin can bind to Met1 and His68 of Ub but it was shown that Met1 was the preferred binding site. This was demonstrated indirectly after oxidation of the Met1 thioether, which decreased the binding, whereas transplatin was unaffected.[78,79] Other high-resolution top-down studies using Fourier transform ion cyclotron resonance and orbitrap MS instruments on Ub adducts with cisplatin and oxaliplatin support the observations of Gibson and Castello.[80]

A comparison of the suitability of MALDI and ESI as an ion introduction method for MS was carried out by studying the binding behavior of cisplatin and oxaliplatin to Ub, which detected largely the same species.[76] However, the softer ESI method in combination with a quadrupole time-of-flight (QTOF) analyzer gave higher-resolution spectra, while in MALDI-MS species were detected where the Pt had lost all of its original ligands. This is most probably related to MALDI being a harder ionization source and therefore increased fragmentation was induced. Besides the ionization, it was found that the mass analyzer also influences the abundance of metal–protein adducts and that QTOFs are superior to ion traps for this purpose.[81]

By mixing more than one protein with a metallodrug, the selectivity of metallodrug-protein binding can be investigated.[82] A high-resolution ESI-linear ion trap quadrupole-orbitrap-MS study with cisplatin, transplatin, or RAPTA-C in a mixture of cyt *c*, Ub, and SOD revealed that cisplatin and transplatin bind indiscriminately to proteins, while RAPTA-C is more selective with affinity for Ub and cyt *c*. The binding of cisplatin to SOD and cyt *c* was also studied by X-ray crystallographic analysis. Cisplatin was found to bind to bovine erythrocyte-SOD at His19 with high selectivity as no other binding sites were identified.[83] Surprisingly, amine ligands as opposed to the chlorido ligand underwent ligand exchange reactions with His19 of SOD. This may be influenced by the cocrystallization conditions employed in the study and the accessibility of Cys residues. In contrast, when the interaction of cisplatin with human SOD1 was studied, cisplatin coordinated to the free thiol of Cys111, which may affect the biological function of the protein.[84] The binding sites of cisplatin on cyt *c* were reported by Merlino et al.[85] The crystal structure featured monofunctional $Pt(NH_3)_2Cl$ and bifunctional $Pt(NH_3)_2$ adducts with Met65 and Met65/Glu61, respectively.

When comparing the binding of RAPTA derivatives with cyt *c*, the RAPTA derivatives containing only halido ligands presented extensive protein metalation while those featuring bidentate chelating ligands proved less reactive.[86] The rate of aquation correlated well with cytotoxicity while in earlier studies no correlation between cytotoxicity and binding to oligonucleotides was found.[87] This may indicate that DNA binding is not part or plays a less important role in the mechanism of action of RAPTA derivatives. The interaction between NAMI-A and cyt *c* was analyzed with a variety of complementary methods.[88] Inductively coupled plasma optical emission spectrometry (ICP-OES) data show that 1 h after the addition of NAMI-A, 40% of the protein is loaded with ruthenium. ESI-MS supported the ICP-OES data and this was echoed by ^1H NMR spectroscopy data showing the characteristic NAMI-A signals disappear far more rapidly in the presence of cyt *c*.[88]

The binding of metal complexes to model proteins has only minor relevance for the biological activity despite giving information on the general reactivity and the types of adducts formed. However, a significant number of studies have been reported to elucidate the interaction with metallodrugs, especially cisplatin, with more physiologically relevant proteins. Cisplatin resistance is often linked to copper homeostasis proteins. ICP-AES analysis showed that stoichiometric binding between cisplatin and the copper chaperone protein Atox1 occurred and X-ray crystallographic data analysis gave insight into the binding sites. The Pt^{II} ion was attached to the thiols of Cys12 and Cys15 and the amide nitrogen of Cys12.[89]

Notable is the work of Meggers et al. who have designed specific inhibitors for kinases that fit into their ATP binding site. Inspired by the structure of the organic molecule staurosporine, organometallic inhibitors based on metals such as Ru and Os were studied.[90] Importantly, the metal centers do not interact directly with the protein and only have a structural function to fill the binding site.[91,92] This was proven in a series of X-ray crystallographic studies where minor changes in the structures of the inhibitors allowed the introduction of selectivity for a specific human kinase, while changing the metal center from Ru to Os gave similar results.[90,93,94] The same group also developed a small molecule inhibitor of the repair enzyme MTH1 with low nM inhibitory activity. The native enzyme interacts with 8-oxo-dGMP and the organometallic inhibitor was designed to mimic this ligand and to more effectively fit the active site.[95]

4.1.4 Target Profiling: Toward Metalloproteomics

The identification of the intracellular target(s) of metallodrugs presents a needle-in-a-haystack problem. The distribution of the metal between the cell compartments is often known and determined with element-specific analysis methods, but the number of potential biomolecular targets is enormous. Target profiling on the whole matrix of the cell requires more advanced multidimensional analytical techniques and method development must consider the properties of metal complexes such as their tendency to undergo ligand exchange reactions.

Cisplatin-resistant ovarian A2780 cells were treated with Au-based anticancer drugs and after lysis the proteins were separated using 2D PAGE.[96] Variations in the protein expression profiles were detected and the proteins were characterized by MALDI-TOF-MS and validated by Western blot. While the well-known Au complex auranofin (Fig. 10.1) altered the amount of proteasome proteins expressed, an investigational Au compound impacted proteins related to mRNA splicing, trafficking, and stability, and both Au compounds reduced the level of thioredoxin-like protein 1, which is involved in oxidative stress. However, such a proteome profiling approach does not give direct information on the protein target of anticancer agents. This issue was addressed in a pull-down study with a specifically designed arene-functionalized RAPTA pharmacophore loaded on streptavidin-modified beads.[97] The loaded beads were exposed to cell lysate of ovarian cancer CH1 cells. Proteomic analysis of the proteins attached to the beads allowed the identification of 184 proteins, out of which 15 were associated with cancer. Notably, many of these proteins could be related to effects observed in complementary biological assays.[97] A complementary MudPIT-based method was used to analyze trypsinized cell lysate from *Escherichia coli* that was treated with cisplatin[98] or $[Ru(\eta^6\text{-}p\text{-cymene})(DMSO)Cl_2]$.[99] Analysis of the cisplatin MS/MS data showed that it bound to 31 targets, including five being specific low-abundance DNA binding proteins and one involved in DNA mismatch repair. The other important targets were efflux proteins and a topoisomerase regulator. This is another indication that metal complexes are prone to reacting with a wide range of biomolecular targets rather than being specific. However, this may be an advantage, considering that cancer cells are responsive to cytostatics upregulating their defense mechanisms. The $Ru^{II}(\eta^6\text{-}p\text{-cymene})$ fragment was found bound to five protein targets. Four of the five proteins were identified as stress response proteins and the other was a DNA repair protein.

4.2 *Helicobacter pylori* – Metallodrugs to Treat Other Diseases

Colloidal bismuth subcitrate (CBS) is a commonly used Bi^{III}-based antiulcer drug targeting *H. pylori* infection of the human gastric mucosa. Native- and SDS-PAGE were used to analyze the proteome of *H. pylori* after treatment with CBS.[100] Higher separation efficiency was observed under denaturing conditions. LA-ICP-MS of the gels allowed for the detection of seven bismuth-binding protein targets in treated *H. pylori*. Continuous monitoring of the Zn-proteome in *H. pylori* during treatment of CBS showed that an overexpression of Zn-binding proteins occurred.[100] To make proteome analysis easier, Sun and coworkers introduced a novel analysis technique termed continuous-flow gel electrophoresis (GE)-ICP-MS.[101,102] After injecting the sample and applying a voltage the proteins separate as on a slab gel, except the eluent can be connected online to an ICP-MS and fractions for analysis with other methods can be collected (Fig. 10.11). The system was used to characterize the Bi-binding profile in cell lysates of *H. pylori* treated with CBS. This method produced quantitative data and proved to be more sensitive than LA-ICP-MS and synchrotron XRF.

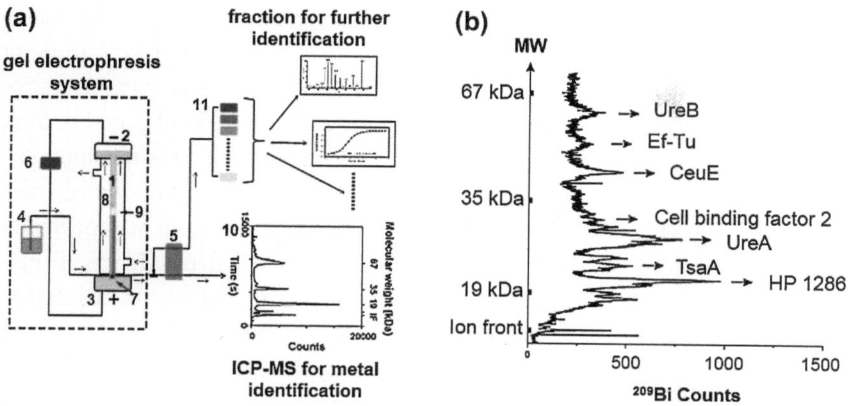

Figure 10.11 (a) Column gel system hyphenated to the inductively coupled plasma mass spectrometry (ICP-MS). 1. Gel column; 2. upper running buffer; 3. lower running buffer; 4. elution buffer; 5. peristaltic pump; 6. power supply; 7. dialysis membrane; 8. cooling system; 9. loaded sample; 10. proteins are separated according to molecular weight (MW) and analyzed online by ICP-MS; 11. fractions are taken for further analysis. (b) The continuous flow gel electrophoresis (GE)-ICP-MS profile of Bi-binding proteins in *Helicobacter pylori* treated with 20 μg/mL colloidal bismuth subcitrate (CBS). *Reproduced from Hu, L.; Cheng, T.; He, B.; Li, L.; Wang, Y.; Lai, Y. T.; Jiang, G.; Sun, H. Identification of Metal-Associated Proteins in Cells by Using Continuous-Flow Gel Electrophoresis and Inductively Coupled Plasma Mass Spectrometry.* Angew. Chem. Int. Ed. Engl. ***2013,*** 52, *4916–4920 with permission from Wiley © 2013.*

5. CONCLUSIONS

Metal-based drugs have been used in the treatment of a variety of diseases. Their notable chemical and physical properties based on the ability for ligand exchange and the resulting reactivity to biological nucleophiles are the reason why they are in some cases superior to organic drugs. However, their use in medicine requires suitable analytical methods to study their fate in biological systems. One of the advantages of metallodrugs is that they are often based on metals that are not endogenous to living organisms. Therefore element-specific methods can be employed to sensitively detect these species. A multitude of element-specific methods has been developed and these efforts have resulted in more easily accessible instrumentation and methods, including the elemental mapping of entire organisms, organs, cells, and cell compartments.

The complex matrices in the biological environment give rise to an enormous number of potential metal–biomolecule conjugates, and especially protein interaction and the structural characterization at the molecular level have become a research focus. Proteins have been suggested as carriers for metallodrugs for a long time but increasingly bioactive metal complexes are being specifically developed to selectively inhibit enzymes. Structural characterization at the molecular level is required, including the identification of metal binding sites and speciation, to develop meaningful structure–activity relationships.

The enormous complexity of biological systems makes it obvious that the application of a wide variety of complementary methods is required to get a full picture of the mode of action of metal-based drugs. Each of these methods has specific strengths and disadvantages, and the combined dataset guides the development of the next generation of metallodrugs.

ACKNOWLEDGMENTS

We thank the organizations and foundations that have supported our research efforts in this area, especially the Universities of Auckland (University of Auckland Doctoral Scholarship to H.H. and Frederick Douglas Brown Postgraduate Science Research Scholarship to M.P.S.) and Vienna, the India–New Zealand Education Council, Education New Zealand, the India–New Zealand Research Institute, the Royal Society of New Zealand and COST CM1105.

REFERENCES

1. Arnesano, F.; Natile, G. Mechanistic Insight into the Cellular Uptake and Processing of Cisplatin 30 Years after Its Approval by FDA. *Coord. Chem. Rev.* **2009,** *253,* 2070–2081.
2. Jakupec, M. A.; Galanski, M.; Arion, V. B.; Hartinger, C. G.; Keppler, B. K. Antitumour Metal Compounds: More Than Theme and Variations. *Dalton Trans.* **2008,** 183–194.
3. Leijen, S.; Burgers, S.; Baas, P.; Pluim, D.; Tibben, M.; van Werkhoven, E.; Alessio, E.; Sava, G.; Beijnen, J.; Schellens, J. M. Phase I/II Study with Ruthenium Compound NAMI-A and Gemcitabine in Patients with Non-Small Cell Lung Cancer after First Line Therapy. *Invest. New Drugs* **2015,** *33,* 201–214.
4. Groessl, M.; Dyson, P. J. Bioanalytical and Biophysical Techniques for the Elucidation of the Mode of Action of Metal-Based Drugs. *Curr. Top. Med. Chem.* **2011,** *11,* 2632–2646.
5. Groessl, M.; Hartinger, C. G. Anticancer Metallodrug Research Analytically Painting the "omics" Picture–Current Developments and Future Trends. *Anal. Bioanal. Chem.* **2013,** *405,* 1791–1808.
6. Kratz, F.; Elsadek, B. Clinical Impact of Serum Proteins on Drug Delivery. *J. Control. Release* **2012,** *161,* 429–445.
7. Kratz, F. Drug Conjugates with Albumin and Transferrin. *Expert Opin. Therap. Pat.* **2002,** *12,* 433–439.
8. Sun, H.; Li, H.; Sadler, P. J. Transferrin as a Metal Ion Mediator. *Chem. Rev.* **1999,** *99,* 2817–2842.
9. Timerbaev, A. R.; Hartinger, C. G.; Aleksenko, S. S.; Keppler, B. K. Interactions of Antitumor Metallodrugs with Serum Proteins: Advances in Characterization Using Modern Analytical Methodology. *Chem. Rev.* **2006,** *106,* 2224–2248.
10. Bytzek, A. K.; Hartinger, C. G. Capillary Electrophoretic Methods in the Development of Metal-Based Therapeutics and Diagnostics: New Methodology and Applications. *Electrophoresis* **2012,** *33,* 622–634.
11. Holtkamp, H.; Grabmann, G.; Hartinger, C. G. Electrophoretic Separation Techniques and Their Hyphenation to Mass Spectrometry in Biological Inorganic Chemistry. *Electrophoresis* **2016.** http://dx.doi.org/10.1002/elps.201500502.

12. Holtkamp, H.; Hartinger, C. G. Capillary Electrophoresis in Metallodrug Development. *Drug Discov. Today Technol.* **2015,** *16,* 16–22.
13. Groessl, M.; Terenghi, M.; Casini, A.; Elviri, L.; Lobinski, R.; Dyson, P. J. Reactivity of Anticancer Metallodrugs with Serum Proteins: New Insights from Size Exclusion Chromatography-ICP-MS and ESI-MS. *J. Anal. At. Spectrom.* **2010,** *25,* 305–313.
14. Sulyok, M.; Hann, S.; Hartinger, C. G.; Keppler, B. K.; Stingeder, G.; Koellensperger, G. Two Dimensional Separation Schemes for Investigation of the Interaction of an Anticancer Ruthenium(III) Compound with Plasma Proteins. *J. Anal. At. Spectrom.* **2005,** *20,* 856–863.
15. Groessl, M.; Hartinger, C. G.; Polec-Pawlak, K.; Jarosz, M.; Keppler, B. K. Capillary Electrophoresis Hyphenated to Inductively Coupled Plasma-Mass Spectrometry: A Novel Approach for the Analysis of Anticancer Metallodrugs in Human Serum and Plasma. *Electrophoresis* **2008,** *29,* 2224–2232.
16. Brabec, V.; Kasparkova, J. Molecular Aspects of Resistance to Antitumor Platinum Drugs. *Drug Resist. Updat.* **2002,** *5,* 147–161.
17. Holding, J. D.; Lindup, W. E.; van Laer, C.; Vreeburg, G. C.; Schilling, V.; Wilson, J. A.; Stell, P. M. Phase I Trial of a Cisplatin-albumin Complex for the Treatment of Cancer of the Head and Neck. *Br. J. Clin. Pharmacol.* **1992,** *33,* 75–81.
18. Sooriyaarachchi, M.; Narendran, A.; Gailer, J. Comparative Hydrolysis and Plasma Protein Binding of *Cis*-Platin and Carboplatin in Human Plasma In Vitro. *Metallomics* **2011,** *3,* 49–55.
19. Antonarakis, E. S.; Emadi, A. Ruthenium-Based Chemotherapeutics: Are They Ready for Prime Time? *Cancer Chemother. Pharmacol.* **2010,** *66,* 1–9.
20. Cetinbas, N.; Webb, M. I.; Dubland, J. A.; Walsby, C. J. Serum-Protein Interactions with Anticancer Ru(III) Complexes KP1019 and KP418 Characterized by EPR. *J. Biol. Inorg. Chem.* **2009,** *15,* 131–145.
21. Webb, M. I.; Walsby, C. J. Control of Ligand-Exchange Processes and the Oxidation State of the Antimetastatic Ru(III) Complex NAMI-a by Interactions with Human Serum Albumin. *Dalton Trans.* **2011,** *40,* 1322–1331.
22. Webb, M. I.; Chard, R. A.; Al-Jobory, Y. M.; Jones, M. R.; Wong, E. W.; Walsby, C. J. Pyridine Analogues of the Antimetastatic Ru(III) Complex NAMI-A Targeting Non-Covalent Interactions with Albumin. *Inorg. Chem.* **2012,** *51,* 954–966.
23. Moller, C.; Sturup, S.; Hansen, H.; Gammelgaard, B. Comparison of Two CE-ICP-MS Interfaces and Quantitative Measurements of Carboplatin in Plasma Samples Using an Internal Standard. *J. Anal. At. Spectrom.* **2009,** *24,* 1208–1212.
24. Bytzek, A. K.; Boeck, K.; Hermann, G.; Hann, S.; Keppler, B. K.; Hartinger, C. G.; Koellensperger, G. LC- and CE-ICP-MS Approaches for the In Vivo Analysis of the Anticancer Drug Candidate Sodium *Trans*-[Tetrachloridobis(1*H*-Indazole) Ruthenate(III)] (KP1339) in Mouse Plasma. *Metallomics* **2011,** *3,* 1049–1055.
25. Bytzek, A. K.; Koellensperger, G.; Keppler, B. K.; Hartinger, C. G. Biodistribution of the Novel Anticancer Drug Sodium *Trans*-[Tetrachloridobis(1*H*-Indazole)Ruthenate(III)] KP-1339/IT139 in Nude BALB/c Mice and Implications on Its Mode of Action. *J. Inorg. Biochem.* **2016.** http://dx.doi.org/10.1016/j.jinorgbio.2016.1002.1037.
26. Dömötör, O.; Hartinger, C. G.; Bytzek, A. K.; Kiss, T.; Keppler, B. K.; Enyedy, E. A. Characterization of the Binding Sites of the Anticancer Ruthenium(III) Complexes KP1019 and KP1339 on Human Serum Albumin Via Competition Studies. *J. Biol. Inorg. Chem.* **2013,** *18,* 9–17.
27. Khalaila, I.; Bergamo, A.; Bussy, F.; Sava, G.; Dyson, P. J. The Role of Cisplatin and NAMI-a Plasma-Protein Interactions in Relation to Combination Therapy. *Int. J. Oncol.* **2006,** *29,* 261–268.
28. Durrant, S. F. Laser Ablation Inductively Coupled Plasma Mass Spectrometry: Achievements, Problems, Prospects. *J. Anal. At. Spectrom..* **1999,** *14,* 1385–1403.

29. Becker, J. S.; Lobinski, R.; Becker, J. S. Metal Imaging in Non-Denaturating 2D Electrophoresis Gels by Laser Ablation Inductively Coupled Plasma Mass Spectrometry (LA-ICP-MS) for the Detection of Metalloproteins. *Metallomics* **2009,** *1,* 312–316.
30. O'Farrell, P. H. High Resolution Two-Dimensional Electrophoresis of Proteins. *J. Biol. Chem.* **1975,** *250,* 4007–4021.
31. Hanif, M.; Nazarov, A. A.; Legin, A.; Groessl, M.; Arion, V. B.; Jakupec, M. A.; Tsybin, Y. O.; Dyson, P. J.; Keppler, B. K.; Hartinger, C. G. Maleimide-Functionalised Organoruthenium Anticancer Agents and Their Binding to Thiol-Containing Biomolecules. *Chem. Commun.* **2012,** *48,* 1475–1477.
32. Li, H.; Snelling, J. R.; Barrow, M. P.; Scrivens, J. H.; Sadler, P. J.; O'Connor, P. B. Mass Spectrometric Strategies to Improve the Identification of Pt(II)-Modification Sites on Peptides and Proteins. *J. Am. Soc. Mass Spectrom.* **2014,** *25,* 1217–1227.
33. Gonias, S. L.; Pizzo, S. V. Complexes of Serum Albumin and Cis-Dichlorodiammine-platinum(II). The Role of Cysteine 34 as a Nucleophilic Entering Group and Evidence for Reaction between Bound Platinum and a Second Macromolecule. *J. Biol. Chem.* **1983,** *258,* 5764–5769.
34. Ivanov, A. I.; Christodoulou, J.; Parkinson, J. A.; Barnham, K. J.; Tucker, A.; Woodrow, J.; Sadler, P. J. Cisplatin Binding Sites on Human Albumin. *J. Biol. Chem.* **1998,** *273,* 14721–14730.
35. Allardyce, C. S.; Dyson, P. J.; Coffey, J.; Johnson, N. Determination of Drug Binding Sites to Proteins by Electrospray Ionisation Mass Spectrometry: The Interaction of Cisplatin with Transferrin. *Rapid Commun. Mass Spectrom.* **2002,** *16,* 933–935.
36. Hu, W.; Luo, Q.; Wu, K.; Li, X.; Wang, F.; Chen, Y.; Ma, X.; Wang, J.; Liu, J.; Xiong, S.; Sadler, P. J. The Anticancer Drug Cisplatin Can Cross-Link the Interdomain Zinc Site on Human Albumin. *Chem. Commun.* **2011,** *47,* 6006–6008.
37. Will, J.; Wolters, D. A.; Sheldrick, W. S. Characterisation of Cisplatin Binding Sites in Human Serum Proteins Using Hyphenated Multidimensional Liquid Chromatography and ESI Tandem Mass Spectrometry. *ChemMedChem* **2008,** *3,* 1696–1707.
38. Ferraro, G.; Massai, L.; Messori, L.; Merlino, A. Cisplatin Binding to Human Serum Albumin: A Structural Study. *Chem. Commun.* **2015,** *51,* 9436–9439.
39. Kratz, F.; Keppler, B. K.; Messori, L.; Smith, C.; Baker, E. N. Protein-Binding Properties of Two Antitumour Ru(III) Complexes to Human Apotransferrin and Apolactoferrin. *Met. Based Drugs* **1994,** *1,* 169–173.
40. Zhang, Y.; Ho, A.; Yue, J.; Kong, L.; Zhou, Z.; Wu, X.; Yang, F.; Liang, H. Structural Basis and Anticancer Properties of Ruthenium-Based Drug Complexed with Human Serum Albumin. *Eur. J. Med. Chem.* **2014,** *86,* 449–455.
41. Groessl, M.; Bytzek, A.; Hartinger, C. G. The Serum Protein Binding of Pharmacologically Active Gallium(III) Compounds Assessed by Hyphenated CE-MS Techniques. *Electrophoresis* **2009,** *30,* 2720–2727.
42. Moreno-Gordaliza, E.; Giesen, C.; Lázaro, A.; Esteban-Fernández, D.; Humanes, B.; Cañas, B.; Panne, U.; Tejedor, A.; Jakubowski, N.; Gómez-Gómez, M. M. Elemental Bioimaging in Kidney by LA–ICP–MS as a Tool to Study Nephrotoxicity and Renal Protective Strategies in Cisplatin Therapies. *Anal. Chem.* **2011,** *83,* 7933–7940.
43. Fuertes, M. A.; Castilla, J.; Alonso, C.; Perez, J. M. Cisplatin Biochemical Mechanism of Action: From Cytotoxicity to Induction of Cell Death through Interconnections between Apoptotic and Necrotic Pathways. *Curr. Med. Chem.* **2003,** *10,* 257–266.
44. Egger, A. E.; Theiner, S.; Kornauth, C.; Heffeter, P.; Berger, W.; Keppler, B. K.; Hartinger, C. G. Quantitative Bioimaging by LA-ICP-MS: a Methodological Study on the Distribution of Pt and Ru in Viscera Originating from Cisplatin- and KP1339-Treated Mice. *Metallomics* **2014,** *6,* 1616–1625.
45. Egger, A. E.; Kornauth, C.; Haslik, W.; Hann, S.; Theiner, S.; Bayer, G.; Hartinger, C. G.; Keppler, B. K.; Pluschnig, U.; Mader, R. M. Extravasation of Pt-Based Chemotherapeutics – Bioimaging of Their Distribution in Resectates Using Laser Ablation-Inductively Coupled Plasma-Mass Spectrometry (LA-ICP-MS). *Metallomics* **2015,** 7, 508–515.

46. Bianga, J.; Bouslimani, A.; Bec, N.; Quenet, F.; Mounicou, S.; Szpunar, J.; Bouyssiere, B.; Lobinski, R.; Larroque, C. Complementarity of MALDI and LA ICP Mass Spectrometry for Platinum Anticancer Imaging in Human Tumor. *Metallomics* **2014,** *6,* 1382–1386.
47. Theiner, S.; Kornauth, C.; Varbanov, H. P.; Galanski, M.; Van Schoonhoven, S.; Heffeter, P.; Berger, W.; Egger, A. E.; Keppler, B. K. Tumor Microenvironment in Focus: LA-ICP-MS Bioimaging of a Preclinical Tumor Model upon Treatment with Platinum(IV)-Based Anticancer Agents. *Metallomics* **2015,** *7,* 1256–1264.
48. Pongratz, M.; Schluga, P.; Jakupec, M. A.; Arion, V. B.; Hartinger, C. G.; Allmaier, G.; Keppler, B. K. Transferrin Binding and Transferrin-mediated Cellular Uptake of the Ruthenium Coordination Compound KP1019, Studied by Means of AAS, ESI-MS and CD Spectroscopy. *J. Anal. At. Spectrom.* **2004,** *19,* 46–51.
49. Aitken, J. B.; Antony, S.; Weekley, C. M.; Lai, B.; Spiccia, L.; Harris, H. H. Distinct Cellular Fates for KP1019 and NAMI-A Determined by X-Ray Fluorescence Imaging of Single Cells. *Metallomics* **2012,** *4,* 1051–1056.
50. Crossley, E. L.; Aitken, J. B.; Vogt, S.; Harris, H. H.; Rendina, L. M. Selective Aggregation of a Platinum-Gadolinium Complex within a Tumor-cell Nucleus. *Angew. Chem. Int. Ed. Engl.* **2010,** *49,* 1231–1233.
51. Legin, A. A.; Schintlmeister, A.; Jakupec, M. A.; Galanski, M.; Lichtscheidl, I.; Wagner, M.; Keppler, B. K. NanoSIMS Combined with Fluorescence Microscopy as a Tool for Subcellular Imaging of Isotopically Labeled Platinum-Based Anticancer Drugs. *Chem. Sci.* **2014,** *5,* 3135–3143.
52. Garcia Sar, D.; Montes-Bayon, M.; Blanco Gonzalez, E.; Sanz-Medel, A. Speciation Studies of Cis-platin Adducts with DNA Nucleotides Via Elemental Specific Detection (P and Pt) Using Liquid Chromatography-Inductively Coupled Plasma-Mass Spectrometry and Structural Characterization by Electrospray Mass Spectrometry. *J. Anal. At. Spectrom.* **2006,** *21,* 861–868.
53. Saris, C. P.; van de Vaart, P. J.; Rietbroek, R. C.; Blommaert, F. A. In Vitro Formation of DNA Adducts by Cisplatin, Lobaplatin and Oxaliplatin in Calf Thymus DNA in Solution and in Cultured Human Cells. *Carcinogenesis* **1996,** *17,* 2763–2769.
54. Iijima, H.; Patrzyc, H. B.; Dawidzik, J. B.; Budzinski, E. E.; Cheng, H. C.; Freund, H. G.; Box, H. C. Measurement of DNA Adducts in Cells Exposed to Cisplatin. *Anal. Biochem.* **2004,** *333,* 65–71.
55. Le Pla, R. C.; Ritchie, K. J.; Henderson, C. J.; Wolf, C. R.; Harrington, C. F.; Farmer, P. B. Development of a Liquid Chromatography-Electrospray Ionization Tandem Mass Spectrometry Method for Detecting Oxaliplatin-DNA Intrastrand Cross-Links in Biological Samples. *Chem. Res. Toxicol.* **2007,** *20,* 1177–1182.
56. Marzilli, L. G.; Saad, J. S.; Kuklenyik, Z.; Keating, K. A.; Xu, Y. Relationship of Solution and Protein-Bound Structures of DNA Duplexes with the Major Intrastrand Cross-Link Lesions Formed on Cisplatin Binding to DNA. *J. Am. Chem. Soc.* **2001,** *123,* 2764–2770.
57. Gelasco, A.; Lippard, S. J. NMR Solution Structure of a DNA Dodecamer Duplex Containing a *Cis*-Diammineplatinum(II) d(GpG) Intrastrand Cross-Link, the Major Adduct of the Anticancer Drug Cisplatin. *Biochemistry* **1998,** *37,* 9230–9239.
58. Chaney, S. G.; Campbell, S. L.; Temple, B.; Bassett, E.; Wu, Y.; Faldu, M. Protein Interactions with Platinum-DNA Adducts: From Structure to Function. *J. Inorg. Biochem.* **2004,** *98,* 1551–1559.
59. Zenker, A.; Galanski, M.; Bereuter, T. L.; Keppler, B. K.; Lindner, W. Kinetics of Binding Properties of 5'-GMP with Cisplatin under Simulated Physiological Conditions by Capillary Electrophoresis. *J. Chromatogr. B* **2000,** *745,* 211–219.
60. Warnke, U.; Rappel, C.; Meier, H.; Kloft, C.; Galanski, M.; Hartinger, C. G.; Keppler, B. K.; Jaehde, U. Analysis of Platinum Adducts with DNA Nucleotides and Nucleosides by Capillary Electrophoresis Coupled to ESI-MS: Indications of Guanosine 5'-Monophosphate *O6-n7* Chelation. *ChemBioChem* **2004,** *5,* 1543–1549.

61. Zenker, A.; Galanski, M.; Bereuter, T. L.; Keppler, B. K.; Lindner, W. Time-Dependent Interactions of Platinum(II) Complexes with 5'-GMP under Simulated Physiological Conditions Studied by Capillary Electrophoresis. *J. Biol. Inorg. Chem.* **2000,** *5,* 498–504.
62. Groessl, M.; Hartinger, C. G.; Dyson, P. J.; Keppler, B. K. CZE-ICP-MS as a Tool for Studying the Hydrolysis of Ruthenium Anticancer Drug Candidates and Their Reactivity towards the DNA Model Compound DGMP. *J. Inorg. Biochem.* **2008,** *102,* 1060–1065.
63. Grabmann, G.; Keppler, B. K.; Hartinger, C. G. A Systematic Capillary Electrophoresis Study on the Effect of the Buffer Composition on the Reactivity of the Anticancer Drug Cisplatin to the DNA Model 2'-Deoxyguanosine 5'-Monophosphate (dGMP). *Anal. Bioanal. Chem.* **2013,** *405,* 6417–6424.
64. Zabel, R.; Kullmann, M.; Kalayda, G.V.; Jaehde, U.; Weber, G. Optimized Sample Preparation Strategy for the Analysis of Low Molecular Mass Adducts of a Fluorescent Cisplatin Analogue in Cancer Cell Lines by CE-Dual-LIF. *Electrophoresis* **2015,** *36,* 509–517.
65. Zorbas-Seifried, S.; Hartinger, C. G.; Meelich, K.; Galanski, M.; Keppler, B. K.; Zorbas, H. DNA Interactions Of pH-Sensitive, Antitumor Bis(Aminoalcohol)Dichloroplatinum(II) Complexes. *Biochemistry* **2006,** *45,* 14817–14825.
66. Todd, R. C.; Lippard, S. J. Structure of Duplex DNA Containing the Cisplatin 1,2-$\{Pt(NH_3)_2\}^{2+}$-d(GpG) Cross-Link at 1.77 Å Resolution. *J. Inorg. Biochem.* **2010,** *104,* 902–908.
67. Jamieson, E. R.; Lippard, S. J. Structure, Recognition, and Processing of Cisplatin-DNA Adducts. *Chem. Rev.* **1999,** *99,* 2467–2498.
68. Hartinger, C. G.; Groessl, M.; Meier, S. M.; Casini, A.; Dyson, P. J. Application of Mass Spectrometric Techniques to Delineate the Modes-of-Action of Anticancer Metallodrugs. *Chem. Soc. Rev.* **2013,** *42,* 6186–6199.
69. Wang, F.; Xu, J.; Wu, K.; Weidt, S. K.; Mackay, C. L.; Langridge-Smith, P. R.; Sadler, P. J. Competition between Glutathione and DNA Oligonucleotides for Ruthenium(II) Arene Anticancer Complexes. *Dalton Trans.* **2013,** *42,* 3188–3195.
70. Chen, H.; Parkinson, J. A.; Morris, R. E.; Sadler, P. J. Highly Selective Binding of Organometallic Ruthenium Ethylenediamine Complexes to Nucleic Acids: Novel Recognition Mechanisms. *J. Am. Chem. Soc.* **2003,** *125,* 173–186.
71. Murray, B. S.; Babak, M.V.; Hartinger, C. G.; Dyson, P. J. The Development of RAPTA Compounds for the Treatment of Tumors. *Coord. Chem. Rev.* **2016,** *306,* 86–114.
72. Hartinger, C. G.; Jakupec, M. A.; Zorbas-Seifried, S.; Groessl, M.; Egger, A.; Berger, W.; Zorbas, H.; Dyson, P. J.; Keppler, B. K. KP1019, a New Redox-Active Anticancer Agent–Preclinical Development and Results of a Clinical Phase I Study in Tumor Patients. *Chem. Biodivers.* **2008,** *5,* 2140–2155.
73. Todd, R. C.; Lippard, S. J. Consequences of Cisplatin Binding on Nucleosome Structure and Dynamics. *Chem. Biol.* **2010,** *17,* 1334–1343.
74. Wu, B.; Ong, M. S.; Groessl, M.; Adhireksan, Z.; Hartinger, C. G.; Dyson, P. J.; Davey, C. A. A Ruthenium Antimetastasis Agent Forms Specific Histone Protein Adducts in the Nucleosome Core. *Chem. Eur. J.* **2011,** *17,* 3562–3566.
74a. Mohideen, K.; Muhammad, R.; Davey, C. A. Perturbations in Nucleosome Structure from Heavy Metal Association. *Nucl. Acids Res.* **2010,** *38* (18), 6301–6311. http://dx.doi.org/10.1093/nar/gkq420.
75. Adhireksan, Z.; Davey, G. E.; Campomanes, P.; Groessl, M.; Clavel, C. M.; Yu, H.; Nazarov, A. A.; Yeo, C. H.; Ang, W. H.; Droge, P.; Rothlisberger, U.; Dyson, P. J.; Davey, C. A. Ligand Substitutions between Ruthenium-Cymene Compounds Can Control Protein Versus DNA Targeting and Anticancer Activity. *Nat. Commun.* **2014,** *5,* 3462.
76. Hartinger, C. G.; Ang, W. H.; Casini, A.; Messori, L.; Keppler, B. K.; Dyson, P. J. Mass Spectrometric Analysis of Ubiquitin-Platinum Interactions of Leading Anticancer Drugs: MALDI Versus ESI. *J. Anal. At. Spectrom.* **2007,** *22,* 960–967.
77. Gibson, D.; Costello, C. E. A Mass Spectral Study of the Binding of the Anticancer Drug Cisplatin to Ubiquitin. *Eur. Mass Spectrom.* **1999,** *5,* 501–510.

78. Peleg-Shulman, T.; Najajreh, Y.; Gibson, D. Interactions of Cisplatin and Transplatin with Proteins: Comparison of Binding Kinetics, Binding Sites and Reactivity of the Pt-protein Adducts of Cisplatin and Transplatin towards Biological Nucleophiles. *J. Inorg. Biochem.* **2002,** *91,* 306–311.
79. Peleg-Shulman, T.; Gibson, D. Cisplatin−Protein Adducts Are Efficiently Removed by Glutathione but Not by 5'-Guanosine Monophosphate. *J. Am. Chem. Soc.* **2001,** *123,* 3171–3172.
80. Meier, S. M.; Tsybin, Y. O.; Dyson, P. J.; Keppler, B. K.; Hartinger, C. G. Fragmentation Methods on the Balance: Unambiguous Top-down Mass Spectrometric Characterization of Oxaliplatin-Ubiquitin Binding Sites. *Anal. Bioanal. Chem.* **2012,** *402,* 2655–2662.
81. Meier, S. M.; Babak, M. V.; Keppler, B. K.; Hartinger, C. G. Efficiently Detecting Metallodrug-Protein Adducts: Ion Trap Versus Time-of-Flight Mass Analyzers. *ChemMedChem* **2014,** *9,* 1351–1355.
82. Casini, A.; Gabbiani, C.; Michelucci, E.; Pieraccini, G.; Moneti, G.; Dyson, P. J.; Messori, L. Exploring Metallodrug-Protein Interactions by Mass Spectrometry: Comparisons between Platinum Coordination Complexes and an Organometallic Ruthenium Compound. *J. Biol. Inorg. Chem.* **2009,** *14,* 761–770.
83. Calderone, V.; Casini, A.; Mangani, S.; Messori, L.; Orioli, P. L. Structural Investigation of Cisplatin-Protein Interactions: Selective Platination of His19 in a Cuprozinc Superoxide Dismutase. *Angew. Chem. Int. Ed. Engl.* **2006,** *45,* 1267–1269.
84. Banci, L.; Bertini, I.; Blazevits, O.; Calderone, V.; Cantini, F.; Mao, J.; Trapananti, A.; Vieru, M.; Amori, I.; Cozzolino, M.; Carri, M. T. Interaction of Cisplatin with Human Superoxide Dismutase. *J. Am. Chem. Soc.* **2012,** *134,* 7009–7014.
85. Ferraro, G.; Messori, L.; Merlino, A. The X-ray Structure of the Primary Adducts Formed in the Reaction between Cisplatin and Cytochrome C. *Chem. Commun.* **2015,** *51,* 2559–2561.
86. Casini, A.; Mastrobuoni, G.; Ang, W. H.; Gabbiani, C.; Pieraccini, G.; Moneti, G.; Dyson, P. J.; Messori, L. ESI-MS Characterisation of Protein Adducts of Anticancer Ruthenium(II)-arene PTA (RAPTA) Complexes. *ChemMedChem* **2007,** *2,* 631–635.
87. Scolaro, C.; Geldbach, T. J.; Rochat, S.; Dorcier, A.; Gossens, C.; Bergamo, A.; Cocchietto, M.; Tavernelli, I.; Sava, G.; Rothlisberger, U.; Dyson, P. J. Influence of Hydrogen-Bonding Substituents on the Cytotoxicity of RAPTA Compounds. *Organometallics* **2006,** *25,* 756–765.
88. Casini, A.; Mastrobuoni, G.; Terenghi, M.; Gabbiani, C.; Monzani, E.; Moneti, G.; Casella, L.; Messori, L. Ruthenium Anticancer Drugs and Proteins: A Study of the Interactions of the Ruthenium(III) Complex Imidazolium *Trans*-[Tetrachloro(Dimethyl Sulfoxide)(Imidazole)Ruthenate(III)] with Hen Egg White Lysozyme and Horse Heart Cytochrome C. *J. Biol. Inorg. Chem.* **2007,** *12,* 1107–1117.
89. Boal, A. K.; Rosenzweig, A. C. Crystal structures of Cisplatin Bound to a Human Copper Chaperone. *J. Am. Chem. Soc.* **2009,** *131,* 14196–14197.
90. Maksimoska, J.; Williams, D. S.; Atilla-Gokcumen, G. E.; Smalley, K. S.; Carroll, P. J.; Webster, R. D.; Filippakopoulos, P.; Knapp, S.; Herlyn, M.; Meggers, E. Similar Biological Activities of Two Isostructural Ruthenium and Osmium Complexes. *Chem. Eur. J.* **2008,** *14,* 4816–4822.
91. Meggers, E. Exploring Biologically Relevant Chemical Space with Metal Complexes. *Curr. Opin. Chem. Biol.* **2007,** *11,* 287–292.
92. Meggers, E. Targeting Proteins with Metal Complexes. *Chem. Commun.* **2009,** 1001–1010.
93. Atilla-Gokcumen, G. E.; Di Costanzo, L.; Meggers, E. Structure of Anticancer Ruthenium Half-Sandwich Complex Bound to Glycogen Synthase Kinase 3β. *J. Biol. Inorg. Chem.* **2011,** *16,* 45–50.
94. Debreczeni, J. E.; Bullock, A. N.; Atilla, G. E.; Williams, D. S.; Bregman, H.; Knapp, S.; Meggers, E. Ruthenium Half-Sandwich Complexes Bound to Protein Kinase Pim-1. *Angew. Chem. Int. Ed. Engl.* **2006,** *45,* 1580–1585.

95. Streib, M.; Kraling, K.; Richter, K.; Xie, X.; Steuber, H.; Meggers, E. An Organometallic Inhibitor for the Human Repair Enzyme 7,8-Dihydro-8-Oxoguanosine Triphosphatase. *Angew. Chem. Int. Ed. Engl.* **2014,** *53,* 305–309.
96. Guidi, F.; Landini, I.; Puglia, M.; Magherini, F.; Gabbiani, C.; Cinellu, M. A.; Nobili, S.; Fiaschi, T.; Bini, L.; Mini, E.; Messori, L.; Modesti, A. Proteomic Analysis of Ovarian Cancer Cell Responses to Cytotoxic Gold Compounds. *Metallomics* **2012,** *4,* 307–314.
97. Babak, M.V.; Meier, S. M.; Huber, K.V. M.; Reynisson, J.; Legin, A. A.; Jakupec, M. A.; Roller, A.; Stukalov, A.; Gridling, M.; Bennett, K. L.; Colinge, J.; Berger, W.; Dyson, P. J.; Superti-Furga, G.; Keppler, B. K.; Hartinger, C. G. Target Profiling of an Antimetastatic RAPTA Agent by Chemical Proteomics: Relevance to the Mode of Action. *Chem. Sci.* **2015,** *6,* 2449–2456.
98. Will, J.; Sheldrick, W. S.; Wolters, D. Characterisation of Cisplatin Coordination Sites in Cellular *Escherichia coli* DNA-Binding Proteins by Combined Biphasic Liquid Chromatography and ESI Tandem Mass Spectrometry. *J. Biol. Inorg. Chem.* **2008,** *13,* 421–434.
99. Will, J.; Kyas, A.; Sheldrick, W. S.; Wolters, D. Identification of (η^6-arene)Ruthenium(II) Protein Binding Sites in *E. coli* Cells by Combined Multidimensional Liquid Chromatography and ESI Tandem Mass Spectrometry: Specific Binding of [(η^6-p-cymene)$RuCl_2$(DMSO)] to Stress-Regulated Proteins and to Helicases. *J. Biol. Inorg. Chem.* **2007,** *12,* 883–894.
100. Tsang, C. N.; Bianga, J.; Sun, H.; Szpunar, J.; Lobinski, R. Probing of Bismuth Antiulcer Drug Targets in *H. pylori* by Laser Ablation-Inductively Coupled Plasma Mass Spectrometry. *Metallomics* **2012,** *4,* 277–283.
101. Hu, L.; Cheng, T.; He, B.; Li, L.; Wang, Y.; Lai, Y. T.; Jiang, G.; Sun, H. Identification of Metal-Associated Proteins in Cells by Using Continuous-Flow Gel Electrophoresis and Inductively Coupled Plasma Mass Spectrometry. *Angew. Chem. Int. Ed. Engl.* **2013,** *52,* 4916–4920.
102. Wang, Y.; Hu, L.; Yang, X.; Chang, Y. Y.; Hu, X.; Li, H.; Sun, H. On-line Coupling of Continuous-Flow Gel Electrophoresis with Inductively Coupled Plasma-Mass Spectrometry to Quantitatively Evaluate Intracellular Metal Binding Properties of Metallochaperones *Hp*HypA and *Hp*HspA in *E. coli* Cells. *Metallomics* **2015,** *7,* 1399–1406.

INDEX

'*Note*: Page numbers followed by "f" indicate figures, "t" indicate tables.'

FSC
www.fsc.org
MIX
Papier | Fördert
gute Waldnutzung
FSC® C083411

Kolibri 360 GmbH
Ferdinand-Jühlke-Straße 7
99095 Erfurt, Deutschland
produktsicherheit@kolibri360.de